Michael S. Zhdanov

Integral Transforms in Geophysics

With 71 Figures

Springer-Verlag
Berlin Heidelberg New York
London Paris Tokyo

Prof. Dr. MICHAEL S. ZHDANOV
Institute of Terrestrial Magnetism
Ionosphere and Radiowave Propagation
Academy of Sciences
Izmiran, Troitsk, Moscow Region 142092
USSR

Translated from Russian by:

Dr. Tamara M. Pyankova

Title of the original Russian edition:
Analogi integrala tipa Koshi v teorii geofizicheskikh poleĭ
© by Nauka, Moscow 1984

ISBN-13: 978-3-642-72630-9 e-ISBN-13: 978-3-642-72628-6
DOI: 10.1007/978-3-642-72628-6

Library of Congress Cataloging-in-Publication Data. Zhdanov, Mikhail Semenovich. Integral transforms in geophysics. Rev. and enl. translation of: Analogi integrala tipa Koshi v teorii geofizicheskikh poleĭ. Bibliography: p. Includes index. 1. Magnetism, Terrestrial–Measurement. 2. Electromagnetic fields–Measurement. 3. Integral transforms. 4. Cauchy problem. 5. Mathematical analysis. I. Title. QC820.Z4313 1987 551 87-20491

Softcover reprint of the hardcover 2nd edition 1988

2132/3130-543210

This book
is dedicated to
Olga and Helena

Foreword

Integral transforms of geophysical fields serve as one of the major instruments for processing and interpreting geophysical data. These transforms enjoy great favor in various fields of geophysics where they are used in gravimetric, magnetometric, geoelectrical, and seismic investigations. Different as the properties of geophysical fields are, behind the theory of their integral transforms there are general principles which hold true both for geopotential (gravitational and permanent magnetic) and for electromagnetic and seismic (wave) fields. Integral transforms permit the solution of a wide range of geophysical problems, including construction of images of geophysical sections as well as field filtering and separation.

In this book we have endeavored to provide a unified treatment of the theory of integral transforms ranging from the techniques of transformation of plane potential fields to the theory of migration of electromagnetic and seismic fields.

Most achievements attained over the last two decades in the theory of integral transforms of two-dimensional potential fields involve the body of functions of a complex variable. This approach has received the greatest effort at development in works by G. M. Voskoboinikov, G. Ya. Golizdra, A. V. Tsirulsky, and by V. N. Strakhov.

Of overriding importance in regard to the logarithmic potential theory (and generally to the theory of functions of a complex variable) are the Cauchy-type integrals. The latter are used to elaborate methods for analytical continuation of fields, establish the location of their singular points, and study non-single-valued solutions of inverse problems.

The significance and necessity of extending the results of the two-dimensional theory to a three-dimensional case seem to be obvious. It is shown in the works published by the author since 1973 through 1976 that a variety of results of the logarithmic potential theory can be extended to the three-dimensional case by constructing Cauchy-type integral analogs for three-dimensional fields, which are modifications of the integrals introduced by Gr. C. Moisil and N. Theodoresco (1931), as well as by A. V. Bitsadze (1953, 1972). A similar approach to the solution of this problem is found in studies by T. Kolbenheyer (1976, 1978), E. Vargova (1977), and A. Sitarova (1977), who examine three-dimensional problems of potential theory in terms of four-component vectors which are analytical, according to Moisil and Theodoresco.

Serving as a further extension and generalization of the author's earlier research, this book suggests a new and, in our view, simpler way (compared to the Moisil-Theodoresco-Bitzadze approach) to construct three-dimensional analogs of the Cauchy-type integral. It applies to three-dimensional problems of gravimetry and magnetometry both for homogeneous disturbing bodies and for arbitrarily distributed density and magnetization. The derived relations possess a remarkable property – when used with two-dimensional fields, they are reduced automatically to the corresponding formulas of the theory of a complex potential. This feature of the suggested approach to three-dimensional problems of the potential theory appears to be particularly important and useful in regard to further possible extensions of the methods and practices of the theory of functions of a complex variable to the three-dimensional case.

It turns out at the same time that the body of Cauchy-type integral analogs is applicable not only to the potential theory but to many electromagnetic problems as well. Natural extensions of the Cauchy-type integral to electromagnetic fields are the Stratton-Chu type integrals whose theoretical fundamentals have been developed by M. N. Berdichevsky and the author (Zhdanov 1975b, 1976a, b, Berdichevsky and Zhdanov 1981, 1984). This book outlines, in a fairly comprehensive manner, the theory of the Stratton-Chu type integrals for various models of a field (nonstationary, quasi-stationary, wave, etc.) and a medium (homogeneous and inhomogeneous, say, for a horizontally layered model which is of critical importance in geoelectrical problems). All this is utilized to elaborate methods of transformation (separation, analytical continuation) and interpretation of electromagnetic fields dealt with in geomagnetism and geoelectrics.

In the theory of elastic oscillations, the analogs of the Cauchy-type integral are the Kirchhoff-type integrals, which are used in the book to develop the theory of seismic migration.

Thus, the principal goal of this book is to provide the reader with new ideas and methods of application of integral transforms, drawing upon the Cauchy-type integrals as well as their three-dimensional, electromagnetic, and seismic (wave) analogs to geophysical field theory.

We should like to note here that the most significant results on application of the classical Cauchy-type integrals to problems of mathematical physics have been obtained by Soviet mathematicians N. I. Muskhelishvily (1954, 1962), F. D. Gakhov (1963), and I. N. Vekua (1959). A brilliant example of the application of the Cauchy-type integral to the theory of a two-dimensional geopotential field is found in the works of V. N. Strakhov (1970c, d, etc.). The simplicity and elegance of the employed methods as well as the profundity and abundance of theoretical findings have won overall approval.

The author of the book has endeavored to show that an adequate body of mathematics can be developed for studying three-dimensional geophysical (potential and rotational – electromagnetic and seismic wave) fields, the mathematics ensuring as complete and ready a solution to complicated problems as that provided in the two-dimensional case.

Now a few words about organization of the subject matter.

This book is arranged in four parts. The first part deals with essential information about the classical theory of the Cauchy-type integral and its applications to a plane geopotential field. While preparing this part, the author followed mainly the works of V. N. Strakhov and A. V. Tsirulsky to whom he is pleased, on this occasion, to express his deep gratitude for fruitful discussions. It was not his aim to cover completely all recent findings of the plane field theory. He takes up only the aspects significant for transition to three-dimensional fields.

The second part is a comprehensive exposition of the theory of three-dimensional Cauchy-type integral analogs and its application to direct and inverse problems of the three-dimensional potential theory, as well as to integral transforms of three-dimensional geopotential fields.

The third part of the book is devoted to the theory of the Stratton-Chu type integrals. In the terms of this theory, thought is given to various types of integral transforms of the field and to the related solution of inverse problems of geomagnetism and geoelectrics based on analytical continuation and migration of electromagnetic fields.

And finally, the fourth part concerns the theory of integral transforms of seismic (wave) fields. Basic to these problems are the Kirchhoff integral formulas and the Kirchhoff-type integrals. The latter are used to consider the migration techniques for seismic (wave) fields, involving their analytical continuation into the lower half-space. This transformation is one of the most effective means of producing seismic sections of the Earth.

All the subject matter, including designations, is organized so as to point out the close linkage between the plane field theory and the three-dimensional theory of potential, electromagnetism, and elasticity. This is particularly important for demonstrating that the established relationship between the Cauchy-type integral analogs and the classical theory of functions of a complex variable makes it feasible to extend the majority of results of the two-dimensional theory to the three-dimensional case. In this way, the long-lived qualitative distinction between the approaches to a plane field and to three-dimensional potential, electromagnetic, and seismic (wave) fields becomes obsolete.

All the formulas are written in the International System of Units (SI).

The Russian book entitled *Analoghy Integrala Tipa Cauchy in Teoriy Gheofizicheskikh Poley* (Cauchy-Type Integral Analogs in Geophysical Field Theory) was issued in 1984 by the Nauka publishing house. While preparing the English version, the author expanded significantly the range of topics and added many new data. Accordingly, the book has been given a new name. In particular, the fourth part dealing with the theory of integral transforms of seismic fields was prepared specially for the English edition. Moreover, for a better insight into the underlying principles of construction of three-dimensional Cauchy-type integral analogs, a brief exposition of the quaternion theory is included in a specialized appendix.

In conclusion, we wish to extend our acknowledgments. While preparing the book, the author got much assistance from scientific editor Prof. M. N. Berdichevshy, who is also a co-author of Sect. 8.4 and Chap. 9. Much valuable advice, contributing to the quality of the book, has come from Prof. V. I. Dmitriev who has kindly taken the trouble of reviewing the manuscript. The recommendations given by Prof. P. Weidelt and Prof. U. Schmucker also proved extremely helpful in preparing the English version. Collaborators on this book are post-graduates doing research under the guidance of the author. Particularly, the theory for the Stratton-Chu type integrals for inhomogeneous media (Sect. 8.5) has been developed jointly with V. V. Spichak, the theory of the electromagnetic field migration (Chap. 10) has been elaborated together with M. A. Frenkel, while the theory of migration of seismic (wave) fields has been resolved in cooperation with V. Yu. Matusevich. Dr. T. M. Pyankova devoted much effort and talent to the translation of the book into English.

The author is grateful to all of them.

In conclusion, the author takes the opportunity to express his sincere gratitude to Olga N. Zhdanova for her constant encouragement and help during work on this manuscript.

Moscow, October 1987 M. S. Zhdanov

Contents

Notations

D	domain of complex plane or space
$\bar{D}$	closed domain
$C\bar{D}$	compliment of the closed domain $\bar{D}$ with respect to the total plane or space
$C(\zeta')$	Cauchy-type integral
ζ	point of a complex plane
$C(\zeta_0)$	singular integral in terms of the Cauchy pricipal value
$\hat{G}$	Hilbert operator
grad, grad_Γ	three- and two-dimensional gradient
div, div_Γ	three- and two-dimensional divergence
rot, rot_Γ	three- and two-dimensional rotation
∇, ∇_Γ	three- and two-dimensionale Hamilton's operator
Δ, Δ_Γ	three- and two-dimensional Laplace's operator
$\mathbf{d}_x$, $\mathbf{d}_y$, $\mathbf{d}_z$	unit vectors of Cartesian basis
$\mathbf{r}$	radius vector
δ	Dirac function; symmetric Kronecker symbol
ε	skew-symmetric Kronecker symbol
i	imaginary unit
$\mathbf{g}$, $g(\zeta)$	gravitational field and its complex intensity
ρ	density of masses, resistivity of a medium
γ	universal constant of gravitation
$\mathbf{I}$, $I(\zeta)$	intensity of magnetization, complex intensity of magnetization
$\mathbf{H}$, $H(\zeta)$	magnetic field and its complex intensity
W, $W(\zeta)$	magnetic potential, complex magnetic potential
$\mathbf{F}^e$, $\mathbf{F}^i$; $\mathbf{f}^e$, $\mathbf{f}^i$	external and internal potential fields and their spectra
U, U^e, U^i, u, u^e, u^i	scalar potential, its external and internal parts and their spectra
$\text{sign}\, x =$	$\begin{cases} +1, & x>0 \\ 0, & x=0 \\ -1, & x<0 \end{cases}$
P_L^+, P_L^-	upper and lower half-planes
S	surface in space
$\mathbf{n}$	unit normal vector
$\mathbf{C}^s(\mathbf{r}';\boldsymbol{\varphi})$	three-dimensional analog of the Cauchy-type integral, $\boldsymbol{\varphi}$ – its vector density
$\hat{G}_x$, $\hat{G}_y$	three-dimensional Hilbert operators
$W_{,a}$	derivative of the Scalar field W with respect to a

$A_{a,\beta}$	derivative of the α-component of the fector field **A** with respect to β
$\mathbf{A}\cdot\mathbf{B}$	scalar product of the vectors **A** and **B**
$\mathbf{A}\times\mathbf{B}$	vector product of the vectors **A** and **B**
U, U_D	Newton potentials of simple and polarized field sources
U^L, U_D^L	Newton potentials of line (L)
U^S, U_D^S	and surface (S) sources
$\mathbf{U}$	displacement vector field
$\mathbf{H}$, $\mathbf{B}$	vectors of the magnetic field
$\mathbf{E}$, $\mathbf{D}$	vectors of the electric field
$\mathbf{j}$	conduction current density, volume density of rotations of the vector field
q	spatial density of free electric charges, volume density of sources of the vector field
$\mathbf{j}^e$, q^e	densities of extraneous electric currents and charges
μ	magnetic permeability
ε	dielectric constant (permittivity)
σ	electric conductivity of a medium, Young modulus
t	time
$\mathbf{j}_S$	density of the surface electric current
η	surface density of the electric charge
ω	cyclic frequency of oscillations
ε^*	complex permittivity of a medium
σ^*	complex electric conductivity of a medium
k^*	wave number of a medium
k	wave number of a medium for a quasi-stationary model of the field
λ	electromagnetic wave length
$\mathbf{j}^m$, q^m	densities of extrameous magnetic currents and magentic charges
G^t	fundamental Green function for the telegraph equation
J_1	Bessel function
$\varkappa$	Heavyside unit function
G^d	fundamental Green function for the diffusion equation
G^w	fundamental Green function for the wave equation
G	fundamental Green function for the Laplace equation
G^h	fundamental Green function for the Helmholtz equation
$H_0^{(1)}$	zero-order Hankel function of the first kind
$\mathbf{C}^E$, $\mathbf{C}^H$	Stratton-Chu type integrals
$\mathbf{M}^E$, $\mathbf{M}^H$	modified Stratton-Chu type integrals
ij $(i,j=1,2)$	Green electromagnetic tensors
$\mathbf{AB}$	dyad product of the vectors A and B
$\mathbf{I}$	idemfactor (unit tensor)
$\hat{\mathbf{G}}^e$	electromagnetic tensor Green functions of the "electric" types, elastic oscillation tensor
$\hat{\mathbf{G}}^m$	electromagenetic tensor Green function of the "magnetic" type
$\mathbf{M}^e$, $\mathbf{M}^m$	Stratton-Chu type integrals for an inhomogeneous medium
$\mathbf{E}^e$, $\mathbf{H}^e$	external electromagnetic field
$\mathbf{E}^i$, $\mathbf{H}^i$	internal electromagnetic field
$\mathbf{E}^n$, $\mathbf{H}^n$	normal electromagnetic field

$\mathbf{E}^a$, $\mathbf{H}^a$	anomalous electromagnetic field
$\overset{*}{\sigma}_n$	normal complex electric conductivity of a medium
τ	reversed time
$\mathbf{E}^{mi}$, $\mathbf{H}^{mi}$ $(i = 1, 2, 3, 4)$	migration electromagnetic fields
$\hat{\mathbf{U}}$	elastic medium deformation tensor
$\hat{\tau}$	stress tensor
Δ^*	operator of the elastic oscillation equation
$\mathbf{U}^t$, $\mathbf{U}^l$	rotational and non-rotational parts of the displacement vector (transverse and longitudinal waves)
C_t	transverse wave velocity
C_l	longitudinal wave velocity
$\hat{\mathbf{G}}^t$, $\hat{\mathbf{G}}^l$	transverse and longitudinal elastic oscillation tensors
$\hat{\mathbf{G}}^w$	Green tensor for the wave equation
$\mathbf{K}^w$	Kirchhoff-type integral for a wave field
$\mathbf{K}^e$	Kirchhoff-type integral for elastic displacement field
$\underline{\mathbf{a}}$	quaternion
$\underline{\mathbf{a}} * \underline{\mathbf{b}}$	quaternion product of the quaternions $\underline{\mathbf{a}}$ and $\underline{\mathbf{b}}$
$\mathbf{A} * \mathbf{B}$	quaternion product of the vectors $\mathbf{A}$ and $\mathbf{B}$

Part I
Cauchy-Type Integrals in the Theory of a Plane Geopotential Field

1 Cauchy-Type Integral

1.1 Definition

1.1.1 Cauchy Integral Formula

For functions of a complex variable there is a known *Cauchy integral theorem* which is formulated as follows. Let L be a smooth closed curve (contour). Hereinafter the term smooth curve (contour) covers a simple (i.e., having no points of self-intersection) closed or open line with a slip tangent and with no cusps. If the function $f(\zeta)$ is analytical in the domain D bounded by the contour L and continuous on the contour itself (i.e., continuous in the closed domain $\bar{D} = D \cup L$), we deal with the following Cauchy formula:

$$\frac{1}{2\pi i} \int_L \frac{f(\zeta)}{\zeta - \zeta'} d\zeta = \begin{cases} f(\zeta'), & \zeta' \in D \,, \\ 0, & \zeta' \in C\bar{D} \,, \end{cases} \tag{1.1.1}$$

where $C\bar{D}$ is the complement of the closed domain $\bar{D}$ with respect to the total complex plane. (From now on it is assumed that the contour is traversed in the positive sense for which the inner domain is always on the left.)

Thus, knowing values of the function $f(\zeta)$ on the boundary L of the domain D we can reconstruct, using the Cauchy formula, its values everywhere inside the domain. A concisely worded statement of this fact is: *the Cauchy formula solves a boundary value problem for analytical functions.* The integral in the left-hand side of the Cauchy formula is called the *Cauchy integral.* For points ζ' lying in the outer domain ($C\bar{D}$), the Cauchy integral vanishes.

If the function $f(\zeta)$ is analytical in an infinite domain $C\bar{D}$, including a point at infinity, and if it is continuous on the contour L, the following *Cauchy formula for an infinite domain* is valid:

$$\frac{1}{2\pi i} \int_L \frac{f(\zeta)}{\zeta - \zeta'} d\zeta = \begin{cases} f(\zeta') + f(\infty), & \zeta' \in C\bar{D}, \\ f(\infty), & \zeta' \in D, \end{cases} \tag{1.1.2}$$

where the contour L is traversed in the same sense as in formula (1.1.1).

Note that Eq. (1.1.1) holds good, provided L is a piecewise smooth closed curve, i.e., it is composed of a finite set of open smooth curves a_1a_2, a_2a_3, a_3a_4, $\ldots$, $a_{n-1}a_n$ arranged so that the terminal point of each previous arc coincides with the initial point of a subsequent arc. In principle, these arcs can have, in addition to initial and terminal points, a finite number of other points in common (points of intersection).

1.1.2 Concept of the Cauchy-Type Integral

It is important that the Cauchy integral retains its meaning also in the case where an arbitrary closed or open smooth curve is specified on a complex plane and located entirely in its finite part, being superimposed by an arbitrary continuous function $\varphi(\zeta)$ (ζ – is the complex coordinate of points on the curve L):

$$C(\zeta') = \frac{1}{2\pi i} \int_L \frac{\varphi(\zeta)}{\zeta - \zeta'} d\zeta . \tag{1.1.3}$$

Integral (1.1.3) is called the *Cauchy-type integral.* The function $\varphi(\zeta)$ is known as its *density,* while $1/(\zeta - \zeta')$ as its *kernel.* It is readily seen that Cauchy-type integral (1.1.3) fits a function which is analytical over the entire plane of a complex variable, except for the points on the curve L itself. We will provide two proofs of this claim.

The first proof, usually quoted in courses on the theory of functions of a complex variable, lies in establishing differentiability with respect to ζ' (parameter) under the integral.

For the sake of proof, let $\Phi(\zeta')$ designate the following integral:

$$\Phi(\zeta') = \frac{1}{2\pi i} \int_L \frac{\varphi(\zeta)}{(\zeta - \zeta')^2} d\zeta ,$$

yielded by formal differentiation of the Cauchy-type integral kernel with respect to ζ'. Now we will consider the difference:

$$\frac{C(\zeta' + \Delta\zeta') - C(\zeta')}{\Delta\zeta'} - \Phi(\zeta') = \frac{1}{2\pi i} \int_L \varphi(\zeta) \left[\frac{1}{(\zeta - \zeta' - \Delta\zeta')(\zeta - \zeta')} - \frac{1}{(\zeta - \zeta')^2} \right] d\zeta .$$

By virtue of the analyticity of the kernel with respect to ζ' (for $\zeta \neq \zeta'$), the bracketed value can be rendered arbitrarily small, for a sufficiently small $|\Delta\zeta'|$. Since the contour L is assumed to be of a finite length, we obtain, proceeding to the limit, that

$$C^{(1)}(\zeta') = \lim_{\Delta\zeta' \to 0} \frac{C(\zeta' + \Delta\zeta') - C(\zeta')}{\Delta\zeta'} = \Phi(\zeta') = \frac{1}{2\pi i} \int_L \frac{\varphi(\zeta)}{(\zeta - \zeta')^2} d\zeta . \tag{1.1.4}$$

In a similar way, it is proved that $C(\zeta')$ has derivatives of all orders and

$$C^{(n)}(\zeta') = \frac{n!}{2\pi i} \int_L \frac{\varphi(\zeta) d\zeta}{(\zeta - \zeta')^{(n+1)}} , \quad n = 1, 2, \ldots, \tag{1.1.5}$$

where $\zeta' \notin L$.

The second proof reduces to direct checking whether the functions $\operatorname{Re} C(\zeta')$ and $\operatorname{Im} C(\zeta')$ satisfy the Cauchy-Riemann conditions:

$$\frac{\partial \operatorname{Re} C(\zeta')}{\partial x'} = \frac{\partial \operatorname{Im} C(\zeta')}{\partial z'} , \quad \frac{\partial \operatorname{Re} C(\zeta')}{\partial z'} = -\frac{\partial \operatorname{Im} C(\zeta')}{\partial x'} , \tag{1.1.6}$$

where $\zeta' = x' + iz'$.

Indeed, relations (1.1.6) follow automatically from the fact that the kernel $1/(\zeta-\zeta')$ is a function analytical with respect to ζ' (for $\zeta'\notin L$). Hence, its real and imaginary parts do satisfy the Cauchy-Riemann conditions. We call the attention of the reader to this proof because in subsequent sections of the book dealing with various analogs of the Cauchy-type integral we shall see that these analogs satisfy the corresponding system of equations in partial derivatives by means of direct differentiation.

Then, it is evident that once ζ' tends to infinity, $C(\zeta')$ tends to zero so that $C(\infty)=0$; or better, at high values $|\zeta'|$ we have:

$$C(\zeta')=O(1/\zeta') \ , \tag{1.1.7}$$

since L is of a finite length.

Thus, the Cauchy-type integral with a continuous density, $\varphi(\zeta)$, is a function analytical everywhere except for the path of integration, L, and it vanishes at infinity. The path of integration, L, is a singular line of the function $C(\zeta')$. The behavior of the Cauchy-type integral on the path L and in its neighborhood will be treated closely in the next subsection.

If L is a closed curve dividing the complex plane into an inner, D, and an outer, $C\bar{D}$, domain (as a rule, the sense of traversal of the path L is considered positive when the inner domain, D, remains on the left), the function $C(\zeta')$ falls into two analytical functions: $C^+(\zeta')$ defined in D, and $C^-(\zeta')$ defined in $C\bar{D}$. Note that generally these functions do not continue one another analytically[1]. For example, we will use the Cauchy-type integral taken around the circle $|\zeta|=2$ (Fig. 1) with a density of $\varphi(\zeta)=2i/(\zeta-i)(\zeta-3i)$. It is evident that

$$C(\zeta')=\frac{1}{\pi}\int_L\frac{1}{(\zeta-i)(\zeta-3i)}\frac{d\zeta}{(\zeta-\zeta')}=\frac{1}{2\pi i}\int_L\frac{1}{(\zeta-3i)}\frac{d\zeta}{(\zeta-\zeta')}-\frac{1}{2\pi i}\int_L\frac{1}{(\zeta-i)}\frac{d\zeta}{(\zeta-\zeta')} \ . \tag{1.1.8}$$

The function $1/(\zeta-3i)$ is analytical in the domain D, while the function $1/(\zeta-i)$ is analytical in the domain $C\bar{D}$ (see Fig. 1), and it vanishes at infinity. Hence, according to the Cauchy integral formula (1.1.1), the first integral for $\zeta'\in D$ is $1/(\zeta'-3i)$, and for $\zeta'\in C\bar{D}$ it is zero. To the contrary, the second integral, according to formula (1.1.2), is zero for $\zeta'\in D$, and is equal to $-1/(\zeta'-i)$ for $\zeta'\in C\bar{D}$. Hence, we have that

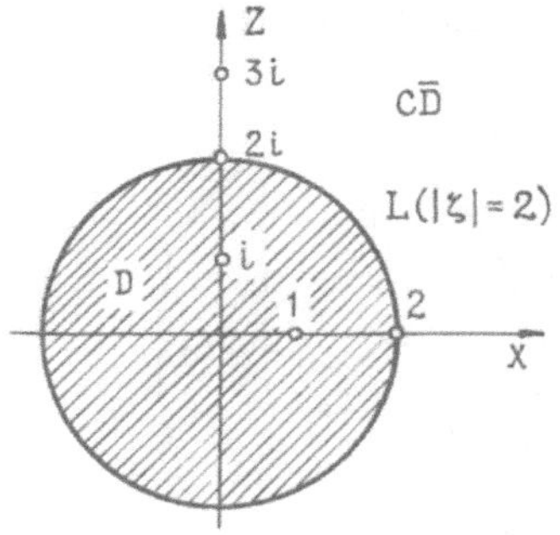

Fig. 1. Analysis of the properties of the Cauchy-type integral taken around the circle $|\zeta|=2$

[1] The fundamentals of analytical continuation are outlined in Sect. 4.1.

$$C^{+}(\zeta') = \frac{1}{\zeta' - 3i}, \quad C^{-}(\zeta') = \frac{-1}{\zeta' - i}. \tag{1.1.9}$$

As is seen, integral (1.1.8) describes two absolutely different analytical functions in the domains D and $C\bar{D}$.

In concluding this subsection, we will note that the concept of the Cauchy-type integral applies readily to the case where the path of integration, L, tends to infinity. The only additional requirement here is consideration of the behavior of functions (either specified or sought) in the neighborhood of a point at infinity (Muskhelishvily 1962; Gakhov 1963) (for greater detail see Sect. 1.2.5).

1.1.3 Piecewise Analytical Functions

Let $D_1, D_2, \ldots, D_N$ be several simply connected domains having no inner points in common but adjoining along smooth curves $L_1, L_2, \ldots, L_M$ constituting the common part of their bounding lines. We will also assume that $\Phi_1(\zeta), \Phi_2(\zeta), \ldots, \Phi_N(\zeta)$ are functions analytical in $D_1, D_2, \ldots, D_N$, respectively, and are extendible continuously to the corresponding boundary curves $L_1, L_2, \ldots, L_M$.

Let $\Phi(\zeta)$ stand for the function defined by the functions $\Phi_1(\zeta), \Phi_2(\zeta), \ldots, \Phi_N(\zeta)$ in each of the domains $D_1, D_2, \ldots, D_N$. This function will be referred to as a *piecewise analytical function with a line of discontinuities* $\mathrm{L} = L_1 \cup L_2 \cup \ldots L_M$.

The results of the previous subsection yield a conclusion that once L is a closed contour, *the Cauchy-type integral is a piecewise analytical function with a discontinuity line* L. Let us recall an important property of piecewise analytical functions, which will be often used in subsequent discussion.

Consider a piecewise analytical function $\Psi(\zeta)$ defined in two domains, D_1 and D_2, adjoining along a smooth curve L (Fig. 2) (the ends of L are not assumed to belong to L itself), that is

$$\Psi(\zeta) = \begin{cases} \Psi_1(\zeta), & \zeta \in D_1, \\ \Psi_2(\zeta), & \zeta \in D_2, \end{cases} \tag{1.1.10}$$

where $\Psi_1(\zeta)$ and $\Psi_2(\zeta)$ are the functions continuously extendible to L, with their boundary values along L equal:

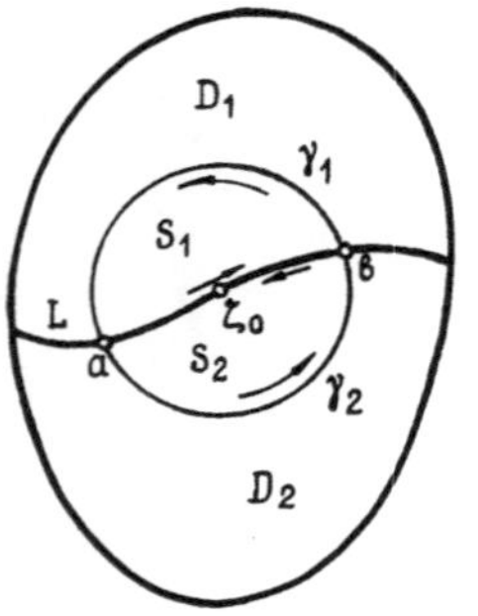

Fig. 2. Scheme explaining the properties of a piecewise analytical function $\Psi(\zeta)$

$$\Psi_1(\zeta) = \Psi_2(\zeta) \ , \quad \zeta \in L \ . \tag{1.1.11}$$

Then the function $\Psi(\zeta)$ redefined on L in the following manner: $\Psi(\zeta) = \Psi_1(\zeta) = \Psi_2(\zeta)$, $\zeta \in L$ is analytical in the domain $D_1 \cup D_2 \cup L$. To prove it, it seems sufficient to establish the analyticity of the function $\Psi(\zeta)$ in the neighborhood of any point ζ_0 on the line L not coinciding with its ends. Let us draw a circle γ from a point ζ_0 of such a small radius that the circle might cross the curve L only at two points, a and b (this is possible since L is a smooth curve). Let S_1 and S_2 designate portions of the circle bounded by γ, which lie in the domains D_1 and D_2, and γ_1 and γ_2 stand for the corresponding portions of the circle γ. Then, ∂S_1 and ∂S_2 are assumed to be boundaries of the domains S_1 and S_2 described in the positive sense. Evidently, they have a common part, ab, traversed in the opposite directions: $\partial S_1 = \gamma_1 \cup ab$, $\partial S_2 = \gamma_2 \cup ba$ (see Fig. 2). On the basis of Cauchy integral formula (1.1.1) for all points ζ' found inside S_1 or S_2 we have

$$\Psi(\zeta') = \frac{1}{2\pi i} \int_{\partial S_1} \frac{\Psi_1(\zeta) d\zeta}{\zeta - \zeta'} + \frac{1}{2\pi i} \int_{\partial S_2} \frac{\Psi_2(\zeta) d\zeta}{\zeta - \zeta'} \ . \tag{1.1.12}$$

Indeed, the first integral is equal to $\Psi_1(\zeta')$ for $\zeta' \in S_1$ and to zero for $\zeta' \in S_2$, while the second integral, on the contrary, is zero for $\zeta' \in S_1$ and equals $\Psi_2(\zeta')$ for $\zeta' \in S_2$. At the same time, expression (1.1.12) reduces to a single integral taken on γ, since the boundary values of the functions $\Psi_1(\zeta)$ and $\Psi_2(\zeta)$ on L coincide and, hence, the integrals taken over the arc ab (in opposite directions) cancel one another

$$\Psi(\zeta') = \frac{1}{2\pi i} \int_{\gamma} \frac{\Psi(\zeta)}{\zeta - \zeta'} d\zeta \ . \tag{1.1.13}$$

Expression (1.1.13) is, in turn, an analytical function everywhere inside γ, in accordance with the property of the Cauchy-type integral, which was to be proved.

1.2 Main Properties

The properties of the Cauchy-type integrals have been systematically exposed in many handbooks of the theory of functions of a complex variable, e.g., in the excellent books of N.I. Muskhelishvily (1962), F.D. Gakhov (1963), and A.V. Bitsadze (1972). They will be discussed but briefly in this book for the convenience of the reader wishing to acquire only the information actually needed for understanding the considerations given below. Of prime concern is the behavior of the Cauchy-type integral on the path of integration itself or near it. Prior to taking up the problem proper, we will examine a subsidiary issue of classes of functions describing densities of the Cauchy-type integral.

1.2.1 Hölder Condition

In introducing the concept of the Cauchy-type integral in the previous subsection, we required, first of all, that its density, $\varphi(\zeta)$ be a continuous function of $\zeta \in L$.

As is known, function continuity lies in the fact that the modulus $|\varphi(\zeta_2)-\varphi(\zeta_1)|$ can be made arbitrarily small, provided the modulus $|\zeta_2-\zeta_1|$ is sufficiently small, i.e., the increments in the argument and in the function tend to zero simultaneously. The order of smallness of an increment of the function with respect to that of the argument is not treated at all, it can be of any value. But it should be used as a basis of classification of a set of continuous functions in the study of the Cauchy-type integral properties. The most important among them is a *class of Hölder continuous functions.*

Definition 1.2.1. The function $\varphi(\zeta)$ of a variable (generally complex) ζ specified for a certain connected set E of values of this variable is said to satisfy the Hölder condition, provided there exist positive constants A and λ so that

$$|\varphi(\zeta_2)-\varphi(\zeta_1)| \leqslant A\,|\zeta_2-\zeta_1|^{\lambda} \tag{1.2.1}$$

for any pair of points ζ_1 and ζ_2 out of E. A is called the Hölder constant, while λ is termed the Hölder indicator.

Functions satisfying the Hölder condition are also known as *Hölder continuous functions.* We should note again that the requirement on Hölder continuity is more severe than that of ordinary continuity. This is evidenced at least by the following example.

Consider the function

$$\varphi(x) = \begin{cases} 1/\ln x, & 0<x\leqslant 1/2, \\ 0, & x=0. \end{cases}$$

Evidently, the function $\varphi(x)$ is continuous over a segment [0, 1/2]. We will assume that within the segment [0, 1/2] the function $\varphi(x)$ also satisfies the Hölder condition with certain specific values of constant quantities A and λ, and write (1.2.1) for two points, namely 0 and $x\in(0, 1/2]$:

$$|\varphi(x)-\varphi(0)| = \left|\frac{1}{\ln x}\right| \leqslant A\,|x|^{\lambda}\ .$$

Hence we have that $|x|^{\lambda}|\ln x| \geqslant 1/A$, where $x\in(0, 1/2]$.

On the other hand, $\lim_{x\to 0} x^{\lambda}\ln x = 0$ *for any* $\lambda>0$.

As a result, for any A there is such $x_0\in(0, 1/2]$ that $|x_0|^{\lambda}\ln x_0 < 1/A$. This inconsistency shows that the function $\varphi(x)$ over the segment concerned, continuous as it is, does not meet the Hölder condition.

We have imposed so far no restrictions on the range of the Hölder indicator, λ. It is evident, however, that if the value of λ were in excess of unity, the derivative $\varphi'(\zeta)$, according to condition (1.2.1), would be everywhere zero and the function $\varphi(\zeta)$ would be identically constant. It is usually assumed therefore that $0<\lambda\leqslant 1$. When $\lambda = 1$, the Hölder condition goes over into the known Lipschitz condition.

If ζ_1 and ζ_2 are such that $|\zeta_1-\zeta_2|<1$ and the Hölder condition is satisfied for some value of λ_1, it will be obviously satisfied also for any $\lambda<\lambda_1$. Thus, the higher the value of λ, the more narrow the class of Hölder continuous functions. The most narrow class of functions meets the Lipschitz condition.

It is readily established from the latter property that once the functions $\varphi_1(\zeta)$ and $\varphi_2(\zeta)$ satisfy the Hölder condition with indicators λ_1 and λ_2, respectively, then $\varphi_1 \pm \varphi_2$, $\varphi_1\varphi_2$, as well as φ_1/φ_2 (provided that the denominator does not vanish) do satisfy the Hölder condition with an indicator $\lambda = \min(\lambda_1, \lambda_2)$.

Simple and *sufficient evidence that the function $\varphi(\zeta)$ satisfies the Lipschitz condition is the differentiability of $\varphi(\zeta)$ on E and the limitedness of its derivative.* Indeed, the above assumptions and taking due account of the mean value theorem ensure fulfilment of condition (1.2.1) for $\lambda = 1$. But this feature is not necessary, which is seen at least from the following example.

Let $\varphi(x)$ be a function specified in the interval $(-1, 1)$, $\varphi(x) = |x|$. Evidently it satisfies the Lipschitz condition everywhere over this interval, although it is not differentiable at $x = 0$.

The concept of the Hölder condition can be extended to a function of any number of variables.

Definition 1.2.2. The function $\varphi(\zeta_1, \zeta_2, \ldots, \zeta_n)$ of variables $\zeta_1, \zeta_2, \ldots, \zeta_n$ specified for a certain connected set E is said to satisfy the Hölder condition within this set, provided there exist positive constants $A_1, A_2, \ldots, A_n$; $\lambda_1, \lambda_2, \ldots, \lambda_n$ $(0<\lambda_1, \lambda_2, \ldots, \lambda_n \leqslant 1)$ so that

$$|\varphi(\zeta_1'', \zeta_2'', \ldots, \zeta_n'') - \varphi(\zeta_1', \zeta_2', \ldots, \zeta_n')| \leqslant A_1|\zeta_1'' - \zeta_1'|^{\lambda_1} + A_2|\zeta_2'' - \zeta_2'|^{\lambda_2} + \ldots + A_n|\zeta_n'' - \zeta_n'|^{\lambda_n} \tag{1.2.2}$$

for any two populations $\zeta_1', \zeta_2', \ldots, \zeta_n'$ and $\zeta_1'', \zeta_2'', \ldots, \zeta_n''$ of values of the variables, $\zeta_1, \zeta_2, \ldots, \zeta_n$ out of the set E.

Clearly, once the function $\varphi(\zeta_1, \zeta_2, \ldots, \zeta_n)$ meets the Hölder condition for the total of variables $\zeta_1, \zeta_2, \ldots, \zeta_n$, it is Hölder continuous for each of these variables separately and, moreover, it is uniformly continuous relative to other variables (i.e., the relevant Hölder constant and indicator are independent of values of these parameters).

In what follows, of particular concern is the case where functions specified for some smooth or piecewise smooth lines meet the Hölder condition. Precisely for this class of functions it is possible to establish the most critical properties of the Cauchy-type integral.

1.2.2 Calculation of Singular Integrals in Terms of the Cauchy Principal Value

Refer again to the direct examination of Cauchy-type integral (1.1.3). As shown earlier, the integral $C(\zeta')$ is an analytical function everywhere in the complex plane outside the curve L. When the point ζ' lies on L, $\zeta' = \zeta_0 \in L$, the integral in the right-hand side of formula (1.1.3) in its ordinary sense does not exist. But this integral can be made to convey quite a definite meaning for a wide and important class of functions $\varphi(\zeta)$ satisfying the Hölder condition by introducing the concept of the *Cauchy principal value.* It comes about as follows.

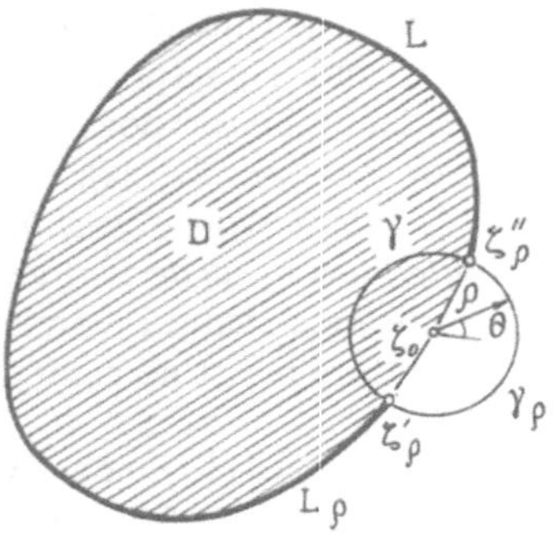

Fig. 3. Evaluation of a singular integral in terms of the Cauchy principal value for a closed line L

For the sake of simplicity, we assume that L is a closed smooth curve and the point $\zeta' = \zeta_0$ lies on L. Let us draw from the point ζ_0, as a center, a circle of such a small radius ϱ that it might intersect L exactly at two points, ζ'_ϱ and ζ''_ϱ (Fig. 3). The portion of the curve L lying outside the circle $|\zeta-\zeta_0| \leqslant \varrho$ bounded by γ will be designated by L_ϱ.

The integral

$$I_\varrho(\zeta_0) = \frac{1}{2\pi i} \int_{L_\varrho} \frac{\varphi(\zeta)}{\zeta-\zeta_0}\, d\zeta \ , \tag{1.2.3}$$

seems to make sense in its ordinary terms.

Definition 1.2.3. If there is

$$\lim_{\varrho\to 0} I_\varrho(\zeta_0) = C(\zeta_0) \ , \tag{1.2.4}$$

this limit is called an integral in terms of the Cauchy principal value or a Cauchy singular integral.

In the definition of a principal value, it is essential that the ends ζ'_ϱ and ζ''_ϱ of the arc eliminated from L in the calculation of integral (1.2.3) are equally spaced from ζ_0.

The principal value of an integral is denoted by the same symbol as is the ordinary integral[2] implying that if the integral has no ordinary sense we deal with its principal value:

$$C(\zeta_0) = \frac{1}{2\pi i} \int_L \frac{\varphi(\zeta)}{\zeta-\zeta_0}\, d\zeta \ . \tag{1.2.5}$$

We will show that for integral (1.2.5) to exist in terms of the Cauchy principal value for any $\zeta_0 \in L$, it is sufficient that the function $\varphi(\zeta)$ be Hölder continuous everywhere on L. We will also derive an expression for a singular integral from an ordinary integral.

---

[2] Sometimes, in mathematical literature, the symbol for an integral in terms of principal value is provided by two Latin letters VP ("Valeur Principale") or by a star or prime.

Indeed, using Cauchy integral formula (1.1.1) we cast expression (1.2.3) in the following form:

$$I_\varrho(\zeta_0) = \frac{1}{2\pi i} \int_{L_\varrho} \frac{\varphi(\zeta)}{\zeta-\zeta_0}\, d\zeta + \varphi(\zeta_0) - \frac{1}{2\pi i} \int_{\Delta_\varrho \cup \gamma_\varrho} \frac{\varphi(\zeta_0)}{\zeta-\zeta_0}\, d\zeta \ , \tag{1.2.6}$$

where γ_ϱ is the portion of the circle γ lying outside the finite domain D bounded by the curve L (Fig. 3). The latter formula can be rewritten as:

$$I_\varrho(\zeta_0) = \frac{1}{2\pi i} \int_{L_\varrho} \frac{\varphi(\zeta)-\varphi(\zeta_0)}{\zeta-\zeta_0}\, d\zeta + \varphi(\zeta_0) - \frac{\varphi(\zeta_0)}{2\pi i} \int_{\gamma_\varrho} \frac{d\zeta}{\zeta-\zeta_0} \ . \tag{1.2.7}$$

The Hölder condition ensures existence of an ordinary uniformly convergent improper integral

$$\int_L \frac{\varphi(\zeta)-\varphi(\zeta_0)}{\zeta-\zeta_0}\, d\zeta = \lim_{\varrho\to 0} \int_{L_\varrho} \frac{\varphi(\zeta)-\varphi(\zeta_0)}{\zeta-\zeta_0}\, d\zeta \ . \tag{1.2.8}$$

The integral extended along the portion of the circle, γ_ϱ, in formula (1.2.7) can be readily evaluated in the frame of polar coordinates (Fig. 3) $\zeta-\zeta_0=\varrho e^{i\theta}$. Then we have:

$$\int_{\gamma_\varrho} \frac{d\zeta}{\zeta-\zeta_0} = i \int_{\gamma_\varrho} d\theta = i[\theta(\zeta_\varrho'')-\theta(\zeta_\varrho')] \ .$$

It follows therefrom that on account of the smoothness of the curve L one has

$$\lim_{\varrho\to 0} \int_{\gamma_\varrho} \frac{d\zeta}{\zeta-\zeta_0} = i\pi \ . \tag{1.2.9}$$

Thus, proceeding to the limit in Eq. (1.2.7) and allowing for (1.2.8) and (1.2.9), we obtain:

$$C(\zeta_0) = \frac{1}{2\pi i} \int_L \frac{\varphi(\zeta)}{\zeta-\zeta_0}\, d\zeta = \frac{1}{2\pi i} \int_L \frac{\varphi(\zeta)-\varphi(\zeta_0)}{\zeta-\zeta_0}\, d\zeta + \frac{1}{2}\varphi(\zeta_0) \ . \tag{1.2.10}$$

It is not difficult at all to analyze a more general case where L is an open smooth line connecting the points a and b in a complex plane (Fig. 4) (in the assumption

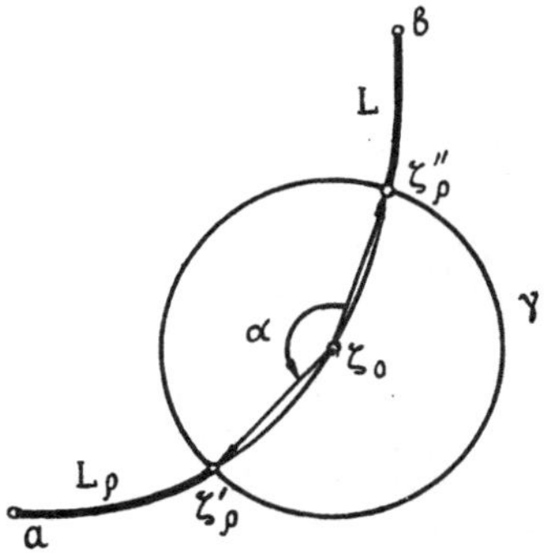

Fig. 4. Evaluation of a singular integral in terms of the Cauchy principal value for an open line L

that the point $\zeta_0 \in L$ is not an end point). In this case, expression (1.2.3) can be represented in the form

$$I_\varrho(\zeta_0) = \frac{1}{2\pi i} \int_{L_\varrho} \frac{\varphi(\zeta)-\varphi(\zeta_0)}{\zeta-\zeta_0} d\zeta + \frac{\varphi(\zeta_0)}{2\pi i} \int_{L_\varrho} \frac{d\zeta}{\zeta-\zeta_0} . \tag{1.2.11}$$

Calculations have revealed that

$$\int_{L_\varrho} \frac{d\zeta}{\zeta-\zeta_0} = \ln(\zeta-\zeta_0)\Big|_a^{\zeta'_\varrho} + \ln(\zeta-\zeta_0)\Big|_{\zeta''_\varrho}^{b} = \ln\frac{b-\zeta_0}{a-\zeta_0} - \ln\frac{\zeta''_\varrho-\zeta_0}{\zeta'_\varrho-\zeta_0} ,$$

where $\ln(\zeta-\zeta_0)$ is taken to be a contour value of the analytical function $\ln(\zeta-\zeta')$, which is single-valued in a plane cut along the curve connecting the branch points ζ_0 and ∞; for concreteness, it is assumed that the cut is on the right of the curve L (moving from a to b along L).

However, we have that

$$\ln\frac{\zeta''_\varrho-\zeta_0}{\zeta'_\varrho-\zeta_0} = \ln\left|\frac{\zeta''_\varrho-\zeta_0}{\zeta'_\varrho-\zeta_0}\right| + i[\arg(\zeta''_\varrho-\zeta_0) - \arg(\zeta'_\varrho-\zeta_0)] .$$

By construction, $|\zeta''_\varrho-\zeta_0| = |\zeta'_\varrho-\zeta_0|$. As a result, the first summand becomes zero. The expression in brackets is equal to an angle made by the vectors $\zeta_0\zeta'_\varrho$ and $\zeta_0\zeta''_\varrho$, this angle being counted off on the left of the curve L in view of the selected cut. Hence,

$$\lim_{\varrho\to 0} \ln\frac{\zeta''_\varrho-\zeta_0}{\zeta'_\varrho-\zeta_0} = -i\pi .$$

As a consequence, we obtain:

$$\lim_{\varrho\to 0} \int_{L_\varrho} \frac{d\zeta}{\zeta-\zeta_0} = \int_L \frac{d\zeta}{\zeta-\zeta_0} = \ln\frac{b-\zeta_0}{a-\zeta_0} + i\pi . \tag{1.2.12}$$

In particular, for a closed curve L at $a=b$ we have:

$$\int_L \frac{d\zeta}{\zeta-\zeta_0} = i\pi . \tag{1.2.13}$$

Thus, if the function $\varphi(\zeta)$ satisfies the Hölder condition, there is a limit of expression (1.2.11), which is equal, in accordance with (1.2.8) and (1.2.12), to

$$C(\zeta_0) = \frac{1}{2\pi i} \int_L \frac{\varphi(\zeta)}{\zeta-\zeta_0} d\zeta = \frac{1}{2\pi i} \int_L \frac{\varphi(\zeta)-\varphi(\zeta_0)}{\zeta-\zeta_0} d\zeta + \frac{\varphi(\zeta_0)}{2\pi i} \ln\frac{b-\zeta_0}{a-\zeta_0} + \frac{1}{2}\varphi(\zeta_0) . \tag{1.2.14}$$

It is evident, in particular, that for a closed contour, i.e., for $a=b$, formula (1.2.14) goes over into (1.2.10).

We should like to note that in the subsequent discussion the Cauchy singular integral will always imply its principal value.

1.2.3 Sokhotsky-Plemelj Formulas

As indicated earlier, for an analytical function defined by the Cauchy-type integral, the curve of integration itself, L, is a singular line. Here, of eminent concern and importance is the behavior of the Cauchy-type integral on approaching the line L. The main result to be obtained here and to be of fundamental significance for the Cauchy-type integral theory is as follows. *The Cauchy-type integral with a density satisfying the Hölder condition has continuous limit values on approaching the line L on each of its sides, but these limit values are different, so that on transition through L they suddenly change.*

Now we assume to have a Cauchy-type integral (1.1.3) where $\zeta' \notin L$ and $\varphi(\zeta)$ meet the Hölder condition. We also assume that the contour L is closed. If it happens to be open, we will complement it by a smooth curve to make it closed, taking that $\varphi(\zeta)=0$ on this complementary curve. Let ζ_0 designate an arbitrary fixed point on L and expression (1.1.3) be cast in the form:

$$C(\zeta') = \psi(\zeta') + \frac{1}{2\pi i}\varphi(\zeta_0)\int_L \frac{d\zeta}{\zeta-\zeta'} , \qquad (1.2.15)$$

where $\psi(\zeta')$ is an auxiliary analytical function equal to

$$\psi(\zeta') = \frac{1}{2\pi i}\int_L \frac{\varphi(\zeta)-\varphi(\zeta_0)}{\zeta-\zeta'}\, d\zeta . \qquad (1.2.16)$$

We will designate the limit values of the analytical functions $C(\zeta')$ and $\psi(\zeta')$ by $C^+(\zeta_0)$ and $\psi^+(\zeta_0)$ for a point ζ' tending from the inside of a finite L-bounded domain D to the point $\zeta_0 \in L$ and by $C^-(\zeta_0)$ and $\psi^-(\zeta_0)$ if this point tends from the outside of the above domain (for an open curve they correspond to the limit values on the left and on the right when L is traversed in the positive sense). To point out the direction of passage to the limit, we will write $\zeta' \to \zeta_0^+$ and $\zeta' \to \zeta_0^-$, respectively.

Using Cauchy integral formula (1.1.1) and relations (1.2.13) we can write:

$$\frac{1}{2\pi i}\int_L \frac{d\zeta}{\zeta-\zeta'} = \begin{cases} 1, & \zeta' \in D , \\ 0, & \zeta' \in C\bar{D} , \\ \frac{1}{2}, & \zeta' \in L . \end{cases} \qquad (1.2.17)$$

This directly implies that the integral in the left-hand side of expression (1.2.17) has (different) limit values for $\zeta' - \zeta_0$ from the outside and inside of the domain D:

$$\lim_{\zeta' \to \zeta_0^+} \frac{1}{2\pi i}\int_L \frac{d\zeta}{\zeta-\zeta'} = 1 , \qquad (1.2.18)$$

$$\lim_{\zeta' \to \zeta_0^-} \frac{1}{2\pi i}\int_L \frac{d\zeta}{\zeta-\zeta'} = 0 . \qquad (1.2.19)$$

If the density $\varphi(\zeta)$ meets the Hölder condition, it is easy to show that *the function $\psi(\zeta')$ fitted by formula (1.2.16) behaves as a continuous function on transi-*

tion through the point $\zeta' = \zeta_0$ of the curve L, i.e., it has a single definite limit value for ζ' approaching ζ_0 from any side of the contour along any path and this limit value is (with due reference to (1.2.10)) as follows:

$$\lim_{\zeta'\to\zeta_0^\pm} \psi(\zeta') = \psi(\zeta_0) = \frac{1}{2\pi i} \int_L \frac{\varphi(\zeta)}{\zeta-\zeta_0} d\zeta - \frac{\varphi(\zeta_0)}{2\pi i} \int_L \frac{d\zeta}{\zeta-\zeta_0} = C(\zeta_0) - \frac{1}{2}\varphi(\zeta_0) \ , \quad (1.2.20)$$

where $C(\zeta_0)$ is a singular integral in terms of the Cauchy principal value. Detailed proof of this claim is found in the relevant handbooks of the theory of functions of a complex variable, as in books by F. D. Gakhov (1963) and A. V. Bitsadze (1972). Hence, there are also limit values for the Cauchy-type integral, which are, according to formulas (1.2.15), (1.2.18)–(1.2.20), as follows:

$$C^+(\zeta_0) = C(\zeta_0) + \tfrac{1}{2}\varphi(\zeta_0) \ , \quad (1.2.21\text{a})$$

$$C^-(\zeta_0) = C(\zeta_0) - \tfrac{1}{2}\varphi(\zeta_0) \ . \quad (1.2.21\text{b})$$

The latter formulas go under the name of *Sokhotsky-Plemelj formulas.*

Subtracting and summing up formulas (1.2.21 a) and (1.2.21 b) we arrive at a pair of tantamount formulas:

$$C^+(\zeta_0) - C^-(\zeta_0) = \varphi(\zeta_0) \ , \quad (1.2.22\text{a})$$

$$C^+(\zeta_0) + C^-(\zeta_0) = 2C(\zeta_0) = \frac{1}{\pi i} \int_L \frac{\varphi(\zeta)}{\zeta-\zeta_0} d\zeta \ . \quad (1.2.22\text{b})$$

N. I. Muskehlishvily proves in his book (1962) that formula (1.1.22a) is valid not only for functions $\varphi(\zeta)$ satisfying the Hölder condition, but also for simply continuous functions. It should be borne in mind, however, that the difference $C^+(\zeta_0) - C^-(\zeta_0)$ implies a limit $\lim\{C(\zeta) - C(\zeta')\}$ for $\zeta\to\zeta_0^+$ and $\zeta'\to\zeta_0^-$ so that $|\zeta-\zeta_0| = |\zeta'-\zeta_0|$, i.e., the points ζ and ζ' tending to ζ_0 are equally spaced from the latter. This yields, in particular, the following important corollary.

Let the function $\varphi(\zeta)$ be continuous over some smooth portion, $\tilde{L}$, of the line L ($\tilde{L}$ may be coincident with L) and let $C(\zeta')$ be extendible continuously over L on the left (right), then the function $C(\zeta)$ is also extendible continuously on the right (left) of $\tilde{L}$, except maybe for the ends of $\tilde{L}$.

The Sokhotsky-Plemelj formulas (1.2.21 a), (1.2.21 b) or (1.2.22a), (1.2.22b) lead to exceedingly wide-ranging and significant consequences both in the theory of functions of a complex variable and in its multinumerous applications.

1.2.4 Generalization of the Sokhotsky-Plemelj Formulas for Piecewise Smooth Curves

It is pertinent to note that the above considerations have been concerned with the existence of singular integrals and limit values of the Cauchy-type integral for smooth curves L. It is evident, however, that these results can be readily extended also to piecewise smooth curves.

Indeed, the property of contour smoothness has been virtually utilized only twice: in the calculation of the limits

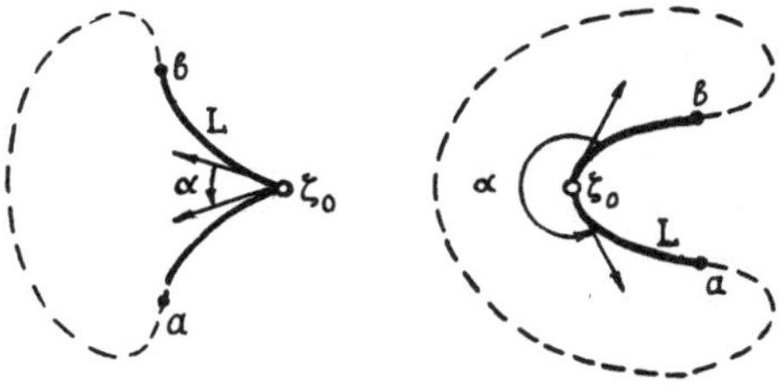

Fig. 5. Generalization of the Sokhotsky-Plemelj formulas for a corner ζ_0 of a piecewise smooth curve L

$$\lim_{\varrho\to 0}\int_{\gamma_\varrho}\frac{d\zeta}{\zeta-\zeta_0}\ \text{[formula (1.2.9)]} \quad\text{and}\quad \lim_{\varrho\to 0}\int_{L_\pi}\frac{d\zeta}{\zeta-\zeta_0}\ \text{[formula (1.2.12)]} .$$

But on examining closely the whole reasoning leading to formulas (1.2.9) and (1.2.12), one is readily convinced of the following. Let ζ_0 be a corner of a piecewise smooth curve L and α be an angle made by two lines tangent to L at a point ζ_0 and counted off on the left of the curve L (inside the domain D if L is a closed contour) (Fig. 5). Then, the following formulas

$$\lim_{\varrho\to 0}\int_{\gamma_\varrho}\frac{d\zeta}{\zeta-\zeta_0}=i(2\pi-\alpha) \tag{1.2.23}$$

and

$$\lim_{\varrho\to 0}\int_{L_\varrho}\frac{d\zeta}{\zeta-\zeta_0}=\int_L\frac{d\zeta}{\zeta-\zeta_0}=\ln\frac{b-\zeta_0}{a-\zeta_0}+i\alpha \tag{1.2.24}$$

seem to hold good.

As a result, expression (1.2.14) for the singular Cauchy-type integral in the case of an open curve L is written in the form:

$$C(\zeta_0)=\frac{1}{2\pi i}\int_L\frac{\varphi(\zeta)-\varphi(\zeta_0)}{\zeta-\zeta_0}d\zeta+\frac{\varphi(\zeta_0)}{2\pi i}\ln\frac{b-\zeta_0}{a-\zeta_0}+\frac{\alpha}{2\pi}\varphi(\zeta_0) . \tag{1.2.25}$$

In a similar way, passage to the limit in formula (1.2.7) with due account of (1.2.23) yields for a closed curve L:

$$C(\zeta_0)=\frac{1}{2\pi i}\int_L\frac{\varphi(\zeta)-\varphi(\zeta_0)}{\zeta-\zeta_0}d\zeta+\frac{\alpha}{2\pi}\varphi(\zeta_0) , \tag{1.2.26}$$

which is consistent with the result obtained directly from (1.2.25) for $a=b$. The Sokhotsky-Plemelj formulas for corners of the curve L are changed accordingly. Indeed, formula (1.2.20) for a corner ζ_0 with due reference to (1.2.24) will be cast in the form:

$$\lim_{\zeta'\to\zeta_0^\pm}\psi(\zeta')=\psi(\zeta_0)=\frac{1}{2\pi i}\int_L\frac{\varphi(\zeta)}{\zeta-\zeta_0}d\zeta-\frac{\varphi(\zeta_0)}{2\pi i}\int_L\frac{d\zeta}{\zeta-\zeta_0}=C(\zeta_0)-\frac{\alpha}{2\pi}\varphi(\zeta_0) . \tag{1.2.27}$$

Hence, proceeding to the limit in formula (1.2.15) and allowing for (1.2.17), just as in the above considerations, we derive:

$$C^+(\zeta_0) = C(\zeta_0) + \left(1 - \frac{\alpha}{2\pi}\right)\varphi(\zeta_0)\ , \quad C^-(\zeta_0) = C(\zeta_0) - \frac{\alpha}{2\pi}\varphi(\zeta_0)\ . \tag{1.2.28}$$

Note that formulas (1.2.28) remain valid also for cusps, under the assumption that $\alpha = 0$ or $\alpha = 2\pi$, depending on whether the cusp faces the right or the left side (outward or inward) of the curve L. Proof of this is provided in the book by N. I. Muskhelishvily (1962).

In concluding the present subsection, we should like to call the reader's attention to the following circumstance. In view of the Sokhotsky-Plemelj formulas, the limit values of the Cauchy-type integral (the $C^+(\zeta_0)$ and $C^-(\zeta_0)$ functions) are continuous functions of ζ_0 everywhere on a piecewise smooth contour L. Moreover, one can see that *if L is a smooth curve and $\varphi(\zeta)$ satisfies the Hölder condition with an indicator λ on L, the limit values of the Cauchy-type integral, $C^+(\zeta_0)$ and $C^-(\zeta_0)$ will also satisfy this condition, indicator being the same for $\lambda < 1$ or lower than λ by an arbitrarily small value for $\lambda = 1$.*

A similar property is exhibited by a singular integral in terms of the Cauchy principal value $C(\zeta_0)$. Thanks to these properties, we can employ values $C^+(\zeta_0)$, $C^-(\zeta_0)$, and $C(\zeta_0)$ as densities of some other Cauchy-type integrals, i.e., we can integrate repeatedly with the Cauchy kernel.

1.2.5 Cauchy-Type Integrals Along the Real Axis

Subsequently, in applications we will often refer to Cauchy-type integrals along the real axis. Let $\varphi(x)$ be a complex function of a real variable x satisfying the Hölder condition for any finite x and tending to a certain limit $\varphi(\infty)$ when $x \to \pm\infty$. We assume also that for high values of x the following inequality

$$|\varphi(x) - \varphi(\infty)| < \frac{A}{|x|^\mu}\ , \quad \mu > 0\ , \quad A > 0 \tag{1.2.29}$$

is valid. Relying on these assumptions we will consider the Cauchy-type integral

$$C(\zeta) = \frac{1}{2\pi i}\int_{-\infty}^{+\infty} \frac{\varphi(x)}{x - \zeta}\,dx\ , \tag{1.2.30}$$

where ζ is not on the real axis.

If $\varphi(\infty) \neq 0$, improper integral (1.2.30) is divergent, i.e., the expression $\frac{1}{2\pi i}\int_a^b \frac{\varphi(x)}{x-\zeta}\,dx$ has no limit when $a \to -\infty$ and $b \to +\infty$ independently of each other.

But if $|a| = |b| = M$, there is a limit

$$\lim_{M\to\infty} \frac{1}{2\pi i}\int_{-M}^{M} \frac{\varphi(x)}{x-\zeta}\,dx\ ,$$

and this limit is known as the principal value of integral (1.2.30) taken over an infinite axis. For the sake of proof, we will write

$$\int_{-M}^{M} \frac{\varphi(x)}{x-\zeta}\, dx = \int_{-M}^{M} \frac{\varphi(x)-\varphi(\infty)}{x-\zeta}\, dx + \varphi(\infty) \int_{-M}^{M} \frac{dx}{x-\zeta} \; . \tag{1.2.31}$$

The integrand in the first integral of the right-hand side is of an order of $|x|^{-1-\mu}$ for large $|x|$, which is due to (1.2.29). As a consequence, the corresponding improper integral is convergent. The second integral is evaluated readily:

$$\int_{-M}^{M} \frac{dx}{x-\zeta} = \ln |M-\zeta| - \ln|-M-\zeta| \pm ia \; . \tag{1.2.32}$$

Here a is the angle made by the straight lines connecting the point ζ with the points M and $(-M)$; it is taken with a positive sign if the point ζ lies in the upper half-plane, and it caries a negative sign when the point ζ is in the lower half-plane. Evidently, when $M \to \infty$ the angle $a \to \pi$ and $\ln \left| \frac{M-\zeta}{-M-\zeta} \right| \to \ln 1 = 0$.

Thus, we derive that

$$\lim_{M\to\infty} \frac{1}{2\pi i} \int_{-M}^{M} \frac{\varphi(x)}{x-\zeta}\, dx = \frac{1}{2\pi i} \int_{-\infty}^{+\infty} \frac{\varphi(x)-\varphi(\infty)}{x-\zeta}\, dx \pm \tfrac{1}{2}\varphi(\infty) \; . \tag{1.2.33}$$

In particular, once the density $\varphi(x)$ vanishes at infinity, $\varphi(\infty) = 0$, then

$$\lim_{M\to\infty} \frac{1}{2\pi i} \int_{-M}^{M} \frac{\varphi(x)}{x-\zeta}\, dx = \frac{1}{2\pi i} \int_{-\infty}^{+\infty} \frac{\varphi(x)}{x-\zeta}\, dx \; . \tag{1.2.34}$$

Expressions (1.2.33) and (1.2.34) define the principal value of integral (1.2.30) taken over infinite limits.

Hereinafter, infinite integrals will always cover the concept of their principal values.

By reasoning similar to that used above one can see that all the described properties of the Cauchy-type integral are also valid for integrals taken over the real axis. These results hold for the Cauchy-type integral taken over an arbitrary smooth curve L tending to infinity also.

1.3 Cauchy and Hilbert Integral Transforms

This subsection is concerned with some significant results following directly from the Cauchy-type integral theory and relating to the properties of boundary values of functions. These results permit us to construct pairs of inversible Cauchy and Hilbert integral transforms used extensively later.

1.3.1 Integral Boundary Conditions for Analytical Functions

Let us consider a domain D in the plane bounded by a smooth contour L. We assume a specified continuous complex function $\varphi(\zeta)$ on L. The question arises whether there exists an analytical function in the domain $D(C\bar{D})$ for which the specified function $\varphi(\zeta)$ will be its limit value on the contour L. Generally, this question should be answered negatively. Indeed, on the basis of the values of the real part, $\operatorname{Re}\varphi(\zeta)$, we can construct a function $U(x, z)$ harmonic in the domain D (or $C\bar{D}$) whose limit values on the contour L will coincide with $\operatorname{Re}\varphi(\zeta)$ (the known Dirichlet problem). Similarly, from the imaginary part $\operatorname{Im}\varphi(\zeta)$ we can construct another function $V(x, z)$ harmonic in D (or $C\bar{D}$). For the complex-valued function $f(\zeta) = U(x, z)+iV(x, z)$ formed from the above two to be analytical in D (or $C\bar{D}$), it is necessary and sufficient that U and V be interrelated by the Cauchy-Riemann relations. These conditions, however, may be not fulfilled under arbitrary specification of the functions $\operatorname{Re}\varphi(\zeta)$ and $\operatorname{Im}\varphi(\zeta)$. In other words, if $\varphi(\zeta)$ is a boundary value on the contour L of some function $f(\zeta) = U(x, z)+iV(x, z)$ analytical in the domain D (or $C\bar{D}$), the function $\varphi(\zeta)$ should satisfy certain conditions ensuring the Cauchy-Riemann conditions for the functions $U(x, z)$ and $V(x, z)$. These conditions will be referred to as *boundary conditions for an analytical function.* Let us establish the boundary conditions.

We will write the Cauchy-type integrals with a continuous density $\varphi(\zeta)$:

$$C(\zeta') = \frac{1}{2\pi i}\int_L \frac{\varphi(\zeta)}{\zeta-\zeta'}\,d\zeta\ . \tag{1.3.1}$$

If $\varphi(\zeta)$ is a boundary value of the function $f(\zeta)$ analytical in D, we have, according to Cauchy formula (1.1.1) that

$$C(\zeta') = 0 \quad \text{for any} \quad \zeta' \in C\bar{D}\ . \tag{1.3.2}$$

Conversely, it is easy to see that if (1.3.2) is the case, $\varphi(\zeta)$ is a boundary value of the function $f(\zeta)$ defined by the expression:

$$f(\zeta') = C(\zeta') = \frac{1}{2\pi i}\int_L \frac{\varphi(\zeta)}{\zeta-\zeta'}\,d\zeta\ , \quad \zeta' \in D\ , \tag{1.3.3}$$

analytical in D and extendible continuously from D onto L. Indeed, by virtue of (1.3.2), the function $C(\zeta')$ is extendible continuously onto L from the outside of the contour and

$$C^-(\zeta_0) = \lim_{\zeta'\to\zeta_0^-} C(\zeta') = 0\ . \tag{1.3.4}$$

As a consequence, allowing for the Sokhotsky-Plemelj formula (1.2.22a) and in accordance with the considerations of Sect. 1.2.3, the function $C(\zeta')$ is also extendible continuously onto L from the inside of the contour and

$$\lim_{\zeta'\to\zeta_0^+} f(\zeta') = C^+(\zeta_0) = \varphi(\zeta_0)\ . \tag{1.3.5}$$

In this way, we proved the following theorem.

Theorem 1.3.1. For the continuous complex function $\varphi(\zeta)$ specified on a closed smooth contour L to be a boundary value of the function analytical in the inner domain D, it is necessary and sufficient to meet the condition

$$\frac{1}{2\pi i}\int_L \frac{\varphi(\zeta)}{\zeta-\zeta'}\,d\zeta = 0 \quad \text{for any} \quad \zeta' \in C\bar{D}\ . \tag{1.3.6}$$

By similar reasoning based on the Cauchy formula for an infinite domain (1.1.2) we can prove the following theorem.

Theorem 1.3.2. The necessary and sufficient prerequisite for the function $\varphi(\zeta)$ to be a boundary value of the function $f(\zeta)$ analytical in the outer domain $C\bar{D}$ is the relation:

$$\frac{1}{2\pi i}\int_L \frac{\varphi(\zeta)}{\zeta-\zeta'}\,d\zeta = f(\infty) \quad \text{for any} \quad \zeta' \in D\ . \tag{1.3.7}$$

Conditions (1.3.6) and (1.3.7) can be made somewhat more compact if the function $\varphi(\zeta)$ is required to satisfy the Hölder condition on L.

Theorem 1.3.3. For the Hölder continuous function $\varphi(\zeta)$ specified on a closed smooth contour L to be a boundary value of the function analytical everywhere in D, it is necessary and sufficient to meet the condition

$$\varphi(\zeta_0) = \frac{1}{\pi i}\int_L \frac{\varphi(\zeta)}{\zeta-\zeta_0}\,d\zeta \quad \text{for any} \quad \zeta_0 \in L\ . \tag{1.3.8}$$

Proof. The necessity of condition (1.3.8) is proved by proceeding to the limit in (1.3.6) for $\zeta \to \zeta_0^+$ and by applying the Sokhotsky-Plemelj formula (1.2.21 b)

$$\frac{1}{2\pi i}\int_L \frac{\varphi(\zeta)}{\zeta-\zeta_0}\,d\zeta - \tfrac{1}{2}\varphi(\zeta_0) = 0\ .$$

Conversely, if condition (1.3.8) is the case, for Cauchy-type integral (1.3.1) we obtain, according to the Sokhotsky-Plemelj formula (1.2.21 b), that $C^-(\zeta_0) = 0$. Hence, in accordance with formula (1.2.22b), we have that $\varphi(\zeta_0) = C^+(\zeta_0)$, which was to be proved.

This theorem, just as theorem (1.3.1) earlier, is extended to the case of an infinite domain $C\bar{D}$.

Theorem 1.3.4. The necessary and sufficient prerequisite for the Hölder continuous function $\varphi(\zeta)$ to be a boundary value of the function $f(\zeta)$ analytical in the outer domain $C\bar{D}$ is the relation:

$$\varphi(\zeta_0) = -\frac{1}{\pi i}\int_L \frac{\varphi(\zeta)}{\zeta-\zeta_0}\,d\zeta + 2f(\infty) \quad \text{for any} \quad \zeta_0 \in L\ . \tag{1.3.9}$$

Finally, it is noteworthy that using formulas (1.3.8) and (1.3.9) we can evaluate directly singular integrals where the integral density is a boundary value of an analytical function.

1.3.2 Determination of a Piecewise Analytical Function from a Specified Discontinuity

Now we will solve a problem of critical importance for further considerations. Let a function $\varphi(\zeta)$ satisfying the Hölder condition be specified on a closed smooth curve (contour) L. It is necessary to define a piecewise analytical function $\Phi(\zeta')$ vanishing at infinity and experiencing a discontinuity on transition through the contour L, which is equal to $\varphi(\zeta)$, i.e.,

$$\Phi^+(\zeta_0) - \Phi^-(\zeta_0) = \varphi(\zeta_0) \ , \quad \zeta_0 \in L \ .$$

A solution to this problem in terms of the Sokhotsky-Plemelj formula (1.2.22a) seems to be the following Cauchy-type integral:

$$\Phi(\zeta') = \frac{1}{2\pi i} \int_L \frac{\varphi(\zeta)}{\zeta - \zeta'} d\zeta \ . \tag{1.3.10}$$

We will show that this solution is unique. Indeed, assuming two available solutions $\Phi(\zeta')$ and $\tilde{\Phi}(\zeta')$ and considering their difference $\Psi(\zeta') = \Phi(\zeta') - \tilde{\Phi}(\zeta')$, we find that the discontinuity on the line L is zero for this difference. Hence, $\Psi(\zeta')$ is a function analytical throughout the plane (if it is redefined appropriately on L) and vanishing at infinity. This implies, by the Liouville theorem, that it is identically zero, so the two solutions coincide.

It is remarkable that the above result is readily extended to the case where L is an arbitrary piecewise smooth curve (closed or open). It is evident from the fact that the Sokhotsky-Plemelj formulas (1.2.22a) and (1.2.22b) hold good also for piecewise smooth lines. And finally, the above solution is valid for a discontinuity line L approaching infinity (Muskhelishvily 1962).

1.3.3 Inversion Formulas for the Cauchy-Type Integral (Cauchy Integral Transforms)

Let L be a closed smooth contour and $\varphi(\zeta)$ be an arbitrary Hölder continuous function specified on L. We will designate by $\psi(\zeta_0)$ a singular integral:

$$\psi(\zeta_0) = \frac{1}{\pi i} \int_L \frac{\varphi(\zeta)}{\zeta - \zeta_0} d\zeta \ , \quad \zeta_0 \in L \ . \tag{1.3.11}$$

According to the remark made at the end of Sect. 1.2.4, the function $\psi(\zeta)$ is also Hölder continuous on L. In a particular case, where $\varphi(\zeta)$ is a boundary value of the function analytical in the domain D, we have, in accordance with (1.3.8):

$$\psi(\zeta_0) \equiv \varphi(\zeta_0) \ , \quad \zeta_0 \in L \ . \tag{1.3.12}$$

Generally, however, $\psi(\zeta_0)$ is not equal to $\varphi(\zeta_0)$.

In many applications, the following problem arises: *how to define $\varphi(\zeta)$ from specified $\psi(\zeta)$, i.e., to solve integral equation (1.3.11).*

Equation (1.3.11) is one of the simple singular integral equations whose theory is outlined in detail in the cited books by N. I. Muskhelishvily (1962) and F. D.

Gakhov (1963). Since it is of particular interest in conjunction with the subject matter of the present book, one of its solutions presented in the former study (Muskhelishvily 1962) will be given here. At the same time, having solved Eq. (1.3.11) we arrive at the so called *inversion formulas for the Cauchy-type integral,* which are extremely useful in many transformations of geophysical fields.

Let us introduce into our considerations a piecewise analytical function (1.3.1) vanishing at infinity. Then, on the basis of the Sokhotsky-Plemelj formula (1.2.22b), Eq. (1.3.11) is rewritten in the form

$$C^+(\zeta_0)+C^-(\zeta_0)=\psi(\zeta_0)\ , \quad \zeta_0\in L\ . \tag{1.3.13}$$

The second piecewise analytical function $\Phi(\zeta')$ is defined as follows:

$$\Phi(\zeta')=\begin{cases} C(\zeta') & \text{for} \quad \zeta'\in D\ , \\ -C(\zeta') & \text{for} \quad \zeta'\in C\bar{D}\ . \end{cases} \tag{1.3.14}$$

Then condition (1.3.13) is cast as:

$$\Phi^+(\zeta_0)-\Phi^-(\zeta_0)=\psi(\zeta_0)\ , \quad \zeta_0\in L\ . \tag{1.3.15}$$

As a result, $\Phi(\zeta')$ is a piecewise analytical function with a specified jump $\psi(\zeta_0)$ and, hence, we obtain from the result of the previous subsection [formula (1.3.10)] that

$$\Phi(\zeta')=\frac{1}{2\pi i}\int_L\frac{\psi(\zeta)}{\zeta-\zeta'}\,d\zeta\ . \tag{1.3.16}$$

Applying the Sokhotsky-Plemelj formula (1.2.22b) to the Cauchy-type integral (1.3.16), we find

$$\Phi^+(\zeta_0)+\Phi^-(\zeta_0)=\frac{1}{\pi i}\int_L\frac{\psi(\zeta)}{\zeta-\zeta_0}\,d\zeta\ . \tag{1.3.17}$$

On the other hand, by virtue of (1.3.14) and the Sokhotsky-Plemelj formula (1.2.22a), for the Cauchy-type integral $C(\zeta')$ we have

$$\Phi^+(\zeta_0)+\Phi^-(\zeta_0)=C^+(\zeta_0)-C^-(\zeta_0)=\varphi(\zeta_0)\ . \tag{1.3.18}$$

Comparing (1.3.17) and (1.3.18) we write finally:

$$\varphi(\zeta_0)=\frac{1}{\pi i}\int_L\frac{\psi(\zeta)}{\zeta-\zeta_0}\,d\zeta\ . \tag{1.3.19}$$

Formula (1.3.19) is just the (unique) solution to Eq. (1.3.11).

In a similar way, we can derive (1.3.11) from (1.3.19). Thus, we have *a pair of mutually invertible integral transforms*

$$\psi(\zeta_0)=\frac{1}{\pi i}\int_L\frac{\varphi(\zeta)}{\zeta-\zeta_0}\,d\zeta\ , \tag{1.3.20}$$

$$\varphi(\zeta_0)=\frac{1}{\pi i}\int_L\frac{\psi(\zeta)}{\zeta-\zeta_0}\,d\zeta\ . \tag{1.3.21}$$

Formulas (1.3.20) and (1.3.21) will be also referred to as *Cauchy integral transforms.*

1.3.4 Hilbert Transforms

Of great value for applications are Cauchy integral transforms written for the case where L is a real axis.

Let $\varphi(x)$ be a complex function of a real variable x satisfying the Hölder condition for any finite x and vanishing at infinity. Under these assumptions, as indicated in Sect. 1.2.5, there are principal values of an infinite Cauchy-type integral and relations (1.3.20) and (1.3.21) take the form:

$$\psi(x_0) = \frac{1}{\pi i}\int_{-\infty}^{+\infty} \frac{\varphi(x)}{x-x_0}\,dx \;, \quad \varphi(x_0) = \frac{1}{\pi i}\int_{-\infty}^{+\infty} \frac{\psi(x)}{x-x_0}\,dx \;. \tag{1.3.22}$$

Separating the real and imaginary parts of functions $\varphi(x)$ and $\psi(x)$ in formulas (1.3.22) we derive two pairs of the following formulas

$$\begin{aligned} &\mathrm{Re}\,\psi(x_0) = \frac{1}{\pi}\int_{-\infty}^{+\infty} \frac{\mathrm{Im}\,\varphi(x)}{x-x_0}\,dx \;, \quad \mathrm{Im}\,\varphi(x_0) = -\frac{1}{\pi}\int_{-\infty}^{+\infty} \frac{\mathrm{Re}\,\psi(x)}{x-x_0}\,dx \;; \\ &\mathrm{Re}\,\varphi(x_0) = \frac{1}{\pi}\int_{-\infty}^{+\infty} \frac{\mathrm{Im}\,\psi(x)}{x-x_0}\,dx \;, \\ &\mathrm{Im}\,\psi(x_0) = -\frac{1}{\pi}\int_{-\infty}^{+\infty} \frac{\mathrm{Re}\,\varphi(x)}{x-x_0}\,dx \;. \end{aligned} \tag{1.3.23}$$

Relations (1.3.23) are called *Hilbert integral transforms.* Using the symbol $\hat{G}$ to designate the following integral operator (Hilbert operator):

$$\hat{G}f(x) = \frac{1}{\pi}\int_{-\infty}^{+\infty} \frac{f(x)}{x-x_0}\,dx \;, \tag{1.3.24}$$

these relations can be rewritten as

$$\begin{aligned} &\mathrm{Re}\,\psi(x_0) = \hat{G}\,\mathrm{Im}\,\varphi(x) \;, \quad \mathrm{Im}\,\varphi(x_0) = -\hat{G}\,\mathrm{Re}\,\psi(x) \;; \\ &\mathrm{Re}\,\varphi(x_0) = \hat{G}\,\mathrm{Im}\,\psi(x) \;, \quad \mathrm{Im}\,\psi(x_0) = -\hat{G}\,\mathrm{Re}\,\varphi(x) \;. \end{aligned} \tag{1.3.25}$$

Note that the function $\varphi(x)$ is provided by an arbitrary Hölder continuous function vanishing at infinity. Hence, we can also select arbitrary real and imaginary parts of this function, if only they satisfy the above requirements. Thus, relation (1.3.25) permits formulation of the following significant finding.

Let $U(x)$ be an arbitrary real function of a real variable x satisfying the Hölder condition for any finite x and vanishing at infinity. The result of Hilbert transformation of $U(x)$ is designated by $V(x_0)$:

$$V(x_0) = \hat{G}U(x) = \frac{1}{\pi}\int_{-\infty}^{+\infty} \frac{U(x)}{x-x_0}\,dx \;. \tag{1.3.26}$$

Relation (1.3.26) can be treated, for a specified function $V(x)$, as an integral equation in $U(x)$ [positing that $V(x)$ satisfies the same conditions on the real axis as $U(x)$ does]. Then, in accordance with (1.3.25), we can solve this integral equation by applying the Hilbert operator to $V(x)$:

$$U(x_0) = -\hat{G}V(x) = -\frac{1}{\pi}\int_{-\infty}^{+\infty}\frac{V(x)}{x-x_0}\,dx\ . \tag{1.3.27}$$

In a similar way, we can obtain (1.3.26) by inverting (1.3.27).

Thus, relations (1.3.26) and (1.3.27) form a pair of mutually invertible integral transforms of real functions specified on the real axis.

In conclusion, we will consider an important particular case where $\varphi(x)$ is a boundary value on the real axis of the function $\varphi(\zeta)$ analytical everywhere in the upper half-plane and vanishing at infinity. Then, according to (1.3.12) we have that $\psi(x)\equiv\varphi(x)$, $-\infty<x<+\infty$. Hence,

$$\operatorname{Re}\psi(x)\equiv\operatorname{Re}\varphi(x)\ ,\quad \operatorname{Im}\psi(x)\equiv\operatorname{Im}\varphi(x)\ . \tag{1.3.28}$$

Therefore, formulas (1.3.25) take the form

$$\begin{aligned}\operatorname{Re}\varphi(x_0) &= \hat{G}\operatorname{Im}\varphi(x) = \frac{1}{\pi}\int_{-\infty}^{+\infty}\frac{\operatorname{Im}\varphi(x)}{x-x_0}\,dx\ ,\\ \operatorname{Im}\varphi(x_0) &= -\hat{G}\operatorname{Re}\varphi(x) = -\frac{1}{\pi}\int_{-\infty}^{+\infty}\frac{\operatorname{Re}\varphi(x)}{x-x_0}\,dx\ .\end{aligned} \tag{1.3.29}$$

In other words, *the boundary values on the real axis of a real and imaginary parts of a function analytical everywhere in the upper half-plane are interrelated by Hilbert transforms (1.3.29).*

Similar formulas for the boundary values of a function analytical everywhere in the lower half-plane and vanishing at infinity are derived from (1.3.29) by reversing the sign of the Hilbert operator (since moving along the X-axis from $-\infty$ to $+\infty$ we leave the region of analyticity on the right of the path of integration):

$$\begin{aligned}\operatorname{Re}\varphi(x_0) &= -\hat{G}\operatorname{Im}\varphi(x) = -\frac{1}{\pi}\int_{-\infty}^{+\infty}\frac{\operatorname{Im}\varphi(x)}{x-x_0}\,dx\ ,\\ \operatorname{Im}\varphi(x_0) &= \hat{G}\operatorname{Re}\varphi(x) = \frac{1}{\pi}\int_{-\infty}^{+\infty}\frac{\operatorname{Re}\varphi(x)}{x-x_0}\,dx\ .\end{aligned} \tag{1.3.30}$$

The Hilbert operator finds much favor in techniques used to separate geophysical fields (see Chap. 3).

2 Representation of Plane Geopotential Fields in the Form of the Cauchy-Type Integral

2.1 Plane Potential Fields and Their Governing Equations

This chapter is devoted to two-dimensional (plane) potential fields. The main geopotential fields dealt with in geophysics are the gravitational and constant magnetic fields. The physical nature of these fields is outlined comprehensively in many treatises on general and applied geophysics. That is why, within the scope of this book, we will dwell only on the problems of mathematical theory of analysis, integral transforms, and interpretation of geopotential fields. Moreover, we shall confine ourselves just to the issues resolved most efficiently in terms of the Cauchy-type integral. How wide the range of these issues is will be shown below.

2.1.1 Vector Field Equations

We will start with a description of the basic equations of the vector field. Let a vector field $\mathbf{F}(\mathbf{r})$ be specified in a region P of three-dimensional space, i.e., each point in this region with a radius vector $\mathbf{r}$ (drawn from the origin of a given Cartesian system of coordinates xyz) is associated with a particular vector $\mathbf{F}$. Assume also that the vector function $\mathbf{F}(\mathbf{r})$ is continuously differentiable in $\mathbf{P}$, i.e., it has all continuous partial derivatives in this region. Then, *everywhere in P the vector field $F(r)$ satisfies the following equations*

$$\operatorname{div}\mathbf{F}(\mathbf{r}) = q(\mathbf{r}) \ , \quad \operatorname{rot}\mathbf{F}(\mathbf{r}) = \mathbf{j}(\mathbf{r}) \ . \tag{2.1.1}$$

The scalar function $q(\mathbf{r})$ is known as *volume density of sources of the field* $\mathbf{F}$, while the vector function $\mathbf{j}(\mathbf{r})$ is termed *volume density of rotations of the field* $\mathbf{F}$.

In the investigation of geophysical fields, the range of the vector field $\mathbf{F}$ is normally taken to be the whole of infinite space, while the field sources and rotations are assumed to be localized within a finite region V. Under these assumptions, the problem of establishing the field from Eqs. (2.1.1) [i.e., from specified volume densities of sources, $q(\mathbf{r})$, and rotations, $\mathbf{j}(\mathbf{r})$] has a unique solution defined by

$$\mathbf{F}(\mathbf{r}') = -\frac{1}{4\pi}\operatorname{grad}'\iiint_V \frac{q(\mathbf{r})}{|\mathbf{r}-\mathbf{r}'|}\,dv + \frac{1}{4\pi}\operatorname{rot}'\iiint_V \frac{\mathbf{j}(\mathbf{r})}{|\mathbf{r}-\mathbf{r}'|}\,dv \ , \tag{2.1.2}$$

where integration is carried out with respect to the variable $\mathbf{r}$, and the prime appearing with the symbols grad and rot indicates that differentiation is accomplished with respect to the variable $\mathbf{r}'$. The expressions

$$U(\mathbf{r}') = -\frac{1}{4\pi} \iiint_V \frac{q(\mathbf{r})}{|\mathbf{r}-\mathbf{r}'|} dv \tag{2.1.3}$$

and

$$\mathbf{A}(\mathbf{r}') = \frac{1}{4\pi} \iiint_V \frac{\mathbf{j}(\mathbf{r})}{|\mathbf{r}-\mathbf{r}'|} dv \tag{2.1.4}$$

are referred to as a *scalar and a vector potential* of the field $\mathbf{F}(\mathbf{r})$. Using them, we can rewrite formula (2.1.2) as

$$\mathbf{F}(\mathbf{r}') = \mathrm{grad}'U(\mathbf{r}') + \mathrm{rot}'\mathbf{A}(\mathbf{r}') \ . \tag{2.1.5}$$

Definition 2.1. The field $\mathbf{F}_q$ satisfying the equations

$$\mathrm{div}\,\mathbf{F}_q(\mathbf{r}) = q(\mathbf{r}) \ , \quad \mathrm{rot}\,\mathbf{F}_q(\mathbf{r}) = 0 \ , \tag{2.1.6}$$

is called an irrotational or a potential field in the whole infinite space. It is defined unequivocally by specifying a scalar potential $U(\mathbf{r})$:

$$\mathbf{F}_q(\mathbf{r}') = \mathrm{grad}'U(\mathbf{r}') \ . \tag{2.1.7}$$

Substituting (2.1.7) into (2.1.6) we derive the following *Poisson equation for a scalar potential:*

$$\Delta U(\mathbf{r}) = q(\mathbf{r}) \ . \tag{2.1.8}$$

In a region devoid of field sources, the scalar potential is fitted by the *Laplace equation:*

$$\Delta U(\mathbf{r}) = 0 \ . \tag{2.1.8a}$$

Definition 2.2. The vector field $\mathbf{F}_j$ satisfying the equations

$$\mathrm{div}\,\mathbf{F}_j(\mathbf{r}) = 0 \ , \quad \mathrm{rot}\,\mathbf{F}_j(\mathbf{r}) = \mathbf{j}(\mathbf{r}) \ , \tag{2.1.9}$$

is called a rotational or a solenoidal field in the whole infinite space. It is defined unequivocally by specifying a vector potential $\mathbf{A}(\mathbf{r})$:

$$\mathbf{F}_j(\mathbf{r}') = \mathrm{rot}'\mathbf{A}(\mathbf{r}') \ . \tag{2.1.10}$$

Below are given some important definitions to be employed in the subsequent discussion.

Definition 2.3. If expression (2.1.7) holds good for a given vector field $\mathbf{F}_q$ within a region V of space, the field $\mathbf{F}_q$ is called potential in the region V.

Definition 2.4. If the vector field $\mathbf{F}_j$ in a region V of space can be represented in the form of (2.1.10), it is termed solenoidal in the region V.

Definition 2.5. The vector field $\mathbf{F}$ which is potential and solenoidal simultaneously in a region V of space is called the Laplace vector field in the region V.

It is easy to recognize that the Laplace field everywhere in V meets the Laplace vector equation $\Delta \mathbf{F} = \mathbf{0}$. Substituting Eqs. (2.1.10) into (2.1.9) we arrive at the *Poisson vector equation for a vector potential:*

$$\Delta \mathbf{A}(\mathbf{r}) = -\mathbf{j}(\mathbf{r}) \quad [\text{for } \operatorname{div} \mathbf{A}(\mathbf{r}) = 0] \ , \tag{2.1.11}$$

which goes, outside rotations, also into the Laplace equation

$$\Delta \mathbf{A}(\mathbf{r}) = \mathbf{0} \ . \tag{2.1.11 a}$$

In conjunction with the Laplace vector field, it is worth remembering the definition of a harmonic function.

Definition 2.6. The function $U(\mathbf{r})$, specified in a region V of space is called harmonic in this region, provided it is twice continuously differentiable in this region and satisfies the Laplace equation $\Delta U = 0$ therein.

Thus, the Cartesian components of the Laplace vector field are harmonic functions.

And finally, we should like to note that by virtue of formulas (2.1.7) and (2.1.8a), *the potential field everywhere outside sources is a Laplace field and its scalar potential is a harmonic function.*

2.1.2 Concept of a Plane Field

This subsection is concerned only with irrotational fields, i.e., with potential fields in the whole space. Furthermore, we assume that the volume density of their sources is independent of one of the space coordinates, say, y (Fig. 6). In other words, it is postulated that field sources are concentrated within regions bounded by cylindrical surfaces, with their elements parallel to the y-axis, and that their density is constant along this axis. Let us examine the behavior of the above general formulas, as applied to this particular case.

We should note, first of all, that here the field $\mathbf{F}$ itself and its scalar potential are obviously independent of the variable y, too. As a result, all the derivatives of the field and potential with respect to y are zero. Therefore, the terms related to differentiation with respect to y disappear from Eqs. (2.1.6) and (2.1.8), while the y-component of the vector field also vanishes, in accordance with (2.1.7).

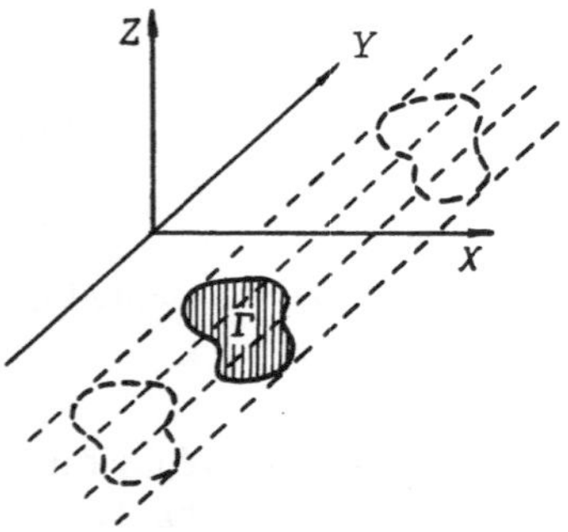

Fig. 6. Definition of the concept of a plane field. Γ is the domain occupied with field sources

Thus, the vector field becomes *plane-parallel,* since all the field vectors are always parallel to the *XOZ* coordinate plane; at all points on any straight line going in parallel with the y-axis, the field vectors are equal in magnitude and direction. Evidently, this field is defined completely by a plane field of vectors in any vertical plane, say, in the *XOZ* plane. From now on it is therefore referred to as a *plane field.*

2.1.3 Plane Field Equations

Now we will write plane field equations obtained on appropriate reduction of Eqs. (2.1.6), (2.1.7), and (2.1.8):

$$\mathrm{div}_\Gamma \mathbf{F}(\boldsymbol{\varrho}) = q(\boldsymbol{\varrho}) \ , \quad \mathrm{rot}_\Gamma \mathbf{F}(\boldsymbol{\varrho}) = \mathbf{0} \ ; \tag{2.1.12}$$

$$\mathbf{F}(\boldsymbol{\varrho}) = \mathrm{grad}_\Gamma U(\boldsymbol{\varrho}) \ ; \tag{2.1.13}$$

$$\Delta_\Gamma U(\boldsymbol{\varrho}) = q(\boldsymbol{\varrho}) \ . \tag{2.1.14}$$

The latter formulas include the following symbols: $\boldsymbol{\varrho}$ is the radius vector of a point (X, Z) of the *XOZ* plane; grad_Γ, div_Γ, rot_Γ, and Δ_Γ are the operators of two-dimensional differentiation. These operators are defined as follows. Take a closed smooth contour L in the *XOZ* plane (Fig. 7) embracing an observation point (x, z) with a radius vector $\boldsymbol{\varrho}_0$. Let $\Delta\Gamma$ designate the domain bounded by the contour L. Assume that $\mathbf{l}$ is a unit vector of the line tangent to L (in the positive sense of contour traversal), $\mathbf{v}$ is the unit vector of an outward normal to L lying in the *XOZ* plane, and $\mathbf{n}$ is the unit vector of a normal to the *XOZ* plane pointing along the y-axis (Fig. 7). Now let $\Delta\Gamma \to 0$, so that the contour L might tighten continuously to the point (x, z). Then the *two-dimensional differentiation operators* grad_Γ, div_Γ, and rot_Γ, are defined by the following equivalent formulas:

$$\mathrm{grad}_\Gamma U(\boldsymbol{\varrho}) = \frac{\partial U}{\partial x}\mathbf{d}_x + \frac{\partial U}{\partial z}\mathbf{d}_z = \lim_{\Delta\Gamma\to 0} \frac{1}{\Delta\Gamma} \int_\Gamma U\mathbf{v}\,dl \ , \tag{2.1.15}$$

$$\mathrm{div}_\Gamma \mathbf{F}(\boldsymbol{\varrho}) = \frac{\partial F_x}{\partial x} + \frac{\partial F_z}{\partial z} = \lim_{\Delta\Gamma\to 0} \frac{1}{\Delta\Gamma} \int_L \mathbf{F}\cdot\mathbf{v}\,dl \ , \tag{2.1.16}$$

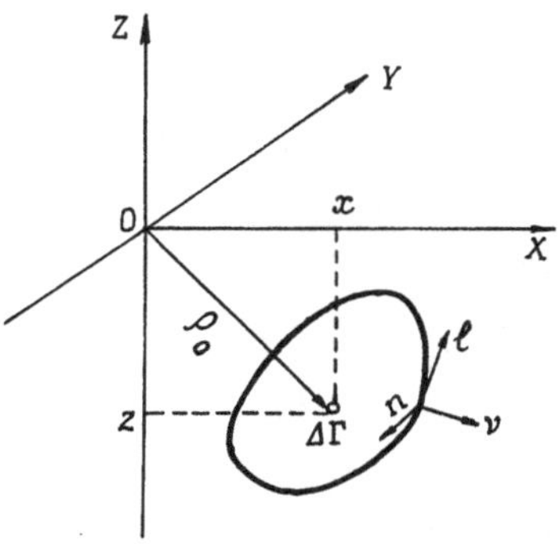

Fig. 7. Scheme explaining the definition of two-dimensional differentiation operators; **n**, **v**, and **l** are the right triad of mutually orthogonal vectors

$$\mathrm{rot}_\Gamma \mathbf{F}(\varrho) = \left(\frac{\partial F_x}{\partial z} - \frac{\partial F_z}{\partial x}\right) \mathbf{d}_y = \mathbf{n} \lim_{\Delta\Gamma \to 0} \frac{1}{\Delta\Gamma} \int_L \mathbf{F}\cdot\mathbf{l}\,dl \ , \tag{2.1.17}$$

where $\mathbf{d}_x$, $\mathbf{d}_y$, and $\mathbf{d}_z$ are the unit vectors of Cartesian coordinates, F_x and F_z are the scalar components of a plane vector field **F**.

Equations (2.1.15) through (2.1.17) follow directly from the two-dimensional Ostrogradsky-Gauss formulas

$$\iint_{\Delta\Gamma} \left(\frac{\partial U}{\partial x}\mathbf{d}_x + \frac{\partial U}{\partial z}\mathbf{d}_z\right) ds = \int_L U\mathbf{v}\,dl \ , \tag{2.1.18}$$

$$\iint_{\Delta\Gamma} \left(\frac{\partial F_x}{\partial x} + \frac{\partial F_z}{\partial z}\right) ds = \int_L \mathbf{F}\cdot\mathbf{v}\,dl \ , \tag{2.1.19}$$

and from the two-dimensional Stokes formula

$$\iint_{\Delta\Gamma} \left(\frac{\partial F_x}{\partial z} - \frac{\partial F_z}{\partial x}\right) ds = \int_L \mathbf{F}\cdot\mathbf{l}\,dl \tag{2.1.20}$$

(where $ds = dx\,dz$) which are valid for functions having continuous partial derivatives in $\Delta\Gamma$ and which are continuous in the closed domain $\bar{\Delta\Gamma}$. The latter formulas will be also extensively used subsequently.

2.1.4 Plane Field Flow Function

Let Γ represent the projection of a cylindrical region V filled with field sources onto the XOZ plane (see Fig. 6). We will consider the space domain $C\bar{\Gamma}$ devoid of field sources. In this domain, Eqs. (2.1.12) with due reference to (2.1.16) and (2.1.17) are written in the form

$$\frac{\partial F_x}{\partial x} + \frac{\partial F_z}{\partial z} = 0 \ , \quad \frac{\partial F_x}{\partial z} - \frac{\partial F_z}{\partial x} = 0 \ . \tag{2.1.21}$$

The former equation indicates that the expression $-F_z dx + F_x dz$ in the domain V is an exact differential of a function V:

$$dV = -F_z dx + F_x dz \ , \tag{2.1.22}$$

so that

$$\partial V/\partial x = -F_z \ , \quad \partial V/\partial z = F_x \ . \tag{2.1.23}$$

The function $V(x, z)$ is reconstructed in the domain $C\bar{\Gamma}$ from its exact differential (2.1.22) by means of the formula

$$V(x, z) = \int_{(x_0, z_0)}^{(x, z)} -F_z dx + F_x dz + a \ , \tag{2.1.24}$$

where the integral is taken around any path L connecting, within $C\bar{\Gamma}$, a fixed point (x_0, z_0) with a variable point (x, z), and a is an arbitrary constant.

Let us consider a line M of equal values of the function $V(xz)$: $V(x, z) = C = \text{const}$. The vector of a tangent line $\mathbf{T}$ at each point of this line has the form: $\mathbf{T} = (\partial V/\partial z - \partial V/\partial x)$, i.e., by virtue of (2.1.23) it coincides with the vector $\mathbf{F}: \mathbf{T} = (F_x, F_z) = \mathbf{F}$. In other words, the line M is a *vector line* of the field $\mathbf{F}$ or a *current line.* That is why the function $V(x, z)$ is given the name of a *flow function of a plane vector field* $\mathbf{F}$. Knowing the flow function, we can determine the plane vector field $\mathbf{F}$ itself on the basis of (2.1.23).

2.2 Logarithmic Potentials and the Cauchy-Type Integral

Plane potential fields introduced in the previous subsection are expressed in terms of the so called logarithmic potentials. We will consider the main properties of logarithmic potentials and establish their relationship to the Cauchy-type integral.

2.2.1 Logarithmic Potentials

The expression for an irrotational (potential over the whole space) vector field $\mathbf{F}$ is, according to (2.1.3) and (2.1.7):

$$\mathbf{F}(\mathbf{r}') = -\frac{1}{4\pi} \iiint_V q(\mathbf{r}) \frac{\mathbf{r}-\mathbf{r}'}{|\mathbf{r}-\mathbf{r}'|^3} dv \;. \tag{2.2.1}$$

Upon transition to densities of sources $q(\mathbf{r})$ independent of the y-coordinate and concentrated in the domain V bounded by a cylindrical surface (see Fig. 6), expression (2.2.1) is transformed as follows:

$$\mathbf{F}(\mathbf{r}') = -\frac{1}{4\pi} \iint_\Gamma q(\mathbf{r}) \int_{-\infty}^{+\infty} \frac{\mathbf{r}-\mathbf{r}'}{|\mathbf{r}-\mathbf{r}'|^3} dy ds \;, \tag{2.2.2}$$

where $\mathbf{r} = (x, y, z)$; $\mathbf{r}' = (x', y', z')$; $|\mathbf{r}-\mathbf{r}'| = [(x-x')^2+(y-y')^2+(z-z')^2]^{1/2}$; Γ is the cross section of a cylindrical volume V by the XOZ plane (see Fig. 6); $ds = dx dz$ is the differential element of the area.

We will rewrite expression (2.2.2) in components:

$$F_x(x', z') = -\frac{1}{4\pi} \iint_\Gamma q(\mathbf{r})(x-x') \int_{-\infty}^{+\infty} \frac{dy}{|\mathbf{r}-\mathbf{r}'|^3} ds \;,$$

$$F_y(x', z') = 0 \;, \tag{2.2.3}$$

$$F_z(x', z') = -\frac{1}{4\pi} \iint_\Gamma q(\mathbf{r})(z-z') \int_{-\infty}^{+\infty} \frac{dy}{|\mathbf{r}-\mathbf{r}'|^3} ds \;.$$

Direct integration indicates that

$$\int_{-\infty}^{+\infty} \frac{dy}{|\mathbf{r}-\mathbf{r}'|^3} = \frac{2}{|\boldsymbol{\varrho}-\boldsymbol{\varrho}'|^2} \;, \tag{2.2.4}$$

where $\boldsymbol{\varrho} = (x, z)$; $\boldsymbol{\varrho}' = (x', z')$; $|\boldsymbol{\varrho}-\boldsymbol{\varrho}'| = [(x-x')^2+(z-z')^2]^{1/2}$.

Substituting (2.2.4) into (2.2.3) and allowing for

$$\mathbf{F}(\mathbf{r}) = \mathbf{F}(\varrho)\ ,\quad q(\mathbf{r}) = q(\varrho)\ ,$$

$$\frac{x-x'}{|\varrho-\varrho'|^2} = \frac{\partial}{\partial x'}\ln\frac{1}{|\varrho-\varrho'|}\ ;\quad \frac{z-z'}{|\varrho-\varrho'|^2} = \frac{\partial}{\partial z'}\ln\frac{1}{|\varrho-\varrho'|}\ ,$$

we write

$$\mathbf{F}(\varrho') = \operatorname{grad}'_\Gamma\left(-\frac{1}{2\pi}\iint_\Gamma q(\varrho)\ln\frac{1}{|\varrho-\varrho'|}\,ds\right)\ . \tag{2.2.5}$$

The function

$$U(\varrho') = -\frac{1}{2\pi}\iint_\Gamma q(\varrho)\ln\frac{1}{|\varrho-\varrho'|}\,ds \tag{2.2.6}$$

is termed a *logarithmic potential of a simple two-dimensional source* distributed within the domain Γ.

One can readily see that the logarithmic potential satisfies the equation

$$\Delta_\Gamma U(\varrho) = q(\varrho)\ . \tag{2.2.7}$$

In the case of a two-dimensional point source (point mass or charge) of intensity Q located at a point $\varrho_0(x_0, z_0)$ in the XOZ plane, its density is known to be fitted by $q(\varrho) = Q\delta(\varrho-\varrho_0)$, where $\delta(\varrho-\varrho_0)$ is the two-dimensional Dirac singular function defined by the relation

$$\iint_\Gamma \delta(\varrho-\varrho_0)f(\varrho)\,ds = \begin{cases} f(\varrho_0)\ , & \varrho_0\in\Gamma\ ,\\ 0\ , & \varrho_0\in C\bar{\Gamma}\ .\end{cases} \tag{2.2.8}$$

Here $f(\varrho)$ is an arbitrary function continuous in Γ.

Substituting the density of a point source into (2.2.6) and allowing for (2.2.8), we arrive at a *logarithmic potential of a point source*

$$U_P(\varrho') = -\frac{Q}{2\pi}\ln\frac{1}{|\varrho_0-\varrho'|}\ . \tag{2.2.9}$$

A two-dimensional dipole source is formed from two approaching point sources of intensities $(+Q)$ and $(-Q)$, respectively, spaced at a short distance Δl from one another. Setting $\Delta l\to 0$ and $Q\to\infty$, but with $\lim Q\Delta l = M = \text{const}$, we obtain a *two-dimensional dipole source* whose density is defined by the function

$$q(\varrho) = M\frac{\partial\delta(\varrho-\varrho_0)}{\partial\mathbf{v}} = \mathbf{M}\cdot\operatorname{grad}_\Gamma^{\rho_0}\delta(\varrho-\varrho_0)\ . \tag{2.2.10}$$

Here ϱ_0 is the dipole location point; $\mathbf{v}$ is the unit vector pointing from the point source $(-Q)$ toward the point source $(+Q)$; $\mathbf{M} = M\mathbf{v}$ is the dipole moment [differentiation in formula (2.2.10) is carried out with respect to the variable ϱ_0 which is designated by the corresponding superscript].

Substituting (2.2.10) into (2.2.6) and taking account of (2.2.8) we derive a *logarithmic potential of a dipole source*

$$U_D(\varrho') = -\frac{1}{2\pi} M \frac{\partial}{\partial \mathbf{v}} \ln \frac{1}{|\varrho_0 - \varrho'|} = -\frac{1}{2\pi} \mathbf{M} \cdot \operatorname{grad}_\Gamma^{\rho_0} \ln \frac{1}{|\varrho_0 - \varrho'|} . \quad (2.2.11)$$

Of great importance in logarithmic potential theory are *linear distributions of point sources and dipoles.* Let L be an arbitrary smooth curve lying in the XOZ plane along which point sources are distributed continuously with linear density $q_L(\varrho)$ being a continuous function of points $\varrho \in L$. This source is called a *simple linear source (two-dimensional simple layer).* Its potential is specified by the formula

$$U_P^L(\varrho') = -\frac{1}{2\pi} \int_L q_L(\varrho) \ln \frac{1}{|\varrho - \varrho'|} dl , \quad (2.2.12)$$

derived by integrating (2.2.9) along the line L (upon substitution of $Q = Q(\varrho) = q_L(\varrho) dl$).

Now assume that along the curve L there are continuously distributed dipoles with a linear density of moments $\mathbf{m}_L(\varrho) = m_L(\varrho) . \mathbf{v}(\varrho)$, where $\mathbf{v}(\varrho)$ stands for the unit vectors normal to L at points $\varrho \in L$. A set of these dipoles makes up a *linear dipole source (two-dimensional double layer),* whose potential is evaluated by the formula

$$U_D^L(\varrho') = -\frac{1}{2\pi} \int_L m_L(\varrho) \frac{\partial}{\partial \mathbf{v}} \ln \frac{1}{|\varrho - \varrho'|} dl$$

$$= -\frac{1}{2\pi} \int_L \mathbf{m}_L(\varrho) \cdot \operatorname{grad}_\Gamma \ln \frac{1}{|\varrho - \varrho'|} dl , \quad (2.2.13)$$

where differentiation under the integral is carried out with respect to the integration variable ϱ. This formula is derived by integrating (2.2.11) (upon substitution $\mathbf{M} = \mathbf{M}(\varrho) = \mathbf{m}_L(\varrho)\, dl$).

If point sources with a density $q(\varrho)$ fill continuously a domain Γ of the XOZ plane, we deal with a *simple two-dimensional source of a plane field.* Its potential is fitted by a formula obtained by integrating (2.2.9) (upon substitution of $Q = Q(\varrho) = q(\varrho)\, ds$) and coincident with (2.2.6):

$$U_P^\Gamma(\varrho') = -\frac{1}{2\pi} \iint_\Gamma q(\varrho) \ln \frac{1}{|\varrho - \varrho'|} ds . \quad (2.2.14)$$

Finally, if the domain Γ is continuously filled with dipoles of a two-dimensional density of moments $\mathbf{m}(\varrho)$, this source will be referred to as a *two-dimensional polarized source of a plane field.* Its potential is defined by integrating formula (2.2.11) over the domain Γ (note that $\mathbf{M} = \mathbf{M}(\varrho) = \mathbf{m}(\varrho)\, ds$):

$$U_D^\Gamma(\varrho') = -\frac{1}{2\pi} \iint_\Gamma m(\varrho) \frac{\partial}{\partial \mathbf{v}} \ln \frac{1}{|\varrho - \varrho'|} ds$$

$$= -\frac{1}{2\pi} \iint_\Gamma \mathbf{m}(\varrho) \cdot \operatorname{grad}_\Gamma \ln \frac{1}{|\varrho - \varrho'|} ds , \quad (2.2.15)$$

Expression (2.2.15) can be transformed by means of a two-dimensional Ostrogradsky-Gauss formula (2.1.19), as follows:

$$-\frac{1}{2\pi}\iint_{\Gamma} \mathbf{m}(\varrho)\,\mathrm{grad}_{\Gamma}\ln\frac{1}{|\varrho-\varrho'|}\,ds = -\frac{1}{2\pi}\int_{L} \mathbf{m}(\varrho)\cdot\mathbf{v}(\varrho)\ln\frac{1}{|\varrho-\varrho'|}\,dl + \frac{1}{2\pi}\iint_{\Gamma} \mathit{div}_{\Gamma}\mathbf{m}(\varrho)\ln\frac{1}{|\varrho-\varrho'|}\,ds\,. \quad (2.2.16)$$

Introducing designations

$$\mathbf{m}(\varrho)\cdot\mathbf{v}(\varrho) = q_L(\varrho)\;, \quad -\mathrm{div}_{\Gamma}\mathbf{m}(\varrho) = q(\varrho)\;, \quad (2.2.17)$$

we write in the final form

$$U_D^{\Gamma}(\varrho') = -\frac{1}{2\pi}\int_{\Gamma} q_L(\varrho)\ln\frac{1}{|\varrho-\varrho'|}\,dl - \frac{1}{2\pi}\iint_{\Gamma} q(\varrho)\ln\frac{1}{|\varrho-\varrho'|}\,ds = U_P^L(\varrho') + U_P^{\Gamma}(\varrho')\;. \quad (2.2.18)$$

Hence, a *two-dimensional polarized source is equivalent to a sum of a simple two-dimensional source and a simple layer.*

2.2.2 Logarithmic Potentials in Complex Coordinates

Let us introduce a complex variable $\zeta = x+iz$ in the *XOZ* plane. Then the points of this plane with radius vectors ϱ and ϱ' are associated with complex numbers ζ and ζ', respectively. Within these designations, expressions (2.2.12) through (2.2.15) are rewritten as follows:

$$U_P^L(\zeta') = -\frac{1}{2\pi}\int_{L} q_L(\zeta)\ln\frac{1}{|\zeta-\zeta'|}\,dl\;, \quad (2.2.19\,\mathrm{a})$$

$$U_D^L(\zeta') = -\frac{1}{2\pi}\int_{L} m_L(\zeta)\frac{\partial}{\partial\mathbf{v}}\ln\frac{1}{|\zeta-\zeta'|}\,dl = -\frac{1}{2\pi}\int_{L} \mathbf{m}_L(\zeta)\cdot\mathrm{grad}\ln\frac{1}{|\zeta-\zeta'|}\,dl\;, \quad (2.2.19\,\mathrm{b})$$

$$U_P^{\Gamma}(\zeta') = -\frac{1}{2\pi}\iint_{\Gamma} q(\zeta)\ln\frac{1}{|\zeta-\zeta'|}\,ds\;, \quad (2.2.19\,\mathrm{c})$$

$$U_D^{\Gamma}(\zeta') = -\frac{1}{2\pi}\iint_{\Gamma} m(\zeta)\frac{\partial}{\partial\mathbf{v}}\ln\frac{1}{|\zeta-\zeta'|}\,ds = -\frac{1}{2\pi}\iint_{\Gamma} \mathbf{m}(\zeta)\cdot\mathrm{grad}\ln\frac{1}{|\zeta-\zeta'|}\,ds\;. \quad (2.2.19\,\mathrm{d})$$

The resultant relations are useful in the study of complex intensities of a plane field. At this point, however, we will confine ourselves to an examination of formula (2.2.19b) for the potential of a double layer. This expression can be cast in a somewhat different form if we employ the Cauchy-Riemann relations for the real and imaginary parts of an analytical function.

$$\ln(\zeta-\zeta') = \ln|\zeta-\zeta'| + i\theta \ , \tag{2.2.20}$$

where

$$\theta = \arg(\zeta-\zeta') \ .$$

Let us introduce an auxiliary system of axes in a complex plane, a system consisting of a positive tangent vector **l** and a normal **ν** to the line L. This system is oriented just as the system of OZ and OX, i.e., the axis **ν** is on the right of the axis **l** (Fig. 8). Within the frame of this coordinate system, the Cauchy-Riemann relations for the functions $\ln|\zeta-\zeta'|$ and θ (with respect to the variable ζ) take the form

$$\frac{\partial \ln|\zeta-\zeta'|}{dl} = -\frac{\partial\theta}{\partial\nu} \ , \tag{2.2.21 a}$$

$$\frac{\partial \ln|\zeta-\zeta'|}{d\nu} = \frac{\partial\theta}{\partial l} \ , \tag{2.2.21 b}$$

where differentiation is carried out with respect to the variable ζ.

Transformation of the integrand in formula (2.2.19b) by means of (2.2.21a) and (2.2.21b) yields a simple expression for the double layer potential:

$$U_D^L(\zeta') = -\frac{1}{2\pi}\int_L m_L(\zeta)\frac{\partial\theta}{\partial l}\,dl = -\frac{1}{2\pi}\int_L m_L(\zeta)\,d\theta \ . \tag{2.2.22}$$

In a particular case, provided $m_L(\zeta) = m_0 = \text{const}$, we find

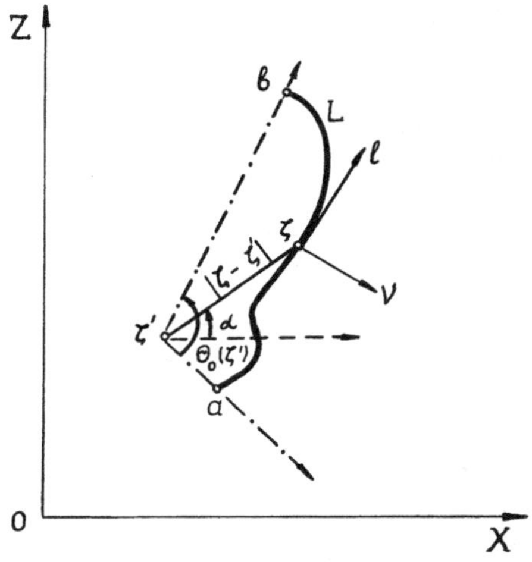

Fig. 8. Scheme explaining the notation of logarithmic potentials in terms of complex coordinates: **l** and **ν** are the positive vectors of the tangent and normal to the line L

$$U_D^L(\zeta') = -\frac{1}{2\pi} m_0 \int_L d\theta = -\frac{1}{2\pi} m_0 \theta_0(\zeta') \ , \tag{2.2.23}$$

where $\theta_0(\zeta')$ is the angle of vision of the curve L from the point ζ'.

2.2.3 Cauchy-Type Integral as a Sum of the Logarithmic Potentials of Simple and Double Layers

For the sake of simplicity, it is assumed that L is a closed smooth curve and $\varphi(\zeta)$ is an arbitrary continuously differentiable real function specified on L. Under these assumptions, we will show that the Cauchy-type integral over the curve L with a density $\varphi(\zeta)$ is expressed in terms of the above logarithmic potentials. Indeed,

$$C(\zeta') = \operatorname{Re} C(\zeta') + i \operatorname{Im} C(\zeta') = \frac{1}{2\pi i} \int_L \frac{\varphi(\zeta)}{\zeta - \zeta'} d\zeta \ . \tag{2.2.24}$$

Using (2.2.20) we will transform the integrand

$$\frac{dl}{\zeta - \zeta'} = d \ln (\zeta - \zeta') = d \ln |\zeta - \zeta'| + i d\theta \ . \tag{2.2.25}$$

Hence,

$$\operatorname{Re} C(\zeta') = \frac{1}{2\pi} \int_L \varphi(\zeta) d\theta \ , \tag{2.2.26}$$

$$\operatorname{Im} C(\zeta') = -\frac{1}{2\pi} \int_L \varphi(\zeta) d \ln |\zeta - \zeta'| \ . \tag{2.2.27}$$

Comparing (2.2.26) and (2.2.22) one can see that the *real part of the Cauchy-type integral is a logarithmic potential of a double layer with a linear density of dipole moments equal* (with a negative sign) *to the density of the Cauchy-type integral:* $m_L(\zeta) = -\varphi(\zeta)$. Using partial differentiation and noting that the integrated part is zero by virtue of the closed nature of the contour L, we can rewrite formula (2.2.27) as follows:

$$\operatorname{Im} C(\zeta') = \frac{1}{2\pi} \int_L \ln |\zeta - \zeta'| d\varphi = -\frac{1}{2\pi} \int_L \frac{\partial \varphi}{\partial l} \ln \frac{1}{|\zeta - \zeta'|} dl \ . \tag{2.2.28}$$

A comparison of (2.2.28) and (2.2.19a) reveals that the *imaginary part of the Cauchy-type integral is a logarithmic potential of a simple layer with a linear density*

$$q_L(\zeta) = \frac{\partial \varphi}{\partial l} (\zeta) \ . \tag{2.2.29}$$

Total intensity (mass or charge) of this linear (closed) source is zero:

$$Q = \int_L q_L(\zeta)\,dl = \int_L \frac{\partial \varphi}{\partial l}\,dl = 0 \ ,$$

since L is a closed contour.

Thus, the Cauchy-type integral can be represented as an algebraic sum of logarithmic potentials of a double layer, U_D^L, (with a negative sign) and of a simple layer, U_P^L, (with a factor i):

$$C(\zeta') = -U_D^L(\zeta') + iU_P^L(\zeta') \ . \tag{2.2.30}$$

Representation (2.2.30) makes it possible to reduce the examination of the Cauchy-type integral virtually to a consideration of the properties of logarithmic potentials of double and simple layers. For instance, it can be used to derive the Sokhotsky-Plemelj formulas, since U_D^L is a discontinuous function on transition through L [the magnitude of a discontinuity is just equal to the density $\varphi(\zeta)$], whereas U_P^L behaves as a continuous function on transition through the path of integration. We should like to stress, however, that this way is far from being optimal since it involves cumbersome calculations and hardly comprehensible results. More efficient and fruitful (particularly in terms of applications) is an approach relying on direct study of the Cauchy-type integral. Within this approach which was developed to the highest degree by Soviet mathematicians N. I. Muskhelishvily (1954, 1962), F. D. Gakhov (1963), and I. N. Vekua, the simplicity and elegance of the methods employed as well as the profundity and abundance of theoretical findings have now general appreciation. Precisely these considerations have stimulated us in our endeavors to develop a body of three-dimensional analogs of the Cauchy- and Stratton-Chu type integrals, which will be outlined in subsequent sections of the book.

2.3 Complex Intensity and Potential of a Plane Field

Plane fields can be described not only in terms of vector functions of a vector argument but also by means of complex-valued functions of a complex argument. These complex functions are known as *complex intensities of a plane field.* They permit the most effective application of the theory of functions of a complex variable to plane fields.

2.3.1 Concept of Complex Intensity of a Plane Field

Consider a plane vector field $\mathbf{F}(\boldsymbol{\varrho}) = F_x(x, z), F_z(x, z))$, which meets, in accordance with (2.1.12), (2.1.16), and (2.1.17), the equations

$$\frac{\partial F_x}{\partial x} + \frac{\partial F_z}{\partial x} = q(x,z) \ , \quad \frac{\partial F_x}{\partial z} - \frac{\partial F_z}{\partial x} = 0 \ . \tag{2.3.1}$$

We will introduce, just as in Sect. 2.2, a complex variable $\zeta = x + iz$ in the XOZ plane and designate by $F(\zeta)$ the following complex function of a complex argument:

$$F(\zeta) = -F_x(x, z) + iF_z(x, z) \ . \qquad (2.3.2)$$

The function $F(\zeta)$ will be referred to as *complex intensity of a plane field* **F**.

It is pertinent to mention, first of all, that *outside-sources complex intensity of a plane field is an analytical function vanishing at infinity.* Indeed, outside-sources Eqs. (2.3.1) take the form

$$\frac{\partial F_x}{\partial x} + \frac{\partial F_z}{\partial z} = 0 \ , \quad \frac{\partial F_x}{\partial z} - \frac{\partial F_z}{\partial x} = 0 \ . \qquad (2.3.3)$$

Since

$$\operatorname{Re} F(\zeta) = -F_x \ , \quad \operatorname{Im} F(\zeta) = F_z \ , \qquad (2.3.4)$$

on substitution of (2.3.4) into (2.3.3) it is evident that the real and imaginary parts of the function $F(\zeta)$ satisfy the Cauchy-Riemann conditions:

$$\frac{\partial \operatorname{Re} F}{\partial x} = \frac{\partial \operatorname{Im} F}{\partial z} \ , \quad \frac{\partial \operatorname{Re} F}{\partial z} = -\frac{\partial \operatorname{Im} F}{\partial x} \ . \qquad (2.3.5)$$

It is stated in Sect. 2.1.1 that thought is given only to sources concentrated in finite domains of space, hence, $F_x \to 0$, $F_z \to 0$ when $\varrho \to \infty$ and $F(\zeta) \to 0$ when $|\zeta| \to \infty$.

2.3.2 Complex Intensity Equations

Now we will write equations fitting the complex intensity of a field. The shortest way to derive these equations is to proceed from real variables x, z to complex ones

$$\zeta = x + iz \ , \quad \bar{\zeta} = x - iz \qquad (2.3.6)$$

and to employ differentiation operators

$$\frac{\partial}{\partial \zeta} = \frac{1}{2}\left(\frac{\partial}{\partial x} - i\frac{\partial}{\partial z}\right) , \qquad (2.3.7)$$

$$\frac{\partial}{\partial \bar{\zeta}} = \frac{1}{2}\left(\frac{\partial}{\partial x} + i\frac{\partial}{\partial z}\right) . \qquad (2.3.8)$$

Indeed, multiplying the first Eq. (2.3.1) by -1 and the second by $-i$, and summing them up we obtain

$$-\left(\frac{\partial F_x}{\partial x} + \frac{\partial F_z}{\partial z}\right) - i\left(\frac{\partial F_x}{\partial z} - \frac{\partial F_z}{\partial x}\right) = \left(\frac{\partial}{\partial x} + i\frac{\partial}{\partial z}\right)(-F_x + iF_z) = -q(x, z) \ . \qquad (2.3.9)$$

Introducing complex coordinates and using operator (2.3.8) we write

$$\frac{\partial F(\zeta)}{\partial \bar{\zeta}} = -\frac{1}{2}q(\zeta) \ . \qquad (2.3.10)$$

We have arrived at a *complex differential equation fitting the complex intensity of a plane field.*

It is noteworthy that once $\psi(\zeta)$ is an arbitrary function analytical in the domain D, the Cauchy-Riemann relations imply that

$$\partial\psi(\zeta)/\partial\bar{\zeta}=0 \; . \tag{2.3.11}$$

Formula (2.3.11) is often called a *complex expression of the Cauchy-Riemann system.* At the same time, application of the operator $\partial/\partial\zeta$ yields a derivative of a complex function with respect to a complex argument

$$\partial\psi(\zeta)/\partial\zeta=\psi'(\zeta) \; . \tag{2.3.12}$$

In particular, by virtue of the analyticity of the complex intensity outside the domain Γ occupied by sources we have

$$\partial F(\zeta)/\partial\bar{\zeta}=0 \; , \quad \zeta\in C\bar{\Gamma} \; . \tag{2.3.13}$$

2.3.3 Representation of Complex Intensity in Terms of Field Source Density

Let us derive an expression for complex intensity of a plane field in terms of density of its sources $q(\zeta)$. To this end, we will use formulas (2.2.5) and (2.2.6) relating the field $\mathbf{F}$ to the logarithmic potential $U(\varrho)$:

$$F_x=\partial U/\partial x \; , \quad F_z=\partial U/\partial z \; , \tag{2.3.14}$$

and substitute corresponding formulas (2.2.19a) through (2.2.19d) into the right-hand side of (2.3.14). For example, in the case of a simple layer potential, we have

$$F_x(\zeta')=-\frac{1}{2\pi}\int_L q_L(\zeta)\frac{x-x'}{|\zeta-\zeta'|^2}\,dl \; , \tag{2.3.15}$$

$$F_z(\zeta')=-\frac{1}{2\pi}\int_L q_L(\zeta)\frac{z-z'}{|\zeta-\zeta'|^2}\,dl \; . \tag{2.3.16}$$

Substituting (2.3.15) and (2.3.16) into (2.3.2) we write

$$F(\zeta')=\frac{1}{2\pi}\int_L q_L(\zeta)\frac{(x-x')-i(z-z')}{|\zeta-\zeta'|^2}\,dl \; . \tag{2.3.17}$$

Allowing for the fact that

$$(x-x')-i(z-z')=(\overline{\zeta-\zeta'}) \; , \quad |\zeta-\zeta'|^2=(\zeta-\zeta')\,(\overline{\zeta-\zeta'}) \tag{2.3.18}$$

(the bar denotes complex conjugate values), we find finally

$$F(\zeta')=\frac{1}{2\pi}\int_L \frac{q_L(\zeta)}{\zeta-\zeta'}\,dl \; . \tag{2.3.19}$$

Formula (2.3.19) is the most convenient expression for complex intensity of a simple layer. By similar reasoning in conjunction with the potential of a two-dimensional simple source defined by one of the formulas (2.2.19a) through (2.2.19d) we establish its complex intensity as

$$F(\zeta') = \frac{1}{2\pi} \iint_{\Gamma} \frac{q(\zeta)}{\zeta - \zeta'} \, ds \; . \tag{2.3.20}$$

Formula (2.3.20) suggests directly that $F(\zeta')$ is a function analytical outside Γ and vanishing at infinity [since $1/(\zeta-\zeta')$ is a function analytical everywhere when $\zeta-\zeta'$ is not zero].

Now we will represent in a similar way complex intensity of the field of a double layer and of a two-dimensional polarized source. A short but unwieldy calculation in accordance with formula (2.2.19b) for a double layer field yields

$$F(\zeta') = -F_x(\zeta') + iF_z(\zeta') = -\frac{\partial U_D^L(\zeta')}{\partial x'} + i\,\frac{\partial U_D^L(\zeta')}{\partial z'}$$

$$= \frac{1}{2\pi} \int_L \left[\frac{[m_{Lx}(\zeta) + i m_{Lz}(\zeta)]\;\overline{(\zeta-\zeta')}}{|\zeta-\zeta'|^2\,(\zeta-\zeta')} \right] dl = \frac{1}{2\pi} \int_L \frac{\mu_L(\zeta)}{(\zeta-\zeta')^2}\, dl \; , \tag{2.3.21}$$

where $\mu_L(\zeta) = m_{Lx}(\zeta) + i m_{Lz}(\zeta)$ is the linear *complex density of dipole moments of a double layer.*

Similarly, complex density of a two-dimensional polarized source is defined

$$F(\zeta') = \frac{1}{2\pi} \iint_{\Gamma} \frac{\mu(\zeta)}{(\zeta-\zeta')^2}\, ds \; , \tag{2.3.22}$$

where $\mu(\zeta) = m_x(\zeta) + i m_z(\zeta)$ is the *complex density of dipole moments of a polarized source or its complex polarization.*

We should like to note here that in geophysical applications simple two-dimensional sources are used to describe distributions of gravitational masses, while polarized two-dimensional sources are useful for those of magnetized masses.

2.3.4 Complex Potential

It will be shown that complex intensity of a field can be represented as a derivative of another complex function, the so-called *complex potential.*

Indeed, let us take into consideration the following complex function:

$$P(\zeta) = U(x,\, z) + iV(x,\, z) \; , \tag{2.3.23}$$

where $U(x, z)$ is the scalar potential of a vector field $\mathbf{F}$, while $V(x, z)$ is its flow function.

First of all, it is evident that by virtue of (2.1.13) and (2.1.23), the functions U and V satisfy the Cauchy-Riemann conditions:

$$\partial U/\partial x = \partial V/\partial z = F_x \; , \quad \partial U/\partial z = -\partial V/\partial x = F_z \; . \tag{2.3.24}$$

As a consequence, $P(\zeta)$ is a function analytical everywhere in $C\bar{\Gamma}$. Let us calculate the derivative of the function $P(\zeta)$:

$$\frac{\partial}{\partial \zeta} P(\zeta) = \frac{1}{2}\left(\frac{\partial}{\partial x} - i\frac{\partial}{\partial z}\right)(U+iV)$$

$$= \frac{1}{2}\left(\frac{\partial U}{\partial x} + \frac{\partial V}{\partial z}\right) + i\,\frac{1}{2}\left(\frac{\partial V}{\partial x} - \frac{\partial U}{\partial z}\right) = F_x - iF_z = -F(\zeta) \ . \qquad (2.3.25)$$

Hence, it is finally written as

$$F(\zeta) = -\frac{\partial}{\partial \zeta} P(\zeta) \ , \quad \zeta \in C\bar{\Gamma} \ . \qquad (2.3.26)$$

The function $P(\zeta)$ is known as a *complex potential* of the field **F**. One can readily see that the real part of the complex potential is an ordinary logarithmic potential, while its imaginary part is a flow function.

2.4 Direct Solution to the Equation for Complex Field Intensity

This subsection is concerned with a method used to solve directly Eq. (2.3.10) for complex intensity of a plane field, a method establishing this value from a specified distribution of source density $q(\zeta)$ in a complex plane. In the foregoing discussion, we outlined a method for solving this problem, which relies on the known analytical expressions for logarithmic potentials. We will see, however, in what follows that it is the complex form of these relations that is most convenient for subsequent applications. To solve the problem in question for a two-dimensional case, just as for a three-dimensional case, the Ostrogradsky-Gauss formula is employed, but it is written in a complex form. Let us begin our exposition with a derivation of this formula.

2.4.1 Two-Dimensional Ostrogradsky-Gauss Formula in Complex Notation

Refer again to the two-dimensional Ostrogradsky-Gauss formula (2.1.19) valid for a plane vector field $\mathbf{F}(F_x, F_z)$ which has continuous partial derivatives in the domain D bounded by a piecewise smooth curve L, and which is continuous in the closed domain $\bar{D}$:

$$\iint_D \left(\frac{\partial F_x}{\partial x} + \frac{\partial F_z}{\partial z}\right) ds = \int_L (F_x \nu_x + F_z \nu_z)\, dl \ , \qquad (2.4.1)$$

where ν_x and ν_z are the Cartesian components of the unit vector $\boldsymbol{\nu}$ of an outward pointing normal to L, related to the differentials dx and dz by simple relations:

$$\nu_x dl = dz \ , \quad \nu_z dl = -dx \ . \qquad (2.4.2)$$

Substituting (2.4.2) into (2.4.1) we write

$$\iint_D \left(\frac{\partial F_x}{\partial x} + \frac{\partial F_z}{\partial z}\right) ds = \int_L (F_x dz - F_z dx) \ . \qquad (2.4.3)$$

If we define a new vector field $\tilde{\mathbf{F}} = (-F_z, F_x)$, relation (2.4.3) is rewritten in the form:

$$\iint_D \left(\frac{\partial F_x}{\partial z} - \frac{\partial F_z}{\partial x}\right) ds = -\int_L (F_x dx + F_z dz) \ . \tag{2.4.4}$$

Multiplying Eq. (2.4.3) by $-i$ and summing it up with (2.4.4), we obtain

$$i \iint_D \left(\frac{\partial}{\partial x} + i\frac{\partial}{\partial z}\right) (-F_x + iF_z)\, ds = \int_L (-F_x + iF_z)\,(dx + i dz) \ . \tag{2.4.5}$$

Passing to complex coordinates ζ and complex intensity of the field $F(\zeta)$ defined by formula (2.3.2) and using operator (2.3.8), we arrive at a *complex notation of the Ostrogradsky-Gauss formula*

$$\iint_D \frac{\partial F(\zeta)}{\partial \bar{\zeta}}\, ds = \frac{1}{2i} \int_L F(\zeta)\, d\zeta \ . \tag{2.4.6}$$

In particular, formula (2.4.6) directly suggests *the Cauchy integral theorem,* asserting that the *integral of an analytical function $F(\zeta)$ along any closed piecewise smooth curve L [lying in a simply connected region of analyticity $F(\zeta)$] is zero.*

Indeed, upon substitution of Eq. (2.3.13) holding for analytical functions into (2.4.6), we find

$$\int_L F(\zeta)\, d\zeta = 0 \ . \tag{2.4.7}$$

Similar reasoning readily yield an equivalent complex form of the Ostrogradsky-Gauss theorem

$$\iint_D \frac{\partial F(\zeta)}{\partial \zeta}\, ds = -\frac{1}{2i} \int_L F(\zeta)\, d\bar{\zeta} \ . \tag{2.4.8}$$

The latter formula indicates that in a two-dimensional case the Ostrogradsky-Gauss theorem provides a means for evaluating the integral of a derivative of a complex function.

2.4.2 Pompei Formulas

Of particular concern in plane field theory is the relation following from the Ostrogradsky-Gauss complex formula which will be referred to Z as a *Pompei formula.*

Assume that inside the domain D there is specified a complex function $w(\zeta)$ satisfying the conditions of applicability of the Ostrogradsky-Gauss theorem, i.e., having continuous partial derivatives with respect to x and z in D and being continuous inside $\bar{D}$. Once $\psi(\zeta)$ is an arbitrary function analytical inside D, the following differential identities seem to be the case by virtue of (2.3.11) and (2.3.12):

$$\frac{\partial}{\partial \bar{\zeta}} [\psi(\zeta)\, w(\zeta)] = \psi(\zeta) \frac{\partial w(\zeta)}{\partial \bar{\zeta}} , \tag{2.4.9}$$

$$\frac{\partial}{\partial \zeta} [\bar{\psi}(\zeta)\, w(\zeta)] = \bar{\psi}(\zeta) \frac{\partial w(\zeta)}{\partial \zeta} . \tag{2.4.10}$$

Specifying $F(\zeta)$ in the Ostrogradsky-Gauss formula (2.4.6) as the function $F(\zeta) = \psi(\zeta)\, w(\zeta)$ we obtain

$$\iint_D \psi(\zeta) \frac{\partial w(\zeta)}{\partial \bar{\zeta}}\, ds = \frac{1}{2i} \int_L \psi(\zeta)\, w(\zeta)\, d\zeta . \tag{2.4.11}$$

Now, selecting the function $F(\zeta) = \bar{\psi}(\zeta)\, w(\zeta)$ for the Ostrogradsky-Gauss formula (2.4.8) we find

$$\iint_D \bar{\psi}(\zeta) \frac{\partial w(\zeta)}{\partial \zeta}\, ds = -\frac{1}{2i} \int_L \bar{\psi}(\zeta)\, w(\zeta)\, \overline{d\zeta} . \tag{2.4.12}$$

In particular, $\psi(\zeta)$ in formula (2.4.11) can be provided by the Cauchy kernel $\psi(\zeta) = 1/(\zeta - \zeta')$ (ζ' being a fixed point within D), and the region of integration can be taken to be domain D_ε produced by extracting a small circle $|\zeta - \zeta'| \leqslant \varepsilon$ with the center at a point ζ' and radius ε from the domain D (Fig. 9). Evidently, function $1/(\zeta - \zeta')$ is analytical inside D, hence formula (2.4.11) is applicable:

$$\iint_{D_\varepsilon} \frac{\partial w(\zeta)}{\partial \bar{\zeta}} \frac{ds}{\zeta - \zeta'} = \frac{1}{2i} \int_L \frac{w(\zeta)}{\zeta - \zeta'}\, d\zeta - \frac{1}{2i} \int_{|\zeta - \zeta'| = \varepsilon} \frac{w(\zeta)}{\zeta - \zeta'}\, d\zeta . \tag{2.4.13}$$

It is easy to show that the limit of the second integral in the right-hand side of (2.4.13) is equal to $\pi\omega(\zeta')$. Indeed, since the equation $\zeta - \zeta' = \varepsilon e^{i\theta}$ is the case everywhere on the circle $|\zeta - \zeta'| = \varepsilon$ (see Fig. 9), we have

$$\lim_{\varepsilon \to 0} \frac{1}{2i} \int_{|\zeta - \zeta'| = \varepsilon} \frac{w(\zeta)}{\zeta - \zeta'}\, d\zeta = \lim_{\varepsilon \to 0} \frac{1}{2} \int_{|\zeta - \zeta'| = \varepsilon} w(\varepsilon e^{i\theta} + \zeta')\, d\theta = \pi\omega(\zeta') . \tag{2.4.14}$$

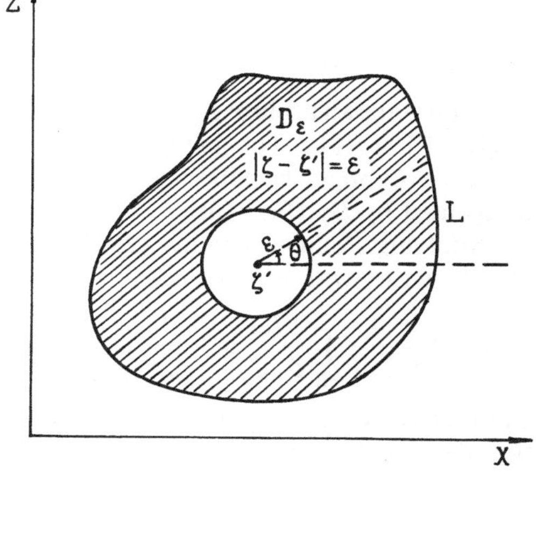

Fig. 9. Derivation of the Pompei formulas: ζ' is a fixed point of the domain D; D_ε is the domain produced by extracting a circle $|\zeta - \zeta'| \leqslant \varepsilon$ from D

Proceeding to the limit for $\varepsilon \to 0$ in (2.4.13) and allowing for (2.4.14) we obtain finally

$$w(\zeta') = \frac{1}{2\pi i} \int_L \frac{w(\zeta)}{\zeta - \zeta'} d\zeta - \frac{1}{\pi} \iint_D \frac{\partial w(\zeta)}{\partial \bar{\zeta}} \frac{ds}{\zeta - \zeta'} . \tag{2.4.15}$$

In a similar way, using formula (2.4.8) and identity (2.4.10) we derive the formula:

$$w(\zeta') = -\frac{1}{2\pi i} \int_L \frac{w(\zeta)}{\bar{\zeta} - \bar{\zeta}'} d\bar{\zeta} - \frac{1}{\pi} \iint_D \frac{\partial w(\zeta)}{\partial \zeta} \frac{ds}{\bar{\zeta} - \bar{\zeta}'} . \tag{2.4.16}$$

Formulas (2.4.15) and (2.4.16) are termed *Pompei formulas.*

If the function $w(\zeta)$ inside the domain D is known to satisfy the equation

$$\partial w(\zeta)/\partial \bar{\zeta} = q(\zeta) , \tag{2.4.17}$$

formula (2.4.15) yields the following solution to this equation:

$$w(\zeta') = \frac{1}{2\pi i} \int_L \frac{w(\zeta)}{\zeta - \zeta'} d\zeta - \frac{1}{\pi} \iint_D \frac{q(\zeta)}{\zeta - \zeta'} ds . \tag{2.4.18}$$

If the function $w(\zeta)$ is analytical inside D, i.e., $\partial w(\zeta)/\partial \bar{\zeta} = 0$, $\zeta \in D$, (2.4.18) yields directly the Cauchy integral formula

$$w(\zeta') = \frac{1}{2\pi i} \int_L \frac{w(\zeta)}{\zeta - \zeta'} d\zeta . \tag{2.4.19}$$

2.4.3 Solution to the Equation for Complex Intensity

We will use the Pompei formulas to examine the complex intensity of a plane field $F(\zeta)$. In accordance with the definition (Sects. 2.3.1 and 2.3.2), complex intensity satisfies Eq. (2.3.10) everywhere in a complex plane

$$\partial F(\zeta)/\partial \bar{\zeta} = -\tfrac{1}{2} q(\zeta) \tag{2.4.20}$$

and vanishes at infinity

$$F(\zeta) \to 0 \quad \text{for} \quad |\zeta| \to \infty . \tag{2.4.21}$$

Let us take an arbitrary point ζ' in a complex plane and draw therefrom a circle $L_R(|\zeta - \zeta'| = R)$ of such a large radius R that the domain Γ occupied by the field sources $q(\zeta)$ might be completely inside L_R. The domain bounded by L_R will be designated by D_R (Fig. 10).

Applying Pompei formula (2.4.15) to the function $F(\zeta)$ and to the domain D_R, we write, with due reference to (2.4.20),

$$F(\zeta') = \frac{1}{2\pi i} \int_{L_R} \frac{F(\zeta)}{\zeta - \zeta'} d\zeta + \frac{1}{2\pi} \iint_\Gamma \frac{q(\zeta)}{\zeta - \zeta'} ds . \tag{2.4.22}$$

Evaluate the first integral in the right-hand side of (2.4.22) using the representation $\zeta - \zeta' = Re^{i\theta}$, then

Fig. 10. Solution to the equation for complex intensity using the Pompei formula: Γ is the domain occupied with field sources

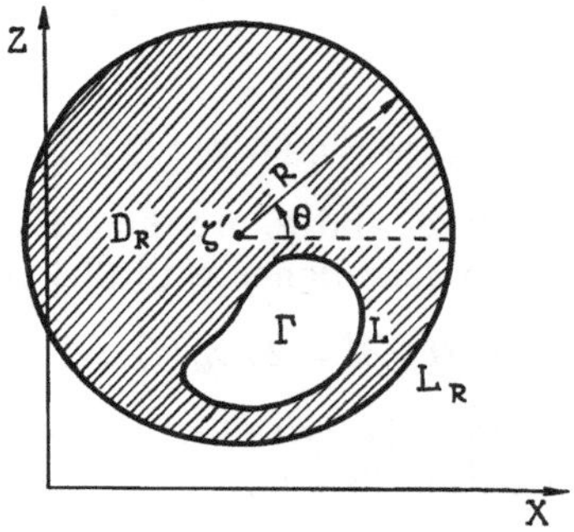

$$\int_{L_R} \frac{F(\zeta)}{\zeta-\zeta'}\,d\zeta = \int_0^{2\pi} F(\zeta'+Re^{i\theta})\,d\theta\ . \tag{2.4.23}$$

Being analytical at infinity, the function $F(\zeta)$ tends to zero uniformly over θ, therefore setting $R\to\infty$ in (2.4.23) we find

$$\int_{L_R} \frac{F(\zeta)}{\zeta-\zeta'}\,d\zeta\to 0 \quad \text{for} \quad R\to\infty\ . \tag{2.4.24}$$

Since Eq. (2.4.22) holds for any R, proceeding to the limit for $R\to\infty$ in (2.4.22) we write finally

$$F(\zeta') = \frac{1}{2\pi}\iint_\Gamma \frac{q(\zeta)}{\zeta-\zeta'}\,ds\ . \tag{2.4.25}$$

Thus, we have found a solution of Eq. (2.4.20) equivalent to the solution of (2.3.20) established earlier in terms of the theory of logarithmic potentials. In view of this, the Pompei formulas can be regarded as a complex notation of the well-known integral equations employed in the logarithmic potential theory.

2.5 Representation of the Gravitational Field in Terms of the Cauchy-Type Integral

The above general concepts of the complex theory of plane fields will be applied to a description of the gravitational field.

2.5.1 Complex Intensity of the Gravitational Field

The gravitational field **g** of a two-dimensional distribution of masses concentrated with a density $\varrho(x, z)$ in the domain Γ of the XOZ plane is known to satisfy the equations

$$\mathrm{div}_\Gamma \mathbf{g} = -4\pi\gamma\varrho\ , \quad \mathrm{rot}_\Gamma \mathbf{g} = \mathbf{0}\ , \tag{2.5.1}$$

where γ is the universal constant of gravitation.

Let us define complex intensity of the gravitational field $g(\zeta)$ by the following formula:

$$g(\zeta) = -g_x(x, z) + ig_z(x, z) \ , \qquad (2.5.2)$$

where g_x and g_z are the scalar components of the gravitational field **g**.

In accordance with (2.3.10), the function $g(\zeta)$ satisfies the equation

$$\partial g(\zeta)/\partial\bar{\zeta} = 2\pi\gamma\varrho(\zeta) \ , \qquad (2.5.3)$$

where $\varrho(\zeta) = \varrho(x+iz) = \varrho(x, z)$, whose solution is governed by a formula following from (2.4.25):

$$g(\zeta') = -2\gamma \iint_{\Gamma} \frac{\varrho(\zeta)}{\zeta-\zeta'} ds \ . \qquad (2.5.4)$$

Thus, formula (2.5.4) represents complex intensity of the gravitational field in terms of the Cauchy-kernel integral taken over the domain occupied by masses. Meanwhile, in many important practically problems (say, analytical continuation of the gravitational field into a domain taken up by masses, examination of the properties of analytically continued fields, etc.), it proves extremely useful to represent complex intensities of the gravitational fields in terms of the Cauchy-type integral. This problem was resolved for the first time by Soviet workers A. V. Tsirulsky (1963, 1964), G. Ya. Golizdra (1966), and V. N. Strakhov (1970d). Below we will outline the practices of simplifying the analytical representations of the gravitational field, following mainly the studies by V. N. Strakhov.

2.5.2 Representation of the Gravitational Field of a Homogeneous Domain in Terms of the Cauchy-Type Integral

We will begin our examination with a simple case, where masses with a constant density $\varrho(\zeta) = \varrho_0 = \text{const}$ are concentrated inside a given domain. The complex intensity of the gravitational field of such a homogeneous domain is defined by

$$g(\zeta') = -2\gamma\varrho_0 \iint_{\Gamma} \frac{ds}{\zeta-\zeta'} \ . \qquad (2.5.5)$$

Let us reduce the double integral in the right-hand side to the Cauchy-type integral. First consider the problem of evaluating the field when the observation point lies outside $\Gamma(\zeta' \in C\bar{\Gamma})$. Here one can directly use the corollary to the Ostrogradsky-Gauss theorem, namely formula (2.4.11) positing therein that $\psi(\zeta) = 1/(\zeta-\zeta')$ and $w(\zeta) = \bar{\zeta}$, since for $\zeta' \in C\bar{\Gamma}$ the Cauchy kernel is a definitely analytical function of $\zeta \in \Gamma$.

$$\iint_{\Gamma} \frac{ds}{\zeta-\zeta'} = \frac{1}{2i} \int_{L} \frac{\bar{\zeta}}{\zeta-\zeta'} d\zeta \ . \qquad (2.5.6)$$

Substitution of (2.5.6) into (2.5.5) yields

$$g(\zeta') = -\frac{1}{2\pi i} \int_{L} \frac{2\pi\gamma\varrho_0\bar{\zeta}}{\zeta-\zeta'} d\zeta \ , \quad \zeta' \in C\bar{\Gamma} \ . \qquad (2.5.7)$$

This is just the *representation of the gravitational field outside masses in terms of the Cauchy-type integral* that was first derived by A. V. Tsirulsky (1963).

To evaluate the gravitational field inside masses (for $\zeta' \in \Gamma$), we will employ one of the Pompei formulas (2.4.15), postulating therein that $w(\zeta) = \bar{\zeta}$:

$$\iint_{\Gamma} \frac{ds}{\zeta - \zeta'} = -\pi\bar{\zeta}' + \frac{1}{2i}\int_{L} \frac{\bar{\zeta}}{\zeta - \zeta'}\, d\zeta \;, \quad \zeta' \in \Gamma \;. \tag{2.5.8}$$

Substituting (2.5.8) into (2.5.5) we obtain

$$g(\zeta') = 2\pi\gamma\varrho_0\bar{\zeta}' - \frac{1}{2\pi i}\int_{L} \frac{2\pi\gamma\varrho_0\bar{\zeta}}{\zeta - \zeta'}\, d\zeta \;, \quad \zeta' \in \Gamma \;. \tag{2.5.9}$$

Formula (2.5.9) is the *representation of the gravitational field inside masses in terms of the Cauchy-type integral.* Using the Sokhotsky-Plemelj formulas for the Cauchy-type integral one can readily see that the right-hand sides of Eqs. (2.5.7) and (2.5.9) are equal to one another on the boundary L, so that $g(\zeta')$ is a continuous function in the whole of space.

The terms in the right-hand side of (2.5.9) have a simple physical meaning. Let us designate

$$\begin{aligned} &2\pi\gamma\varrho_0\bar{\zeta}' = g_0(\zeta') \;, \quad \zeta' \in \Gamma \;, \\ &\frac{1}{2\pi i}\int_{L} \frac{2\pi\gamma\varrho_0\bar{\zeta}}{\zeta - \zeta'}\, d\zeta = g_{C\Gamma}(\zeta') \;, \quad \zeta' \in \Gamma \;. \end{aligned} \tag{2.5.10}$$

Evidently, the function $g_0(\zeta)$ satisfies the equation $\partial g_0(\zeta)/\partial\bar{\zeta} = 2\pi\gamma\varrho_0$ in the whole of the complex plane. Comparing the latter formula with (2.5.3) one sees that the function $g_0(\zeta)$ is complex intensity of the gravitational field produced by masses distributed with a uniform density ϱ_0 over the whole complex plane. Since according to (2.5.9),

$$g_{C\Gamma}(\zeta') = g_0(\zeta') - g(\zeta') \;, \tag{2.5.11}$$

where $\zeta' \in \Gamma$ and $g(\zeta')$ are the field inside masses concentrated in the domain Γ, $g_{C\Gamma}(\zeta')$ is, by virtue of the principle of field superposition, the gravitational field of masses filling, with a uniform density ϱ_0, an infinite domain $C\Gamma$, complementing the domain Γ with respect to the whole complex plane.

Now let us derive a single analytical expression for complex intensities of gravitational fields inside and outside masses. Note, first of all, that the function $2\pi\gamma\varrho_0\bar{\zeta}'$ is constant with respect to ζ, i.e., analytical with respect to ζ. As a result, using the Cauchy integral formula we obtain

$$\frac{1}{2\pi i}\int_{L} \frac{2\pi\gamma\varrho_0\bar{\zeta}'}{\zeta - \zeta'}\, d\zeta = \begin{cases} 2\pi\gamma\varrho_0\bar{\zeta}' \;, & \zeta' \in \Gamma \;, \\ 0 \;, & \zeta' \in C\bar{\Gamma} \;. \end{cases} \tag{2.5.12}$$

Allowing for (2.5.12), formulas (2.5.7) and (2.5.9) can be unified to end up with a single formula valid for any ζ':

$$g(\zeta') = -\frac{1}{2\pi i}\int_{L} \frac{2\pi\gamma\varrho_0(\bar{\zeta} - \bar{\zeta}')}{\zeta - \zeta'}\, d\zeta \;, \tag{2.5.13}$$

where ζ' is any point in a complex plane (including $\zeta' \in L$ due to the continuity of $g(\zeta')$). The latter formula can be cast in the form:

$$g(\zeta') = i\gamma\varrho_0 \int_L \exp(-2i\arg(\zeta-\zeta'))\, d\zeta \ . \tag{2.5.14}$$

Expression (2.5.14) was first obtained by V.N. Strakhov. It turns out exceptionally useful in the solution of direct gravimetric problems.

2.5.3 Representation of the Gravitational Field of a Domain with an Analytical Mass Distribution in Terms of the Cauchy-Type Integral

Now consider a more general case where the density of masses distributed inside the domain Γ, $\varrho(x, z)$ is an analytical function of the variables x and z in the domain, and a continuous function in a closed domain $\bar{\Gamma}$. We will introduce new independent variables $\zeta = x+iz$, $\bar{\zeta} = x-iz$. Evidently, the function $\tilde{\varrho}(\zeta, \bar{\zeta}) = \varrho((\zeta+\bar{\zeta})/2, (\zeta-\bar{\zeta})/2i)$ is also an analytical function of the complex variables ζ and $\bar{\zeta}$ in Γ and it is continuous in $\bar{\Gamma}$.

Let $\Phi(\zeta, \bar{\zeta})$ represent an indefinite integral

$$\Phi(\zeta, \bar{\zeta}) = \int \tilde{\varrho}(\zeta, \bar{\zeta})\, \overline{d\zeta} \ . \tag{2.5.15}$$

Then

$$\partial\Phi(\zeta, \bar{\zeta})/\partial\bar{\zeta} = \tilde{\varrho}(\zeta, \bar{\zeta}) \ . \tag{2.5.16}$$

Allowing for formula (2.5.16), expression (2.5.4) for complex intensity of the gravitational field is cast in the form

$$g(\zeta') = -2\gamma \iint_\Gamma \frac{\partial\Phi/\partial\bar{\zeta}}{\zeta-\zeta'}\, ds \ . \tag{2.5.17}$$

Evaluate the integral in the right-hand side of (2.5.17) employing again Pompei formula (2.4.15) for $\zeta' \in \Gamma$:

$$\iint_\Gamma \frac{\partial\Phi/\partial\bar{\zeta}}{\zeta-\zeta'}\, ds = -\pi\Phi(\zeta', \bar{\zeta}') + \frac{1}{2i}\int_L \frac{\Phi(\zeta, \bar{\zeta})}{\zeta-\zeta'}\, d\zeta \ , \quad \zeta' \in \Gamma \ . \tag{2.5.18}$$

If $\zeta' \in C\bar{\Gamma}$, the Cauchy kernel $1/(\zeta-\zeta')$ is a definitely analytical function and the double integral can be evaluated, just as in the case of a constant density ϱ, directly by formula (2.4.11):

$$\iint_\Gamma \frac{\partial\Phi/\partial\bar{\zeta}}{\zeta-\zeta'}\, ds = \frac{1}{2i}\int_L \frac{\Phi(\zeta, \bar{\zeta}')}{\zeta-\zeta'}\, d\zeta \ , \quad \zeta' \in C\bar{\Gamma} \ . \tag{2.5.19}$$

Substitution of (2.5.18) and (2.5.19) into (2.5.17) yields the following formula:

$$g(\zeta') = \begin{cases} 2\pi\gamma\Phi(\zeta', \bar{\zeta}') - \dfrac{1}{2\pi i}\displaystyle\int_L \frac{2\pi\gamma\Phi(\zeta, \bar{\zeta})}{\zeta-\zeta'}\, d\zeta \ , & \zeta' \in \Gamma \ , \\[2ex] -\dfrac{1}{2\pi i}\displaystyle\int_L \frac{2\pi\gamma\Phi(\zeta, \bar{\zeta})}{\zeta-\zeta'}\, d\zeta \ , & \zeta \in C\bar{\Gamma} \ . \end{cases} \tag{2.5.20}$$

Using the Sokhotsky-Plemelj formulas for the Cauchy-type integral one can see that the right-hand sides of Eqs. (2.5.20) are equal to one another on the boundary L so that $g(\zeta')$ is, just as in the case of a uniform density, a continuous function in the whole space.

Formulas (2.5.20) were first suggested by G. Ya. Golizdra (1966). Their derivation was given in a most comprehensive manner by V.N. Strakhov (1970d). Our exposition relies precisely on this work by V.N. Strakhov and draws upon the method outlined in the book by I.N. Vekua (1959, p. 43).

Now write the right-hand side of formula (2.5.20) as a unified analytical expression. To this end, we will use again Cauchy integral formula (1.1.1), applying it to the function $2\pi\gamma\Phi(\zeta', \bar{\zeta}')$ constant with the variable ζ:

$$\frac{1}{2\pi i}\int_L \frac{2\pi\gamma\Phi(\zeta', \bar{\zeta}')}{\zeta-\zeta'}\,d\zeta = \begin{cases} 2\pi\gamma\Phi(\zeta', \bar{\zeta}')\ , & \zeta' \in \Gamma\ , \\ 0\ , & \zeta' \in C\bar{\Gamma}\ . \end{cases} \tag{2.5.21}$$

Allowing for (2.5.21), we derive the following *unified representation of complex intensity of the gravitational field of a material domain with an analytical mass distribution given in terms of the Cauchy-type integral:*

$$g(\zeta') = -\frac{1}{2\pi i}\int_L \frac{2\pi\gamma[\Phi(\zeta, \bar{\zeta}) - \Phi(\zeta', \bar{\zeta}')]}{\zeta-\zeta'}\,d\zeta\ . \tag{2.5.22}$$

2.5.4 Case of Vertical or Horizontal Variations in the Density

Consider a number of important particular cases of $\varrho(x, z)$ distribution in the domain Γ. A simple but remarkable situation occurs where the density undergoes variations only in the horizontal or only in the vertical. To begin with, let $\varrho = \varrho(x)$. Then, according to (2.5.15),

$$\Phi(\zeta, \bar{\zeta}) = \int \varrho\left(\frac{\zeta+\bar{\zeta}}{2}\right) d\bar{\zeta} = 2\int \varrho\left(\frac{\zeta+\bar{\zeta}}{2}\right) d\left(\frac{\zeta+\bar{\zeta}}{2}\right) = 2R\left(\frac{\zeta+\bar{\zeta}}{2}\right)\ , \tag{2.5.23}$$

where

$$R(x) = \int \varrho(x)\,dx\ . \tag{2.5.24}$$

Substituting (2.5.23) into (2.5.22) we write

$$g(\zeta') = \frac{1}{2\pi i}\int_L \frac{4\pi\gamma[R((\zeta'+\bar{\zeta}')/2) - R((\zeta+\bar{\zeta})/2)]}{\zeta-\zeta'}\,d\zeta\ . \tag{2.5.25}$$

For instance, once $\varrho(x)$ is a quadratic function of x, $\varrho(x) = \varrho_0 + ax^2$, then, in accordance with (2.5.24), we have

$$R(x) = \varrho_0 x + \tfrac{1}{3}ax^3 + c_0\ . \tag{2.5.26}$$

Substitution of (2.5.26) into (2.5.25) yields

$$g(\zeta' = x' + iz') = \frac{1}{2\pi i}\int_L \frac{4\pi\gamma\,[\varrho_0(x'-x) + 1/3\,a(x'^3 - x^3)]}{\zeta-\zeta'}\,d\zeta\ . \tag{2.5.27}$$

Now assume that the density ϱ varies solely in the vertical $\varrho = \varrho(z)$. Then, according to (2.5.15), we have

$$\Phi(\zeta, \bar{\zeta}) = \int \varrho\left(\frac{\zeta-\bar{\zeta}}{2i}\right) d\bar{\zeta} = -2i \int \varrho\left(\frac{\zeta-\bar{\zeta}}{2i}\right) d\left(\frac{\zeta-\bar{\zeta}}{2i}\right)$$

$$= -2i \int \varrho(z)\, dz = -2iQ(z) \ , \tag{2.5.28}$$

where

$$Q(z) = \int \varrho(z)\, dz \ . \tag{2.5.29}$$

Substitution of (2.5.28) into (2.5.22) yields

$$g(\zeta') = -\frac{1}{2\pi i} \int_L \frac{4\pi\gamma[Q((\zeta'-\bar{\zeta}')/2)-Q((\zeta-\bar{\zeta})/2)]}{\zeta-\zeta'}\, d\zeta \ . \tag{2.5.30}$$

For instance, if $\varrho(z)$ is a quadratic function of z: $\varrho(z) = \varrho_0 + az^2$, then we have

$$Q(z) = \varrho_0 z + \tfrac{1}{3}az^3 + c_0 \ . \tag{2.5.31}$$

Hence, the gravitational field of masses whose density varies quadratically with depth is defined according to the formula

$$g(\zeta' = x' + iz') = -\frac{1}{2\pi i} \int_\Gamma \frac{4\pi\gamma[\varrho_0(z'-z) + 1/3\, a(z'^3 - z^3)]}{\zeta-\zeta'}\, d\zeta \ . \tag{2.5.32}$$

2.5.5 Case of Linear Density Variations Along the Coordinate Axis

Of critical importance for applications is the case where the mass density, ϱ, is a linear function of the coordinates:

$$\varrho = \varrho(x, z) = \varrho_0 + ax + bz \ . \tag{2.5.33}$$

Setting $x = (\zeta+\bar{\zeta})/2$ and $z = (\zeta-\bar{\zeta})2i$ into (2.5.33) we have

$$\varrho(x, z) = \varrho_0 + \alpha\zeta + \bar{\alpha}\bar{\zeta} \ , \tag{2.5.34}$$

where $\alpha = (a-ib)/2$. Accordingly, the function $\Phi(\zeta, \bar{\zeta})$ is evaluated as follows:

$$\Phi(\zeta, \bar{\zeta}) = \varrho_0\bar{\zeta} + \alpha\zeta\bar{\zeta} + \tfrac{1}{2}\bar{\alpha}\bar{\zeta}^2 \ . \tag{2.5.35}$$

Substitution of (2.5.35) into (2.5.22) and little transformation yield

$$g(\zeta') = \frac{1}{2\pi i} \int_L \frac{2\pi\gamma\{(\bar{\zeta}'-\bar{\zeta})[\varrho_0 + \alpha\zeta + 1/2\,\bar{\alpha}(\bar{\zeta}'+\bar{\zeta})] + \alpha\bar{\zeta}'(\zeta'-\zeta)\}}{\zeta-\zeta'}\, d\zeta$$

$$= i\gamma \int_\Gamma [\varrho_0 + \alpha\zeta + \tfrac{1}{2}\bar{\alpha}(\bar{\zeta}'+\bar{\zeta})] \frac{\bar{\zeta}-\bar{\zeta}'}{\zeta-\zeta'}\, d\zeta + i\alpha\gamma\bar{\zeta}' \int_L d\zeta \ . \tag{2.5.36}$$

The second integral in the right-hand side of (2.5.36) vanishes, by the Cauchy theorem, while the first integral can be cast in a form which is similar to the one in formula (2.5.14):

$$g(\zeta') = i\gamma \int_L [\varrho_0 + a\zeta + \tfrac{1}{2}\bar{a}(\bar{\zeta}' + \bar{\zeta})] \exp(-2i\arg(\zeta - \zeta'))\, d\zeta . \qquad (2.5.37)$$

2.5.6 General Case of Continuous Density Distribution

We will show that the above results can be extended to the case where the density $\varrho(x, z)$ is arbitrarily continuous in $\bar{\Gamma}$ and continuously differentiable in Γ. Let us redefine the function $\varrho(x, z)$ for some domain Γ^* enclosing completely the domain Γ with its boundary $\Gamma^* \supset \bar{\Gamma}$, so that $\varrho(x, z)$ might be continuous in Γ^*. We will specify an auxiliary function in Γ^* by the following equation:

$$\partial\sigma(\zeta)/\partial\bar{\zeta} = \varrho(\zeta = x + iz) = \varrho(x, z) . \qquad (2.5.38)$$

Note that Γ^* may coincide with a complex plane as a whole.

In accordance with the findings of Sect. 2.4, Eq. (2.5.38) inside Γ admits of a general solution fitted by the formula

$$\sigma(\zeta) = \varphi(\zeta) - \frac{1}{\pi} \iint_\Gamma \varrho(\eta) \frac{ds}{\eta - \zeta} , \qquad (2.5.39)$$

where $\varphi(\zeta)$ is an arbitrary function analytical in Γ. This formula follows from the Pompei formula (2.4.15), provided it is understood that

$$\frac{1}{2\pi i} \int_L \frac{\sigma(\eta)}{\eta - \zeta}\, d\eta = \varphi(\zeta) .$$

A particular solution to (2.5.38) is available on the assumption that $\varphi(\zeta) = 0$:

$$\sigma(\zeta) = -\frac{1}{\pi} \iint_\Gamma \varrho(\eta) \frac{ds}{\eta - \zeta} . \qquad (2.5.40)$$

Substituting (2.5.38) into (2.5.4) we derive the following expression for complex intensity of the gravitational field:

$$g(\zeta') = -2\gamma \iint_\Gamma \frac{\partial\sigma(\zeta)}{\partial\bar{\zeta}} \frac{ds}{\zeta - \zeta'} . \qquad (2.5.41)$$

On the basis of formula (2.5.41) and considerations similar to those of Sect. 2.5.3, we can represent the function $g(\zeta')$ in terms of the Cauchy-type integral on the boundary of the domain Γ.

$$g(\zeta') = \begin{cases} 2\pi\gamma\sigma(\zeta') - \dfrac{1}{2\pi i} \displaystyle\int_L \frac{2\pi\gamma\sigma(\zeta)}{\zeta - \zeta'}\, d\zeta , & \zeta' \in \Gamma , \\[2ex] -\dfrac{1}{2\pi i} \displaystyle\int_L \frac{2\pi\gamma\sigma(\zeta)}{\zeta - \zeta'}\, d\zeta , & \zeta' \in C\bar{\Gamma} . \end{cases} \qquad (2.5.42)$$

This latter representation can also be reduced to a unified formula similar to (2.5.22):

$$g(\zeta') = \frac{1}{2\pi i} \int_L \frac{2\pi\gamma[\sigma(\zeta') - \sigma(\zeta)]}{\zeta - \zeta'}\, d\zeta . \qquad (2.5.43)$$

In the case where $\varrho = \varrho_0 = \text{const}$, a particular solution to (2.5.38) is

$$\sigma(\zeta) = \varrho_0 \bar{\zeta} \ . \tag{2.5.44}$$

Substituting (2.5.44) into (2.5.43) we arrive at (2.5.13).

2.5.7 Calculation of the Gravitational Field of an Infinitely Extended Domain

Of exceptional value for applications is the solution of the problem on the gravitational field of a material domain extending into infinity.

Consider a domain Γ bounded by a horizontal line M

$$z = -h \tag{2.5.45}$$

and a smooth or a piecewise smooth line L (Fig. 11) fitting the equation

$$z = \Delta z(x) - h \ , \tag{2.5.46}$$

with

$$\Delta z(x) \geqslant 0 \quad \text{and} \quad \lim_{x \to \pm\infty} \Delta z(x) = 0 \ . \tag{2.5.47}$$

It is assumed that the domain Γ is filled with masses of a uniform density, $\varrho_0 = \text{const}$. Let us evaluate complex intensity of the gravitational field of this domain. To employ the formulas presented in the foregoing discussion, and valid only for finite domains, we will cut a subdomain Γ_R, bounded by a circle O_R of a radius R with a center at the origin of the coordinates, out of the domain Γ (Fig. 11). The portions of the curve L and straight line M that happen to be inside O_R will be designated by L_R and M_R, while the portions of the circle found between L and M, by O_R^+ and O_R^-. Thus, the boundary of the domain Γ_R consists of four portions of curves: $L_R \cup O_R^- \cup M_R \cup O_R^+$. As a result, in accordance with formula (2.5.7), complex intensity of the gravitational field of the domain Γ_R outside the latter is calculated by the formula

$$\begin{aligned} g(\zeta') = &-\frac{1}{2\pi i}\int_{L_R}\frac{2\pi\gamma\varrho_0\bar{\zeta}}{\zeta-\zeta'}d\zeta - \frac{1}{2\pi i}\int_{O_R^-}\frac{2\pi\gamma\varrho_0\bar{\zeta}}{\zeta-\zeta'}d\zeta \\ &-\frac{1}{2\pi i}\int_{M_R}\frac{2\pi\gamma\varrho_0\bar{\zeta}}{\zeta-\zeta'}d\zeta - \frac{1}{2\pi i}\int_{O_R^+}\frac{2\pi\gamma\varrho_0\bar{\zeta}}{\zeta-\zeta'}d\zeta \ . \end{aligned} \tag{2.5.48}$$

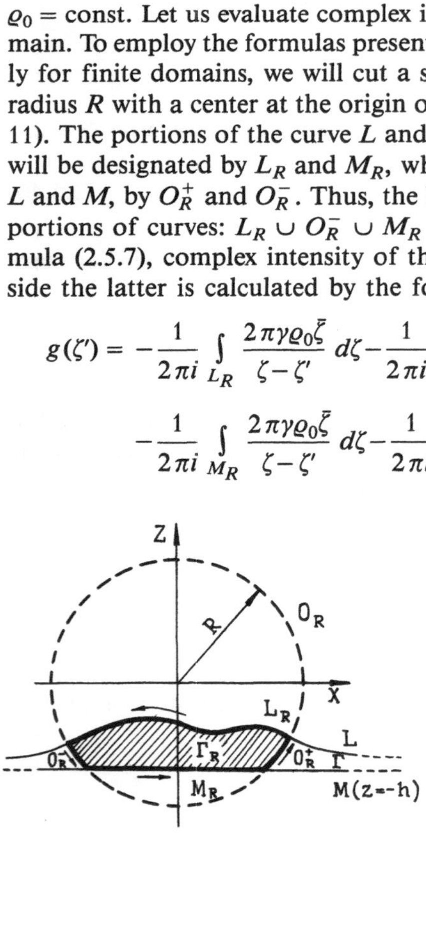

Fig. 11. Scheme explaining the method used to calculate the gravitational field of an infinitely extended domain: L and M are the boundaries of a material domain Γ

Let us transform the integral over the segment of the straight line M. For this purpose, assuming $z=(\zeta-\bar{\zeta})\,2i$ in (2.5.45), we cast the equation for the horizontal line in the form:

$$\bar{\zeta}=\zeta+2ih\ . \tag{2.5.49}$$

(For greater details of this form of the curve equation see Sect. 4.2.2.)

The function appearing in the right-hand side of Eq. (2.5.49) is analytical over the entire finite part of a complex plane. As a consequence, using the Cauchy integral theorem we can transform the integral over the segment M_R into the integral over the curve $O_R^- \cup L_R \cup O_R^+$. These transformations yield

$$\begin{aligned}-\frac{1}{2\pi i}\int_{M_R}\frac{2\pi\gamma\varrho_0\bar{\zeta}}{\zeta-\zeta'}\,d\zeta&=\frac{1}{2\pi i}\int_{M_R}\frac{2\pi\gamma\varrho_0(\zeta+2ih)}{\zeta-\zeta'}\,d\zeta\\&=\frac{1}{2\pi i}\int_{O_R^-\cup L_R\cup O_R^+}\frac{2\pi\gamma\varrho_0(\zeta+2ih)}{\zeta-\zeta'}\,d\zeta\ ,\end{aligned} \tag{2.5.50}$$

preserving the sense of integration along the curves, as is shown in Fig. 11.

Substitution of (2.5.50) into (2.5.48) yields

$$\begin{aligned}g(\zeta')=&-\frac{1}{2\pi i}\int_{L_R}\frac{2\pi\gamma\varrho_0(\bar{\zeta}-\zeta-2ih)}{\zeta-\zeta'}\,d\zeta-\frac{1}{2\pi i}\int_{O_R^-}\frac{2\pi\gamma\varrho_0(\bar{\zeta}-\zeta-2ih)}{\zeta-\zeta'}\,d\zeta\\&-\frac{1}{2\pi i}\int_{O_R^+}\frac{2\pi\gamma\varrho_0(\bar{\zeta}-\zeta-2ih)}{\zeta-\zeta'}\,d\zeta\ .\end{aligned} \tag{2.5.51}$$

By virtue of condition (2.5.47), the last two integrals in (2.5.51) for $R\to\infty$ tend to zero; therefore, passing to the limit in (2.5.51) we find

$$g(\zeta')=-\frac{1}{2\pi i}\int_{L}\frac{2\pi\gamma\varrho_0(\bar{\zeta}-\zeta-2ih)}{\zeta-\zeta'}\,d\zeta\ , \tag{2.5.52}$$

where the line L happens to run to the left. Noting that $\bar{\zeta}-\zeta-2ih=-2i\Delta z(x)=-2i\Delta z[(\zeta+\bar{\zeta})/2]$, we obtain in a finalized form

$$g(\zeta')=2\gamma\varrho_0\int_{L}\frac{\Delta z[(\zeta+\bar{\zeta})/2]}{\zeta-\zeta'}\,d\zeta\ . \tag{2.5.53}$$

It is precisely formula (2.5.53) that yields a solution to the problem of evaluating the gravitational field of an infinitely extended domain outside masses. Referring again to the Cartesian coordinates x, z and expanding the real and imaginary parts in formula (2.5.53), taking due account of (2.5.46), we write

$$-g_x(x',z')+ig_z(x',z')=-2\gamma\varrho_0\int_{-\infty}^{\infty}\frac{\Delta z(x)(1+id\Delta z(x)/dx)}{(x-x')+i(\Delta z-h-z')}\,dx\ .$$

Multiplying the numerator and denominator of the integrand by a complex conjugate to the denominator, we arrive finally at

$$g_x(x', z') = 2\gamma\rho_0 \int_{-\infty}^{+\infty} \frac{\Delta z(x)[(x-x') + (\Delta z - h - z')\, d\Delta z(x)/dx]}{(x-x')^2 + (\Delta z - h - z')^2}\, dx ,$$

$$g_z(x', z') = 2\gamma\rho_0 \int_{-\infty}^{+\infty} \frac{\Delta z(x)[(\Delta z - h - z') - (x-x')\, d\Delta z(x)/dx]}{(x-x')^2 + (\Delta z - h - z')^2}\, dx . \qquad (2.5.54)$$

Assuming that $\max_{-\infty<x<+\infty} \frac{\Delta z(x)}{h} \ll 1$ and $\max_{-\infty<x<+\infty} \left|\frac{d\Delta z(x)}{dx}\right| \ll 1$ in (2.5.46) and setting $z' = 0$, we obtain the so called *linearized representations of the gravitational field:*

$$g_x(x', 0) = 2\gamma\varrho_0 \int_{-\infty}^{+\infty} \frac{\Delta z(x)(x-x')}{(x-x')^2 + h^2}\, dx ,$$

$$g_z(x', 0) = -2\gamma\varrho_0 h \int_{-\infty}^{+\infty} \frac{\Delta z(x)}{(x-x')^2 + h^2}\, dx . \qquad (2.5.55)$$

The extremely simple derivation of formulas (2.5.54) and (2.5.55) presented here has been suggested by V.N. Strakhov (1970d).

2.6 Representation of a Fixed Magnetic Field in Terms of the Cauchy-Type Integral

The complex theory of plane fields will now be applied to a fixed magnetic field. The mathematical model of this field exciters is a two-dimensional polarized source, therefore our investigation starts with a description of complex intensity of a polarized source in terms of the Cauchy-type integral.

2.6.1 Complex Potential of a Polarized Source

Consider a two-dimensional polarized source filling a domain Γ with a complex polarization $\mu(\zeta)$. In accordance with formula (2.3.22), complex intensity of this source is defined as follows:

$$F(\zeta') = \frac{1}{2\pi} \iint_\Gamma \frac{\mu(\zeta)}{(\zeta - \zeta')}\, ds . \qquad (2.6.1)$$

Represented the kernel of integral (2.6.1) as a derivative of the Cauchy kernel:

$$\frac{1}{(\zeta - \zeta')^2} = \frac{\partial}{\partial \zeta'} \frac{1}{(\zeta - \zeta')} . \qquad (2.6.2)$$

Substituting (2.6.2) into (2.6.1) and factoring the derivative with respect to the parameter outside the integral sign, we obtain

$$F(\zeta') = \frac{\partial}{\partial \zeta'} \left(\frac{1}{2\pi} \iint_\Gamma \frac{\mu(\zeta)}{\zeta - \zeta'} \right) ds . \qquad (2.6.3)$$

The complex potential of a polarized source will be designated by $P_D(\zeta')$. Then we find from (2.6.3) that

$$F(\zeta') = -\partial P_D(\zeta')/\partial \zeta' , \tag{2.6.4}$$

where

$$P_D(\zeta') = -\frac{1}{2\pi} \iint_\Gamma \frac{\mu(\zeta)}{\zeta - \zeta'} ds . \tag{2.6.5}$$

Evidently, representation (2.6.5) for the complex potential of a polarized source is structurally identical to that of (2.4.25) for the complex intensity of the field. But unlike the latter, $\mu(\zeta)$ in the former is a complex-valued function, which markedly simplifies the examination of this formula.

2.6.2 Complex Intensities and Potential of a Magnetic Field

Now let the domain Γ be filled with magnetized masses with an intensity of magnetization $\mathbf{I}(x, z) = (I_x(x, z), I_z(x, z))$. As is known, the density of dipole moments of the corresponding two-dimensional polarized source is

$$\mathbf{m}(x, z) = 4\pi \mathbf{I}(x, z) .$$

Accordingly, complex polarization of the source is represented in the form:

$$\mu(\zeta) = 4\pi I_x(x, z) + i4\pi I_z(x, z) = 4\pi I(\zeta) , \tag{2.6.6}$$

where $I(\zeta)$ is the *complex intensity of magnetization* of magnetic masses filling the domain Γ.

Substituting (2.6.6) into (2.6.1) we find the complex intensity of a plane magnetic field $\mathbf{H}(x, z) = (H_x(x, z), H_z(x, z))$:

$$H(\zeta) = -H_x(x, z) + iH_z(x, z) = 2 \iint_\Gamma \frac{I(\zeta)}{(\zeta - \zeta')^2} ds . \tag{2.6.7}$$

The complex magnetic potential $W(\zeta')$ is cast, accordingly, in the form:

$$W(\zeta') = -2 \iint_\Gamma \frac{I(\zeta)}{\zeta - \zeta'} ds . \tag{2.6.8}$$

Formula (2.6.8) is equivalent to formula (2.5.4) for the complex intensity of the gravitational field, provided $\gamma\varrho(\zeta)$ in the latter is replaced by $I(\zeta)$. As previously noted, the basic difference lies in the fact that $I(\zeta)$ is a complex function of a complex variable ζ, while $\varrho(\zeta)$ is a real function. In what follows, we will concentrate mainly on the properties of a complex magnetic potential.

2.6.3 Representation of the Magnetic Potential of a Homogeneous Domain in Terms of the Cauchy-Type Integral

Let us rewrite formula (2.6.8) for the case of a uniformly magnetized body:

$$W(\zeta') = -2I_0 \iint_{\Gamma} \frac{ds}{\zeta-\zeta'} \,, \quad \text{where } I_0 = \text{const.} \tag{2.6.9}$$

Expression (2.6.9) can be transformed by means of the Pompei formulas in the same way as (2.5.5) is for the intensity of the gravitational field of a homogeneous domain.

Hence, analogous to (2.5.7) and (2.5.9), we derive

$$W(\zeta') = \begin{cases} -\dfrac{1}{2\pi i} \displaystyle\int_L \frac{2\pi I_0 \bar{\zeta}}{\zeta-\zeta'} d\zeta \,, & \zeta' \in C\bar{\Gamma} \,, \\ 2\pi I_0 \bar{\zeta}' - \dfrac{1}{2\pi i} \displaystyle\int_L \frac{2\pi I_0 \bar{\zeta}}{\zeta-\zeta'} d\zeta \,, & \zeta' \in \Gamma \,. \end{cases} \tag{2.6.10}$$

Just as with the gravitational field, formulas (2.6.10) can be unified to end up with a single expression:

$$W(\zeta') = iI_0 \int_L \frac{\bar{\zeta}-\bar{\zeta}'}{\zeta-\zeta'} d\zeta = iI_0 \int_L \exp\left[-2i\arg(\zeta-\zeta')\right] d\zeta \,. \tag{2.6.11}$$

2.6.4 General Case of Magnetization Distribution

Now consider a general case where magnetization $I(\zeta)$ is an arbitrary continuous function in $\bar{\Gamma}$, and has continuous partial derivatives in Γ. Let us redefine the function $I(\zeta)$, following the considerations outlined in Sect. 2.5.6, for the domain $\Gamma^* \supset \bar{\Gamma}$ so that $I(\zeta)$ be continuous in Γ^*, and write an equation to solve the function $h(\zeta)$ in Γ^*:

$$\partial h(\zeta)/\partial\bar{\zeta} = I(\zeta) \,. \tag{2.6.12}$$

In accordance with the findings of Sect. 2.4, Eq. (2.6.12) admits of a general solution in Γ, which is fitted by the formula

$$h(\zeta) = \Phi(\zeta) - \frac{1}{\pi} \iint_{\Gamma} I(\eta) \frac{ds}{\eta-\zeta} \,, \tag{2.6.13}$$

where $\Phi(\zeta)$ is an arbitrary function analytical in Γ. In particular, we can assume that $\Phi(\zeta) \equiv 0$, then a particular solution to (2.6.12) has the form

$$h(\zeta) = -\frac{1}{\pi} \iint_{\Gamma} I(\eta) \frac{ds}{\eta-\zeta} \,. \tag{2.6.14}$$

Now substituting (2.6.12) into (2.6.8) we represent the magnetic potential as follows:

$$W(\zeta') = -2 \iint_{\Gamma} \frac{\partial h(\zeta)}{\partial \bar{\zeta}} \frac{ds}{\zeta - \zeta'} , \tag{2.6.15}$$

and transform it, similarly to the practice of (2.5.6), into the expression

$$W(\zeta') = \begin{cases} -\dfrac{1}{2\pi i} \displaystyle\int_L \dfrac{2\pi h(\zeta)}{\zeta - \zeta'} d\zeta , & \zeta' \in \Gamma , \\ 2\pi h(\zeta') - \dfrac{1}{2\pi i} \displaystyle\int_L \dfrac{2\pi h(\zeta)}{\zeta - \zeta'} d\zeta , & \zeta' \in C\bar{\Gamma} . \end{cases} \tag{2.6.16}$$

Finally, just as we have done it for the gravitational field, we unify the right-hand sides of (2.6.16) into a single formula

$$W(\zeta') = \frac{1}{2\pi i} \int_L \frac{2\pi [h(\zeta') - h(\zeta)]}{\zeta - \zeta'} d\zeta . \tag{2.6.17}$$

Thus, the magnetic potential outside and inside a magnetized body is fitted by unified formula (2.6.17) in terms of the Cauchy-type integral over the contour of a body. It is remarkable that *representation (2.6.17) holds good for any distribution of magnetization $I(\zeta)$, which has continuous partial derivatives inside Γ and is continuous in a closed domain $\bar{\Gamma}$.*

In a particular case of $I(\zeta) = I_0 = \text{const}$, the function $h(\zeta)$ seems to be equal to $h(\zeta) = I_0 \bar{\zeta}$ and formula (2.6.17) goes over into formula (2.6.11).

2.6.5 Analytical Distribution of Magnetization

We will dwell upon the most critical case where magnetization $I(\zeta)$ is an analytical function inside Γ and is a continuous function in $\bar{\Gamma}$. Then it is readily comprehensible that a particular solution to (2.6.12) is the function

$$h(\zeta) = I(\zeta)\bar{\zeta} . \tag{2.6.18}$$

Substitution of (2.6.18) into (2.6.16) yields

$$W(\zeta') = \begin{cases} -\dfrac{1}{2\pi i} \displaystyle\int_L \dfrac{2\pi I(\zeta)\bar{\zeta}}{\zeta - \zeta'} d\zeta , & \zeta' \in C\bar{\Gamma} , \\ 2\pi I(\zeta')\bar{\zeta}' - \dfrac{1}{2\pi i} \displaystyle\int_L \dfrac{2\pi I(\zeta)\bar{\zeta}}{\zeta - \zeta'} d\zeta , & \zeta' \in \Gamma , \end{cases} \tag{2.6.19}$$

or a unified analytical expression

$$W(\zeta') = \frac{1}{2\pi i} \int_L \frac{2\pi [I(\zeta')\bar{\zeta}' - I(\zeta)\bar{\zeta}]}{\zeta - \zeta'} d\zeta . \tag{2.6.20}$$

Relations (2.6.19) were first derived by V. N. Strakhov (1970d). They are of fundamental importance in the theory of a complex magnetic potential.

3 Techniques for Separation of Plane Fields

3.1 Separation of Geopotential Fields into External and Internal Parts Using Spectral Decomposition

This chapter is concerned with application of the body of the Cauchy-type integral to the solution of one of the oldest problems encountered in geopotential field theory, namely separation of the field into external and internal parts. This problem was first tackled as early as the last century in Gauss' classical works on geomagnetism. Gauss was interested in the location of sources of the terrestrial magnetic field, whether they are found inside or outside the Earth. Having employed his own concept of spherical harmonic analysis, the great German mathematician demonstrated that the external part of the magnetic potential is negligible, compared to the internal part. The contributions of the magnetic field made by external and internal sources were analyzed once again toward the end of the nineteenth century by the German physicist and mathematician Schmiedt (Chapman and Bartels 1940). Having refined appropriately the technique of spherical harmonic analysis and employing more precise field data on the magnetic field, Schmiedt succeeded in evaluating the contribution of external sources into the terrestrial magnetic field, which proved to be negligible too.

In the mid-twentieth century, the German geophysicists Kertz (1954), Siebert (Siebert and Kertz 1957; Siebert 1958, 1962, 1964), and Hartman (1963) developed the Gauss–Schmidt method further to apply it to the analysis of the variations in the magnetic field of the Earth. This was possible because the transient magnetic field in the air (insulator) as well as the fixed magnetic field can be represented in terms of scalar potential[1]. The works by Siebert (1964) and Hartman (1963) rely extensively on the bulk of the theory of functions of a complex variable. For the sake of a more comprehensive exposition of the subject matter, this chapter will take up both the spectral technique for field separation dating back to the ideas of Gauss-Schmiedt and the current approach illustrating the efficiency in applying the Cauchy-type integral.

Further development of this research relating to the extension of field separation methods of three-dimensional situations and non-potential fields has been effected in the works of Weaver (1964), Nedyalkov (1965) and the author of this monograph (Zhdanov 1973c). These methods will be treated in the subsequent chapters.

---

[1] For a close analysis of the equations for a transient magnetic field see Part III.

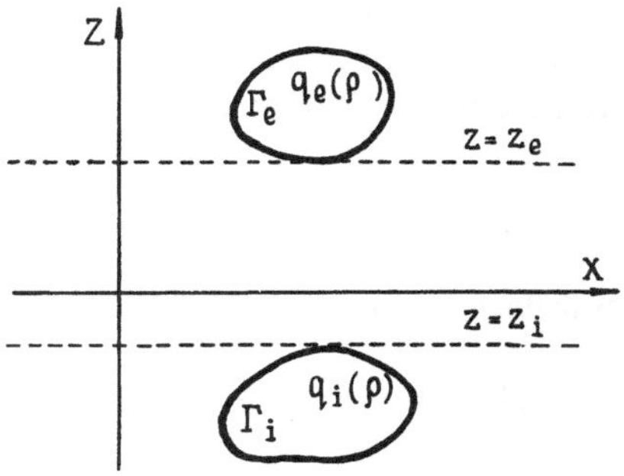

Fig. 12. Distribution of the sources of a plane potential field **F** in the vertical plane *xoz*; q_e is the external source, q_i is the internal source

3.1.1 Statement of the Problem of Plane Field Separation

Consider a plane potential field $\mathbf{F}(x, z) = [F_x(x, z), F_z(x, z)]$. Assume that this field is produced by two sources: an external source $q_e(x, z)$ located in the domain Γ_e, and an internal source $q_i(x, z)$ concentrated in the domain Γ_i, the domains Γ_e and Γ_i being found on the two sides of the x-axis (Fig. 12). Let $\mathbf{F}^e(x, z)$ and $\mathbf{F}^i(x, z)$ designate the fields produced by each of the sources, q_e and q_i, separately. These fields will be referred to as *external* and *internal* fields, respectively. Then, according to the principle of field superposition, we have

$$\mathbf{F}(x, z) = \mathbf{F}^e(x, z) + \mathbf{F}^i(x, z). \tag{3.1.1}$$

The problem of field separation is stated as follows: knowing the total vector field $\mathbf{F}(x, z)$ along the x-axis, define the external and internal fields, $\mathbf{F}^e(x, z)$ and $\mathbf{F}^i(x, z)$ along the same axis.

This problem can be expressed in terms of the scalar potential. Indeed, let $U(x, z)$, $U^e(x, z)$ and $U^i(x, z)$ stand for the potentials of the fields $\mathbf{F}(x, z)$, $\mathbf{F}^e(x, z)$, and $\mathbf{F}^i(x, z)$, respectively. Then we have

$$\mathbf{F} = \operatorname{grad} U \ , \quad \mathbf{F}^e = \operatorname{grad} U^e \ , \quad \mathbf{F}^i = \operatorname{grad} U^i \ . \tag{3.1.2}$$

The problem of potential separation lies in finding an external U^e and internal U^i potentials along the x-axis from the total potential U specified along the same axis.

To solve this problem, we will turn to spectral representations of the field.

3.1.2 Spectral Representations of Plane Fields

It is evident that a plane potential field excited by sources localized in finite domains of a plane is absolutely integrable along any horizontal line. Hence, there is a direct Fourier transform for it:

$$\mathbf{f}(k_x, z) = \int_{-\infty}^{+\infty} \mathbf{F}(x, z) \exp(ik_x x) \, dx \ . \tag{3.1.3}$$

The function $\mathbf{f}(k_x, z)$ will be referred to as a *spatial spectrum* of the field $\mathbf{F}(x, z)$, while the parameter k_x is called a *spatial frequency*. In accordance with the inverse Fourier transform, we have

$$\mathbf{F}(x, z) = \frac{1}{2\pi} \int_{-\infty}^{+\infty} \mathbf{f}(k_x, z) \exp(-ik_x x)\, dk_x \ . \tag{3.1.4}$$

Similarly, a spatial spectrum $u(k_x, z)$ of the scalar potential is introduced:

$$u(k_x, z) = \int_{-\infty}^{+\infty} U(x, z) \exp(ik_x x)\, dx \ , \tag{3.1.5}$$

which is related to $U(x, z)$ by

$$U(x, z) = \frac{1}{2\pi} \int_{-\infty}^{+\infty} u(k_x, z) \exp(-ik_x x)\, dx \ . \tag{3.1.6}$$

Obviously, according to (3.1.2), the spectra of the horizontal and vertical components of the field $\mathbf{F}$ are expressed in terms of the potential spectrum:

$$f_x = -ik_x u \ , \quad f_z = \partial u/\partial z \ . \tag{3.1.7}$$

Let $z = z_e$ and $z = z_i$ stand for the equations of horizontal lines passing through points of domains Γ_e and Γ_i next to the x-axis (Fig. 12). Then, since the scalar potential everywhere outside Γ_e and Γ_i meets the Laplace equation

$$\Delta U = 0 \ , \tag{3.1.8}$$

the potential spectrum in a horizontal band $z_i < z < z_e$ does satisfy the one-dimensional Helmholtz equation

$$\partial^2 u/\partial z^2 - k_x^2 u = 0 \ . \tag{3.1.9}$$

A general solution to Eq. (3.1.9) has the form

$$u = u^+ \exp(|k_x| z) + u^- \exp(-|k_x| z) \ , \tag{3.1.10}$$

where u^+ and u^- are constants.

As is seen, the spectrum u consists of two components. The first component increases with z and, hence, is related to an external field, while the second component, on the contrary, decreases with increasing z and, as a result, fits an internal field. Therefore, we can write

$$u^e = u^+ \exp(|k_x| z) \ , \quad u^i = u^- \exp(-|k_x| z) \ , \tag{3.1.11}$$

where u^e and u^i are the spatial spectra of the external and internal parts of the potential, respectively. Thus, the problem of field separation has been reduced to the determination of unknown constants u^+ and u^-. Let us solve this problem.

3.1.3 Determination of the External and Internal Parts of the Scalar Potential and Field (Gauss-Schmiedt Formulas)

In order to find u^+ and u^-, substitute (3.1.10) into (3.1.7):

$$\begin{aligned} f_x &= -ik_x[u^+\exp(|k_x|z)+u^-\exp(-|k_x|z)] \ , \\ f_z &= |k_x|\,[u^+\exp(|k_x|z)-u^-\exp(-|k_x|z)] \ . \end{aligned} \tag{3.1.12}$$

In particular, along the x-axis, we have

$$f_x|_{z=0} = -ik_x(u^+ + u^-) \ , \quad f_z|_{z=0} = |k_x|(u^+ - u^-) \ . \tag{3.1.13}$$

Hence, we find readily that

$$\begin{aligned} u^+ &= u^e|_{z=0} = \frac{1}{2}\left(\frac{f_x}{-ik_x}+\frac{f_z}{|k_x|}\right)_{z=0} \ , \\ u^- &= u^i|_{z=0} = \frac{1}{2}\left(\frac{f_x}{-ik_x}-\frac{f_z}{|k_x|}\right)_{z=0} \ . \end{aligned} \tag{3.1.14}$$

Substituting (3.1.14) into (3.1.11) and subsequently into (3.1.7) we determine the external and internal parts of a plane field along the x-axis:

$$\begin{aligned} f_x^e|_{z=0} &= \frac{1}{2}\left(f_x - i\frac{k_x}{|k_x|}f_z\right)_{z=0} \ , \\ f_x^i|_{z=0} &= \frac{1}{2}\left(f_x + i\frac{k_x}{|k_x|}f_z\right)_{z=0} \ , \\ f_z^e|_{z=0} &= \frac{1}{2}\left(f_z + i\frac{|k_x|}{k_x}f_x\right)_{z=0} \ , \\ f_z^i|_{z=0} &= \frac{1}{2}\left(f_z - i\frac{|k_x|}{k_x}f_x\right)_{z=0} \ . \end{aligned} \tag{3.1.15}$$

Allowing for the fact that

$$\frac{k_x}{|k_x|} = \frac{|k_x|}{k_x} = \operatorname{sign} k_x = \begin{cases} 1, & k_x > 0 \ , \\ 0, & k_x = 0 \ , \\ -1, & k_x < 0 \ , \end{cases} \tag{3.1.16}$$

we write

$$\begin{aligned} f_x^{e,i}|_{z=0} &= \tfrac{1}{2}\,(f_x \pm i\operatorname{sign} k_x \cdot f_z)_{z=0} \ , \\ f_z^{e,i}|_{z=0} &= \tfrac{1}{2}\,(f_z \pm i\operatorname{sign} k_x \cdot f_x)_{z=0} \ . \end{aligned} \tag{3.1.17}$$

Synthesizing (3.1.17) we obtain finally

$$\mathbf{F}^{e,i}(x, 0) = \frac{1}{2\pi}\int_{-\infty}^{+\infty} \mathbf{f}^{e,i}|_{z=0}\exp(-ik_x x)\, dk_x \ , \tag{3.1.18}$$

or in greater detail

$$F_x^i(x,\,0)=\frac{1}{2}\,F_x(x,\,0)+\frac{i}{4\pi}\int\limits_{-\infty}^{+\infty}\operatorname{sign}k_x\cdot f_z|_{z=0}\exp(-ik_xx)\,dk_x\ ,$$

$$F_z^i(x,\,0)=\frac{1}{2}\,F_z(x,\,0)-\frac{i}{4\pi}\int\limits_{-\infty}^{+\infty}\operatorname{sign}k_x\cdot f_x|_{z=0}\exp(-ik_xx)\,dk_x\ ; \qquad (3.1.19)$$

$$F_x^e(x,\,0)=\frac{1}{2}\,F_x(x,\,0)-\frac{i}{4\pi}\int\limits_{-\infty}^{+\infty}\operatorname{sign}k_x\cdot f_z|_{z=0}\exp(-ik_xx)\,dk_x\ ,$$

$$F_z^e(x,\,0)=\frac{1}{2}\,F_z(x,\,0)+\frac{i}{4\pi}\int\limits_{-\infty}^{+\infty}\operatorname{sign}k_x\cdot f_x|_{z=0}\exp(-ik_xx)\,dk_x\ . \qquad (3.1.20)$$

Following the work of Berdichevsky and Zhdanov (1984), formulas (3.1.19) and (3.1.20) will be referred to as *Gauss-Schmiedt formulas of plane field separation.*

3.2 Kertz-Siebert Technique

Now we will solve the problem stated in the previous subsection in terms of the Cauchy-type integral.

3.2.1 Problem of Separation of Field Complex Intensity

Refer again to the complex coordinates $\zeta=x+iz$ and designate the complex intensities of the total field $\mathbf{F}(x,\,z)$, of its external $\mathbf{F}^e(x,\,z)$ and internal $\mathbf{F}^i(x,\,z)$ parts by $F(\zeta)$, $F^e(\zeta)$, and $F^i(\zeta)$, respectively. Evidently, $F^e(\zeta)$ is a complex analytical function everywhere outside Γ_e, while $F^i(\zeta)$ is a complex analytical function everywhere outside Γ_i.

The situation treated in the above subsection will be made somewhat more complicated by assuming that the total field $F(\zeta)$ is specified on an arbitrary piecewise smooth line passing through a point at infinity and separating a complex plane into two simply connected parts – upper and lower half-planes (Fig. 13).

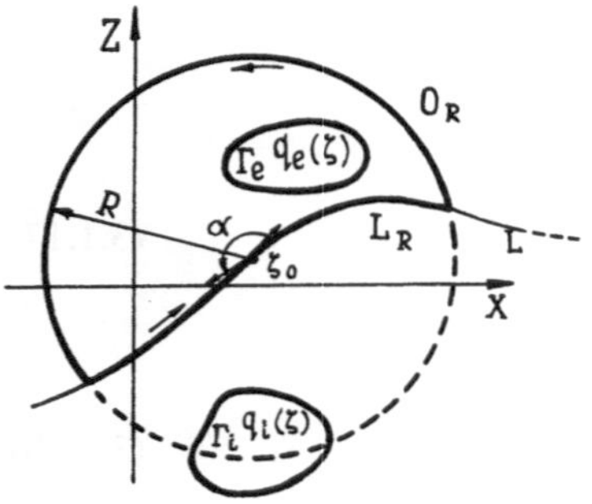

Fig. 13. Substantiation of the Kertz-Siebert technique used to separate plane fields: q_e is the external source, q_i is the internal source

The domain Γ_e lies in the upper half-plane, while the domain Γ_i is in the lower half-plane. *The problem is to define on line L complex analytical functions $F^e(\zeta)$ and $F^i(\zeta)$, these having singularities only in the upper or only in the lower half-plane, respectively, on the basis of a known complex analytical function $F(\zeta)$ on the line L, which function is peculiar both in the upper (in the domain Γ_e) and in the lower (in the domain Γ_i) half-planes.*

To solve this problem, draw from an arbitrary point ζ_0 lying on the line L a circle of such a large radius R that the domain Γ_e can be completely inside this circle (Fig. 13). Close the portion of the circle O_R lying in the upper half-plane by the line L_R, cut off by the circle on L, and consider the Cauchy-type integral along the closed contour $A_R = O_R \cup L_R$:

$$C_R(\zeta') = \frac{1}{2\pi i} \int_{A_R} \frac{F(\zeta)}{\zeta - \zeta'} d\zeta \tag{3.2.1}$$

(the contour is traversed counterclockwise). Since

$$F(\zeta) = F^e(\zeta) + F^i(\zeta) \ , \tag{3.2.2}$$

integral (3.2.1) can be represented as a sum of two integrals:

$$C_R(\zeta') = \frac{1}{2\pi i} \int_{A_R} \frac{F^e(\zeta)}{\zeta - \zeta'} d\zeta + \frac{1}{2\pi i} \int_{A_R} \frac{F^i(\zeta)}{\zeta - \zeta'} d\zeta \ . \tag{3.2.3}$$

Let D_R designate the domain bounded by A_R. The function $F^e(\zeta)$ seems to be analytical and continuous in the domain CD_R, external to D_R, and to vanish at infinity. Hence, in accordance with the Cauchy integral formula for an infinite domain (1.1.2), we have

$$\frac{1}{2\pi i} \int_{A_R} \frac{F^e(\zeta)}{\zeta - \zeta'} d\zeta = \begin{cases} -F^e(\zeta') \ , & \zeta' \in C\bar{D}_R \ , \\ 0 \ , & \zeta' \in D_R \ . \end{cases} \tag{3.2.4}$$

Similarly, the function $F^i(\zeta)$ is analytical and continuous in $\bar{D}_R$, so as a result we have

$$\frac{1}{2\pi i} \int_{A_R} \frac{F^i(\zeta)}{\zeta - \zeta'} d\zeta = \begin{cases} 0 \ , & \zeta' \in C\bar{D}_R \ , \\ F^i(\zeta') \ , & \zeta' \in D_R \ . \end{cases} \tag{3.2.5}$$

Substituting (3.2.4) and (3.2.5) into (3.2.3) we find:

$$C_R(\zeta') = \begin{cases} -F^e(\zeta') \ , & \zeta' \in C\bar{D}_R \ , \\ F^i(\zeta') \ , & \zeta' \in D_R \ . \end{cases} \tag{3.2.6}$$

Formula (3.2.6) is virtually basic to the determination of external and internal fields. However, we are concerned with a situation where F^e and F^i are to be sought on the line L.

3.2.2 Field Separation at Ordinary Points of the Line L

Assume, first, that the point ζ_0 is an ordinary point (rather than a corner) of the line L, i.e., there is a neighborhood of this point where the curve L is a smooth line. Let us determine the limit values of the function $C_R(\zeta')$ on L_R.

According to Sokhotsky-Plemelj formulas (1.2.21 a) and (1.2.21 b), we have

$$C_R^+(\zeta_0) = \lim_{\zeta\to\zeta_0^+} C_R(\zeta) = C_R(\zeta_0) + \tfrac{1}{2}F(\zeta_0) \ ,$$
$$C_R^-(\zeta_0) = \lim_{\zeta\to\zeta_0^-} C_R(\zeta) = C_R(\zeta) - \tfrac{1}{2}F(\zeta_0) \ . \qquad (3.2.7)$$

On the other hand, since the functions $F^e(\zeta)$ and $F^i(\zeta)$ fitting the external and internal fields are continuous in the neighborhood of the point ζ_0, we derive, in accordance with (3.2.6), that

$$C_R^+(\zeta_0) = F^i(\zeta_0) \ , \quad C_R^-(\zeta_0) = -F^e(\zeta_0) \ . \qquad (3.2.8)$$

Substitution of (3.2.8) into (3.2.7) yields

$$F^i(\zeta_0) = C_R(\zeta_0) + \tfrac{1}{2}F(\zeta_0) \ , \quad F^e(\zeta_0) = -C_R(\zeta_0) + \tfrac{1}{2}F(\zeta_0) \ , \qquad (3.2.9)$$

where

$$C_R(\zeta_0) = \frac{1}{2\pi i}\int_{L_R}\frac{F(\zeta)}{\zeta-\zeta_0}d\zeta + \frac{1}{2\pi i}\int_{O_R}\frac{F(\zeta)}{\zeta-\zeta_0}d\zeta \ , \quad \zeta_0 \in L_R \ .$$

Now proceed to the limit for $R\to\infty$. Upon substitution of $\zeta = \zeta_0 + Re^{i\theta}$, the integral over the circle portion O_R is written in the form

$$\int_{O_R}\frac{F(\zeta)}{\zeta-\zeta_0}d\zeta = i\int_{\theta} F(\zeta_0 + Re^{i\theta})\,d\theta \ . \qquad (3.2.10)$$

By virtue of analyticity, the function $F(\zeta)$ tends uniformly over θ to vanish at infinity, therefore, the limit of integral (3.2.10) is zero. As a result, passing to the limit in (3.2.9) for $R\to\infty$, we write finally

$$F^i(\zeta_0) = \frac{1}{2}F(\zeta_0) + \frac{1}{2\pi i}\int_L \frac{F(\zeta)}{\zeta-\zeta_0}d\zeta \ ,$$
$$F^e(\zeta_0) = \frac{1}{2}F(\zeta_0) - \frac{1}{2\pi i}\int_L \frac{F(\zeta)}{\zeta-\zeta_0}d\zeta \ , \qquad (3.2.11)$$

where the integration along L is carried out in such a sense as to leave the upper half-plane on the left (Fig. 13), and the integral implies the Cauchy principal value.

3.2.3 Field Separation at Corners of the Line L

Now consider the case where ζ_0 is a corner of a piecewise smooth curve L and a is the angle made by two lines tangent to L at a point ζ_0 counted off on the left of the curve L [i.e., in the upper half-plane (see Fig. 13)]. Then it is necessary to employ the Sokhotsky-Plemelj formulas for corners (1.2.28):

$$C^+(\zeta_0) = C(\zeta_0) + (1 - a/2\pi)F(\zeta_0) \ ,$$
$$C^-(\zeta_0) = C(\zeta_0) - (a/2\pi)F(\zeta_0 \ . \qquad (3.2.12)$$

Substituting (3.2.8) into (3.2.12) and on the basis of the above considerations, we write

$$F^i(\zeta_0) = \left(1-\frac{a}{2\pi}\right) F(\zeta_0)+\frac{1}{2\pi i}\int_L \frac{F(\zeta)}{\zeta-\zeta_0}\,d\zeta\ , \tag{3.2.13 a}$$

$$F^e(\zeta_0) = \frac{a}{2\pi}F(\zeta_0)-\frac{1}{2\pi i}\int_L \frac{F(\zeta)}{\zeta-\zeta_0}\,d\zeta\ . \tag{3.2.13 b}$$

Evidently, for a smooth curve, $a=\pi$ and formulas (3.2.13 a) and (3.2.13 b) go over into (3.2.11), so that expressions (3.2.13) permit field separation at any point ζ_0 of a piecewise smooth curve L.

Assuming that $F(\zeta)=F^i(\zeta)$ in formula (3.2.13 a) we derive

$$F^i(\zeta_0) = \frac{1}{ia}\int_L \frac{F^i(\zeta_0)}{\zeta-\zeta_0}\,d\zeta\ . \tag{3.2.14}$$

In a similar way, taking $F(\zeta)=F^e(\zeta)$ in (3.2.13 b) we find

$$F^e(\zeta_0) = \frac{-1}{i(2\pi-a)}\int_L \frac{F^e(\zeta_0)}{\zeta-\zeta_0}\,d\zeta\ . \tag{3.2.15}$$

Formulas (3.2.14) and (3.2.15) *specify the boundary conditions for complex intensities of internal and external fields on the line L.* If ζ_0 is an ordinary (rather than a corner) point of the curve L, then $a=\pi$ and formulas (3.2.14) and (3.2.15) are equivalent to conditions (1.3.8) and (1.3.9) for the boundary values of analytical functions.

3.2.4 Kertz-Siebert Formulas

Now assume that the line L coincides with the x-axis. Then $\zeta_0=x_0$ and formulas (3.2.11) are simplified as follows:

$$\begin{aligned} F^i(x_0) &= \frac{1}{2}\,F(x_0)+\frac{1}{2\pi i}\int_{-\infty}^{+\infty} \frac{F(x)}{x-x_0}\,dx\ , \\ F^e(x_0) &= \frac{1}{2}\,F(x_0)-\frac{1}{2\pi i}\int_{-\infty}^{+\infty} \frac{F(x)}{x-x_0}\,dx\ . \end{aligned} \tag{3.2.16}$$

Allowing for the fact that $F=-F_x+iF_z$ and separating the real and imaginary parts in formulas (3.2.16) we obtain

$$\begin{aligned} F^i_x(x_0) &= \frac{1}{2}\,F_x(x_0)-\frac{1}{2\pi}\int_{-\infty}^{+\infty} \frac{F_z(x)}{x-x_0}\,dx\ , \\ F^i_z(x_0) &= \frac{1}{2}\,F_z(x_0)+\frac{1}{2\pi}\int_{-\infty}^{+\infty} \frac{F_x(x)}{x-x_0}\,dx\ ; \end{aligned} \tag{3.2.17}$$

$$F_x^e(x_0) = \frac{1}{2} F_x(x_0) + \frac{1}{2\pi} \int_{-\infty}^{+\infty} \frac{F_z(x)}{x-x_0} dx \ ,$$

$$F_z^e(x_0) = \frac{1}{2} F_z(x_0) - \frac{1}{2\pi} \int_{-\infty}^{+\infty} \frac{F_x(x)}{x-x_0} dx \ . \tag{3.2.18}$$

Formulas (3.2.17) and (3.2.18) are called the *Kertz-Siebert formulas.* If the Hilbert operator (1.3.24) is employed, formulas (3.2.17) and (3.2.18) take the form

$$F_x^i(x_0) = \tfrac{1}{2}[F_x(x_0) - \hat{G}F_z(x)] \ , \quad F_z^i(x_0) = \tfrac{1}{2}[F_z(x_0) + \hat{G}F_x(x)] \ ; \tag{3.2.19}$$

$$F_x^e(x_0) = \tfrac{1}{2}[F_x(x_0) + \hat{G}F_z(x)] \ , \quad F_z^e(x_0) = \tfrac{1}{2}[F_z(x_0) - \hat{G}F_x(x)] \ . \tag{3.2.20}$$

Just as in the above general case, formulas (3.2.19) and (3.2.20) suggest *integral relations between the components of the external and internal fields.*

Indeed, taking a zero external field in (3.2.19), i.e., $F_x = F_x^i$ and $F_z = F_z^i$, we have

$$F_z^i(x_0) = -\hat{G}F_z^i(x_0) \ , \quad F_z^i(x_0) = \hat{G}F_x^i(x_0) \ . \tag{3.2.21}$$

Similarly, setting a zero international field in (3.2.20) we find

$$F_x^e(x_0) = \hat{G}F_z^e(x_0) \ , \quad F_z^e(x_0) = -\hat{G}F_x^e(x_0) \ . \tag{3.2.22}$$

Thus, the vertical and horizontal components of the external and internal fields are interrelated by pairs of Hilbert transforms (taken, however, with the different signs).

3.2.5 Equivalence Between the Kertz-Siebert and the Gauss-Schmiedt Formulas

Concluding this section, we should like to point out the equivalence between the Kertz-Siebert formulas (3.2.17) and (3.2.18) and the Gauss-Schmiedt formulas (3.1.19) and (3.1.20). To this end, it seems sufficient to show that the Hilbert operator can be represented in the form

$$\hat{G}F(x) = -\frac{i}{2\pi} \int_{-\infty}^{+\infty} \operatorname{sign} k_x \cdot f(k_x) \exp(-ik_x x_0) \, dk_x \ , \tag{3.2.23}$$

where

$$f(k_x) = \int_{-\infty}^{+\infty} F(x) \exp(ik_x x) \, dx.$$

Indeed, the integral in the right-hand side of (3.2.23) contains the product of the spectra of functions $F(x)$ and $1/\pi x$ since

$$i \operatorname{sign} k_x = \frac{1}{\pi} \int_{-\infty}^{+\infty} \frac{1}{x} \exp(ik_x x) \, dx \ .$$

Hence, by the convolution spectrum theorem, we have

$$-\frac{i}{2\pi}\int_{-\infty}^{+\infty}\operatorname{sign}k_x\cdot f(k_x)\exp(-ik_xx_0)\,dk_x=\frac{1}{\pi}\int_{-\infty}^{+\infty}\frac{F(x)}{x-x_0}\,dx=\hat{G}F(x)\;,$$

which was to be proved.

Thus, we have derived a set of formulas permitting separation of plane potential fields into external and internal parts. In subsequent chapters of the book these formulas will be extended to three-dimensional situations and to electromagnetic fields.

4 Analytical Continuation of a Plane Field

4.1 Fundamentals of Analytical Continuation

Now let us take up one of the most important concepts of the theory of a plane potential field, namely the concept of analytical continuation. As we have implied, in a two-dimensional case a plane field is described unequivocally by complex intensity. Obviously, the problem of analytical continuation of a plane field reduces to the well-known problem of analytical continuation of complex functions. A detailed treatment of this problem is offered in many treatises and monographs on the theory of functions of a complex variable. Those wishing to delve more deeply into this particular topic will no doubt start with the above literature. In this book, however, we will confine ourselves to the basic theorems and principles underlying the theory of analytical continuation, particularly to those needed for geophysical applications.

4.1.1 Taylor Theorem

First we will prove an important theorem of complex analysis which is known under the name of the *Taylor theorem.*

Theorem 4.1.1. The function $f(\zeta)$ analytical in the domain D is represented in the neighborhood of each point $\zeta_0 \in D$ as a power series

$$f(\zeta) = \sum_{n=0}^{\infty} a_n(\zeta-\zeta_0)^n \ , \tag{4.1.1}$$

whose radius of convergence R is not shorter than the distance d from ζ_0 to the boundary L of the domain D.

Indeed, let us draw a circle γ_a of radius a from a point ζ_0 as a center and express the function $f(\zeta')$ at some point ζ' inside γ_a in terms of its value on γ_a, using the Cauchy integral formula (Fig. 14):

$$f(\zeta') = \frac{1}{2\pi i} \int_{\gamma_a} \frac{f(\zeta)}{\zeta-\zeta'}\, d\zeta \ . \tag{4.1.2}$$

Transform the integrand as follows:

$$\frac{1}{\zeta-\zeta'} = \frac{1}{(\zeta-\zeta_0)[1-(\zeta'-\zeta_0)/(\zeta-\zeta_0)]} \ , \quad \text{where } \zeta \in \gamma_a \ . \tag{4.1.3}$$

Fig. 14. Proof of the Taylor theorem: ζ_0 is the center of the circle of convergence of a power series

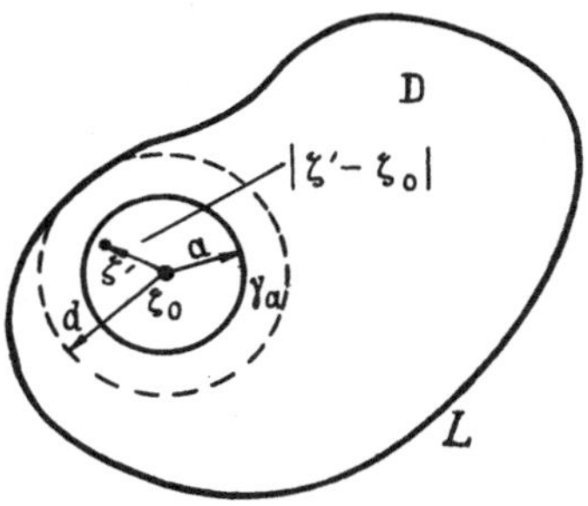

Evidently, for any fixed point ζ' of the circle $|\zeta'-\zeta_0|<a$ the following inequality

$$\left|\frac{\zeta'-\zeta_0}{\zeta-\zeta_0}\right| = |q|<1\ , \quad \zeta\in\gamma_a \tag{4.1.4}$$

is the case. As a result, the function $1/(1-q)$ is expanded into an absolutely and uniformly (with respect to ζ) convergent series:

$$\frac{1}{1-q} = \sum_{n=0}^{\infty} q^n = \sum_{n=0}^{\infty}\left(\frac{\zeta'-\zeta_0}{\zeta-\zeta_0}\right)^n . \tag{4.1.5}$$

Substitution of power series (4.1.5) into (4.1.3) and subsequently into (4.1.2) yields

$$f(\zeta') = \frac{1}{2\pi i}\int_{\gamma_a}\frac{f(\zeta)}{(\zeta-\zeta_0)}\sum_{n=0}^{\infty}\left(\frac{\zeta'-\zeta_0}{\zeta-\zeta_0}\right)^n d\zeta = \sum_{n=0}^{\infty} a_n(\zeta'-\zeta_0)^n\ , \tag{4.1.6}$$

where

$$a_n = \frac{1}{2\pi i}\int_{\gamma_a}\frac{f(\zeta)}{(\zeta-\zeta_0)^{n+1}}\,d\zeta\ , \quad n=0,\ 1,\ 2\ldots \tag{4.1.7}$$

Thus, we have obtained a power series expansion of the function analytical inside D. Taking into consideration formulas (1.1.5), we can write

$$a_n = \frac{1}{n!}f^{(n)}(\zeta_0)\ . \tag{4.1.8}$$

Note finally that since the above considerations are good for any a out of the range $0<a<d$ and the coefficients a_n defined by formulas (4.1.7) are the same, by virtue of the Cauchy theorem, for any a, the radius of convergence of power series (4.1.6) is not shorter than d.

A series (4.1.6) whose coefficients are expressed in terms of the analytical function $f(\zeta)$, according to formulas (4.1.7) and (4.1.8), is called *the Taylor series of the function* $f(\zeta)$. It follows from (4.1.8), in particular, that the analytical function $f(\zeta)$ can be expanded into the Taylor series in only one possible way.

It is pertinent to mention that sometimes in the elaboration of the theory of analytical functions of a complex variable, basic to the definition of an analytical function f in some neighborhood D of a point ζ_0 is the property that the function f is expanded into a power series convergent in the neighborhood D of the point ζ_0. This approach to the theory of analytical functions is known as the *Weierstrass concept.* It should be noted that *in the study of the analytical functions of a real variable* (single or several) *definition of analyticity in Weierstrass' terms is fundamental.* Precisely this definition will be used, in particular in the theory of harmonic functions of three variables, as well as in the electromagnetic field theory. Now we will define, as an example, a real analytical function of a real variable:

A real single-valued function $f(x)$ specified for a segment $a \leqslant x \leqslant b$ is called analytical, provided it is represented in some neighborhood[1] $x_0 - \delta < x < x_0 + \delta$ *of each point $x_0 \in [a, b]$ as a sum of the power series*

$$f(x) = \sum_{n=0}^{\infty} a_n (x - x_0)^n \tag{4.1.9}$$

with real coefficients.

4.1.2 Uniqueness of an Analytical Function

The Taylor theorem suggests, as a corollary, the following claim.

Theorem 4.1.2. Let two analytical function $f(\zeta)$ and $\varphi(\zeta)$ be specified in the domain D. If the functions $f(\zeta)$ and $\varphi(\zeta)$ are equal to one another in a subdomain V of the domain D, then $f(\zeta) \equiv \varphi(\zeta)$ everywhere in the domain D.

To prove this claim, we will introduce into consideration the function $\psi(\zeta) = f(\zeta) - \varphi(\zeta)$. Evidently, $\psi(\zeta)$ is analytical in D and $\psi(\zeta) = 0$ in V. Now let us take an arbitrary point $\eta \in D (\eta \notin V)$ and connect it by a curve L lying entirely inside D with a fixed point $\zeta_0 \in V$ (Fig. 15a).

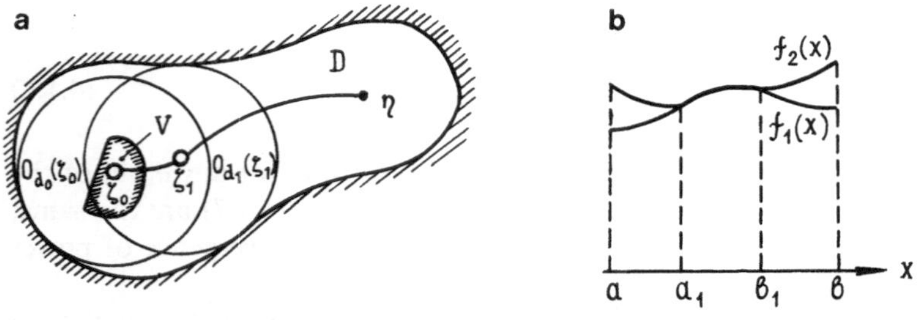

Fig. 15. a Once two analytical functions $f(\zeta)$ and $\varphi(\zeta)$ coincide in a subdomain V, they coincide identically in the whole domain of their analyticity D; **b** Two differentiable functions of a real variable may coincide in part (a_1, b_1) of their definition range (a, b) without coinciding identically within the whole domain (a, b)

[1] At the limiting points of a segment, we naturally take one-side neighborhoods: $a \leqslant x < a + \delta$; $b - \delta < x \leqslant b$.

Choose an arbitrary point $\zeta \in L$ and denote the circle drawn from point ζ as a center and having a radius $d(\zeta)$ equal to the distance between the point ζ and the boundary of the domain D by $O_d(\zeta)$.

We will prove first of all that once $O_{d_0}(\zeta_0)$ is the circle with a center at point ζ_0, we find everywhere in the circle that

$$\psi(\zeta) = 0 \ , \quad \zeta \in O_{d_0}(\zeta_0) \ . \tag{4.1.10}$$

Indeed, expand the function $\psi(\zeta)$ into the Taylor series in this circle:

$$\psi(\zeta) = \sum_{n=0}^{\infty} a_n(\zeta_0)(\zeta - \zeta_0)^n \ . \tag{4.1.11}$$

By virtue of (4.1.8), however, we have

$$a_n = \frac{1}{n!} \psi^{(n)}(\zeta_0) = 0 \ , \quad n = 0, 1, 2, \ldots, \tag{4.1.12}$$

which implies condition (4.1.10).

Now consider a point $\zeta_1 \in L$ lying inside $O_{d_0}(\zeta_0)$ but with the circle $O_{d_1}(\zeta_1)$ going outside $O_{d_0}(\zeta_0)$. Expanding the function $\psi(\zeta)$ into the Taylor series in the neighborhood of ζ_1, we establish, just as it was done earlier, that $\psi(\zeta) \equiv 0$ everywhere in $O_{d_1}(\zeta_1)$. Removing the center of the circle ζ from the point ζ_1 to the point ζ_2, etc. (until η is reached) and repeating the above reasoning, we conclude that $\psi(\eta) = 0$.

Since η is an arbitrary point within D, we have proved thereby that $\psi(\zeta) \equiv 0$ in D. This suggests the equality of the initial functions $f(\zeta) \equiv \varphi(\zeta)$ within D. Hence, the above claim has been proved[2].

It is remarkable that the condition of the theorem can be made less stringent. In fact, for uniqueness it is sufficient that two functions analytical in D coincide in a set of points η_n convergent to the inner point η of the domain D.

Thus, the concept of an analytical function differs markedly from that of differentiability in the sense of real analysis. Indeed, even two infinitely differentiable functions of a real variable may coincide in part of the definition range without coinciding identically (Fig. 15b). At the same time, according to the proved theorem, two analytical functions coinciding in any part of the domain where they are analytical (say, in a small circle or an arc belonging to the domain) do coincide identically within the whole domain.

This theorem could be applied to plane geopotential fields – gravitational or magnetic. As indicated in the previous chapter, the complex intensities of these fields everywhere outside attracting or magnetized masses are fitted by analytical functions. As a result, measuring the gravitational or magnetic fields in any arbitrarily small domain of space outside the masses, we can define, in principle unequivocally these fields in the whole space outside sources.

---

[2] We omit the proof that the point η is virtually achievable in a finite number of steps. This proof relies on the Borel lemma asserting the selection of a finite cover from any cover of a segment by means of open sets (in our case, by circles).

4.1.3 Concept of Analytical Continuation

The concept of analytical continuation will be first studied within the frame of single-valued (analytical) functions. From this standpoint, the concept of *analytical continuation of a function $\varphi_0(\zeta)$ specified in a domain V of a complex plane is ascribed to its redefinition for a function $\varphi(\zeta)$ specified for a larger domain $D \supset V$, so that $\varphi(\zeta)$ is analytical in D and its striction in the domain V coincides with $\varphi_0(\zeta)$*:

$$\varphi(\zeta)|_V \equiv \varphi_0(\zeta) \ .$$

It is evident from the uniqueness theorem (4.1.2) that if analytical continuation is available, it is unique.

The problem of analytical continuation could be solved differently. Below, two basic principles underlying continuation will be presented – those suggested by Weierstrass and Penleve. But first we will consider a more general approach to the problem of analytical continuation implying the concept of a complete analytical function.

The analytical function $\varphi(\zeta)$ taken together with the domain D of its definition is called an *element,* which is generally designated by (φ, D). Of the two elements (φ_1, D_1) and (φ_2, D_2) one is termed a *direct analytical continuation of the other,* provided the intersection $D_1 \cap D_2 = \Delta$ is a domain and $\varphi_1(\zeta) = \varphi_2(\zeta)$ in Δ.

Evidently, the function

$$\varphi(\zeta) = \begin{cases} \varphi_1(\zeta) \ , & \zeta \in D_1 \ , \\ \varphi_2(\zeta) \ , & \zeta \in D_2 \ , \end{cases}$$

is analytical in the domain $d = D_1 \cup D_2$.

The finite set of elements (φ_0, D_0), (φ_1, D_1), $\ldots$, (φ_m, D_m) is called a *chain,* provided each element (φ_j, D_j) is direct analytical continuation of the element (φ_{j-1}, D_{j-1}). The population $\mathscr{H}$ of elements (φ, D), in which one of any two elements out of $\mathscr{H}$ is derived from the other by means of a chain, all elements of which belong to $\mathscr{H}$, is referred to as a *general analytical function.* Analytical functions φ_j pertaining to different elements will be termed *branches* of a general analytical function. A general analytical function containing all analytical continuations of each of its elements is called a *complete analytical function.*

In other words, a complete analytical function is derived upon *all possible analytical continuations* of the initial function φ_0 *along all possible chains of domains* lying in a complex plane.

Analytical continuation may bring us back to the domain where the initial element (φ_0, D_0) has been specified (Fig. 16) and the values of the continued function φ_m and of the initial function φ_0 in the common part Δ_m of the domains D_0 and D_m may not be the same.

Thus, one can see that *general analytical functions* and, accordingly, *complete analytical functions may be multiple-valued.*

Fig. 16. Scheme explaining the nonsingle-valuedness of general analytical functions

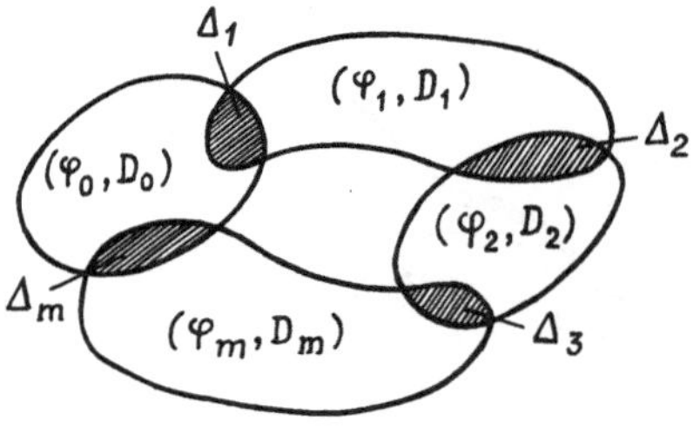

4.1.4 Concept of the Riemann Surface

Obviously, it is very difficult and inconvenient to deal with multiple-valued functions. Even the simplest operations of algebra and analysis involving these functions become rather complicated. For instance, summing up two multiple-valued functions $\Phi(\zeta)$ and $\psi(\zeta)$ one is not sure which result should be ascribed to their sum. In this connection, the theory of functions of a complex variable deals with multiple-sheeted surfaces rather than with plane domains. The former can be represented as lying over the latter and having as many sheets above the point ζ as is the number of values ascribed to this point by an analytical function. Therefore, analytical functions on these surfaces can be regarded as such in an ordinary sense of the word, i.e., as *single-valued functions.*

The multiple-sheeted surface introduced in the above described way is called a *Riemann surface* of a given analytical function. This concept means that any analytical function, as indicated previously, can be treated at its Riemann surface as a function in an ordinary sense of the word. Thus, *a complete analytical function is a function of a point at its Riemann surface* (rather than a function of a point ζ of a complex plane).

The theory of Riemann surfaces is expounded comprehensively in courses on the theory of functions of a complex variable, hence it will not be treated in detail here. We have recalled the concept of a Riemann surface just so as to facilitate the reading of subsequent sections concerned with Riemann spaces of analytical functions of several real variables.

4.1.5 Weierstrass Continuation of an Analytical Function

Weierstrass has suggested a simple way to construct analytical continuation, which occurs as follows. Let the function $\varphi_0(\zeta)$ be analytical in a domain D_0. Take an arbitrary point $\zeta_0 \in D_0$ and expand the function $\varphi_0(\zeta)$ into the Taylor series in the neighborhood of this point:

$$\varphi_0(\zeta) = \sum_{n=0}^{\infty} \frac{\varphi^{(n)}(\zeta_0)}{n!} (\zeta - \zeta_0)^n \ . \tag{4.1.13}$$

Series (4.1.13) is definitely convergent into a circle $O_1(\zeta_0)$ with a center at the point ζ_0 and a radius $d(\zeta_0)$ equal to the distance from ζ_0 to the boundary of the domain D_0, or it may converge also into a larger circle $\tilde{O}_1(\zeta_0)$ (Fig. 17). In the lat-

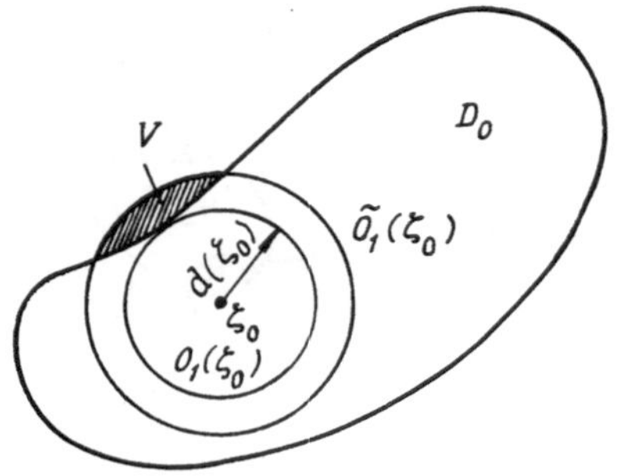

Fig. 17. Weierstrass' continuation of analytical functions: ζ_0 is the center of the circle of convergence of a power series

ter case, it yields analytical continuation of $\varphi(\zeta)$ outside the circle $|\zeta-\zeta|<d(\zeta_0)$ (into the domain V, shaded in Fig. 17), since in the common part of the convergence cycles $O_1(\zeta_0)$ and $\tilde{O}_1(\zeta_0)$ $\varphi(\zeta)\equiv\varphi_0(\zeta)$.

The above operation of analytical continuation can be carried out indefinitely, taking ever new points of the initial domain D_0 for Taylor series centers and new points of analyticity of the functions (elements) obtained upon analytical continuation. All possible continuations into all convergence cycles lead to a complete analytical function, for which the sums of its underlying power series (elements) are its branches.

4.1.6 Singular Points of an Analytical Function

Power series determining the elements of a complete analytical function converge until their cycles of convergence get a *singular point of a function, i.e., a point in whose neighborhood the function cannot be represented by a convergent power series.* Thus, the boundary of the domain into which a given analytical function can be continued ("domain of existence" of a complete analytical function) consists entirely of its singular points. This boundary may be constituted by closed lines or their segments, and it may also contain particular (isolated) points. The latter are either *single-valued or multiple-valued,* depending on whether the function in their rather small neighborhood is single- or multiple-valued. A simple example of a single-valued singular point is the point $\zeta=0$ for the function $1/\zeta$.

Single-valued isolated singular points fall into three types. *The isolated singular point ζ_0 of the function $\varphi(\zeta)$ is called*

a) *a removable singular point* if there is a finite limit $\lim_{\zeta\to\zeta_0}\varphi(\zeta)\neq\infty$;

b) *a pole,* if $\lim_{\zeta\to\zeta_0}\varphi(\zeta)=\infty$;

c) *an essential singular point,* if the limit $\lim_{\zeta\to\zeta_0}\varphi(\zeta)$ does not exist.

Multiple-valued singular points are also termed *branch points of a function.* An example of these is the point $\zeta=0$ for the function $\sqrt[n]{\zeta}$.

4.1.7 Penleve Continuation of an Analytical Function (Principle of Continuity)

Assume that two domains D_1 and D_2 whose intersection is a void set, $D_1 \cap D_2 = \varnothing$, have part of the boundaries in common consisting of a smooth curve L, and consider the domain $D = D_1 \cup D_2 \cup L$.

Principle of continuity. If the functions $\psi_1(\zeta)$ and $\psi_2(\zeta)$ analytical in the domains D_1 and D_2, respectively, are continuous up to L and, moreover,

$$\psi_1(\zeta) = \psi_2(\zeta)\ , \quad \zeta \in L\ ,$$

the function $\psi_2(\zeta)$ is an analytical continuation of the function $\psi_1(\zeta)$ from the domain D_1 through the curve L into the domain D_2.

The proof of the principle of continuity follows directly from the results of Sect. 1.1.3. Indeed, it is shown there that the function

$$\psi(\zeta) = \begin{cases} \psi_1(\zeta)\ , & \zeta \in D_1\ , \\ \psi_2(\zeta)\ , & \zeta \in D_2\ , \\ \psi_1(\zeta) = \psi_2(\zeta)\ , & \zeta \in L\ , \end{cases}$$

is analytical in the domain D, thereby implying that $\psi(\zeta)$ is an analytical continuation of the function $\psi_1(\zeta)$ from D_1 into a larger domain D. Since in D_2 $\psi(\zeta) = \psi_2(\zeta)$, we claim that $\psi_2(\zeta)$ is an analytical continuation of $\psi_1(\zeta)$ from the domain D_1 through L into the domain D_2.

The technique of analytical continuation of a function that relies on the principle of continuity is referred to as the *Penleve continuation*. This approach to analytical continuation will be employed extensively in geophysical applications.

A significant corollary follows from the principle of continuity. Let L – a boundary of the domain Γ – contain a smooth curve L_0, and let the function $\varphi(\zeta)$ be analytical in Γ and continuous up to L_0. Assume also that $\varphi(\zeta) = 0$, $\zeta \in L_0$, then we have $\varphi(\zeta) \equiv 0$ everywhere in Γ.

Indeed, let us add a domain Γ_1, $\Gamma_1 \cap \Gamma_0 = \varnothing$, to the domain Γ along the portion L_0 of its boundary L and consider the function

$$\Phi(\zeta) = \begin{cases} \varphi(\zeta)\ , & \zeta \in \Gamma\ , \\ 0\ , & \zeta \in L_0\ , \\ 0\ , & \zeta \in \Gamma_1\ . \end{cases}$$

By virtue of the principle of continuity, a conclusion is drawn that $\Phi(\zeta)$ is analytical in the domain $D = \Gamma \cup \Gamma_1 \cup L_0$. On the other hand, as $\Phi(\zeta) = 0$ in the domain Γ_1 lying inside D, we have, according to the property of uniqueness of analytical functions, that $\Phi(\zeta) = 0$ everywhere in D, and hence, inside Γ; as a consequence, $\varphi(\zeta) \equiv 0$ in Γ.

4.1.8 Conformal Mapping

Consider a function analytical in the domain D

$$w = f(\zeta) = u(x, z) + iv(x, z)\ , \tag{4.1.14}$$

and satisfying the condition

$$f'(\zeta_0) \neq 0 \tag{4.1.15}$$

at the point $\zeta_0 \in D$.

Inequality (4.1.15) (with due reference to the Cauchy-Riemann conditions) is tantamount to the fact that at the point ζ_0 the Jacobian

$$\frac{\partial(u, v)}{\partial(x, z)} = \begin{vmatrix} \frac{\partial u}{\partial x} & \frac{\partial u}{\partial z} \\ \frac{\partial v}{\partial x} & \frac{\partial v}{\partial z} \end{vmatrix} = \left(\frac{\partial u}{\partial x}\right)^2 + \left(\frac{\partial v}{\partial x}\right)^2 = |f'(\zeta)|^2$$

is nonzero. It is common knowledge that *this condition is, in turn, tantamount to existence of the neighborhood of a point* ζ_0 *where* $w = f(\zeta)$ *is a one-to-one mapping.* It should be borne in mind *that the inverse function* $\zeta = f^{-1}(w)$ *is analytical* in neighborhood of the point $w_0 = f(\zeta_0)$.

A complex analysis reveals that the mapping specified by the analytical function $w = f(\zeta)$ satisfying condition (4.1.15) at the point ζ_0 possesses the following properties:

1) The angle made by any two smooth curves L_1 and L_2 intersecting at a point ζ_0 is equal, in magnitude and reference direction, to the angle between the curves γ_1 and γ_2 into which the curves L_1 and L_2 go upon mapping by $w = f(\zeta)$ (angle conservation).
2) The coefficient of linear extension A at a point ζ_0 upon mapping by $w = f(\zeta)$ of any curve L passing through this point is independent of the shape and direction of the line L, with $A = |f'(\zeta_0)|$.

Definition 4.1.1. The mapping $w = f(\zeta)$ exhibiting the first and the second properties at a point ζ_0 is called conformal.

Thus, we conclude that *mapping performed by the analytical function* $w = f(\zeta)$ is conformal in a rather small neighborhood of each point ζ at which $f'(\zeta) \neq 0$.

Definition 4.1.2. The mapping $w = f(\zeta)$ is called conformal in the domain D if it is conformal at each point of this domain.

It follows from the previous results *that for the mapping to be conformal in the domain D, it is sufficient that the function* $f(\zeta)$ *be analytical in D and that its derivative be nonzero therein.*

The theory of conformal mappings is an essential branch of the theory of functions of a complex variable, which treats in detail the properties of conformal mappings.

4.2 Analytical Continuation of the Cauchy-Type Integral Through a Path of Integration

Refer again to Cauchy-type integral (1.1.3) taken along a (closed or open) piecewise smooth line L. As indicated in Chap. 1, the line L is a singular line for a piecewise analytical function fitted by the Cauchy-type integral.

Meanwhile, in a number of cases, it is possible to continue analytically the Cauchy-type integral through a path of integration. This problem is of particular concern in the study of feasibility of analytical continuation of geopotential fields into domains filled with field sources (attractive or magnetized masses). To solve this problem, one needs additional information on the theory of continuation of analytical functions from analytical curves into domains of a complex plane. A simple result along these lines is related to the problem of analytical continuation of a real analytical function of a real variable.

4.2.1 Analytical Continuation of a Real Analytical Function of a Real Variable

Let a real analytical function $f(x)$ of a real variable be specified for a segment I of a real axis $a \leqslant x \leqslant b$. Then, in accordance with (4.1.9), in the neighborhood $x_0 - \delta < x < x_0 + \delta$ of each point $x_0 \in I$ it is represented by a convergent power series

$$f(x) = \sum_{n=0}^{\infty} a_n (x - x_0)^n \ . \tag{4.2.1}$$

Let $F(\zeta)$ be a complex analytical function in the domain D enclosing completely the segment I. It is claimed that the function $F(\zeta)$ *is an analytical continuation of* $f(\zeta)$ *from the segment* I *into the domain* D, provided that

$$F(x) = f(x) \quad \text{for} \quad x \in I \ . \tag{4.2.2}$$

A simple method could be suggested for this analytical continuation. Indeed, let $F(\zeta)$ stand for the series obtainable from (4.2.1) by substituting a complex variable $\zeta = x + iz$ for x:

$$F(\zeta) = \sum_{n=0}^{\infty} a_n (\zeta - x_0)^n \ . \tag{4.2.3}$$

Evidently, this series is definitely convergent into a cycle $O_\delta(x_0)$ with a center at a point x_0 and a radius δ, $|\zeta - x_0| < \delta$ and its condition is (4.2.2). Thus, the sum $F(\zeta)$ is an analytical continuation of the function $f(x)$ into a complex domain.

We have established thereby the following important fact. *A real analytical function of a real variable is analytically continuable from the segment I of a real axis into a domain D of a complex plane enclosing completely I.*

4.2.2 Concept of an Analytical Arc; the Herglotz-Tsirulsky Equation for the Arc

Take an arc L whose equation is represented parametrically in the form:

$$x = x(s) \ , \quad z = z(s) \ , \quad s_a \leqslant s \leqslant s_b \ , \tag{4.2.4}$$

where s_a and s_b are constants.

Definition 4.2.1. The arc L is called analytical if $x(s)$ and $z(s)$ are analytical functions of a real variable s in a segment I, $s_a \leqslant s \leqslant s_b$ (in terms of the definition provided in Sect. 4.1.1).

The analytical arc is termed *regular* if

$$(dx/ds)^2 + (dz/ds)^2 \neq 0 \tag{4.2.5}$$

everywhere in the segment I, $s_a \leqslant s \leqslant s_b$. Note that the regularity condition is equivalent to the absence of singular points in a differential geometrical sense (Rashevsky 1956) (say, cusps) on the arc L, i.e., it ensures smoothness of the arc L.

Let us take some value of the parameter s_0. It is shown in the previous subsection that the functions $x(s)$ and $z(s)$ are continued analytically from an interval $-\delta_1 + s_0 < s < s_0 + \delta_1 (\delta_1 > 0)$ into a cycle $O_{\delta_1}(s_0)$ $(|\sigma - s_0| < \delta_1)$ of a plane of a complex variable $\sigma = s + it$ (Fig. 18). As a result, the function $\zeta = \varphi(\sigma) = x(\sigma) + iz(\sigma)$ is analytical in the cycle $O_{\delta_1}(s_0)$ and, by virtue of (4.2.5), in a neighborhood $O_\delta(s_0)$ of the point s_0.

$$\varphi'(\sigma) \neq 0 \ . \tag{4.2.6}$$

Hence it is evident that the function $\zeta = \varphi(\sigma)$ performs conformal mapping of the neighborhood $O_\delta(s_0)$ onto a certain domain of the plane ζ enclosing the point $\zeta_0 = \varphi(s_0) \in L_0$.

In view of the fact that the functions $x(s)$ and $z(s)$ are continued analytically along the entire segment I and the regular arc condition (4.2.5) is met everywhere thereon, it is concluded that there is a domain Γ of a complex plane σ which com-

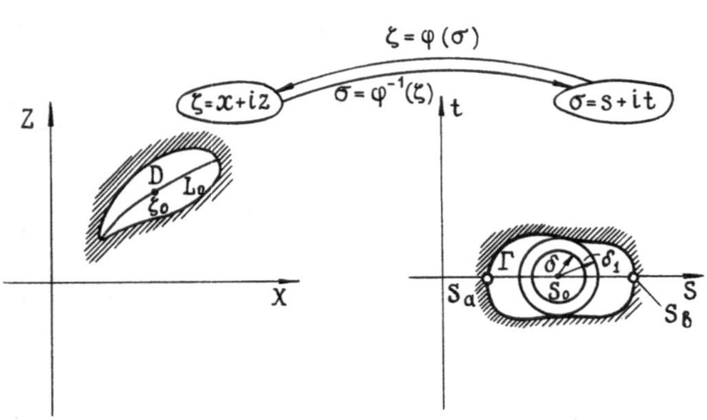

Fig. 18. Definition of the concept of an analytical arc. Conformal mapping of the domain Γ onto the domain D

pletely encloses the segment $[s_a, s_b]$, a domain that the function $\zeta = \varphi(\sigma)$ is analytically continued from the segment I into the domain Γ, mapping the latter conformally onto a domain D containing the arc L_0. Thanks to the one-to-one conformal mappings, there is an inverse mapping

$$\sigma = \varphi^{-1}(\zeta) \ , \tag{4.2.7}$$

bringing the domain D into the domain Γ (see Fig. 18). In the course of this mapping, the arc L_0 goes over into the segment I of a real axis s:

$$\varphi^{-1}(\zeta) = s \ , \quad \zeta \in L_0 \ . \tag{4.2.8}$$

It is to be noted that the parameter s can be provided by a real variable x, without restricting generality. Then (4.2.8) is cast in the form

$$\varphi^{-1}(\zeta) = x \ . \tag{4.2.9}$$

Assuming $x = \frac{1}{2}(\zeta + \bar{\zeta})$ in (4.2.9) we obtain $\varphi^{-1}(\zeta) = \frac{1}{2}(\zeta + \bar{\zeta})$ and hence

$$\bar{\zeta} = 2\varphi^{-1}(\zeta) - \zeta = \psi(\zeta) \ . \tag{4.2.10}$$

It seems that $\psi(\zeta)$ as a sum of two analytical functions is analytical in D.

Thus, the equation for the analytical arc is specified in the form:

$$\bar{\zeta} = \psi(\zeta) \ , \tag{4.2.11}$$

where $\psi(\zeta)$ is the analytical function in a domain D enclosing completely the arc L_0.

The idea of specifying the arc in form (4.2.11) originated with German mathematician G. Herglotz as early as in 1914 (Herglotz 1914). But later on, as is often the case in science, this idea remained in oblivion for a long time, until the Soviet geophysicist A. V. Tsyrulsky (1963, 1964) applied it (independently of the works by Herglotz) to the solution of problems of the theory of a plane potential field. In view of this fact, Eq. (4.2.11) will be called an *equation for a plane curve in the Herglotz-Tsirulsky form.*

It is pertinent to mention that Eq. (4.2.11) could have been derived directly from the implicit curve equation

$$F(x, z) = 0 \ .$$

Indeed, bearing in mind that $x = (\zeta + \bar{\zeta})/2$, $z = (\zeta - \bar{\zeta})/2i$ we write

$$F(x, z) = \tilde{F}(\zeta, \bar{\zeta}) = 0 \ .$$

Solving the latter equation for $\bar{\zeta}$ we obtain (4.2.11).

Thus, we have proved that if the arc L is regular-analytical there exists a domain D enclosing the arc L completely, a domain which is such that inside it the function $\psi(\zeta)$ fitting the Herglotz-Tsirulsky curve equation is analytical. The converse is also true: once there exists a domain $D \supset L$ inside which the function $\psi(\zeta)$ fitting the Herglotz-Tsirulsky curve equation is analytical, the curve L is regular-analytical. We will prove the latter claim. The analyticity of the curve L described by formula (4.2.11) having an analytical right-hand side follows directly from definition 4.2.1. Let us establish the regularity of this curve. Assume the converse.

Let a point $\zeta_0 \in L$ be existent at which the regularity condition (4.2.5) is violated. In other words, ζ_0 is a geometrically singular point of the curve L. We will designate the real and the imaginary parts of $\psi(\zeta)$ by u and v:

$$\psi(\zeta) = u(x, z) + iv(x, z)$$

which are interrelated everywhere in D, including the point $\zeta_0 = (x_0, z_0)$, by the Cauchy-Riemann conditions

$$\partial u/\partial x = \partial v/\partial z \ , \quad \partial u/\partial z = -\partial v/\partial x \ . \tag{4.2.12}$$

Equation (4.2.11) is equivalent to two real equations

$$\begin{aligned} x - u(x, z) &= P(x, z) = 0 \\ z + v(x, z) &= Q(x, z) = 0 \end{aligned} \tag{4.2.13}$$

where $P(x, z)$ and $Q(x, z)$ are the functions specifying implicitly the equation for the arc L.

Since $\zeta_0 = (x_0, z_0)$ is a geometrically singular point of the curve L, at this point we have that

$$\partial P/\partial x = 0 \ , \quad \partial P/\partial z = 0 \ ; \quad \partial Q/\partial x = 0 \ , \quad \partial Q/\partial z = 0$$

(Rashevsky 1956).

As a result, allowing for (4.2.13), we write

$$\left.\frac{\partial u}{\partial x}\right|_{(x_0,z_0)} = 1 \ , \quad \left.\frac{\partial u}{\partial z}\right|_{(x_0,z_0)} = 0 \ ;$$

$$\left.\frac{\partial v}{\partial x}\right|_{(x_0,z_0)} = 0 \ , \quad \left.\frac{\partial v}{\partial z}\right|_{(x_0,z_0)} = -1 \ .$$

Hence, we obtain, in particular,

$$\left.\frac{\partial u}{\partial x}\right|_{(x_0,z_0)} = -\left.\frac{\partial v}{\partial z}\right|_{(x_0,z_0)} \ ,$$

which is in conflict with the Cauchy-Riemann conditions (4.2.12). In this way we proved that L is a regular-analytical arc.

Based on the above considerations, we will define equivalently a regular-analytical arc.

Definition 4.2.2. An arc L is called Herglotz-Tsirulsky analytical if there is a domain D enclosing completely the arc L so that inside D the function $\psi(\zeta)$ describing the curve equation in the Herglotz-Tsirulsky form is analytical.

Evidently, any Herglotz-Tsirulsky analytical arc is regular-analytical and vice versa.

The requirements imposed on the function $\psi(\zeta)$ can be made less stringent by introducing the following:

Fig. 19. Definition of the concept of a one-side analytical arc

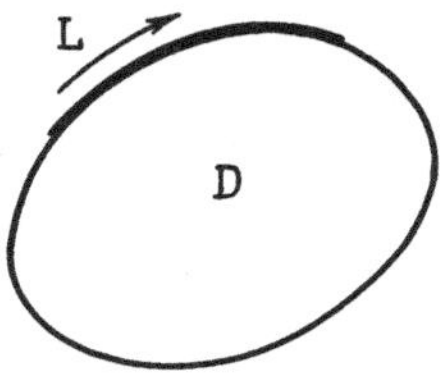

Definition 4.2.3. An arc L is called one-side Herglotz-Tsirulsky analytical if there is a domain D whose boundary coincides partially with L (Fig. 19) so that $\psi(\zeta)$ is analytical in D and continuous in $D \cup L$. The equation for the arc L which is one-side analytical on the left (left-side analytical) will be denoted by $\bar{\zeta} = \psi^+(\zeta)$, while that on the right (right-side analytical) will be represented by $-\bar{\zeta} = \psi^-(\zeta)$.

Note that any one-side Herglotz-Tsirulsky analytical arc is analytical but it is not necessarily regular. Indeed, assume $\bar{\zeta} = \psi^-(\zeta)$ on L, where $\psi^-(\zeta)$ is analytical in D and continuous in $\bar{D}$ (see Fig. 19). Let $\zeta = \zeta(t)$ be a mapping of the circle $|t| < 1$ onto D, then we have

$$\bar{\zeta}(t) = \psi^-[\zeta(t)] \ , \quad t \in l \ ,$$

where l is a portion of the circumference $|t| = 1$. The right-hand side of the latter expression is analytical for $|t| < 1$ and continuous for $|t| \leqslant 1$. Using the function $\zeta(t)$, construct, in a standard way (see Muskelishvily 1962, p. 142), the function $\zeta_*(t)$ analytical for $|t| > 1$ and $\zeta_*(t) = \bar{\zeta}(t)$ for $|t| = 1$. The formula

$$\zeta_*(t) = \psi^-[\zeta(t)] \ , \quad t \in l$$

yields an analytical continuation of $\zeta_*(t)$ from $|t| > 1$ through l into the circle $|t| < 1$. Thus, L is an image of the analytical arc l produced in the analytical mapping $\zeta_*(t)$, i.e., L is an analytical arc. This simple reasoning is due to V. G. Cherednichenko. Meanwhile, it is remarkable that not any analytical arc is one-side Herglotz-Tsirulsky analytical on any side! Indeed, if some analytical arc is one-side Herglotz-Tsirulsky analytical on the two sides, this means that, by definition 4.2.2, it is a regular-analytical arc, i.e., it does not contain any cusps. But, at the same time, the concept of an analytical arc (definition 4.2.1) admits of the presence of such singular points.

4.2.3 Analytical Continuation of a Function Specified Along an Analytical Curve

Let L_0 be a regular-analytical curve in a complex plane with parametric Eq. (4.2.4) and let a complex function $f(\tau)$, $\tau = x + iz \in L_0$, be specified thereon. We also assume that the real and the imaginary parts of the function $f(\tau)$ are analytical functions of the coordinates $x(s)$ and $z(s)$ of the arc L_0, as defined in Sect. 4.1.1.

Now let $F(\zeta)$ be a function analytical in the domain D enclosing completely the arc L_0. It is claimed that *the function $F(\zeta)$ is an analytical continuation of $f(\tau)$ from the curve L_0 into the domain D*, if $F(\tau) = f(\tau)$ for $\tau \in L_0$.

We will show that on the above assumptions the function $f(\tau)$ can be always continued from the analytical arc L_0.

Indeed, let us fix on a point $\tau_0 \in L_0$ and consider the real and the imaginary parts of the function $f(\tau)$:

$$\begin{aligned} \operatorname{Re} f(\tau) &= u(x, z) = u[x(s), z(s)] = \tilde{u}(s) , \\ \operatorname{Im} f(\tau) &= v(x, z) = v[x(s), z(s)] = \tilde{v}(s) . \end{aligned} \tag{4.2.14}$$

Evidently, the functions $\tilde{u}(s)$ and $\tilde{v}(s)$ are real analytical functions of a real variable s. Similarly, as in the previous subsection, it is concluded therefore that these functions are continued analytically from the segment $I(s_a \leqslant x \leqslant s_b)$ into a domain Γ of a complex plane σ enclosing completely the segment I. Hence, the function $f(\tau(s)) = \tilde{u}(s) + i\tilde{v}(s)$ is also continued analytically into the domain Γ, i.e., there is an analytical function $\Phi(\sigma)$ in Γ so that

$$\Phi(s) = f[\tau(s)] . \tag{4.2.15}$$

Now let us employ conformal mapping (4.2.7) $\sigma = \varphi^{-1}(\zeta)$, which performs inverse mapping of the domain D of the plane ζ into the domain Γ of the plane σ, and consider the function

$$F(\zeta) = \Phi[\varphi^{-1}(\zeta)] . \tag{4.2.16}$$

Obviously, the function $F(\zeta)$ is analytical in the domain containing L_0 and, according to (4.2.15) and (4.2.16), we have

$$F(\tau) = \Phi[\varphi^{-1}(\tau)] = \Phi(s) = f(\tau) . \tag{4.2.17}$$

Thus, the function $F(\zeta)$ is an analytical continuation of the function $f(\tau)$ from the arc L_0.

4.2.4 Continuation of the Cauchy-Type Integral Through a Path of Integration; Singular Points of the Continued Field

We have done all the preliminary work needed for analytical continuation of Cauchy-type integral (1.1.3) through a path of integration.

Assume that L is a regular-analytical arc and the real and the imaginary parts of the density of the Cauchy-type integral $\varphi(\zeta)$ are analytical functions of the x- and z-coordinates of the points on the arc L. It is noteworthy that in this case the function $\varphi(\zeta)$ has a definitely limited derivative on L and, hence, it satisfies the Lipschitz condition. This means that we can rightfully employ Sokhotsky-Plemelj formulas (1.2.22a) and (1.2.22b) and write

$$C^-(\tau_0) = C^+(\tau_0) - \varphi(\tau_0) . \tag{4.2.18}$$

In accordance with the findings of the previous subsection, the function $\varphi(\tau_0)$ is a boundary value on L of a function $\varphi(\zeta)$ analytical everywhere in the domain D enclosing L, and particularly in its subdomain D^+ adjoining L on the left [with respect to a given sense of traversal of L (Fig. 20)]. By definition, the function $C^+(\tau_0)$ is a boundary value of the function $C(\zeta')$ – the Cauchy-type in-

Fig. 20. Continuation of the Cauchy-type integral through a path of integration L

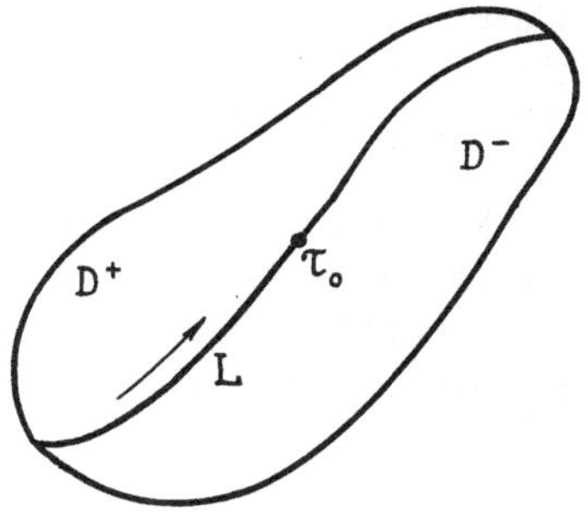

tegral – analytical outside L, if L is approached on the left (from the domain D^+). Thus, the right-hand side of Eq. (4.2.18) is a boundary value on L of some function analytical in D^+.

On the other hand, the left-hand side, $C^-(\tau_0)$ is a boundary value of the function $C(\zeta')$ on approaching L from the right. As a consequence, the right-hand side of Eq. (4.2.18) defines, according to the *principle of continuity* (Sect. 4.1.7), analytical continuation of the left-hand side through the contour L from the right to the left. Designating the result of this continuation by $C_L^-(\zeta')$, we obtain

$$C_L^-(\zeta') = C(\zeta') - \varphi(\zeta') \ , \quad \zeta' \in D^+ \ . \tag{4.2.19}$$

In a similar way, one proves the possibility of analytical continuation of the Cauchy-type integral from the left to the right into a domain D^- adjoining L on the right (see Fig. 20), and the result of analytical continuation is

$$C_L^+(\zeta') = C(\zeta') + \varphi(\zeta') \ , \quad \zeta' \in D^- \ . \tag{4.2.20}$$

Thus, we have proved the following significant claim.

Theorem 4.2.1. If the density $\varphi(\zeta)$ of the Cauchy-type integral $C(\zeta')$ is an analytical function of the coordinates of points of the integration path L and if L is a regular-analytical curve, the integral $C(\zeta')$ can be continued analytically through any point of the integration path from both its sides (except, maybe, for its ends); the result of analytical continuation differs from the Cauchy-type integral itself by a function $\varphi(\zeta')$ defined by analytical continuation of the density $\varphi(\zeta)$ from the curve L into a complex plane taken with a positive sign in the right-to-left continuation and with a negative sign in the left-to-right continuation:

$$C_L^\pm(\zeta') = C(\zeta') \pm \varphi(\zeta') \ . \tag{4.2.21}$$

Two useful corollaries follow from theorem 4.2.1.

Corollary 4.2.1. If L is an analytical arc, the singular points of the functions $C_L^\pm(\zeta')$ coincide with those of the function $\varphi(\zeta')$ derived upon analytical continuation of the Cauchy-type integral density from the line L into a complex plane.

Corollary 4.2.2. If the contour L is a piecewise analytical curve, the corners of this curve in a general case are branch points for a complete analytical function derived upon continuation of the Cauchy-type integral.

The proof of these corollaries follows directly from formula (4.2.21) and it is left to the reader.

Theorem 4.2.1 and its corollary are basic to solution of the problem of continuation of potential fields into domains filled with sources, a problem to be taken up at this point.

4.3 Analytical Continuation of a Plane Magnetic Field into a Domain Occupied by Magnetized Masses

This and the next subsections deal with a very interesting issue of the theory of a plane potential field, namely feasibility of analytical continuation of this field into a domain occupied by sources. This problem has been investigated most thoroughly in fundamental works by A. V. Tsirulsky (1963, 1964), V. N. Strakhov (1970a, c, d) and V. G. Cherednichenko (1978) whose findings are primarily presented here. Our examination starts with a relatively simple problem – continuation of the magnetic potential.

4.3.1 Analytical Continuation of a Magnetic Potential into a Domain of Analytically Distributed Magnetization

Consider the situation treated in detail in Sect. 2.6.5, where magnetization $I(\zeta)$ of masses distributed over a domain Γ of a complex plane is a function analytical inside Γ and continuous in a closed domain $\bar{\Gamma}$. In accordance with (2.6.19), the complex magnetic potential of these masses outside the domain Γ is fitted by the formula

$$W(\zeta') = -\frac{1}{2\pi i}\int_L \frac{2\pi I(\zeta)\bar{\zeta}}{\zeta-\zeta'}\,d\zeta\ ,\quad \zeta'\in C\bar{\Gamma}\ , \tag{4.3.1}$$

where L is a piecewise smooth boundary of the domain Γ.

Examine the feasibility of analytical continuation of the potential $W(\zeta')$ through the boundary L into the domain Γ. Let L_0 denote a regular-analytical arc of the curve L whose equation in the Herglotz-Tsirulsky form is

$$\bar{\zeta} = \varphi_0(\zeta)\ , \tag{4.3.2}$$

where $\varphi_0(\zeta)$ is the function analytical in a domain D enclosing completely L_0 (Fig. 21); D^+ is the part of domain D common with Γ. Assume also that in the arc L_0 the function $I(\zeta)$ (and, hence, $I(\zeta)\bar{\zeta}$) satisfies the Hölder condition (1.2.1). Then, by reasoning similar to that of Sect. 4.2.4, we find for Cauchy-type integral (4.3.1) that the magnetic potential $W(\zeta')$ can be continued into Γ through any point of L_0 (except, maybe, for its ends) and the result of continuation $W^-_{L_0}(\zeta')$ is fitted by the formula following from (4.2.20):

Fig. 21. Substantiation of the possibility for analytical continuation of a magnetic potential into a domain Γ of analytically distributed magnetization

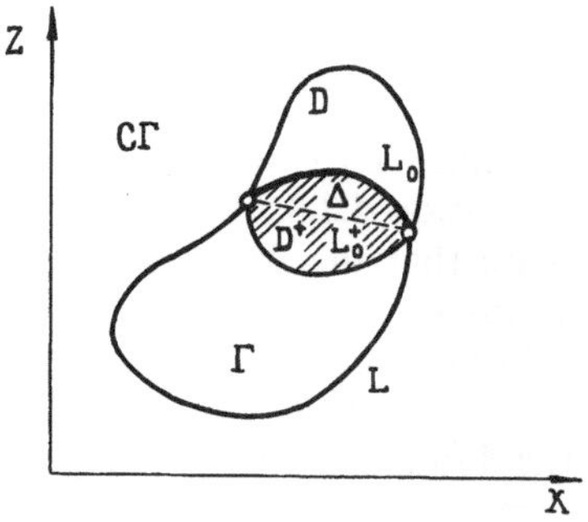

$$W_{L_0}^{-}(\zeta') = -\frac{1}{2\pi i}\int_L \frac{2\pi I(\zeta)\bar{\zeta}}{\zeta-\zeta'}d\zeta + 2\pi I(\zeta')\varphi_0(\zeta') \ . \tag{4.3.3}$$

Thus, the following statement is valid. *A magnetic potential produced by analytically distributed magnetization inside a domain Γ bounded by a piecewise smooth curve L admits analytical continuation from the outer domain $C\bar{\Gamma}$ into* Γ through a regular-analytical arc L_0 *up to which magnetization $I(\zeta)$ is continuous and satisfies the Hölder condition on L_0.*

It is of interest to compare the function $W_{L_0}^{-}$ obtained upon continuation of the magnetic potential into Γ with the magnetic potential itself inside Γ, fitted by (2.6.19):

$$W_{L_0}^{-}(\zeta') - W(\zeta') = 2\pi I(\zeta')[\varphi_0(\zeta') - \bar{\zeta}'] \ . \tag{4.3.4}$$

For instance, once the arc L_0 is a segment of the straight line with an equation:

$$\bar{\zeta} = a_0 + a\zeta \ , \tag{4.3.5}$$

formula (4.3.4) yields the following expression:

$$W_{L_0}^{-}(\zeta') - W(\zeta') = 2\pi I(\zeta')(a_0 + a\zeta' - \bar{\zeta}') \ . \tag{4.3.6}$$

Thus, it is readily seen that *the field continued inside masses always differs from the true field.*

4.3.2 Continuation Through a One-Side Herglotz-Tsirulsky Analytical Arc

It is remarkable that the requirements imposed in Sect. 4.3.1 on the distribution of magnetization $I(\zeta)$ and on the arc L_0 can be rendered less stringent. Namely, the claim of Sect. 4.3.1 holds good, provided L_0 is a left-side analytical arc (from the inside of the domain D) and $I(\zeta)$ is a function analytical in Γ and continuous in $\Gamma \cup L_0$ (i.e., the requirements that $I(\zeta)$ should be Hölder continuous on L_0 and that L_0 be a regular-analytical arc are not necessary!).

Indeed, represent the right-hand side of expression (4.3.1) as a sum of two integrals – one taken over the arc L_0 and the other taken over the remaining part $L - L_0$ of the curve L:

$$W(\zeta') = -\frac{1}{2\pi i}\int_{L_0}\frac{2\pi I(\zeta)\bar{\zeta}}{\zeta-\zeta'}d\zeta - \frac{1}{2\pi i}\int_{L-L_0}\frac{2\pi I(\zeta)\bar{\zeta}}{\zeta-\zeta'}d\zeta\ . \tag{4.3.7}$$

Since L_0 is a left-side Herglotz-Tsirulsky analytical arc, then, according to definition 4.2.3, there is a domain D^+ adjoining L_0 inside the domain Γ, a domain that the function $\varphi_0(\zeta)$ determining the equation for L_0 in the Herglotz-Tsirulsky form is analytical in D^+ (and continuous in $D^+ \cup L_0$). Let us draw an arbitrary smooth arc L_0^+ connecting the ends of the arc L_0 and enclosed completely inside D^+ (see Fig. 21). By virtue of the analyticity of $I(\zeta)$ and $\varphi_0(\zeta)$ in the domain Δ between L_0 and L_0^+, we have, on the basis of the Cauchy theorem, that

$$-\frac{1}{2\pi i}\int_{L_0}\frac{2\pi I(\zeta)\bar{\zeta}}{\zeta-\zeta'}d\zeta = -\frac{1}{2\pi i}\int_{L_0^+}\frac{2\pi I(\zeta)\varphi_0(\zeta)}{\zeta-\zeta'}d\zeta\ ,\quad \zeta'\in C\bar{\Gamma}\ . \tag{4.3.8}$$

Substitution of (4.3.8) into (4.3.7) yields

$$W(\zeta') = -\frac{1}{2\pi i}\int_{L_0^+}\frac{2\pi I(\zeta)\varphi_0(\zeta)}{\zeta-\zeta'}d\zeta - \frac{1}{2\pi i}\int_{L-L_0}\frac{2\pi I(\zeta)\bar{\zeta}}{\zeta-\zeta'}d\zeta\ ,\quad \zeta'\in C\bar{\Gamma}\ . \tag{4.3.9}$$

The integrals in the right-hand side of (4.3.9) are Cauchy-type integrals and define a function definitely analytical in the domain $C\bar{\Gamma}\cup\Delta\cup L_0$, this function coinciding with the magnetic potential $W(\zeta')$ in $C\bar{\Gamma}$. Consequently, for any $\zeta'\in\Delta\cup L_0$ (naturally, except for the ends of the arc L_0) the integrals in the right-hand side of (4.3.9) describe analytical continuation of $W(\zeta')$ into Γ, which was to be proved.

Note, furthermore, that in these "relaxed" conditions on $I(\zeta)$ and L_0 formula (4.3.3) is also retained, because in accordance with the remark made in Sect. 1.2.3, formulas (1.2.22a) and (1.2.22b) hold not only for the Cauchy-type integral densities satisfying the Hölder condition but also for merely continuous densities.

4.3.3 Analyticity Condition for the Boundary of a Domain Occupied by Magnetized Masses

The converse claim permitting us to establish the analytical properties of the boundary of a domain from analytical continuation of a magnetic potential is also valid.

Let the function $I(\zeta)$ fitting the magnetization distribution over Γ be analytical in Γ and continuous in $\Gamma\cup L_0$. Then, if the magnetic potential $W(\zeta')$ admits analytical continuation through L_0 into Γ, L_0 is analytical.

For the sake of proof, we will write for the Cauchy-type integral

$$C(\zeta') = \frac{1}{2\pi i}\int_L\frac{2\pi I(\zeta)\bar{\zeta}}{\zeta-\zeta'}d\zeta \tag{4.3.10}$$

the Sokhotsky-Plemelj formula (1.2.22a) in the form

$$C^-(\zeta_0) = C^+(\zeta_0) - 2\pi I(\zeta_0)\bar{\zeta}_0\ ,\quad \zeta_0\in L_0\ . \tag{4.3.11}$$

Allowing for the fact that $C^-(\zeta_0) = -W^-(\zeta_0)$, according to (4.3.1), we obtain

$$W^-(\zeta_0) = -C^+(\zeta_0) + 2\pi I(\zeta_0)\bar{\zeta}_0 \,, \quad \zeta_0 \in L_0 \,. \tag{4.3.12}$$

If the function $W(\zeta')$ admits analytical continuation into Γ through the arc L_0, one has

$$W^-(\zeta_0) = W^-_{L_0}(\zeta_0) \,. \tag{4.3.13}$$

With due account taken of (4.3.13), if it is required that $I(\zeta_0) \neq 0$ for any point $\zeta_0 \in L_0$ formula (4.3.12) is rewritten as

$$\frac{W^-_{L_0}(\zeta_0) + C^+(\zeta_0)}{2\pi I(\zeta_0)} = \bar{\zeta}_0 \,. \tag{4.3.14}$$

The left-hand side permits, by the theorem, analytical continuation into a part D^+ of the domain Γ through all the points of the arc L_0 (see Fig. 21). This continuation will be denoted by $\varphi(\zeta)$. Then we have

$$\varphi(\zeta_0) = \bar{\zeta}_0 \,, \quad \zeta_0 \in L_0 \,, \tag{4.3.15}$$

where $\varphi(\zeta)$ is the function analytical in D^+. This implies that L_0 is a left-side Herglotz-Tsirulsky analytical arc and, in accordance with Sect. 4.2.2, it is merely analytical. Thus, we have proved the following theorem.

Theorem 4.3.1. Let a domain Γ be specified, which is bounded by a piecewise smooth curve L and within which there are distributed magnetized masses with magnetization $I(\zeta)$ analytical in Γ and continuous in $\Gamma \cup L$. For the magnetic potential $W(\zeta)$ of the domain Γ to be continued into Γ through any point of the arc L_0 of the boundary L (except maybe for its ends and the points where $I(\zeta_0) = 0$), it is necessary and sufficient that the arc L_0 be left-side Herglotz-Tsirulsky analytical.

In concluding, we should like to note that the theorem holds good also where complex intensity of the magnetic field or any of its derivatives, rather than the magnetic potential, is continued.

4.3.4 Singular Points of Analytical Continuation of the Magnetic Potential

The location of singular points of the magnetic potential continued analytically into the domain Γ occupied by magnetized masses is readily established on the basis of formula (4.3.3).

Indeed, the Cauchy-type integral included in the right-hand side of formula (4.3.3) is an analytical function everywhere inside Γ. Hence, singular points of the potential $W^-_{L_0}(\zeta')$ continued through an analytical curve L_0 coincide, inside Γ, with the singular points of the function $\varphi_0(\zeta')$ fitting the equation of the arc L_0 in the Herglotz-Tsirulsky form.

Let us consider an important particular case where the entire line L bounding Γ is a left-side analytical curve with an equation

$$\bar{\zeta} = \varphi^{+}(\zeta) \ . \tag{4.3.16}$$

Here, $\varphi^{+}(\zeta)$ *is sure to have singular points inside* Γ, *the set of its singularities coinciding with the set of singularities of the function* $W(\zeta)$ *continued analytically into* Γ.

Indeed, if $\varphi(\zeta)$ had no singular points inside Γ, analytical continuation of the magnetic potential into Γ would yield, according to (4.3.3), a function analytical in the whole space [outside Γ, the potential $W(\zeta')$ is a definitely analytical function vanishing at infinity]. As a consequence, by the Liouville theorem, $W(\zeta')\equiv 0$, which is not correct.

Now we will elucidate which points on the contour L bounding the magnetized domain Γ are singular for the analytically continued magnetic potential. Let the line L consists of a finite number of various left-side Herglotz-Tsirulsky analytical (from the inside of the domain Γ) arcs $L_1, L_2, \ldots, L_N$ described by equations in the Herglotz-Tsirulsky form

$$\bar{\zeta} = \varphi_n(\zeta) \ , \quad n = 1, 2, \ldots, N, \tag{4.3.17}$$

with

$$\varphi_k(\zeta) \not\equiv \varphi_l(\zeta) \quad \text{for any} \quad k \neq l \ . \tag{4.3.18}$$

Condition (4.3.18) defines precisely the concept of "various" one-side analytical arcs.

Let us designate the points corresponding to the ends of the arcs $L_1, L_2, \ldots, L_N$ by $\zeta_0, \zeta_1, \zeta_2, \ldots, \zeta_N$. It is not difficult to show that *for any analytical distribution of magnetization* $I(\zeta)$ *inside* Γ *satisfying the Hölder condition on L, the corners of the contour L are singular points (branch points) of the function* $W(\zeta')$ *continued analytically into* Γ.

Indeed, continuing $W(\zeta')$ analytically through the arc L_k we find, in accordance with (4.3.3), that

$$W^{-}_{L_k}(\zeta') = -\frac{1}{2\pi i}\int\limits_L \frac{2\pi I(\zeta)\bar{\zeta}}{\zeta-\zeta'}\, d\zeta + 2\pi I(\zeta')\varphi_k(\zeta') \ , \tag{4.3.19}$$

while continuing $W(\zeta')$ through the arc L_{k+1} we define

$$W^{-}_{L_{k+1}}(\zeta') = -\frac{1}{2\pi i}\int\limits_L \frac{2\pi I(\zeta)\bar{\zeta}}{\zeta-\zeta'}\, d\zeta + 2\pi I(\zeta')\varphi_{k+1}(\zeta') \ . \tag{4.3.20}$$

By virtue of condition (4.3.18)

$$W^{-}_{L_k}(\zeta') \not\equiv W^{-}_{L_{k+1}}(\zeta') \ . \tag{4.3.21}$$

Hence, the point ζ_k (the point connecting the arcs L_k and L_{k+1}) is a branch point of the continued potential.

Note, in conclusion, that the results obtained for the magnetic potential are readily extendible to complex intensity of the magnetic field and to any of its derivatives as well.

4.3.5 Determination of Complex Magnetization of a Body from its Magnetic Potential

Consider the following *inverse problem.* Assume that we know a magnetic potential $W(\zeta')$ outside the domain Γ occupied by magnetized masses as well as a smooth boundary of the domain Γ. It is necessary to determine magnetization $I(\zeta)$, which is a function analytical inside Γ and continuous in $\Gamma \cup L$. To solve the problem, continue analytically the magnetic potential $W(\zeta')$ from the domain of its definition to everywhere in $C\bar{\Gamma}$, in particular up to the boundary L. The result of this continuation will be denoted, just as in the above discussion, by $W^-(\zeta_0)$. Then, in accordance with (4.3.1), we have

$$W^-(\zeta_0) = -C^-(\zeta_0) \ , \qquad \zeta_0 \in L \ , \tag{4.3.22}$$

where $C(\zeta_0)$ is defined by formula (4.3.10). Substituting the Sokhotsky-Plemelj formula (1.2.21b) into (4.3.22) and allowing for (4.3.10) we write

$$W^-(\zeta_0) = -\frac{1}{2\pi i}\int_L \frac{2\pi I(\zeta)\bar{\zeta}}{\zeta-\zeta_0}\, d\zeta + \pi I(\zeta_0)\bar{\zeta}_0 \ , \quad \zeta_0 \in L \ . \tag{4.3.23}$$

The resultant expression (4.3.23) is a *singular integral equation for the boundary values on the line L of complex magnetization* $I(\zeta)$.

The theory basic to the solution of this type of equations is outlined comprehensively in the above-cited monograph by N. I. Muskhelishvily (1962) and in other special studies that the reader can look through. Here we will confine ourselves to an issue of supreme importance for geophysics, namely uniqueness of the solution to Eq. (4.3.23).

Obviously, the solution is unique if, and only if, there are no analytical distributions of magnetization in the domain Γ, which would give rise to an identically zero external magnetic potential.

V. N. Strakhov has shown (1970a, c) that once the equation for the contour L bounding the domain L cannot be represented in the form

$$\bar{\zeta} = f(\zeta)/\prod_{i=0}^{K} (\zeta-\zeta_i)^{n_i} \ , \quad n_i \geqslant 0 \ ; \quad \sum_i n_i > 1 \ , \quad \zeta_i \in \Gamma \ , \tag{4.3.24}$$

where $f(\zeta)$ is an arbitrary function analytical in Γ and continuous in $\Gamma \cup L$, then for these domains there are no analytical distributions of magnetization that might give rise to a zero external potential.

Thus, for such domains Γ, a solution to Eq. (4.3.23) is unique.

For instance, for $I(\zeta) = I_0 = \text{const}$, this solution is defined by the following formula derived first by V. N. Strakhov (1970c):

$$I_0 = \frac{W^-(\zeta_0)}{\pi\bar{\zeta}_0 + i\int_L \bar{\zeta}\, d\zeta/(\zeta-\zeta_0)} \ . \tag{4.3.25}$$

If $I(\zeta)$ is an arbitrary analytical function inside Γ we can, having solved Eq. (4.3.23) and established $I(\zeta_0)$ for $\zeta_0 \in L$, employ the Cauchy integral formula (1.1.1) and reconstruct $I(\zeta)$ everywhere inside Γ:

$$I(\zeta') = \frac{1}{2\pi i} \int_L \frac{I(\zeta)}{\zeta - \zeta'} d\zeta \, , \quad \zeta' \in \Gamma \, . \tag{4.3.26}$$

4.4 Analytical Continuation of a Plane Gravitational Field into a Domain Occupied by Attracting Masses

The body of the Cauchy-type integral will be employed to solve the problem of continuing the gravitational field into perturbing masses.

4.4.1 Characteristics of the Gravitational Field of a Homogeneous Domain Bounded by an Analytical Curve

Our investigation will be started with a simple case, where the domain Γ is occupied by masses of a constant density ϱ_0 and is bounded by an analytical curve L whose equation has the form

$$\bar{\zeta} = \varphi_L(\zeta) \, , \tag{4.4.1}$$

where $\varphi_L(\zeta)$ is a function analytical in some, generally not simply connected, domain D enclosing completely Γ (Fig. 22).

Complex intensity of the gravitational field of the domain will be designated by $g_\Gamma(\zeta')$. Let us also define complex intensity $g_{C\Gamma}(\zeta')$ of the domain $C\bar{\Gamma}$ filled with attracting masses of the same density ϱ_0. Then, according to formulas (2.5.7) and (2.5.10), we have

$$g_\Gamma(\zeta') = -\frac{1}{2\pi i} \int_L \frac{2\pi\gamma\varrho_0\bar{\zeta}}{\zeta - \zeta'} d\zeta \, , \quad \zeta' \in C\bar{\Gamma} \, , \tag{4.4.2}$$

$$g_{C\Gamma}(\zeta') = \frac{1}{2\pi i} \int_L \frac{2\pi\gamma\varrho_0\bar{\zeta}}{\zeta - \zeta'} d\zeta \, , \quad \zeta' \in \Gamma \, . \tag{4.4.3}$$

Applying the Sokhotsky-Plemelj formulas (1.2.22 a) and (1.2.22 b) to the Cauchy-type integral contained in the right-hand sides of (4.4.2) and (4.4.3), we arrive at

$$g_{C\Gamma}(\zeta_0) + g_\Gamma(\zeta_0) = 2\pi\gamma\varrho_0\bar{\zeta} = 2\pi\gamma\varrho_0\varphi_L(\zeta_0) \, , \quad \zeta_0 \in L \, . \tag{4.4.4}$$

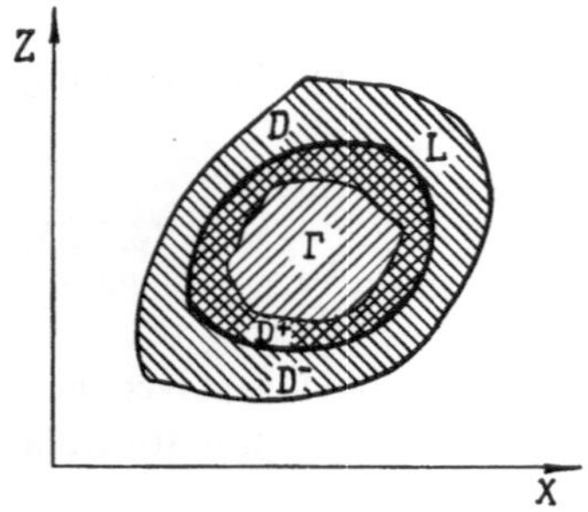

Fig. 22. Proof of theorem 4.4.1 about the analyticity conditions of a boundary L of a domain D filled with masses of a constant density ϱ_0

Relation (4.4.4) permits statement of the following theorem proved for the first time by A. V. Tsirulsky (1963).

Theorem 4.4.1. For a closed smooth curve L to be analytical, it is necessary and sufficient that the function $g_\Gamma(\zeta')$ might be continued analytically inside, while the function $g_{C\Gamma}(\zeta')$ outside the domain Γ bounded by L through any point on L.

We will prove the *sufficiency of the theorem condition.* Assume that the function $g_\Gamma(\zeta')$ can be continued analytically into some domain D^+ adjoining L from the inside, whereas the function $g_{C\Gamma}(\zeta')$ can be analytically continued outside the domain Γ into some domain adjoining L from the outside. Thus, L is a common boundary of the domains D^+ and D^- (see Fig. 22). The letter D will represent the domain including the two domains D^+ and D^- and the curve L: $D = D^+ \cup D^- \cup L$. Derive the function $\varphi_L(\zeta)$ as follows:

$$\varphi_L(\zeta) = \frac{1}{2\pi\gamma\varrho_0}\,[g_{C\Gamma}(\zeta)+g_\Gamma(\zeta)] \ . \tag{4.4.5}$$

Evidently, $\varphi_L(\zeta)$ is a function analytical in D, and, according to (4.4.4), we have

$$\varphi_L(\zeta_0) = \bar{\zeta}_0 \ , \quad \zeta_0 \in L \ . \tag{4.4.6}$$

Equation (4.4.6) is the equation for the contour L in the Herglotz-Tsirulsky form and, by virtue of the analyticity of $\varphi_L(\zeta)$ in D, the contour L is analytical, which was to be proved.

The *necessity of the theorem condition* follows directly from theorem 4.2.1 asserting the possibility of continuation of the Cauchy-type integral of an analytical density through a regular-analytical path of integration (since $\bar{\zeta} = \varphi_L(\zeta)$ is an analytical function and L is a smooth curve). The continued fields are defined, in accordance with theorem 4.2.1, by the formulas:

$$g^-_{\Gamma_L}(\zeta') = 2\pi\gamma\varrho_0\varphi_L(\zeta') - g_{C\Gamma}(\zeta') \ , \quad \zeta' \in \Gamma \ , \tag{4.4.7}$$

$$g^+_{C\Gamma_L}(\zeta') = 2\pi\gamma\varrho_0\varphi_L(\zeta') - g_\Gamma(\zeta') \ , \quad \zeta' \in \bar{\Gamma} \ , \tag{4.4.8}$$

whence it is evident that the set of singular points of the fields $g_\Gamma(\zeta')$ and $g_{C\Gamma}(\zeta')$, continued analytically through the curve L, coincides with the set of singularities of the function $\varphi_L(\zeta')$ inside and outside the domain Γ, respectively.

In particular, of special concern is the case where the function $\varphi_L(\zeta)$ has no singularities outside Γ. Then, by the proved theorem, the complex intensity $g^+_{C\Gamma_L}(\zeta')$ continued to the whole complex plane has no singularities at all and, by the Liouville theorem, it is constant:

$$g^+_{C\Gamma_L}(\zeta') \equiv a = \text{const} \ . \tag{4.4.9}$$

Substituting (4.4.9) into (4.4.8) we find

$$g_\Gamma(\zeta') = 2\pi\gamma\varrho_0\varphi_L(\zeta') - a \ , \quad \zeta' \in C\bar{\Gamma} \ . \tag{4.4.10}$$

Making ζ' in (4.4.10) tend to infinity, we define the unknown constant a:

$$a = 2\pi\gamma\varrho_0 \lim_{\zeta'\to\infty} \varphi_L(\zeta') \ , \tag{4.4.11}$$

whence we write in the final form

$$g_\Gamma(\zeta') = 2\pi\gamma\varrho_0[\varphi_L(\zeta') - \lim_{\zeta'\to\infty} \varphi_L(\zeta')] \ . \tag{4.4.12}$$

Formula (4.4.12) yields an analytical solution to the problem of determination of complex intensity of a homogeneous domain Γ bounded by an analytical curve L [fitted by the function $\varphi_L(\zeta)$ having no singularities outside Γ].

For example, if L is a circle of radius R with its center at the origin of coordinates, its equation in the Herglotz-Tsirulsky form is

$$\bar{\zeta} = \varphi_L(\zeta) = R^2/\zeta \ . \tag{4.4.13}$$

The function $\varphi_L(\zeta) = R^2/\zeta$ has only one singularity – a pole at the origin of coordinates. Substituting (4.4.13) into (4.4.12) we find the field of a circle $|\zeta| \leqslant R$ filled with attracting masses of a constant density ϱ_0:

$$g_\Gamma(\zeta') = 2\pi\gamma\varrho_0 R^2/\zeta' = 2\gamma M/\zeta' \ , \tag{4.4.14}$$

where $M = \pi R^2 \varrho_0$ is the circle mass.

4.4.2 Continuation of the Gravitational Field into a Domain with an Analytical Density Distribution

Complex intensity of the gravitational field of the domain Γ bounded by a piecewise smooth curve L, with an analytical distribution of the mass density $\varrho(x, z)$ inside Γ, is described, according to (2.5.20), by the following Cauchy-type integral:

$$g_\Gamma(\zeta') = -\frac{1}{2\pi i}\int_L \frac{2\pi\gamma\Phi(\zeta, \bar{\zeta})}{\zeta - \zeta'}\, d\zeta \ , \quad \zeta' \in C\bar{\Gamma} \ . \tag{4.4.15}$$

Here $\Phi(\zeta, \bar{\zeta})$ is the indefinite integral of the mass density, which is defined by formula (2.5.15) and it is an analytical function of its arguments in Γ and is Lipschitz-continuous on L. For complex intensity $g_\Gamma(\zeta')$ the following theorem is valid.

Theorem 4.4.2. Complex intensity of the gravitational field produced by the distribution of density $\varrho(x, z)$ inside Γ, which is an analytical function of the x- and z-coordinates (provided $\Phi(\zeta, \bar{\zeta}) \neq 0$ on L_0), permits analytical continuation through any left-side Herglotz-Tsirulsky analytical arc L_0 of the curve L (except maybe for its ends) into Γ. The singular points of the continued field coincide with those of the function $\Psi(\zeta) = \Phi(\zeta, \varphi^+_{L_0}(\zeta))$, where $\varphi^+_{L_0}(\zeta)$ specifies the equation of the arc L_0 in the Herglotz-Tsirulsky form.

Theorem 4.4.2 is proved by reasoning similar to those used in the proof of theorems 4.2.1 and 4.3.1, so that proving will be left to the reader.

In particular, here formula (4.3.19) takes the form:

$$g_{\bar{\Gamma}_{L_0}}(\zeta') = -\frac{1}{2\pi i}\int_L \frac{2\pi\gamma\Phi(\zeta,\bar{\zeta})}{\zeta-\zeta'}\,d\zeta + 2\pi\gamma\Phi(\zeta',\varphi^+_{L_0}(\zeta'))\ , \quad \zeta'\in D^+\ , \tag{4.4.16}$$

where $g_{\bar{\Gamma}_{L_0}}(\zeta')$ is the result of analytical continuation of $g_\Gamma(\zeta')$ through the arc L_0 into $\bar{\Gamma}$.

Note that formula (4.4.16) indicates validity of the inverse claim that *if the function* $g_\Gamma(\zeta')$ *permits analytical continuation into* Γ *through a smooth* arc L_0 *(except maybe for its ends) of its boundary, the* arc L_0 *is regular analytical.*

Indeed, if there is analytical continuation $g_{\bar{\Gamma}_{L_0}}(\zeta')$ of the field $g_\Gamma(\zeta')$ into Γ through an arc L_0, we have for $\zeta' = \zeta_0 \in L_0$

$$g_{\bar{\Gamma}_{L_0}}(\zeta_0) + \lim_{\zeta'\to\zeta_0^+}\left(\frac{1}{2\pi i}\int_L \frac{2\pi\gamma\Phi(\zeta,\bar{\zeta})}{\zeta-\zeta'}\,d\zeta\right) = 2\pi\gamma\Phi[\zeta_0,\varphi^+_{L_0}(\zeta_0)]\ . \tag{4.4.17}$$

The left-hand side of (4.4.17) admits analytical continuation into Γ, hence, analytical continuation is permitted also by its right-hand side. Therefore, the function $\varphi^+_{L_0}(\zeta)$ should be analytical in some domain D^+ adjoining L from the inside of the domain D^+, i.e., L_0 is a left-side Herglotz-Tsirulsky analytical arc, consequently it is merely analytical. But L_0 is a smooth curve, hence, it is regular-analytical.

4.4.3 Case of a Homogeneous Domain Bounded by a Piecewise Analytical Curve

Consider again a particular case where the density ϱ_0 inside Γ is constant but, unlike in Sect. 4.4.1, L is assumed to be piecewise smooth. Then the function $\Phi(\zeta,\bar{\zeta})$, in accordance with (2.5.15), takes the form

$$\Phi(\zeta,\bar{\zeta}) = \varrho_0\bar{\zeta}\ . \tag{4.4.18}$$

Naturally, theorem 4.3.2 is still valid but formula (4.4.16) for the continued field is somewhat simplified:

$$g_{\bar{\Gamma}_{L_0}}(\zeta') = -\frac{1}{2\pi i}\int_L \frac{2\pi\gamma\varrho_0\bar{\zeta}}{\zeta-\zeta'}\,d\zeta + 2\pi\gamma\varrho_0\varphi^+_{L_0}(\zeta')\ , \quad \zeta'\in D^+\ . \tag{4.4.19}$$

The integral in the right-hand side of (4.4.19), according to formula (4.4.3), implies the complex intensity of the domain $C\bar{\Gamma}$ filled with masses of a constant density ϱ_0; therefore, one can rewrite (4.4.19) in the form:

$$g_{\bar{\Gamma}_{L_0}}(\zeta') = 2\pi\gamma\varrho_0\varphi^+_{L_0}(\zeta') - g_{C\Gamma}(\zeta')\ , \quad \zeta'\in D^+\ , \tag{4.4.20}$$

which is consistent with (4.4.7).

Evidently, applying theorem 4.2.1 to the Cauchy-type integral describing the field $g_{C\Gamma}$ [formula (4.4.3)], one can readily see, just as in Sect. 4.4.1, that *the field* $g_{C\Gamma}(\zeta')$ *can be continued analytically beyond the domain* Γ through any point of an arc L_0 belonging to its boundary (except maybe for its ends) *if, and only if, the* arc L_0 *is right-side Herglotz-Tsirulsky analytical.*

The result of this analytical continuation is fitted by the formula

$$g^{+}_{C\Gamma_{L_0}}(\zeta') = 2\pi\gamma\varrho_0\varphi^{-}_{L_0}(\zeta') - g_{\Gamma}(\zeta') \ , \quad \zeta' \in D^{-} \ , \qquad (4.4.21)$$

where $\varphi^{-}_{L_0}(\zeta')$ specifies the equation for a right-side analytical arc [compare with formula (4.4.8)].

Thus, it is evident that *the necessary and sufficient condition that the complex intensity $g_{\Gamma}(\zeta')$ can be continued into Γ, while $g_{C\Gamma}(\zeta')$ beyond Γ through an* arc *L_0 of the boundary of the domain Γ (except for its ends), is the property of left- or right-side Herglotz-Tsirulsky analyticity of the* arc *L_0, respectively.*

4.4.4 Singular Points of the Continued Field, Lying on the Boundary of a Material Body

The problem of singular points of the continued gravitational field, points belonging to the boundary of a domain occupied by attracting masses, is resolved in the same way as that for the magnetic potential, namely the following theorem holds good.

Theorem 4.4.3. Let the boundary L of the domain Γ filled with masses of an analytical density distribution consists of a finite number of various left-side Herglotz-Tsirulsky analytical arcs. Then the points connecting these arcs are branch points of the function $g^{-}_{\Gamma_L}(\zeta')$ continued analytically into Γ.

The proof of theorem 4.4.3 is similar to that given in Sect. 4.3.4.

A nice question is the behavior of the continued field in the neighborhood of the cusps of the curve L bounding Γ.

The thing is that in the neighborhood of the cusp the arc L_0 may be one-side Herglotz-Tsirulsky analytical, although the cusp itself is a singular point ζ_0 of the arc in the differential geometrical meaning (Rashevsky 1956). That is why, once L_0 is a left-side Herglotz-Tsirulsky analytical arc, the field $g_{\Gamma}(\zeta')$ can be continued analytically into Γ through the cuspidal point ζ_0. A more detailed investigation, which is omitted here (Zhdanov 1970), has revealed that it is possible there where the cuspidal point is facing inward the domain Γ. But evidently *in the neighborhood of a cusp the* arc *L_0 will never be regular-analytical.*

This fact permits statement of the following theorem.

Theorem 4.4.4. Any point of the curve L bounding the domain Γ filled with attracting masses of a constant density ϱ_0, which is singular in the differential-geometrical meaning, is singular either for the field $g_{\Gamma}(\zeta')$, or for the field $g_{C\Gamma}(\zeta')$, or for the two fields, $g_{\Gamma}(\zeta')$ and $g_{C\Gamma}(\zeta')$.

Indeed, let ζ_0 be a geometrically singular point of the curve L but at the same time the functions $g_{\Gamma}(\zeta')$ and $g_{C\Gamma}(\zeta')$ permit analytical continuation through any point of the arc L_0 lying in a neighborhood of ζ_0, including ζ_0 itself. Then, according to the result of Sect. 4.4.3, the arc L_0 is one-side analytical on the two sides, i.e., simply regular-analytical, which is impossible.

Thus, theorems 4.4.3 and 4.4.4 define completely singular points of the continued field, which lie on the boundary of a material body.

4.5 Integral Techniques for Analytical Continuation of Plane Fields

In this section we will present constructive methods for analytical continuation of plane geopotential fields. This problem has been actively studied in geophysics ever since the late 1930's. Currently we have at our disposal a fundamental theory of analytical continuation, effective numerical methods for problem solution, and a variety of algorithms and programs for recalculating the gravitational and magnetic fields. It is not our objective to expatiate on the state of the art in the problem, since it would require a separate book. We will dwell just on one approach to the elaboration of methods for analytical continuation, relying on the Cauchy integral formula. This approach is distinguished by its permitting an insight into the most significant features of the problem of field continuation, on the one hand, and by equal applicability to both plane and three-dimensional geopotential fields, on the other. Moreover, it gives the most striking example of integral transforms in geophysics.

4.5.1 Forms of Analytical Continuation of Plane Fields in Geophysics

The problem of continuation of a plane geopotential field in geophysics is normally stated as follows. Let $F(\zeta)$ be complex intensity of a plane field whose sources are found inside a domain Γ. For the sake of simplicity, we will confine ourselves to the case of a finite domain Γ, although all the results presented below are also valid for infinitely extended domains. In other words, we assume that the function $F(\zeta)$ is analytical everywhere outside the finite domain Γ and it vanishes at infinity. It is assumed furthermore that L is an arbitrary piecewise smooth line passing through a point at infinity (Fig. 23). Its equations are

$$x = x(s) \ , \quad z = z(s) \ , \quad s_a \leqslant s \leqslant s_b \ ,$$

and here

$$z_1 \leqslant z(s) \leqslant 0. \tag{4.5.1}$$

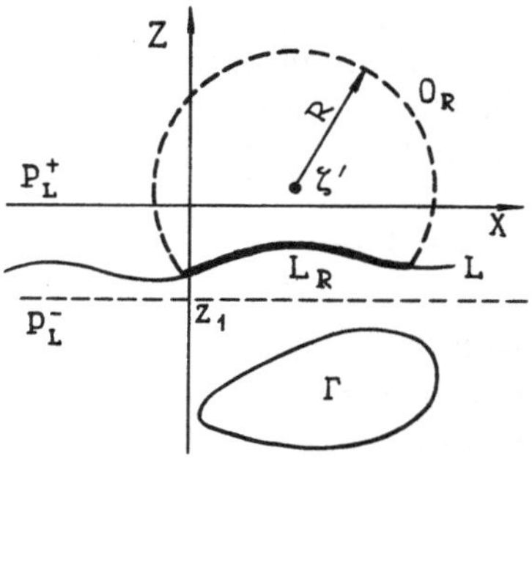

Fig. 23. Analytical continuation of the field into the upper half-plane P_L^+

Let P_L^+ stand for the upper half-plane of a complex plane ζ bounded by the curve L, and P_L^- for the lower half-plane. The field $F(\zeta)$ is known on the curve L, with $\Gamma \subset P_L^-$, i.e., the sources of the field $F(\zeta)$ are concentrated in the lower half-plane. Within this statement two classes of problems involved in analytical continuation can be formulated: (1) *analytical continuation of the field upward* – into the upper half plane P_L^+; (2) *analytical continuation of the field downward* – into the lower half-plane P_L^-.

In geophysical applications, the curve L is identified with the line of observation, i.e., with the Earth's surface. In the upward continuation of the field, the distance from the field sources increases and the minute details of the source structure are no longer discernible. By contrast, continuing the field downwards, we approach the field sources, thereby identifying the effects hardly observed from the surface of the Earth. Moreover, as indicated in the previous subsections, we can penetrate even the domain occupied by field sources and reconstruct the location of sources and, in certain instances, also their density from the singularities of the continued field.

Thus, analytical continuation permits the solution of both qualitative and quantitative problems of interpretation of geopotential fields.

The problems of the first class, i.e., of upward continuation, are among the simplest forms of analytical continuation. These problems are notable for the fact that the function $F(\zeta)$ is continued into the domain where it is analytical. Therefore, to solve this problem we can directly employ the Cauchy integral formula (1.1.1).

Indeed, consider an arbitrary point $\zeta' \in P_L^+$ and draw therefrom a circle of radius R. Part of the curve L that happens to be inside the circle will be represented by L_R, while the part of this circle found inside P_L^+ will be denoted by O_R (see Fig. 23). Evidently, by virtue of Cauchy integral formula (1.1.1), the values of the function $F(\zeta)$ at the point ζ' are calculated from its value on L_R and O_R:

$$F(\zeta') = \frac{1}{2\pi i} \int_{L_R} \frac{F(\zeta)}{\zeta - \zeta'} d\zeta + \frac{1}{2\pi i} \int_{O_R} \frac{F(\zeta)}{\zeta - \zeta'} d\zeta \; . \tag{4.5.2}$$

Now let us proceed to the limit for $R \to \infty$. The integral taken over the part of the circle O_R is written, upon substitution of the variable $\zeta = \zeta' + Re^{i\theta}$, in the form

$$\frac{1}{2\pi i} \int_{O_R} \frac{F(\zeta)}{\zeta - \zeta'} d\zeta = \frac{1}{2\pi} \int_0^{\pi} F(\zeta' + Re^{i\theta}) \, d\theta \; . \tag{4.5.3}$$

Being analytical, the function $F(\zeta)$ tends uniformly over θ to zero at infinity, hence the limit of integral (4.5.3) is zero. The integral extended along the part of the curve L_R for $R \to \infty$ tends to the integral taken over the entire line L. Thus, in the limit, formula (4.5.2) takes the form:

$$F(\zeta') = \frac{1}{2\pi i} \int_L \frac{F(\zeta)}{\zeta - \zeta'} d\zeta \; , \quad \zeta' \in P_L^+ \; , \tag{4.5.4}$$

with the line L traversed in expression (4.5.4) from the left to the right. Formula (4.5.4) does yield a solution for the problem of analytical continuation of complex intensity of the field into the upper half-plane.

4.5.2 Reconstruction of a Function Analytical in the Upper Half-Plane from Its Real or Imaginary Part

Of extreme importance in geophysical applications is the problem of reconstructing the function $F(\zeta)$ from its real or imaginary part. Indeed, in gravimetry, complex intensity of the gravitational field is $g(\zeta) = -g_x + ig_z$, where g_x and g_z are the horizontal and the vertical components of the gravitational field, respectively. As a rule, in practice it is the value of g_z that is measured, hence, the problem is to reconstruct $g(\zeta)$ from g_z, i.e., from its imaginary part. A similar problem arises also in magnetometry.

This problem is readily solved in a simple case, where the line L coincides with the real axis x. Here formula (4.5.4) is written as follows:

$$F(\zeta') = \frac{1}{2\pi i} \int_{-\infty}^{+\infty} \frac{F(x)}{x-\zeta'} dx \ . \tag{4.5.5}$$

Let us identify the real and the imaginary parts of the function $F(\zeta)$:

$$F(\zeta) = \mathrm{Re}F(\zeta) + i\mathrm{Im}F(\zeta) = U(x,\ z) + iV(x,\ z) \ ,$$

and substitute the latter expression into (4.5.5):

$$F(\zeta') = \frac{1}{2\pi i} \int_{-\infty}^{+\infty} \frac{U(x,\ 0)}{x-\zeta'} dx + \frac{1}{2\pi} \int_{-\infty}^{+\infty} \frac{V(x,\ 0)}{x-\zeta'} dx \ . \tag{4.5.6}$$

The above-stated problem is solved on the basis of integral relations between the functions $U(x,\ 0)$ and $V(x,\ 0)$ derived in Sect. 1.3.4 [formulas (1.3.29)]:

$$U(x,\ 0) = \frac{1}{\pi} \int_{-\infty}^{+\infty} \frac{V(\tilde{x},\ 0)}{\tilde{x}-x} d\tilde{x} \ , \tag{4.5.7}$$

$$V(x,\ 0) = -\frac{1}{\pi} \int_{-\infty}^{+\infty} \frac{U(\tilde{x},\ 0)}{\tilde{x}-x} d\tilde{x} \ . \tag{4.5.8}$$

Substituting formula (4.5.7) or (4.5.8) into expression (4.5.6) we can define the function $F(\zeta')$ from its real or imaginary part specified on the real axis.

Indeed, represent, say, the imaginary part of the function $F(\zeta)$ in terms of its real part by means of (4.5.8). Then, (4.5.6) can be rewritten as follows:

$$F(\zeta') = \frac{1}{2\pi i} \int_{-\infty}^{-\infty} \frac{U(x,\ 0)}{x-\zeta'} dx - \frac{1}{2\pi^2} \int_{-\infty}^{+\infty} \frac{1}{x-\zeta'} \int_{-\infty}^{+\infty} \frac{U(\tilde{x},\ 0)}{\tilde{x}-x} d\tilde{x} dx \ . \tag{4.5.9}$$

Since the sources of the field $F(\zeta)$ are concentrated in the finite domain Γ of the plane, the function $F(\zeta)$ itself, as well as its real and imaginary parts, decrease at infinity as $1/|\zeta|$, by virtue of (2.3.20). As a consequence, all the integrals contained in expressions (4.5.6) through (4.5.8) are absolutely and uniformly convergent, so that the order of integration in the second term of (4.5.9) can be exchanged:

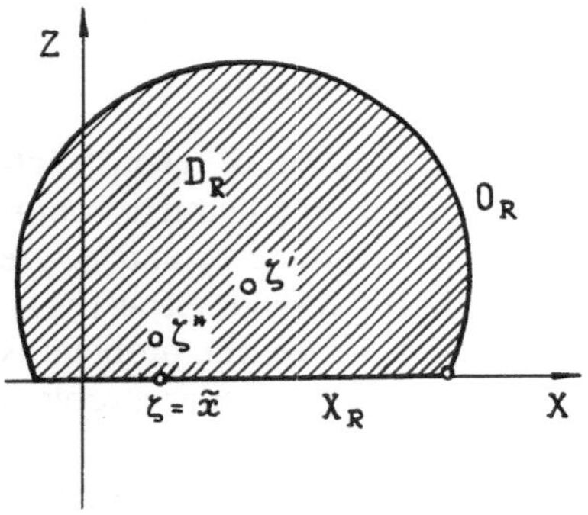

Fig. 24. Scheme explaining the solution to the problem of reconstructing an analytical function from its real or imaginary part

$$-\frac{1}{2\pi^2}\int_{-\infty}^{+\infty}\frac{1}{x-\zeta'}\int_{-\infty}^{+\infty}\frac{U(\tilde{x},0)}{\tilde{x}-x}d\tilde{x}dx$$
$$=\frac{-i}{2\pi}\int_{-\infty}^{+\infty}U(\tilde{x},0)\left[\frac{1}{\pi i}\int_{-\infty}^{+\infty}\frac{1/(x-\zeta')}{\tilde{x}-x}dx\right]d\tilde{x}\ . \qquad (4.5.10)$$

Let us evaluate the inner integral in the right-hand side of (4.5.10). To this end, take a point ζ' and draw therefrom, as a center, a circle of a radius R (Fig. 24). A part of this circle lying in the upper half-plane ($\mathrm{Im}\,\zeta \geqslant 0$) will be designated, as usual, by O_R, while the part of the real axis that has got inside the circle will be denoted by X_R. We will take into consideration the Cauchy-type integral

$$C(\zeta^*)=\frac{1}{2\pi i}\int_{O_R\cup X_R}\frac{1/(\zeta-\zeta')}{\zeta-\zeta^*}d\zeta\ .$$

Evidently, the density of this integral $1(\zeta-\zeta')$ is an analytical function outside the domain D_R bounded by the contour $O_R \cup X_R$ (see Fig. 24) and it tends to zero at infinity. Consequently, we obtain from the Cauchy-type formula for an infinite domain $C\bar{D}_R$ that

$$C(\zeta^*)=\begin{cases}0, & \zeta^*\in D_R\ ,\\ -1/(\zeta^*-\zeta')\ , & \zeta^*\in C\bar{D}_R\ .\end{cases} \qquad (4.5.11)$$

Now we will employ the Sokhotsky-Plemelj formula (1.2.22b):

$$C^+(\zeta_0)+C^-(\zeta_0)=\frac{1}{\pi i}\int_{O_R\cup X_R}\frac{1/(\zeta-\zeta')}{\zeta-\zeta_0}d\zeta\ ,\quad \zeta_0\in O_R\cup X_R\ . \qquad (4.5.12)$$

But, according to (4.5.11), we have

$$C^+(\zeta_0)=\lim_{\zeta^*\to\zeta_0^+}C(\zeta^*)=0\ ,\quad C^-(\zeta_0)=\lim_{\zeta^*\to\zeta_0^-}C(\zeta^*)=-1/(\zeta_0-\zeta')\ . \qquad (4.5.13)$$

Substituting (4.5.13) into (4.5.12) we write

$$\frac{1}{\zeta_0-\zeta'}=-\frac{1}{\pi i}\int_{O_R\in X_R}\frac{1/(\zeta-\zeta')}{\zeta-\zeta_0}d\zeta\ ,\quad \zeta_0\in O_R\cup X_R\ .$$

In particular, for $\zeta_0=\tilde{x}\in X_R$ we obtain

$$\frac{1}{\tilde{x}-\zeta'} = \frac{1}{\pi i} \int_{O_R} \frac{1/(\zeta-\zeta')}{\tilde{x}-\zeta} d\zeta + \frac{1}{\pi i} \int_{X_R} \frac{1/(x-\zeta')}{\tilde{x}-x} dx \ . \tag{4.5.14}$$

Replacing the variables $\zeta = \tilde{x} - Re^{i\theta}$, it is easy to rewrite the first integral in the right-hand side of (4.5.14) as follows:

$$\frac{1}{\pi i} \int_{O_R} \frac{1/(\zeta-\zeta')}{\tilde{x}-\zeta} d\zeta = -\frac{1}{\pi} \int_0^{\pi} [1/(\tilde{x}-\zeta'-Re^{i\theta})] \, d\theta \ ,$$

whence it is obvious that this integral vanishes at $R \to \infty$.

Thus, passing to the limit in (4.5.14) for $R \to \infty$ we obtain

$$\frac{1}{\pi i} \int_{-\infty}^{+\infty} \frac{1/(x-\zeta')}{\tilde{x}-x} dx = \frac{1}{\tilde{x}-\zeta'} \ . \tag{4.5.15}$$

Substitution of (4.5.15) into the right-hand side of (4.5.10) yields

$$-\frac{1}{2\pi^2} \int_{-\infty}^{+\infty} \frac{1}{\tilde{x}-\zeta'} \int_{-\infty}^{+\infty} \frac{U(\tilde{x}, 0)}{\tilde{x}-x} d\tilde{x} dx = \frac{1}{2\pi i} \int_{-\infty}^{+\infty} \frac{U(\tilde{x}, 0)}{\tilde{x}-\zeta'} d\tilde{x} \ . \tag{4.5.16}$$

And finally, substitution of (4.5.16) into (4.5.9) provides an expression for analytical continuation of complex intensity of a plane field into the upper half-plane on the basis of its real part:

$$F(\zeta') = \frac{1}{\pi i} \int_{-\infty}^{+\infty} \frac{U(x, 0)}{x-\zeta'} dx \ , \quad \mathrm{Im}\,\zeta' > 0 \ . \tag{4.5.17}$$

In a similar way, we can represent the function $F(\zeta')$ from its imaginary part. Indeed, substituting (4.5.7) into the first integral of (4.5.6) and exchanging the order of integration we obtain

$$F(\zeta') = \frac{1}{2\pi} \int_{-\infty}^{+\infty} V(\tilde{x}, 0) \left[\frac{1}{\pi i} \int_{-\infty}^{+\infty} \frac{1/(x-\zeta')}{\tilde{x}-x} dx \right] d\tilde{x} + \frac{1}{2\pi} \int_{-\infty}^{+\infty} \frac{V(x, 0)}{x-\zeta'} dx \ . \tag{4.5.18}$$

With due account taken of (4.5.15), formula (4.5.18) is recast in the form

$$F(\zeta') = \frac{1}{\pi} \int_{-\infty}^{+\infty} \frac{V(x, 0)}{x-\zeta'} dx \ . \tag{4.5.19}$$

Expressions (4.5.17) and (4.5.19) are of critical value in the theory of analytical continuation of geopotential fields into the upper half-plane. In particular, expanding the real and imaginary parts in the left- and right-hand sides of formulas (4.5.17) and (4.5.19) we can derive the following *Poisson integral formulas* for continuing conjugate harmonic functions[3] U and V upward:

---

[3] It is worth remembering that the Cauchy-Riemann conditions imply that the real and imaginary parts of any function analytical in D are harmonic in D and are called conjugate harmonic functions.

$$U(x', z') = \frac{z'}{\pi} \int_{-\infty}^{+\infty} \frac{U(x, 0)}{(x-x')^2 + z'^2}\, dx\;,$$
$$V(x', z') = \frac{z'}{\pi} \int_{-\infty}^{+\infty} \frac{V(x, 0)}{(x-x')^2 + z'^2}\, dx\;; \tag{4.5.20}$$

$$U(x', z') = \frac{1}{\pi} \int_{-\infty}^{+\infty} \frac{V(x, 0)(x-x')}{(x-x')^2 + z'^2}\, dx\;,$$
$$V(x', z') = -\frac{1}{\pi} \int_{-\infty}^{+\infty} \frac{U(x, 0)(x-x')}{(x-x')^2 + z'^2}\, dx\;. \tag{4.5.21}$$

Here $z' > 0$.

4.5.3 Analytical Continuation of Plane Fields into a Horizontal Layer Using Spectral Decomposition of the Cauchy Kernel

Now let us turn to the solution of the second class of problems, i.e., to analytical continuation of the field into the lower half-plane. This problem is difficult to solve because we do not know a priori the position of the domain Γ occupied by sources of the field $F(\zeta)$, i.e., we do not know singular points of the latter. Furthermore, one of the objectives of analytical continuation is to determine the location of these singularities. Meanwhile, for the sake of problem simplification, we can assume that the function $F(\zeta)$ is analytical at least up to a horizontal line M: $z = -h$, $h > 0$ (Fig. 25) with $-h < z_1$, i.e., with the line L passing everywhere above the straight line M. Let S designate the region between L and M (Fig. 25). *The problem is to reconstruct the values of $F(\zeta)$ everywhere inside S, knowing its values on L.* Let us solve this problem.

First of all, represent the value of the function $F(\zeta')$ at the point $\zeta' \in S$ through its values on the straight line M, using formula (4.5.4):

$$F(\zeta') = \frac{1}{2\pi i} \int_M \frac{F(\zeta)}{\zeta - \zeta'}\, d\zeta = \frac{1}{2\pi i} \int_{-\infty}^{+\infty} \frac{F(x - ih)}{x - ih - \zeta'}\, dx\;. \tag{4.5.22}$$

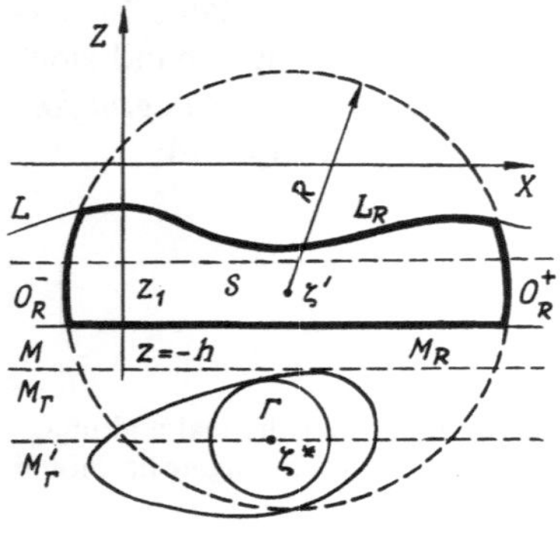

Fig. 25. Analytical continuation of the field into the lower half-plane to a horizontal layer

However, the field $F(\zeta)$ on the line M is not known. Hence, to solve the stated problem, it is necessary to transform the integral extended along the straight line M into the integral taken oven the line L where the field $F(\zeta)$ is specified. The way to this transformation lies in separating the variables in the Cauchy kernel, which can be achieved by its spectral decompositions.

Indeed, the following equation

$$\frac{1}{i(\zeta-\zeta')}=\int_{-\infty}^{0}\exp(ik_x\zeta)\exp(-ik_x\zeta')\,dk_x\ , \tag{4.5.23}$$

where $\mathrm{Im}(\zeta-\zeta')<0$, is valid. In particular, for $\zeta=x-ih$ we have

$$\frac{1}{i(x-ih-\zeta')}=\int_{-\infty}^{0}\exp[ik_x(x-ih)]\exp(-ik_x\zeta')\,dk_x\ . \tag{4.5.24}$$

Substituting this decomposition into formula (4.5.22) and exchanging the order of integration [which is feasible due to the absolute and uniform convergence of integral (4.5.24)], we write

$$F(\zeta')=\frac{1}{2\pi}\int_{-\infty}^{0}\exp(-ik_x\zeta')\int_{-\infty}^{+\infty}F(x-ih)\exp[ik_x(x-ih)]\,dx\,dk_x\ . \tag{4.5.25}$$

Let us transform the inner integral in (4.5.25). To this end, draw a circle of radius R from the point ζ' as a center and denote parts of this circle belonging to S by O_R^- and O_R^+, while the parts of the curve L and straight line M found inside this circle by L_R and M_R, respectively (see Fig. 25). Let us form a closed contour $l_R=L_R\cup O_R^-\cup M_R\cup O_R^+$. Note that the integrand $F(\zeta)\exp ik_x\zeta$ in (4.5.25) is an analytical function of ζ everywhere in the upper half-plane $\mathrm{Im}\,\zeta>-h$ and in particular inside the contour l_R. Consequently, by the Cauchy integral formula we have

$$\int_{l_R}F(\zeta)\exp(ik_x\zeta)\,d\zeta=0\ . \tag{4.5.26}$$

Representing the integral in the left-hand side of (4.5.26) as a sum of four integrals taken over the corresponding parts of the contour l_R we derive

$$\int_{L_R}F(\zeta)\exp(ik_x\zeta)\,d\zeta+\int_{O_R^-}F(\zeta)\exp(ik_x\zeta)\,d\zeta+\int_{M_R}F(x-ih)\exp[ik_x(x-ih)]\,dx$$
$$+\int_{O_R^+}F(\zeta)\exp(ik_x\zeta)\,d\zeta=0\ . \tag{4.5.27}$$

where the line L_R passes to the left and the line M_R runs to the right.

Estimate the integrals extended along the circle portions O_R^- and O_R^+:

$$|\Phi_{O_R^\pm}|=\left|\int_{O_R^\pm}F(\zeta)\exp(ik_x\zeta)\,d\zeta\right|=\left|\int_{O_R^\pm}F(\zeta)\exp(ik_x\mathrm{Re}\,\zeta)\right.$$
$$\left.\times\exp(-k_x\mathrm{Im}\,\zeta)\,d\zeta\right|\leqslant\int_{O_R^\pm}|F(\zeta)||d\zeta|\leqslant\max_{\zeta\in O_R^\pm}|F(\zeta)|\int_{O_R^\pm}|d\zeta|\ , \tag{4.5.28}$$

where it is taken into consideration that $\exp(-k_x\mathrm{Im}\,\zeta)\leqslant 1$, since, in accordance with (4.5.1) $\mathrm{Im}\,\zeta\leqslant 0$ on $O_R^\pm$ and $k_x\leqslant 0$.

For any R, the circle portions $O_R^\pm$ lie invariable between the straight lines $z=0$ and $z=-h$. It is remarkable that for larger R the circle portions $O_R^\pm$ approach the form of vertical portions of height h; therefore, we have at least

$$\int_{O_R^\pm} |d\zeta| < 2h \; . \tag{4.5.29}$$

Substitution of (4.5.29) into (4.5.28) yields

$$|\Phi_{O_R^\pm}| < 2 \max_{\zeta \in O_R^\pm} |F(\zeta)| h \; . \tag{4.5.30}$$

Since $F(\zeta)$ vanishes at infinity, the right-hand side of (4.5.30) for large R becomes zero, thereby making the left-hand side vanish, too. Thus, proceeding to the limit for $R \to \infty$ in (4.5.27), we obtain

$$\int_{-\infty}^{+\infty} F(x-ih) \exp[ik_x(x-ih)]\, dx = \int_L F(\zeta) \exp(ik_x\zeta)\, d\zeta \; , \tag{4.5.31}$$

where the line L, unlike formula (4.5.27), is traversed *from the left to the right.*

Thus, the problem of evaluating the inner integral of (4.5.25) has been reduced to integration over the line L, where the field has been specified. Substituting (4.5.31) into (4.5.25) we write in the final form

$$F(\zeta') = \frac{1}{2\pi} \int_{-\infty}^{0} \exp(-ik_x\zeta') \int_L F(\zeta) \exp(ik_x\zeta)\, d\zeta\, dk_x \; . \tag{4.5.32}$$

Precisely formula (4.5.32) provides an analytical solution to the problem of downward continuation of the field. We should like to stress once again that by construction this formula holds only if $\operatorname{Im}\zeta' > -h$ and all the singularities of the function $F(\zeta)$ are below the straight line $z=-h$. In other words, this formula is valid up to the horizontal line going through the singular point of the field which is the closest to the line L. In accordance with the findings of Sects. 4.3 and 4.4, singular points may be found both on the boundary of the domain Γ containing field sources and inside Γ. As a result, formula (4.5.32) definitely permits field continuation up to the horizontal line M_Γ tangent to the boundary of the domain Γ (see Fig. 25). At the same time, if the point of contact of the line M_Γ and the boundary of the domain Γ are not a singular point of the function $F(\zeta')$, formula (4.5.32) also permits field continuation into Γ – up to the straight line passing through the closest inner singular point of the function $F(\zeta)$. For example, if Γ is a circle of radius a with its center at a point ζ^* filled with masses of uniform density ϱ_0 and $F(\zeta)$ is a complex intensity of the gravitational field of these masses, formula (4.5.32) enables in principle the field $F(\zeta)$ to be continued up to the horizontal line M'_Γ going through the center of the circle, namely the point ζ^* (see Fig. 25).

4.5.4 Case of Field Specification on the Real Axis. The Zamorev Formulas

Consider a particular but for applications extremely important case where the line L coincides with the real axis.

Then formula (4.5.32) is written in the form

$$F(\zeta') = \frac{1}{2\pi}\int_{-\infty}^{0} \exp(-ik_x\zeta')\int_{-\infty}^{+\infty} F(x)\exp(ik_x x)\,dx\,dk_x\,. \qquad (4.5.33\,\mathrm{a})$$

Assume that $\zeta' = x - id$, then

$$F(x'-id) = \frac{1}{2\pi}\int_{-\infty}^{0} \exp(-ik_x x')\exp(|k_x|d)\int_{-\infty}^{+\infty} F(x)\exp(ik_x x)\,dx\,dk_x\,. \qquad (4.5.33\,\mathrm{b})$$

Following Sect. 3.1.2, we will designate the spatial spectrum of the field $F(x-id)$ by $f(k_x, -d)$:

$$f(k_x,-d) = \int_{-\infty}^{+\infty} F(x-id)\exp(ik_x x)\,dx\,. \qquad (4.5.34)$$

Then relations (4.5.33 a) and (4.5.33 b) can be rewritten in spectral terms as follows

$$f(k_x,-d) = \Phi_d(k_x) f(k_x, 0)\,, \qquad (4.5.35)$$

where $\Phi_d(k_x)$ is the frequency characteristic of field continuation to a depth d, defined as

$$\Phi_d(k_x) = \begin{cases} \exp(|k_x|d)\,, & k_x \leqslant 0\,,\\ 0\,, & k_x > 0\,. \end{cases} \qquad (4.5.36)$$

Let $u(k_x, -d)$ and $v(k_x, -d)$ stand for the spatial spectra of the real and the imaginary parts of the function $F(x-id)$ calculated along the horizontal line $z = -d$:

$$u(k_x,-d) = \int_{-\infty}^{+\infty} U(x,-d)\exp(ik_x x)\,dx\,, \qquad (4.5.37\,\mathrm{a})$$

$$v(k_x,-d) = \int_{-\infty}^{+\infty} V(x,-d)\exp(ik_x x)\,dx\,. \qquad (4.5.37\,\mathrm{b})$$

Substitution of (4.5.37 a) and (4.5.37 b) into (4.5.35) yields

$$u(k_x,-d) + iv(k_x,-d) = \Phi_d(k_x)[u(k_x,0) + iv(k_x,0)]\,. \qquad (4.5.38)$$

Obviously, the spectra $u(k_x,0)$ and $v(k_x,0)$ are interrelated by relations following from formulas (4.5.7) and (4.5.8). Indeed, employing representation (3.2.23) for the Hilbert operator, we will cast (4.5.7) in the form

$$U(x,0) = \hat{G}V(\tilde{x},0) = -\frac{i}{2\pi}\int_{-\infty}^{+\infty} \operatorname{sign} k_x \cdot v(k_x,0)\exp(-ik_x x)\,dk_x\,.$$

Hence, allowing for the Fourier transform inversion formula, we obtain

$$u(k_x,0) = -i\,\mathrm{sign}\,k_x \cdot v(k_x,0)\ . \tag{4.5.39a}$$

Similarly, we find

$$v(k_x,0) = i\,\mathrm{sign}\,k_x \cdot u(k_x,0)\ . \tag{4.5.39b}$$

One is readily convinced that formulas (4.5.39a) and (4.5.39b) remain valid also for spectra of fields calculated along any straight line $z = -d$ passing above all singular points of the function $F(\zeta)$, i.e., for $d<h$:

$$u(k_x,-d) = -i\,\mathrm{sign}\,k_x \cdot v(k_x,-d)\ , \tag{4.5.40a}$$

$$v(k_x,-d) = i\,\mathrm{sign}\,k_x \cdot u(k_x,-d)\ . \tag{4.5.40b}$$

Substituting (4.5.40a) and (4.5.40b) into (4.5.38) we derive two equivalent formulas

$$u(k_x,-d)(1-\mathrm{sign}\,k_x) = \Phi_d(k_x)u(k_x,\ 0)(1-\mathrm{sign}\,k_x)\ , \tag{4.5.41a}$$

$$v(k_x,-d)(1-\mathrm{sign}\,k_x) = \Phi_d(k_x)v(k_x,\ 0)(1-\mathrm{sign}\,k_x)\ . \tag{4.5.41b}$$

For $k_x>0$, these formulas turn into a trivial identity, while for $k_x\leqslant 0$ yield, allowing for (4.5.36),

$$\begin{aligned} u(k_x,-d) &= \exp(|k_x|d)u(k_x,\ 0)\ , \\ v(k_x,-d) &= \exp(|k_x|d)v(k_x,\ 0)\ . \end{aligned} \tag{4.5.42}$$

According to formula (4.5.37a), we have

$$u(k_x,-d) = \bar{u}(-k_x,-d)\ ,\quad v(k_x,-d) = \bar{v}(-k_x,-d)\ . \tag{4.5.43}$$

Hence, for $k_x>0$, with due reference to (4.5.42) and (4.5.43), we have

$$u(k_x,-d) = \bar{u}(-k_x,-d) = \exp(|k_x|d)\bar{u}(-k_x,\ 0) = \exp(|k_x|d)u(k_x,\ 0) \tag{4.5.44a}$$

and similarly

$$v(k_x,-d) = \exp(|k_x|d)v(k_x,\ 0)\ ,\quad k_x>0\ . \tag{4.5.44b}$$

Uniting (4.5.42), (4.5.44a) and (4.5.44b) we write finally for any k_x

$$u(k_x,-d) = \exp(|k_x|d)u(k_x,\ 0)\ ,\quad v(k_x,-d) = \exp(|k_x|d)v(k_x,\ 0)\ . \tag{4.5.45}$$

Formula (4.5.45) indicates that the frequency characteristic $\Psi_d(k_x)$ of the operation of downward continuation of the functions harmonic in the upper half-plane has the form

$$\Psi_d(k_x) = \exp(|k_x|d)\ . \tag{4.5.46}$$

Using relations (4.5.39a) and (4.5.39b), one can easily see that the function $F(\zeta)$ can be analytically continued into the lower half-plane on the basis of values of only its real or imaginary part specified on the real axis. Indeed, according to (4.5.34), (4.5.37a), and (4.5.37b) as well as (4.5.39a) and (4.5.39b) we have

$$f(k_x, 0) = u(k_x, 0) + iv(k_x, 0) = (1 - \text{sign}\, k_x) u(k_x, 0) = i(1 - \text{sign}\, k_x) v(k_x, 0) \; . \tag{4.5.47}$$

Substituting (4.5.47) into (4.5.33 a), we derive two equivalent formulas:

$$F(\zeta') = \frac{1}{\pi} \int_{-\infty}^{0} \exp(-ik_x\zeta') \int_{-\infty}^{+\infty} U(x, 0) \exp(ik_x x)\, dx dk_x \;, \tag{4.5.48 a}$$

$$F(\zeta') = \frac{i}{\pi} \int_{-\infty}^{0} \exp(-ik_x\zeta') \int_{-\infty}^{+\infty} V(x, 0) \exp(ik_x x)\, dx dk_x \;, \quad \zeta' \in S \; . \tag{4.5.48 b}$$

Formulas (4.5.48) are known as *the Zamorev formulas*. They permit downward analytical continuation of the function, which is analytical in the upper half-plane, from its real or imaginary part specified on the real axis. These formulas are particularly important in the problems of gravimetry, since only one component, g_z, of the complex intensity of the gravitational field is generally measured in practice. It proves that these measurements are quite sufficient to reconstruct the total complex intensity $g(\zeta) = -g_x(x, z) + ig_z(x, z)$ both in the upper [formula (4.5.19)] and in the lower half-plane right to the singular point of the function $g(\zeta)$ lying closest to the real axis.

4.5.5 Downward Analytical Continuation of Functions Having Singularities Both in the Lower and in the Upper Half-Planes

Now consider a more complicated problem involving downward continuation from the curve L of the function $F(\zeta)$, it having singularities not only below the straight line M but also above the line L (see Fig. 25), i.e., both in the upper and in the lower half-planes. Assume also that the function $F(\zeta)$ satisfies the conditions

$$\lim F(\zeta) = 0 \;, \quad \text{Re}\,\zeta \to \pm\infty \;, \quad \zeta \in S \;, \tag{4.5.49}$$

where S is the infinitely horizontally extended domain lying between L and M.

Evidently, this function can be exemplified by the complex intensity of the field having sources in local domains Γ_e and Γ_i both in the upper and in the lower half-planes (see Sect. 3.1). The problem is to define the value of the function $F(\zeta)$ at an arbitrary point $\zeta' \in S$ from its values on L (see Fig. 25).

To solve the problem, consider, just as in Sect. 4.5.3, the contour $l_R = L_R \cup O_R^- \cup M_R \cup O_R^+$. According to the Cauchy integral formula, we have

$$F(\zeta') = \frac{1}{2\pi i} \int_{l_R} \frac{F(\zeta)}{\zeta - \zeta'} d\zeta \;, \tag{4.5.50}$$

where the contour l_R is traversed counterclockwise.

Integral (4.5.50) can be represented as a sum of four integrals:

$$F(\zeta') = \frac{1}{2\pi i} \int_{L_R} \frac{F(\zeta)}{\zeta - \zeta'} d\zeta + \frac{1}{2\pi i} \int_{O_R^-} \frac{F(\zeta)}{\zeta - \zeta'} d\zeta + \frac{1}{2\pi i} \int_{M_R} \frac{F(\zeta)}{\zeta - \zeta'} d\zeta + \frac{1}{2\pi i} \int_{O_R^+} \frac{F(\zeta)}{\zeta - \zeta'} d\zeta \; . \tag{4.5.51}$$

Let us proceed to the limit for $R \to \infty$ in (4.5.51). Then one can easily see (just as in the above discussion) that by virtue of condition (4.5.49) the integrals taken along the circle arcs O_R^+ and O_R^- tend to zero. Hence, we obtain

$$F(\zeta') = \frac{1}{2\pi i} \int_L \frac{F(\zeta)}{\zeta - \zeta'} d\zeta + \frac{1}{2\pi i} \int_M \frac{F(\zeta)}{\zeta - \zeta'} d\zeta \ , \tag{4.5.52}$$

where the line L is traversed *from the right to the left,* while the line M *from the left to the right.*

The integral extended along M in formula (4.5.52) is identical to that of (4.5.22), which can be reduced, as indicated earlier, to form (4.5.32). Substituting the latter formula into (4.5.52) we find

$$F(\zeta') = -\frac{1}{2\pi i} \int_L \frac{F(\zeta)}{\zeta - \zeta'} d\zeta + \frac{1}{2\pi} \int_{-\infty}^{0} \exp(-ik_x\zeta') \int_L F(\zeta) \exp(ik_x\zeta)\, d\zeta\, dk_x \ , \quad \zeta' \in S \ , \tag{4.5.53}$$

where the integration over L in the two integrals is carried out *from the left to the right.*

It is formula (4.5.53) that yields a solution to the stated problem of downward continuation. Transform this formula by means of spectral decomposition of the Cauchy kernel:

$$\frac{1}{i(\zeta - \zeta')} = -\int_0^{\infty} \exp[ik_x(\zeta - \zeta')]\, dk_x = -\int_0^{\infty} \exp(ik_x\zeta) \exp(-ik_x\zeta')\, dk_x \ , \tag{4.5.54}$$

where $\mathrm{Im}(\zeta - \zeta') > 0$.

Substitution of (4.5.54) into (4.5.53) and a little simplification yield

$$F(\zeta') = \frac{1}{2\pi} \int_{-\infty}^{+\infty} \exp(-ik_x\zeta') \int_L F(\zeta) \exp(ik_x\zeta)\, d\zeta\, dk_x \ , \tag{4.5.55}$$

where the integration along L is carried out *from the left to the right.*

In a particular case, where the line L coincides with the real axis, formula (4.5.55) is written in the form

$$F(\zeta') = \frac{1}{2\pi} \int_{-\infty}^{+\infty} \exp(-ik_x\zeta') \int_{-\infty}^{+\infty} F(x) \exp(ik_x x)\, dx\, dk_x \ , \tag{4.5.56}$$

or, if it is posited that $\zeta' = x' - id$,

$$F(x' - id) = \frac{1}{2\pi} \int_{-\infty}^{+\infty} \exp(-ik_x x') \exp(-k_x d) \int_{-\infty}^{+\infty} F(x) \exp(ik_x x)\, dx\, dk_x \ . \tag{4.5.57}$$

Passing on to spatial spectra, we obtain

$$f(k_x, -d) = \Psi_d(k_x) f(k_x, 0) \ , \tag{4.5.58}$$

where $\Psi_d(k_x)$ is the frequency characteristic of field continuation to a depth d, equal in this case to

$$\Psi_d(k_x) = \exp(-k_x d) \ , \quad -\infty < k_x < +\infty \ . \tag{4.5.59}$$

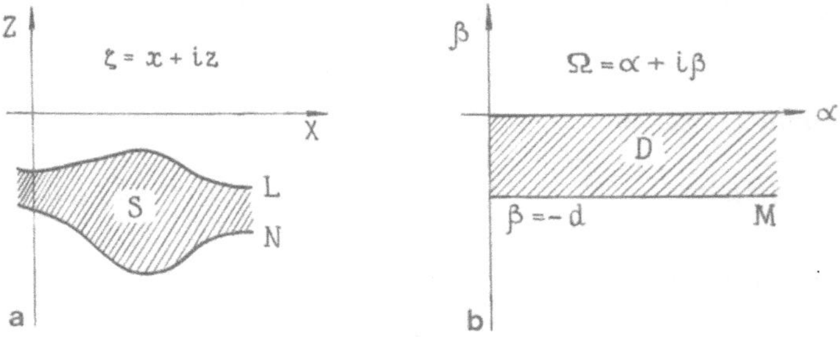

Fig. 26. a Domain S with curvilinear boundaries; **b** Subsidiary plane of a complex variable with a horizontal band D

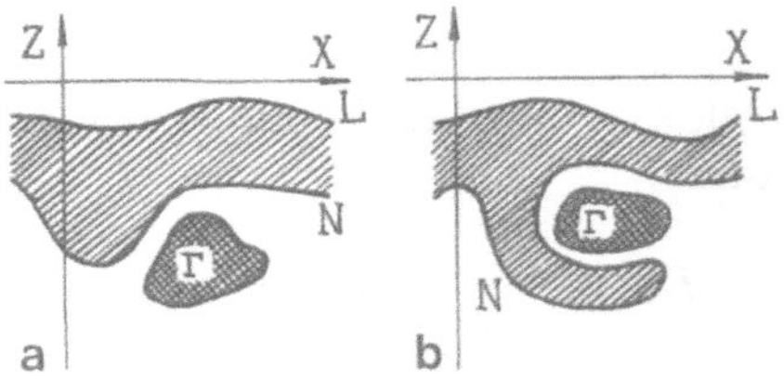

Fig. 27 a, b. Examples of curvilinear-boundary domains. Γ denotes the domains occupied with field sources

4.5.6 Analytical Continuation into Domains with Curvilinear Boundaries

So far we have considered the problem of continuing the function $F(\zeta)$ from the line L to a horizontal straight line M that goes above the singularities of the function $F(\zeta)$ lying in the lower half-plane and closest to the line L. It is of greater interest, however, to continue the field into a domain bounded from below by a curvilinear line N (Fig. 26a). Indeed, in this case we can continue the field not only to a horizontal layer down to a level corresponding to the uppermost point of the domain Γ occupied by field sources, but also below this level (at the side of the domain Γ) and even under it (Fig. 27a, b). Methods used to solve the above-stated problem are dealt with extensively in the literature [see, for instance, studies by V. N. Strakhov (1970e, 1971 a, b) or M. S. Zhdanov (1971, 1973 a)]. It is noteworthy that almost all the suggested methods rely on the natural idea of application of conformal mappings, thus reducing the problem of analytical continuation into curvilinear-boundary domains to that for domains with horizontal boundaries.

Now we will dwell upon one of the simplest methods used to solve this problem. Consider an auxiliary plane of the complex variable $\Omega = a + i\beta$ (Fig. 26b). Let D stand for a horizontal band in the plane Ω, which is bounded by the axis a and by the straight line $\beta = -d$.

Assume that the function $\zeta = \sigma(\Omega)$ maps conformly the horizontal band D onto the domain S, so that the axis a goes over into the line L, and the straight line $\beta = -d$ – into the line N. Consider a function $\tilde{F}(\Omega) = F(\sigma(\Omega))$ resulting from anti-pulling of the function $F(\zeta)$ onto the plane Ω by means of mapping $\sigma(\Omega)$. The function $\tilde{F}(\Omega)$ is analytical in the band D by virtue of the analyticity of the initial function in the domain S with curvilinear boundaries; here the function $\tilde{F}(\Omega)$ is known on the axis a, since the function $F(\zeta)$ is specified on the line L. As a consequence, to continue the function $\tilde{F}(\Omega)$ from the axis a into the domain D, we can employ formula (4.5.56) and write

$$\tilde{F}(\Omega') = \frac{1}{2\pi} \int_{-\infty}^{+\infty} \exp(-ik_a\Omega') \int_{-\infty}^{+\infty} \tilde{F}(a) \exp(ik_a a)\, da\, dk_a \ , \tag{4.5.60}$$

where Ω' is the inverse image of the point ζ' in the plane Ω:

$$\Omega' = \sigma^{-1}(\zeta') \tag{4.5.61}$$

where $\sigma^{-1}(\zeta)$ is the inverse transform of $\sigma(\Omega)$.

Evidently, the functions $F(\zeta)$ and $\tilde{F}(\Omega)$ are interrelated by

$$\tilde{F}(\Omega) = F(\sigma(\Omega)) \ , \quad F(\zeta) = \tilde{F}(\sigma^{-1}(\zeta)) \ . \tag{4.5.62}$$

Allowing for these relations as well as for the fact that $a = \sigma^{-1}(\zeta)$, $\zeta \in L$, we obtain

$$F(\zeta') = \frac{1}{2\pi} \int_{-\infty}^{+\infty} \exp[-ik_a\sigma^{-1}(\zeta')] \int_L \frac{F(\zeta) \exp[ik_a\sigma^{-1}(\zeta)]}{(d\sigma/da)\sigma^{-1}(\zeta)} d\zeta\, dk_a \ . \tag{4.5.63}$$

The resultant formula permits evaluation of the complex-analytical function $F(\zeta)$ everywhere within the domain S (with a curvilinear boundary N), where $F(\zeta)$ is analytical, from its values specified on the line L.

In a particular case, where L coincides with the x-axis, formula (4.5.63) is simplified:

$$F(\zeta') = \frac{1}{2\pi} \int_{-\infty}^{+\infty} \exp[-ik_a\sigma^{-1}(\zeta')] \int_{-\infty}^{+\infty} \frac{F(x) \exp[ik_a\sigma^{-1}(x)]}{(d\sigma/da)\sigma^{-1}(x)} dx\, dk_a \ . \tag{4.5.64}$$

4.5.7 Bateman Formula; Continuation of Complex Intensity of the Field into the Lower Half-Plane Using Its Real Part

For a field $F(\zeta)$ having singularities only in the lower half-plane, formula (4.5.64) permits its analytical continuation downward using only its real (or imaginary) part specified on the x-axis.

Indeed, let us derive function

$$U(\zeta) = \tfrac{1}{2}[F(\zeta) + \bar{F}(\bar{\zeta})] \ , \tag{4.5.65}$$

where $F(\zeta)$ is the complex intensity of a plane field whose sources are found in a finite domain Γ of the lower half-plane (Fig. 28).

The function $U(\zeta)$ seems to be analytical everywhere in the complex plane, except of the domain Γ and the domain Γ_m formed upon mirror reflection of Γ with respect to the real axis (see Fig. 28). Moreover, it is evident that

$$U(x) = \tfrac{1}{2}[U(x, 0) + iV(x, 0) + U(x, 0) - iV(x, 0)] = U(x, 0) \ , \tag{4.5.66}$$

i.e., $U(\zeta)$ is an analytical continuation of the real part $U(x, 0)$ of the function $F(\zeta)$ from the real axis into a complex domain.

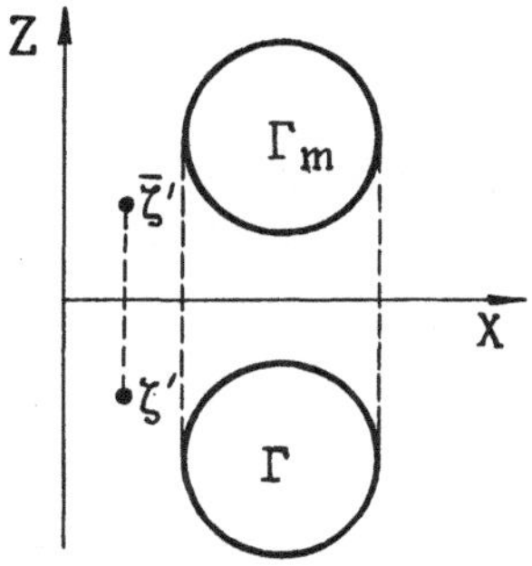

Fig. 28. Mirror reflection of the domain Γ with respect to the real axis

On the other hand, according to (4.5.65), we have

$$F(\zeta') = 2\,U(\zeta') - \bar{F}(\bar{\zeta}') \ , \tag{4.5.67}$$

where ζ' is an arbitrary point of the lower half-plane ($\operatorname{Im}\zeta' < 0$) lying outside Γ (see Fig. 28). Evidently, in this case the point $\bar{\zeta}'$ is found in the upper half-plane and the problem of continuing the function $F(\zeta)$ from the real axis downward is therefore reduced to downward continuation of its real part $U(\zeta)$ and to continuation of the function $\bar{F}(\zeta)$ conjugate to $F(\zeta)$ into the upper half-plane. The latter problem is readily solved using formula (4.5.17):

$$\bar{F}(\zeta') = -\frac{1}{\pi i}\int_{-\infty}^{+\infty} \frac{U(x,\ 0)}{x-\zeta'}\,dx \ . \tag{4.5.68}$$

Substituting (4.5.68) into (4.5.67) we write in the final form

$$F(\zeta') = 2\,U(\zeta') + \frac{1}{\pi i}\int_{-\infty}^{+\infty} \frac{U(x,\ 0)}{x-\zeta'}\,dx \ . \tag{4.5.69}$$

Formula (4.5.69) is called the Bateman formula (Bateman 1946). It permits reconstruction of the function $F(\zeta')$ from analytical continuation $U(\zeta')$ of its real part, specified along the horizontal axis, into a complex domain. In this case, the function $U(\zeta')$ in formula (4.5.69) can be calculated according to formula (4.5.56) (with continuation into a horizontal layer) or formula (4.5.64) (with continuation into the domain S with a curvilinear boundary).

For instance, substitution (4.5.56) into (4.5.69) yields

$$F(\zeta') = \frac{1}{\pi}\int_{-\infty}^{+\infty} \exp(-ik_x\zeta') \int_{-\infty}^{+\infty} U(x,\ 0)\exp(ik_x x)\,dx\,dk_x + \frac{1}{\pi i}\int_{-\infty}^{+\infty} \frac{U(x,\ 0)}{x-\zeta'}\,dx \ . \tag{4.5.70}$$

On the other hand, in accordance with (4.5.54), we have

$$\frac{1}{i(x-\zeta')} = -\int_{0}^{\infty} \exp(ik_x x)\exp(-ik_x\zeta')\,dk_x \ ,$$

since

$$\operatorname{Im}(x-\zeta') = -z' > 0 \ .$$

Allowing for this relation, we can simplify formula (4.5.70) as follows:

$$F(\zeta') = \frac{1}{\pi} \int_{-\infty}^{0} \exp(-ik_x\zeta') \int_{-\infty}^{+\infty} U(x, 0) \exp(ik_x x)\, dx\, dk_x, \tag{4.5.71}$$

which coincides with formula (4.5.48a) derived in a different way.

Now we will employ formula (4.5.64) to calculate $U(\zeta')$ in (4.5.69):

$$F(\zeta') = \frac{1}{\pi} \int_{-\infty}^{+\infty} \exp[-ik_a\sigma^{-1}(\zeta')] \int_{-\infty}^{+\infty} \frac{U(x, 0) \exp[ik_a\sigma^{-1}(x)]}{(d\sigma/da)\sigma^{-1}(x)}\, dx\, dk_a$$

$$+ \frac{1}{\pi i} \int_{-\infty}^{+\infty} \frac{U(x, 0)}{x - \zeta'}\, dx\ . \tag{4.5.72}$$

The latter formula permits continuation of complex intensity of the field into domains of the lower half-plane, having curvilinear boundaries, from the values of its real part $U(x, 0)$ on the x-axis. According to the remark in Sect. 4.5.4, this finding is particularly important for problems of gravimetry.

In concluding this section, we should like to draw the reader's attention to the following significant fact. All the above integral formulas of analytical continuation ensure a rigorous solution to the problem, provided accurate and continuous values of complex intensity of the field on the observation line L are available. However, any numerical realization of these formulas applied directly to experimental data known with some uncertainty only for a discrete set of points on the line L will not yield a satisfactory result. This is due to the fact that analytical continuation of functions into the lower half-plane is an ill-posed problem for which small variations in the initial data may cause arbitrarily large variations in the solution. To solve ill-posed problems numerically, it is necessary to employ special regularizing algorithms. In particular, in downward analytical continuation of plane fields, these regularizing algorithms can be synthesized on the basis of the integral formulas derived in this subsection, but these issues are beyond the scope of this book.

On the whole, the theory of solution of ill-posed problems is nowadays a separate branch of mathematics. There are special papers and monographs that deal with methods of regularization (see, for instance, Tikhonov and Arsenin 1974; Strakhov 1969a; Dmitriev and Zhdanov 1982, etc.). Some examples of developing regularizing algorithms for analytical continuation will be given below for electromagnetic (Sect. 9.3.4) and seismic (Sect. 12.1.4) fields.

Part II
Cauchy-Type Integral Analogs in the Theory of a Three-Dimensional Geopotential Field

5 Three-Dimensional Cauchy-Type Integral Analogs

5.1 Three-Dimensional Analog of the Cauchy Integral Formula

The previous section was concerned with the main concepts of analytical theory of two-dimensional geopotential fields based on the body of the Cauchy-type integral. But two-dimensional (plane) fields can serve just as an approximate model of actual geophysical fields dependent on three space coordinates. In view of this, an extremely important problem of theoretical geophysics is how to extend the results of a two-dimensional theory to a three-dimensional case. This section aims precisely at solving this problem.

5.1.1 Vector Statements of the Ostrogradsky-Gauss Theorem

The basis for the analytical body used to study three-dimensional geopotential fields is provided by the Ostrogradsky-Gauss formula stated for space domains bounded by smooth or piecewise smooth surfaces. It is worth remembering the concept of a smooth surface.

Definition 5.1. Surface S in space is called smooth, provided that:

a) a tangential plane and, hence, a normal can be passed to it at any point of the surface S, and on transition from point to point this normal changes continuously (without jumps);
b) there is a fixed positive number $\delta_0 > 0$ so that for any point A on the surface S a sphere $O_\delta(A)$ can be drawn therefrom as a center with a radius $\delta (\delta < \delta_0)$ and exhibiting the following property: all straight lines parallel to the normal passed to the surface S at the point A lying inside the sphere $O_\delta(A)$ intersect the surface S only once (Fig. 29).

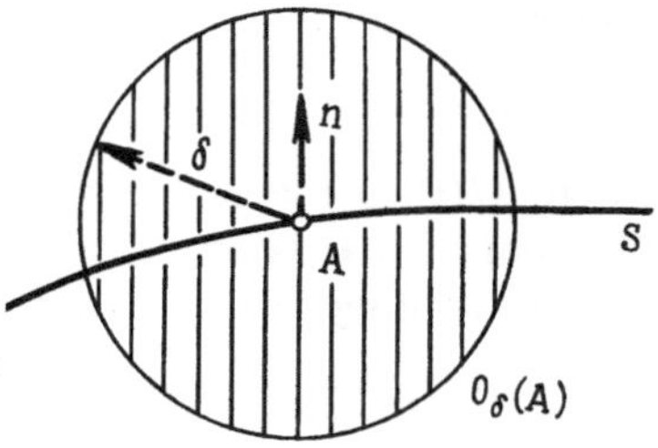

Fig. 29. At a smooth surface, all straight lines parallel to the normal **n** passed to the surface S at a point A lying with some neighborhood $O_\delta(A)$ intersect the surface S only once

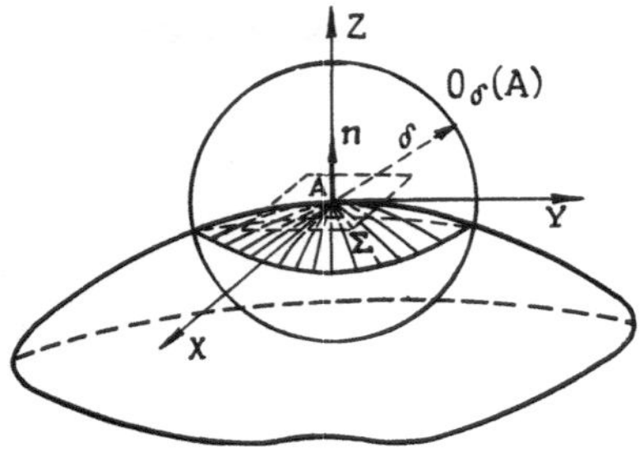

Fig. 30. Coordinate system adjusted to the tangential plane and the normal to a smooth surface

The condition of smoothness of the surface S permits, according to requirement (a), to relate each point A of the surface S to a special orthogonal system of coordinates x, y, z, whose origin is at point A, the z-axis is directed along the normal $\mathbf{n}$, while the x- and y-axes lie in the tangential plane (Fig. 30). Requirement (b) indicates that within this coordinate system, the equation for the part of surface Σ which is found inside the sphere $O_\delta(A)$ can be represented in a form resolvable for $z: z = f(x, y)$, where the function f is continuously differentiable (since, according to (a), S has everywhere a normal and it is continuous).

It is essential that in the solution of certain specific problems the function $f(x, y)$ describing a smooth surface S should be continuously differentiable, several times rather than once. Therefore, in our subsequent discussion of a smooth surface it will be understood that the function $f(x, y)$ is continuously differentiable a desired number of times.

Definition 5.2. Surface S is called piecewise smooth if it is made up of a finite number of smooth parts.

Assume now that the piecewise smooth surface S bounds a finite domain D of space. Let $\mathbf{F}(\mathbf{r})$ be a vector function continuously differentiable everywhere in the domain D up to its boundary. Then, *according to the Ostrogradsky-Gauss theorem, the following formula holds good:*

$$\iiint_D \operatorname{div} \mathbf{F}\, dv = \iint_S \mathbf{F} \cdot \mathbf{n}\, ds \;, \tag{5.1.1}$$

where $\mathbf{n}$ is the unit vector of an outward pointing normal to S. Proof of this formula is available in any treatise on mathematical analysis. It is known as *the first vector statement of the Ostrogradsky-Gauss theorem.*

Other equivalent vector statements of the Ostrogradsky-Gauss theorem can be also presented here. Assume, for example, that

$$\mathbf{F}(\mathbf{r}) = U(\mathbf{r})\mathbf{C} \;, \tag{5.1.2}$$

where $U(\mathbf{r})$ is the scalar function continuously differentiable in D (up to its boundary S); $\mathbf{C}$ is an arbitrary constant vector.

Substituting (5.1.2) into (5.1.1) we find

$$\mathbf{C} \cdot \iiint_D \operatorname{grad} U\, dv = \mathbf{C} \cdot \iint_S U \mathbf{n}\, ds \;, \tag{5.1.3}$$

whence we have that

$$\mathbf{C}\cdot\left(\iiint\limits_{D} \operatorname{grad} U dv - \iint\limits_{S} U\mathbf{n} ds\right) = 0 \ . \tag{5.1.4}$$

Since the vector **C** is chosen arbitrarily, Eq. (5.1.4) directly implies that

$$\iiint\limits_{D} \operatorname{grad} U dv = \iint\limits_{S} U\mathbf{n} ds \ . \tag{5.1.5}$$

Formula (5.1.5) is referred to as *the second vector statement of the Ostrogradsky-Gauss theorem. The third vector statement* of the Ostrogradsky-Gauss formula results from the following substitution:

$$\mathbf{F}(\mathbf{r}) = \mathbf{C}\times\mathbf{P}(\mathbf{r}) \ , \tag{5.1.6}$$

where $\mathbf{P}(\mathbf{r})$ is the vector function continuously differentiable in D (up to its boundary S); **C** is, as in the above case, an arbitrary constant vector.

Substituting (5.1.6) into (5.1.1) and omitting simple transformations similar to those presented above, we write

$$\iiint\limits_{D} \operatorname{rot} \mathbf{P} dv = -\iint\limits_{S} \mathbf{P}\times\mathbf{n} ds \ . \tag{5.1.7}$$

These three equivalent vector statements of the Ostrogradsky-Gauss theorem are extremely important in the theory of geophysical fields.

5.1.2 Vector Statements of the Stokes Theorem

Let S be a piecewise smooth two-side surface bounded by a piecewise smooth contour L, and $\mathbf{F}(\mathbf{r})$ be a vector function continuously differentiable everywhere on S up to its boundary L. We will specify a positive direction of the unit vector of the normal $\mathbf{n}$ to the surface S, and a positive direction of the unit vector of the tangent $\boldsymbol{\tau}$ to the contour L, so that these directions might be consistent with the corkscrew rule, i.e., with a corkscrew rotating in the direction $\boldsymbol{\tau}$ the point might move in the direction $\mathbf{n}$. Then, *according to the Stokes theorem,* the following formula

$$\iint\limits_{S} \operatorname{rot} \mathbf{F}\cdot\mathbf{n} ds = \int\limits_{L} \mathbf{F}\cdot\boldsymbol{\tau} dl \tag{5.1.8}$$

is valid. Formula (5.1.8) is termed *the first vector statement of the Stokes theorem.* Its proof is provided in the corresponding courses of vector analysis.

Let us specify the field **F** by relation (5.1.2):

$$\mathbf{F}(\mathbf{r}) = U(\mathbf{r})\cdot\mathbf{C} \ , \tag{5.1.9}$$

where **C** is an arbitrary constant vector. Substituting (5.1.9) into (5.1.8) and performing a little transformation, we find

$$-\mathbf{C}\iint\limits_S (\operatorname{grad} U\times\mathbf{n})\,ds = \mathbf{C}\int\limits_L U\boldsymbol{\tau}\,dl\ , \tag{5.1.10}$$

whence, by virtue of the arbitrariness of the vector $\mathbf{C}$ selection, we have

$$-\iint\limits_S (\operatorname{grad} U\times\mathbf{n})\,ds = \int\limits_L U\boldsymbol{\tau}\,dl\ . \tag{5.1.11}$$

Formula (5.1.11) is called *the second vector statement of the Stokes theorem.*

5.1.3 Analog of the Cauchy-Type Integral

Now let us derive a relation that is a three-dimensional analog of the Cauchy integral formula. To this end, we will represent the vector function $\boldsymbol{\Phi}(\mathbf{n})$ in the form

$$\boldsymbol{\Phi}(\mathbf{r}) = (\mathbf{C}\cdot\operatorname{grad} h)\operatorname{grad} P + \operatorname{grad} P\times[\operatorname{grad} h\times\mathbf{C}]\ , \tag{5.1.12}$$

where $h(\mathbf{r})$ and $P(\mathbf{r})$ are arbitrary functions twice differentiable continuously in D (up to its boundary S); $\mathbf{C}$ is an arbitrary constant vector. Calculations have revealed that

$$\operatorname{div}\boldsymbol{\Phi} = \mathbf{C}\cdot(\Delta P\operatorname{grad} h + \Delta h\operatorname{grad} P)\ , \tag{5.1.13a}$$

$$\begin{aligned}\boldsymbol{\Phi}\cdot\mathbf{n} &= \mathbf{C}\cdot[(\mathbf{n}\cdot\operatorname{grad} P)\operatorname{grad} h + (\mathbf{n}\times\operatorname{grad} P)\times\operatorname{grad} h]\\ &= \mathbf{C}\cdot[(\mathbf{n}\cdot\operatorname{grad} h)\operatorname{grad} P + (\mathbf{n}\times\operatorname{grad} h)\times\operatorname{grad} P]\ .\end{aligned} \tag{5.1.13b}$$

Applying formula (5.1.1) to the vector function $\boldsymbol{\Phi}(\mathbf{r})$ and allowing for the fact, just as it was done previously, that the selection of the vector $\mathbf{C}$ is arbitrary, we arrive at the following two corollaries to the Ostrogradsky-Gauss theorem:

$$\begin{aligned}&\iiint\limits_D (\Delta P\operatorname{grad} h + \Delta h\operatorname{grad} P)\,dv\\ &\quad = \iint\limits_S [(\mathbf{n}\cdot\operatorname{grad} P)\operatorname{grad} h + (\mathbf{n}\times\operatorname{grad} P)\times\operatorname{grad} h]\,ds\ ,\end{aligned} \tag{5.1.14a}$$

$$\begin{aligned}&\iiint\limits_D (\Delta P\operatorname{grad} h + \Delta h\operatorname{grad} P)\,dv\\ &\quad = \iint\limits_S [(\mathbf{n}\cdot\operatorname{grad} h)\operatorname{grad} P + (\mathbf{n}\times\operatorname{grad} h)\times\operatorname{grad} P]\,ds\ .\end{aligned} \tag{5.1.14b}$$

We take in (5.1.14b) that h is the function harmonic in D and continuous, together with its derivative, up to the boundary S, and $P = G(\mathbf{r}-\mathbf{r}') = -1/4\pi\,|\mathbf{r}-\mathbf{r}'|$ is the Green fundamental function for the Laplace equation, i.e.,

$$\Delta h = 0 \quad \text{where} \quad \mathbf{r}\in D\ , \tag{5.1.15a}$$

$$\Delta P = \Delta G(\mathbf{r}-\mathbf{r}') = \delta(\mathbf{r}-\mathbf{r}')\ , \tag{5.1.15b}$$

where $\delta(\mathbf{r}-\mathbf{r}')$ is the singular delta-function of Dirac. Then we obtain

$$\frac{-1}{4\pi}\iint_S\left[(\mathbf{n}\cdot\operatorname{grad}h)\operatorname{grad}\frac{1}{|\mathbf{r}-\mathbf{r}'|}+(\mathbf{n}\times\operatorname{grad}h)\times\operatorname{grad}\frac{1}{|\mathbf{r}-\mathbf{r}'|}\right]ds$$
$$=\begin{cases}\operatorname{grad}h(\mathbf{r}')\,, & \mathbf{r}'\in D\,,\\ 0, & \mathbf{r}'\in C\bar{D}\,.\end{cases} \tag{5.1.16}$$

Here the differentiation and integration in the left-hand side of (5.1.16) are carried out with respect to $\mathbf{r}$, and $\mathbf{r}'$ is a fixed point. This formula is structurally similar to the Cauchy integral formula (1.1.1). Indeed, just as the latter, it enables us to reconstruct values of the fields inside the domain D from their values on the boundary of the domain, and outside D the integral vanishes.

This similarity makes it possible to call formula (5.1.16) a *three-dimensional analog of the Cauchy integral formula.* Moreover, it turns out that this seemingly formal resemblance reflects the deep inherent unity of these relations, since in a two-dimensional situation formula (5.1.16) reduces to (1.1.1).

5.1.4 Relationship Between the Three-Dimensional Analog and the Classical Cauchy Integral Formula

Let us introduce a right-handed Cartesian system of coordinates xyz. Assume that S is a cylindrical surface with a generator parallel to the y-axis and the function h is independent of y. Then, $\mathbf{n}=(n_x, 0, n_z)$, $\operatorname{grad}h=(h_x, 0, h_z)$ and formula (5.1.16) is cast in the form

$$-\frac{1}{4\pi}\iint_S\left[(\mathbf{n}\cdot\operatorname{grad}h)\operatorname{grad}\frac{1}{|\mathbf{r}-\mathbf{r}'|}+(\mathbf{n}\times\operatorname{grad}h)\times\operatorname{grad}\frac{1}{|\mathbf{r}-\mathbf{r}'|}\right]ds$$
$$=\frac{-\mathbf{d}_x}{4\pi}\int_L[(n_xh_x+n_zh_z)(x'-x)+(n_zh_x-n_xh_z)(z'-z)]$$
$$\times\int_{-\infty}^{+\infty}\frac{dy}{|\mathbf{r}-\mathbf{r}'|^3}\,dl-\frac{\mathbf{d}_z}{4\pi}\int_L[(n_xh_x+n_zh_z)(z'-z)+(n_xh_z-n_zh_x)(x'-x)]$$
$$\times\int_{-\infty}^{+\infty}\frac{dy}{|\mathbf{r}-\mathbf{r}'|^3}\,dl=B_x(\mathbf{r}')\mathbf{d}_x+B_z(\mathbf{r}')\mathbf{d}_z\,, \tag{5.1.17}$$

where $\mathbf{d}_x$ and $\mathbf{d}_z$ are the unit vectors of the x- and z-axes, respectively; L is the contour of the cross section of the surface by the plane xz;

$$B_{x(z)}(\mathbf{r}')=\begin{cases}h_{x(z)}(\mathbf{r}')\,, & \mathbf{r}'\in D\,,\\ 0, & \mathbf{r}'\in C\bar{D}\,.\end{cases} \tag{5.1.18}$$

As already noted in Sect. 2.4.1, the Cartesian components of the unit vector $\mathbf{n}(n_x, 0, n_z)$ of an outward normal to L are linked with the differentials dx and dz by simple relations (2.4.2):

$$n_xdl=dz\,,\quad n_zdl=-dx\,. \tag{5.1.19}$$

Besides, by virtue of (2.2.4), we have

$$\int_{-\infty}^{+\infty} \frac{dy}{|\mathbf{r}-\mathbf{r}'|^3} = \frac{2}{(x-x')^2+(z-z')^2} \ . \tag{5.1.20}$$

Substituting (5.1.19) and (5.1.20) into (5.1.17) we write

$$-\frac{1}{2\pi} \int_L \frac{(x-x')(h_z dx - h_x dz) + (z-z')(h_x dx + h_z dz)}{(x-x')^2+(z-z')^2}$$

$$= \begin{cases} h_x(x', z') \ , & (x', z') \in D \ , \\ 0, & (x', z') \in C\bar{D} \ ; \end{cases} \tag{5.1.21 a}$$

$$-\frac{1}{2\pi} \int_L \frac{(z-z')(h_z dx - h_x dz) + (x-x')(h_x dx + h_z dz)}{(x-x')^2+(z-z')^2}$$

$$= \begin{cases} h_z(x', z') \ , & (x', z') \in D \ , \\ 0, & (x', z') \in C\bar{D} \ . \end{cases} \tag{5.1.21 b}$$

Let us multiply the left- and the right-hand sides of Eq. (5.1.21 a) by i and add it up with (5.1.21 b) then a little transformation yields

$$-\frac{1}{2\pi} \int_L \frac{(-h_x + ih_z)[(x-x') - i(z-z')](dx+idz)}{(x-x')^2+(z-z')^2}$$

$$= \begin{cases} -i[-h_x(x', z') + ih_z(x', z')] \ , & (x', z') \in D \ , \\ 0 \ , & (x', z') \in C\bar{D} \ . \end{cases} \tag{5.1.22}$$

Introduce a complex variable $\zeta = x + iz$ on the plane xz and designate the following complex function by $f(\zeta)$:

$$f(\zeta) = -h_x(x, z) + ih_z(x, z) \ . \tag{5.1.23}$$

Evidently, the function $f(\zeta)$ is analytical in D and continuous in $\bar{D}$, since h_x and h_z are interrelated, due to the harmonicity of h, by the Cauchy-Riemann conditions.

Substitution of (5.1.23) into (5.1.22) yields the classical Cauchy integral formula

$$\frac{1}{2\pi i} \int_L \frac{-h_x + ih_z}{(x-x') + i(z-z')} d(x+iz) = \frac{1}{2\pi i} \int_L \frac{f(\zeta)}{\zeta - \zeta'} d\zeta = \begin{cases} f(\zeta') \ , & \zeta' \in D \ , \\ 0 \ , & \zeta' \in C\bar{D} \ . \end{cases} \tag{5.1.24}$$

Thus, (5.1.16) is a natural extension of the Cauchy formula to a three-dimensional case, hence, it can be termed a *three-dimensional analog of the Cauchy integral formula.*

We should like to note in conclusion of this subsection that (5.1.16) can be rewritten as follows: let $\mathbf{F}(\mathbf{r})$ be a potential vector field in a finite domain D, all sources of which are outside D, i.e., $\mathbf{F}$ is a Laplace field in D continuously differentiable up to the boundary S. Then postulating in (5.1.16) that

$$\operatorname{grad} h = \mathbf{F} \ , \tag{5.1.25}$$

we write

$$-\frac{1}{4\pi}\iint_S \left[(\mathbf{n}\cdot\mathbf{F})\operatorname{grad}\frac{1}{|\mathbf{r}-\mathbf{r}'|}+[\mathbf{n}\times\mathbf{F}]\times\operatorname{grad}\frac{1}{|\mathbf{r}-\mathbf{r}'|}\right]ds = \begin{cases}\mathbf{F}(\mathbf{r}') \ , & \mathbf{r}'\in D \ , \\ 0 \ , & \mathbf{r}'\in C\bar{D} \ ,\end{cases} \tag{5.1.26}$$

where $\mathbf{n}$ is the outward pointing normal with respect to the domain D.

Thus, formula (5.1.26) *solves the boundary problem for the Laplace vector field* $\mathbf{F}$.

5.1.5 Gauss Harmonic Function Theorem

Consider a particular case where $S = S_R$ is a spherical surface of radius R with its center at a point $\mathbf{r}_0$. Then formula (5.1.26) suggests

$$\mathbf{F}(\mathbf{r}_0) = -\frac{1}{4\pi}\iint_{S_R}\left[(\mathbf{n}\cdot\mathbf{F})\operatorname{grad}\frac{1}{|\mathbf{r}-\mathbf{r}_0|}+(\mathbf{n}\times\mathbf{F})\times\operatorname{grad}\frac{1}{|\mathbf{r}-\mathbf{r}_0|}\right]ds \ . \tag{5.1.27}$$

We will transform the integrand in (5.1.27), bearing in mind that on the surface of the sphere S_R

$$\mathbf{n} = \frac{\mathbf{r}-\mathbf{r}_0}{|\mathbf{r}-\mathbf{r}_0|} = \frac{1}{R}(\mathbf{r}-\mathbf{r}_0) \ , \tag{5.1.28 a}$$

$$\operatorname{grad}\frac{1}{|\mathbf{r}-\mathbf{r}_0|} = \frac{\mathbf{r}_0-\mathbf{r}}{|\mathbf{r}-\mathbf{r}_0|^3} = \frac{1}{R^3}(\mathbf{r}_0-\mathbf{r}) \ . \tag{5.1.28 b}$$

Then, we have

$$(\mathbf{n}\cdot\mathbf{F})\operatorname{grad}\frac{1}{|\mathbf{r}-\mathbf{r}_0|}+[\mathbf{n}\times\mathbf{F}]\times\operatorname{grad}\frac{1}{|\mathbf{r}-\mathbf{r}_0|}$$
$$= \frac{1}{R^4}\{[(\mathbf{r}-\mathbf{r}_0)\cdot\mathbf{F}](\mathbf{r}_0-\mathbf{r})-(\mathbf{r}-\mathbf{r}_0)[\mathbf{F}\cdot(\mathbf{r}_0-\mathbf{r})]$$
$$+\mathbf{F}[(\mathbf{r}-\mathbf{r}_0)\cdot(\mathbf{r}_0-\mathbf{r})]\} = -\frac{\mathbf{F}}{R^2} \ . \tag{5.1.29}$$

Consequently, formula (5.1.27) is written in the form

$$\mathbf{F}(\mathbf{r}_0) = \frac{1}{4\pi R^2}\iint_{S_R}\mathbf{F}(\mathbf{r})\,ds \ , \tag{5.1.30}$$

whence for the Cartesian components of the vector field **F** we obtain

$$F_{x(yz)}(\mathbf{r}_0)=\frac{1}{4\pi R^2}\iint_{S_R} F_{x(yz)}(\mathbf{r})\,ds\ . \tag{5.1.31}$$

Allowing for the fact that all the scalar components of the field **F** are, by virtue of (5.1.25), harmonic functions, relation (5.1.31) can be cast as follows:

$$U(\mathbf{r}_0)=\frac{1}{4\pi R^2}\iint_{S_R} U(\mathbf{r})\,ds\ , \tag{5.1.32}$$

where U is any function harmonic inside the sphere S_R, which is continuously differentiable in a closed domain $\bar{D}_R$ bounded by the sphere.

Relation (5.1.32), *implying the fact that the value of a harmonic function at the center of a sphere coincides with the arithmetic mean of its values over the sphere itself, is called the Gauss harmonic function theorem.*

5.1.6 Cauchy Formula Analog for an Infinite Domain

Assume now that $\mathbf{F}(\mathbf{r})$ is a Laplace field everywhere in an infinite domain $C\bar{D}$ (including a point at infinity) which is continuously differentiable up to the boundary S. Then it is easy to derive a formula of type (5.1.26) also for an infinite domain $C\bar{D}$. Indeed, let us fix on an arbitrary point $\mathbf{r}'$ and draw therefrom, as a center, a sphere $O_R(\mathbf{r}')$ of such a large radius R that the domain D might prove completely enclosed in $O_R(\mathbf{r}')$. Let T_R stand for the domain found between $O_R(\mathbf{r}')$ and S (Fig. 31). We will apply formula (5.1.26) to the domain T_R:

$$-\frac{1}{4\pi}\iint_S\left[(\mathbf{v}\cdot\mathbf{F})\,\mathrm{grad}\,\frac{1}{|\mathbf{r}-\mathbf{r}'|}+(\mathbf{v}\times\mathbf{F})\times\mathrm{grad}\,\frac{1}{|\mathbf{r}-\mathbf{r}'|}\right]ds$$

$$-\frac{1}{4\pi}\iint_{O_R(\mathbf{r}')}\left[(\mathbf{v}\cdot\mathbf{F})\,\mathrm{grad}\,\frac{1}{|\mathbf{r}-\mathbf{r}'|}+(\mathbf{v}\cdot\mathbf{F})\times\mathrm{grad}\,\frac{1}{|\mathbf{r}-\mathbf{r}'|}\right]ds$$

$$=\begin{cases}\mathbf{F}(\mathbf{r}')\ , & \mathbf{r}'\in T_R\ ,\\ \mathbf{0}\ , & \mathbf{r}'\in D\ ,\end{cases} \tag{5.1.33}$$

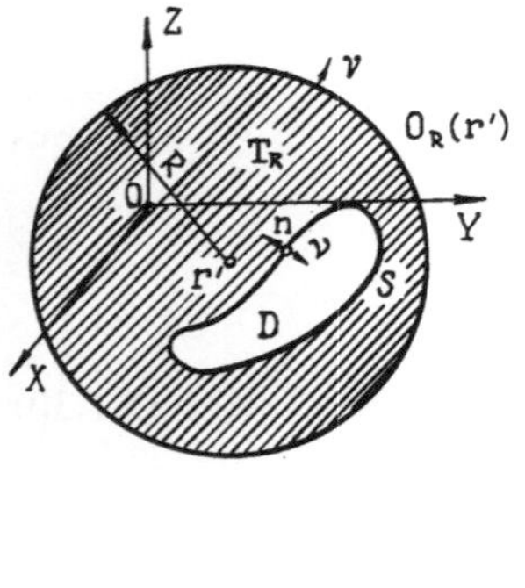

Fig. 31. Derivation of a three-dimensional analog of the Cauchy formula for an infinite domain

where $\mathbf{v}$ is the normal to the surfaces $O_R(\mathbf{r}')$ and S which is outward pointing with respect to the domain T_R. On the surface of the sphere $O_R(\mathbf{r}')$, relations similar to (5.1.28a) and (5.1.28b)

$$\mathbf{v} = (1/R)(\mathbf{r}-\mathbf{r}') \ , \tag{5.1.34a}$$

$$\operatorname{grad} \frac{1}{|\mathbf{r}-\mathbf{r}'|} = \frac{1}{R^3}(\mathbf{r}'-\mathbf{r}) \tag{5.1.34b}$$

are fulfilled.

With due references to these relations, just as it was done in Sect. 5.1.5, the integral taken over $O_R(\mathbf{r}')$ can be transformed as follows:

$$-\frac{1}{4\pi}\iint\limits_{O_R(\mathbf{r}')}\left[(\mathbf{v}\cdot\mathbf{F})\operatorname{grad}\frac{1}{|\mathbf{r}-\mathbf{r}'|}+[\mathbf{v}\times\mathbf{F}]\times\operatorname{grad}\frac{1}{|\mathbf{r}-\mathbf{r}'|}\right]ds$$

$$=\frac{1}{4\pi R^2}\iint\limits_{O_R(\mathbf{r}')}\mathbf{F}(\mathbf{r})\,ds=\mathbf{I}_R \ . \tag{5.1.35}$$

Now we will assume that $\mathbf{F}(\mathbf{r})$ for $|\mathbf{r}|\to\infty$ tends to a limit $\mathbf{F}(\infty)$ uniformly in $\mathbf{r}$: $\lim\limits_{|\mathbf{r}|\to\infty}\mathbf{F}(\mathbf{r})=\mathbf{F}(\infty)$. Then we have

$$\lim_{R\to\infty}\mathbf{I}_R=\mathbf{F}(\infty) \ . \tag{5.1.36}$$

Thus, passing to the limit for $R\to\infty$ in (5.1.33) we derive in the final form

$$-\frac{1}{4\pi}\iint\limits_{S}\left[(\mathbf{n}\cdot\mathbf{F})\operatorname{grad}\frac{1}{|\mathbf{r}-\mathbf{r}'|}+(\mathbf{n}\times\mathbf{F})\times\operatorname{grad}\frac{1}{|\mathbf{r}-\mathbf{r}'|}\right]ds$$

$$=\begin{cases}-\mathbf{F}(\mathbf{r}')+\mathbf{F}(\infty) \ , & \mathbf{r}'\in C\bar{D} \ ,\\ \mathbf{F}(\infty) \ , & \mathbf{r}'\in D \ ,\end{cases} \tag{5.1.37}$$

where it is taken into account that $\mathbf{v}=-\mathbf{n}$. Formula (5.1.37) is a *three-dimensional analog of the Cauchy-formula* (1.1.2) *for an infinite domain.*

5.1.7 Three-Dimensional Analog of the Pompei Formulas

In conclusion of this subsection, we will derive an important relation for a potential field which can be regarded as an extension of the Pompei formulas to a three-dimensional case (see Sect. 2.4.2).

Let $\mathbf{F}$ be a potential field in the domain D, satisfying the equations

$$\operatorname{div}\mathbf{F}=q \ , \quad \operatorname{rot}\mathbf{F}=0 \ , \quad \mathbf{r}\in D \ . \tag{5.1.38}$$

Positing in (5.1.14a) and (5.1.14b) that

$$\operatorname{grad}h=\mathbf{F} \ , \quad P=-1/4\pi|\mathbf{r}-\mathbf{r}'| \ , \tag{5.1.39}$$

we obtain

$$-\frac{1}{4\pi}\iint_S\left[(\mathbf{n}\cdot\mathbf{F})\,\mathrm{grad}\,\frac{1}{|\mathbf{r}-\mathbf{r}'|}+(\mathbf{n}\times\mathbf{F})\times\mathrm{grad}\,\frac{1}{|\mathbf{r}-\mathbf{r}'|}\right]ds$$

$$+\frac{1}{4\pi}\iiint_D \mathrm{div}\,\mathbf{F}\cdot\mathrm{grad}\,\frac{1}{|\mathbf{r}-\mathbf{r}'|}\,dv=\begin{cases}\mathbf{F}(\mathbf{r}')\,, & \mathbf{r}'\in D\,,\\ \mathbf{0}\,, & \mathbf{r}'\in C\bar{D}\,;\end{cases} \tag{5.1.40a}$$

or

$$-\frac{1}{4\pi}\iint_S\left[\left(\mathbf{n}\cdot\mathrm{grad}\,\frac{1}{|\mathbf{r}-\mathbf{r}'|}\right)\mathbf{F}+\left(\mathbf{n}\times\mathrm{grad}\,\frac{1}{|\mathbf{r}-\mathbf{r}'|}\right)\times\mathbf{F}\right]ds$$

$$+\frac{1}{4\pi}\iiint_D \mathrm{div}\,\mathbf{F}\cdot\mathrm{grad}\,\frac{1}{|\mathbf{r}-\mathbf{r}'|}\,dv=\begin{cases}\mathbf{F}(\mathbf{r}')\,, & \mathbf{r}'\in D\,,\\ \mathbf{0}\,, & \mathbf{r}'\in C\bar{D}\,.\end{cases} \tag{5.1.40b}$$

Formulas (5.1.40a) *and* (5.1.40b) *are just three-dimensional analogs of the Pompei formulas.* These formulas yield a solution to Eqs. (5.1.38) in the domain *D*:

$$\mathbf{F}(\mathbf{r}')=-\frac{1}{4\pi}\iint_S\left[(\mathbf{n}\cdot\mathbf{F})\,\mathrm{grad}\,\frac{1}{|\mathbf{r}-\mathbf{r}'|}+[\mathbf{n}\times\mathbf{F}]\times\mathrm{grad}\,\frac{1}{|\mathbf{r}-\mathbf{r}'|}\right]ds$$

$$+\frac{1}{4\pi}\iiint_D q(\mathbf{r})\,\mathrm{grad}\,\frac{1}{|\mathbf{r}-\mathbf{r}'|}\,dv\,,\quad \mathbf{r}'\in D\,, \tag{5.1.41a}$$

$$\mathbf{F}(\mathbf{r}')=-\frac{1}{4\pi}\iint_S\left[\left(\mathbf{n}\cdot\mathrm{grad}\,\frac{1}{|\mathbf{r}-\mathbf{r}'|}\right)\mathbf{F}+\left(\mathbf{n}\times\mathrm{grad}\,\frac{1}{|\mathbf{r}-\mathbf{r}'|}\right)\times\mathbf{F}\right]ds$$

$$+\frac{1}{4\pi}\iiint_D q(\mathbf{r})\,\mathrm{grad}\,\frac{1}{|\mathbf{r}-\mathbf{r}'|}\,dv\,,\quad \mathbf{r}'\in D\,. \tag{5.1.41b}$$

If the field $\mathbf{F}$ is a Laplace field in *D*, i.e., $\mathrm{div}\,\mathbf{F}=0$, $\mathrm{rot}\,\mathbf{F}=\mathbf{0}$, formula (5.1.40a) directly yields a three-dimensional Cauchy integral formula (5.1.16).

5.2 Definition and Properties of the Three-Dimensional Cauchy Integral Analog

In this subsection we will define the concept of a three-dimensional analog of the Cauchy-type integral and show that its properties are similar in many respects to those of the classical Cauchy integral studied in Sect. 1.2. Just as in the classical case, of critical importance is the behavior of the Cauchy integral analog at the integration surface itself or near it.

5.2.1 Concept of a Three-Dimensional Cauchy Integral Analog

A three-dimensional analog of the Cauchy-type integral is elaborated on the basis of formulas (5.1.16). We should like to recall, first of all, how the operators of

Fig. 32. Right triad of mutually orthogonal vectors **n**, **v**, and **l**

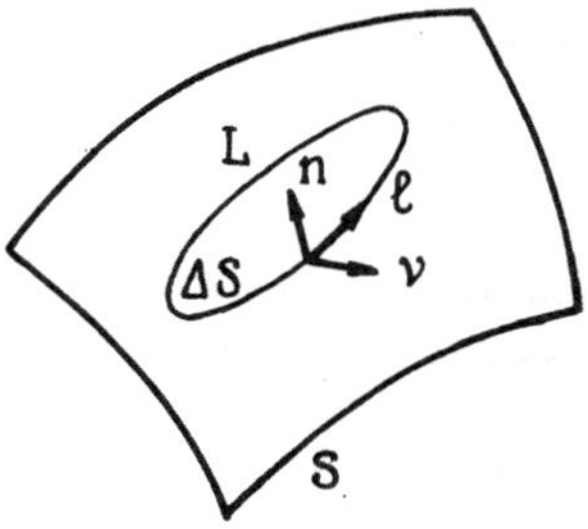

two-dimensional differentiation (grad_S, div_S, and rot_S) of fields specified at the surface are defined [compare with formulas (2.1.5) through (2.1.17) of Sect. 2.1.3]:

$$\mathrm{grad}_S h = \lim_{\Delta S \to 0} \frac{1}{\Delta S} \oint_L h \mathbf{v}\, dl \, ,$$

$$\mathrm{div}_S \mathbf{h}_\tau = \lim_{\Delta S \to 0} \frac{1}{\Delta S} \oint_L \mathbf{h}_\tau \cdot \mathbf{v}\, dl \, , \tag{5.2.1}$$

$$\mathrm{rot}_S \mathbf{h}_\tau = \lim_{\Delta S \to 0} \frac{\mathbf{n}}{\Delta S} \oint_L \mathbf{h}_\tau \cdot \mathbf{l}\, dl \, ,$$

where h is the scalar field; $\mathbf{h}_\tau$ is the vector field tangential with respect to the surface S; L is the contour bounding the domain ΔS of the surface S: $\mathbf{n}$ is the unit vector of the normal to S; $\mathbf{v}$ is the unit vector of the normal to L lying on S; $\mathbf{l}$ is the unit vector tangent to L; vectors $\mathbf{n}$, $\mathbf{v}$, and $\mathbf{l}$ form a right triad, i.e., $\mathbf{n} \times \mathbf{v} = \mathbf{l}$ (Fig. 32).

Let a continuously differentiable vector field $\boldsymbol{\varphi} = \boldsymbol{\varphi}_n + \boldsymbol{\varphi}_\tau$ be specified on S and satisfy the conditions[1]

$$\mathrm{rot}_S \boldsymbol{\varphi}_\tau \equiv 0 \, , \tag{5.2.2}$$

where $\boldsymbol{\varphi}_\tau$ is the tangential part of $\boldsymbol{\varphi}$; $\boldsymbol{\varphi}_n$ is the normal part of $\boldsymbol{\varphi}$. We will write the following expression:

$$\mathbf{C}^S(\mathbf{r}', \boldsymbol{\varphi}) = \frac{-1}{4\pi} \iint_S \left[(\mathbf{n} \cdot \boldsymbol{\varphi}) \,\mathrm{grad} \frac{1}{|\mathbf{r}-\mathbf{r}'|} + (\mathbf{n} \times \boldsymbol{\varphi}) \times \mathrm{grad} \frac{1}{|\mathbf{r}-\mathbf{r}'|} \right] ds \, . \tag{5.2.3}$$

Evidently, (5.2.3) has meaning at all points of space not belonging to S. Let us transform (5.2.3) using the obvious equation

$$\mathrm{grad} \frac{1}{|\mathbf{r}-\mathbf{r}'|} = -\mathrm{grad}' \frac{1}{|\mathbf{r}-\mathbf{r}'|}$$

[1] As shown in Sect. A.3 of Appendix A, to define the Cauchy-type integral analog as a function $\boldsymbol{\varphi}$ we can take any function continuous on S and satisfying the condition

$$\iint_S \mathrm{grad} \frac{1}{|\mathbf{r}-\mathbf{r}'|} \cdot [\mathbf{n} \times \boldsymbol{\varphi}] \, ds = 0 \, .$$

and factoring grad′ outside the integral sign[2]:

$$\mathbf{C}^S(\mathbf{r}', \boldsymbol{\varphi}) = \frac{1}{4\pi}\,\mathrm{grad}' \iint_S \frac{\mathbf{n}\cdot\boldsymbol{\varphi}_n}{|\mathbf{r}-\mathbf{r}'|}\,ds - \frac{1}{4\pi}\,\mathrm{rot}' \iint_S \frac{\mathbf{n}\times\boldsymbol{\varphi}_\tau}{|\mathbf{r}-\mathbf{r}'|}\,ds\,. \tag{5.2.4}$$

Expression (5.2.4) makes it possible to write equations for the vector field $\mathbf{C}^S$. Indeed, the divergence of $\mathbf{C}^S$ is

$$\mathrm{div}'\mathbf{C}^S = \frac{1}{4\pi}\iint_S (\mathbf{n}\cdot\boldsymbol{\varphi}_n)\Delta' \frac{1}{|\mathbf{r}-\mathbf{r}'|}\,ds \equiv 0 \quad \text{for} \quad \mathbf{r}' \notin S\,. \tag{5.2.5}$$

Now calculate the $\mathrm{rot}\,\mathbf{C}^S$:

$$\mathrm{rot}'\mathbf{C}^S = -\frac{1}{4\pi}\,\mathrm{rot}'\mathrm{rot}' \iint_S \frac{\mathbf{n}\times\boldsymbol{\varphi}_\tau}{|\mathbf{r}-\mathbf{r}'|}\,ds$$

$$= -\frac{1}{4\pi}\,\mathrm{grad}' \iint_S \mathrm{div}' \frac{\mathbf{n}\times\boldsymbol{\varphi}_\tau}{|\mathbf{r}-\mathbf{r}'|}\,ds \quad \text{for} \quad \mathbf{r}' \notin S\,. \tag{5.2.6}$$

At the same time, we have

$$\mathrm{div}'\frac{(\mathbf{n}\times\boldsymbol{\varphi}_\tau)}{|\mathbf{r}-\mathbf{r}'|} = \mathrm{grad}'\frac{1}{|\mathbf{r}-\mathbf{r}'|}\cdot(\mathbf{n}\times\boldsymbol{\varphi}_\tau) = -\mathrm{grad}_S\frac{1}{|\mathbf{r}-\mathbf{r}'|}\cdot(\mathbf{n}\times\boldsymbol{\varphi}_\tau)$$

$$= \frac{\mathrm{div}_S(\mathbf{n}\times\boldsymbol{\varphi}_\tau)}{|\mathbf{r}-\mathbf{r}'|} - \mathrm{div}_S\left[\mathbf{n}\times\frac{\boldsymbol{\varphi}_\tau}{|\mathbf{r}-\mathbf{r}'|}\right]. \tag{5.2.7}$$

By virtue of (5.2.1) and (5.2.2), we have

$$\mathrm{div}_S(\mathbf{n}\times\boldsymbol{\varphi}_\tau) = \lim_{\Delta S\to 0}\frac{1}{\Delta S}\oint_L (\mathbf{n}\times\boldsymbol{\varphi}_\tau)\cdot\mathbf{v}\,dl$$

$$= -\lim_{\Delta S\to 0}\frac{1}{\Delta S}\oint_L \boldsymbol{\varphi}_\tau\cdot\mathbf{l}\,dl = -\mathrm{rot}_S\boldsymbol{\varphi}_\tau\cdot\mathbf{n} \equiv 0\,. \tag{5.2.8}$$

In a similar way,

$$\mathrm{div}_S\left(\mathbf{n}\times\frac{\boldsymbol{\varphi}_\tau}{|\mathbf{r}-\mathbf{r}'|}\right) = -\mathrm{rot}_S\left(\frac{\boldsymbol{\varphi}_\tau}{|\mathbf{r}-\mathbf{r}'|}\right)\cdot\mathbf{n} = -\mathrm{rot}\left(\frac{\boldsymbol{\varphi}_\tau}{|\mathbf{r}-\mathbf{r}'|}\right)\cdot\mathbf{n}\,. \tag{5.2.9}$$

Substitution of (5.2.8) and (5.2.9) into (5.2.7), and subsequently into (5.2.6), yields

$$\mathrm{rot}'\mathbf{C}^S = \frac{1}{4\pi}\,\mathrm{grad}' \iint_S \mathrm{rot}\,\frac{\boldsymbol{\varphi}_\tau}{|\mathbf{r}-\mathbf{r}'|}\cdot\mathbf{n}\,ds = \mathbf{0} \quad \text{for} \quad \mathbf{r}' \notin S\,, \tag{5.2.10}$$

since, by virtue of the Stokes theorem (5.1.8), the surface integral in the right-hand side of (5.2.10) reduces to a curvilinear integral taken over the boundary S, with S being a closed surface.

Thus, everywhere outside S, the function $\mathbf{C}^S$ fits the Laplace vector field

---

[2] Hereinafter the prime used with the operators grad, div, rot, denotes differentiation with respect to $\mathbf{r}'$.

$$\operatorname{div}\mathbf{C}^S = 0 \ , \quad \operatorname{rot}\mathbf{C}^S = \mathbf{0} \tag{5.2.11}$$

and the scalar components of $\mathbf{C}^S$ are harmonic functions (generally different on the two sides of S). This property of expression (5.2.3) as well as the fact that in a two-dimensional situation the right-hand side of (5.2.3) reduces to a complex Cauchy-type integral enable (5.2.3) to be called a *three-dimensional analog of the Cauchy-type integral* and the function $\boldsymbol{\varphi}$ – its *vector density.*

In a particular case, where $\boldsymbol{\varphi}(\mathbf{r})$ stands for boundary values on S of the gradient of a function harmonic inside the domain D, we have, according to (5.1.26), that

$$\mathbf{C}^S(\mathbf{r}'; \boldsymbol{\varphi}) = \begin{cases} \boldsymbol{\varphi}(\mathbf{r}') \ , & \mathbf{r}' \in D \ , \\ \mathbf{0} \ , & \mathbf{r}' \in C\bar{D} \ , \end{cases} \tag{5.2.12}$$

where $C\bar{D}$ is an infinite domain complementing the closed domain $\bar{D}$ with respect to the whole space.

It is remarkable that the concept of the Cauchy integral analog is readily extendible to the case where the surface of integration tends to infinity. But it is necessary, just as in a two-dimensional case, (see Sect. 1.2.5), to study the behavior of functions – densities of the Cauchy-type integral – at infinity. The extension of the results of Sect. 1.2.5 to a three-dimensional case is presented in Sect. 5.3.4. It should be stressed that in the above definition of the Cauchy integral analog it is essential that S is a closed surface (which can be closed also through a point at infinity).

Below, there are given some simple properties of the Cauchy-type integral directly following from its definition (5.2.3):

a) If S_1 and S_2 are two surfaces, then we have
$\mathbf{C}^{S_1}(\mathbf{r}'; \boldsymbol{\varphi}) + \mathbf{C}^{S_2}(\mathbf{r}'; \boldsymbol{\varphi}) = \mathbf{C}^{S_1 \cup S_2}(\mathbf{r}'; \boldsymbol{\varphi})$;

b) If $\boldsymbol{\varphi}$ and $\boldsymbol{\psi}$ are two vector functions specified on S and satisfying condition (5.2.2), then we have
$\mathbf{C}^S(\mathbf{r}'; \boldsymbol{\varphi}) + \mathbf{C}^S(\mathbf{r}'; \boldsymbol{\psi}) = \mathbf{C}^S(\mathbf{r}'; \boldsymbol{\varphi} + \boldsymbol{\psi})$;

c) $\mathbf{C}^S(\mathbf{r}'; \lambda\boldsymbol{\varphi}) = \lambda\mathbf{C}^S(\mathbf{r}'; \boldsymbol{\varphi})$, where $\lambda = \text{const}$.

The properties (b) and (c) reflect the linearity of the Cauchy-type integral with respect to its density.

Concluding this subsection, we should like to note that the concept of a three-dimensional Cauchy integral analog was first introduced by Moisil and Theodoresco (1931) and Bitsadze (1953, 1972). The method of elaborating the Cauchy-type integral analogs outlined in the present book has been developed in works by the author (Zhdanov 1973b, 1974, 1975a, 1976a, b, 1979, 1980b). A close approach to this problem is also found in studies by Kolbenheyer (1976, 1978), Vargova (1977), and Sitarova (1977), where three-dimensional problems of potential theory are investigated in terms of four-component vectors, which are analytical, according to Moisil and Theodoresco (1931), (see also Bitsadze 1972).

Note finally that three-dimensional analogs of the Cauchy-type integrals can be also introduced in terms of the quaternion theory. This approach is outlined in Appendix A.

5.2.2 Evaluation of Singular Integrals in Terms of the Cauchy Principal Value

Cauchy-type integral (5.2.3) in cases where the point $\mathbf{r}' = \mathbf{r}_0$ lies on the surface S is an improper integral, since at $\mathbf{r}' = \mathbf{r}_0$ the integrand becomes infinity. This improper integral can be evaluated using a special procedure. Let S_ϱ stand for the part of the surface S outside the sphere O_ϱ of radius ϱ having its center at a point $\mathbf{r}_0$. The improper integral extended along the surface S can be evaluated as a limit (if any) of the corresponding integrals taken over surfaces S_ϱ with $\varrho \to 0$:

$$\mathbf{C}^S(\mathbf{r}_0; \varphi) = \lim_{\varrho \to 0} \mathbf{C}^{S_\varrho}(\mathbf{r}_0; \varphi) \ . \tag{5.2.13}$$

This limit will be called, following the two-dimensional case (Sect. 1.2.2), a *singular integral in terms of the Cauchy principal value.* The principal value of the integral will be denoted by the same symbol, thereby implying that if the integral has no ordinary sense we deal with its principal value. In this case, calculation formulas are determined by the type of point $\mathbf{r}_0 \in S$ at which the singular integral is evaluated.

We will assume in what follows that a piecewise smooth surface S consists of N smooth parts Γ_i $i = 1, 2, \dots$ N bounded by closed contours γ_i, respectively. Then all the points belonging to the interior of Γ_i seem to be ordinary points of the surface S, i.e., at these points there exist planes tangent to the surface. As to the points belonging to the contours γ_i, they may prove to be singular points of the surface S, i.e., at these points the surface may happen to have no tangent planes.

Consider some singular point $\mathbf{r}_0$ of the surface S which is a point of contact of smooth parts $\Gamma_{j_1}, \Gamma_{j_2}, \Gamma_{j_3}, \dots, \Gamma_{j_m}$. Within each of the parts Γ_{j_i} one can draw portions of smooth curves lying on S and ending at the point $\mathbf{r}_0$. Having constructed one-side tangent lines to all the drawn segments of smooth curves at the point $\mathbf{r}_0$, we obtain a conical surface with the vertex at the point $\mathbf{r}_0$. Designate the value of a solid angle (in steradians) formed by the obtained conical surface by $\theta(\mathbf{r}_0)$ (Fig. 33).

Let us formulate the following definitions. A singular point $\mathbf{r}_0$ will be called a *corner* if the angle $\theta(\mathbf{r}_0)$ differs from 0 or 4π. A singular point $\mathbf{r}_0$ will be termed a *cusp* if the angle $\theta(\mathbf{r}_0)$ is equal to zero or 4π. In the subsequent discussion, we will confine ourselves only to these two types of singular points. Note that sets of corners and cusps may form angular lines and edges of regression, respectively, on the surface S.

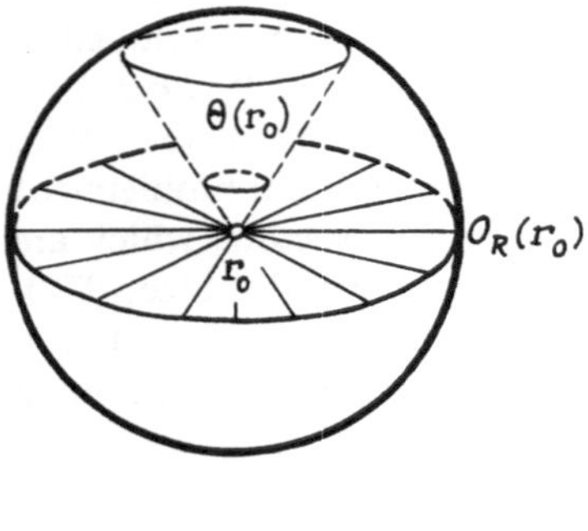

Fig. 33. Solid angle formed by one-side tangents to the surface S at a singular point $\mathbf{r}_0$

Fig. 34. Evaluation of a singular integral in terms of the principal Cauchy value

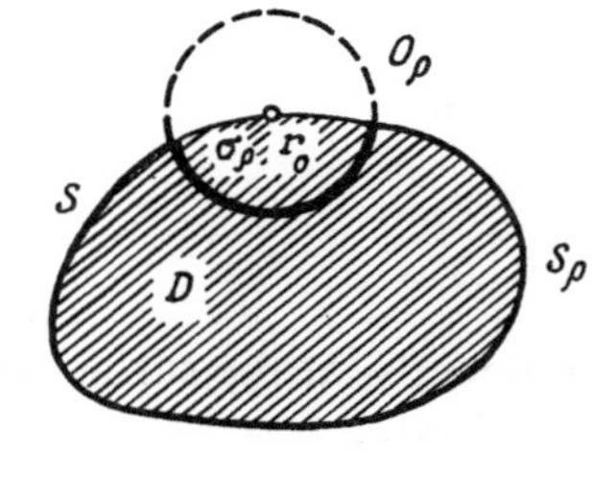

Now refer again to the problem of evaluating singular integral (5.2.13). It will be shown that for integral (5.2.13) in terms of the Cauchy principal value to exist for any $\mathbf{r}_0 \in S$ it is sufficient that the scalar components of the vector function $\boldsymbol{\varphi}(\mathbf{r})$, as functions of three variables, should satisfy the Hölder condition everywhere on S (see Sect. 1.2.1), i.e., there should exist positive constants $A_{\alpha\beta}$ and $\lambda_{\alpha\beta}$ ($\alpha, \beta = x, y, z$; $0 < \lambda_{\alpha,\beta} \leqslant 1$) so that

$$|\varphi_\alpha(x'', y'', z'') - \varphi_\alpha(x', y', z')| \leqslant A_{\alpha x}|x''-x'|^{\lambda_{\alpha x}} + A_{\alpha y}|y''-y'|^{\lambda_{\alpha y}} + A_{\alpha z}|z''-z'|^{\lambda_{\alpha z}} , \tag{5.2.14}$$

where $\alpha = x, y, z$; φ_x, φ_y, and φ_z are the Cartesian components of $\boldsymbol{\varphi}$.

Indeed, let us designate the part of the sphere O_ϱ belonging to the domain D (Fig. 34, the domain D is shaded) by σ_ϱ. The density $\boldsymbol{\psi}(\mathbf{r})$ of the Cauchy-type integral will be provided by a constant vector field equal to the value of $\boldsymbol{\varphi}$ at the point $\mathbf{r}_0$: $\boldsymbol{\psi}(\mathbf{r}) \equiv \boldsymbol{\varphi}(\mathbf{r}_0)$. Then, by virtue of (5.2.12) the Cauchy-type integral of a specified density $\boldsymbol{\psi}$, evaluated along the closed surface formed by S_ϱ and part of the sphere σ_ϱ, (Fig. 34) is zero: $\mathbf{C}^{S_\varrho \cup \sigma_\varrho}(\mathbf{r}_0; \boldsymbol{\psi}) = \mathbf{C}^{S_\varrho \cup \sigma_\varrho}(\mathbf{r}_0; \boldsymbol{\varphi}(\mathbf{r}_0)) = 0$.

As a result, the expression under the limit sign in (5.2.13) can be transformed, using the properties (a) and (b) of Sect. 5.2.1, as follows

$$\begin{aligned}\mathbf{C}^{S_\varrho}[\mathbf{r}_0; \boldsymbol{\varphi}(\mathbf{r})] &= \mathbf{C}^{S_\varrho}[\mathbf{r}_0; \boldsymbol{\varphi}(\mathbf{r})] - \mathbf{C}^{S_\varrho \cup \sigma_\varrho}[\mathbf{r}_0; \boldsymbol{\varphi}(\mathbf{r}_0)] \\ &= \mathbf{C}^{S_\varrho}[\mathbf{r}_0; \boldsymbol{\varphi}(\mathbf{r}) - \boldsymbol{\varphi}(\mathbf{r}_0)] - \mathbf{C}^{\sigma_\varrho}[\mathbf{r}_0; \boldsymbol{\varphi}(\mathbf{r}_0)] .\end{aligned} \tag{5.2.15}$$

The Hölder condition (5.2.14) ensures the existence of an ordinary uniformly convergent improper integral

$$\mathbf{C}^{S}[\mathbf{r}_0; \boldsymbol{\varphi}(\mathbf{r}) - \boldsymbol{\varphi}(\mathbf{r}_0)] = \lim_{\varrho \to 0} \mathbf{C}^{S_\varrho}[\mathbf{r}_0; \boldsymbol{\varphi}(\mathbf{r}) - \boldsymbol{\varphi}(\mathbf{r}_0)] . \tag{5.2.16}$$

Indeed, once $\mathbf{r}_0$ belongs to the interior of Γ_i, limit (5.2.16) exists as an ordinary improper integral, since $S_\varrho \to S$ for $\varrho \to 0$. But if the point $\mathbf{r}_0$ is a point of contact of m smooth parts $\Gamma_{j_1}, \Gamma_{j_2}, \Gamma_{j_3}, \ldots, \Gamma_{j_m}$, we will break up the integral under the limit sign into m integrals over different smooth parts $\Gamma_{j_1}, \Gamma_{j_2}, \ldots, \Gamma_{j_m}$ and, allowing for the Hölder condition, evaluate m ordinary improper integrals whose sum yields again relation (5.2.16).

Now evaluate the limit of the second integral in (5.2.15):

$$\lim_{\varphi\to 0} \mathbf{C}^{\sigma_\varrho}[\mathbf{r}_0;\ \varphi(\mathbf{r}_0)] = -\frac{1}{4\pi}\iint_{\sigma_\varrho}\left\{[\mathbf{n}\cdot\boldsymbol{\varphi}(\mathbf{r}_0)]\,\mathrm{grad}\,\frac{1}{|\mathbf{r}-\mathbf{r}_0|} + [\mathbf{n}\times\boldsymbol{\varphi}(\mathbf{r}_0)]\times\mathrm{grad}\,\frac{1}{|\mathbf{r}-\mathbf{r}_0|}\right\}ds\ . \qquad (5.2.17)$$

Evidently, at the surface of the part of the sphere σ_ϱ the unit vector of the outward normal is

$$\mathbf{n} = (\mathbf{r}-\mathbf{r}_0)/\varrho\ . \qquad (5.2.18\,\mathrm{a})$$

At the same time,

$$\mathrm{grad}\,\frac{1}{|\mathbf{r}-\mathbf{r}_0|} = \frac{\mathbf{r}_0-\mathbf{r}}{|\mathbf{r}-\mathbf{r}_0|^3} = \frac{1}{\varrho^3}(\mathbf{r}_0-\mathbf{r})\ . \qquad (5.2.18\,\mathrm{b})$$

Allowing for (5.2.18a) and (5.2.18b), transform the integrand in (5.2.17):

$$\begin{aligned}
&(\mathbf{n}\cdot\boldsymbol{\varphi}(\mathbf{r}_0))\,\mathrm{grad}\,\frac{1}{|\mathbf{r}-\mathbf{r}_0|} + [\mathbf{n}\times\boldsymbol{\varphi}(\mathbf{r}_0)]\times\mathrm{grad}\,\frac{1}{|\mathbf{r}-\mathbf{r}_0|}\\
&\quad= (\mathbf{n}\cdot\boldsymbol{\varphi}(\mathbf{r}_0))\,\mathrm{grad}\,\frac{1}{|\mathbf{r}-\mathbf{r}_0|} - \mathbf{n}\left(\boldsymbol{\varphi}(\mathbf{r}_0)\cdot\mathrm{grad}\,\frac{1}{|\mathbf{r}-\mathbf{r}_0|}\right)\\
&\qquad+ \boldsymbol{\varphi}(\mathbf{r}_0)\left(\mathbf{n}\cdot\mathrm{grad}\,\frac{1}{|\mathbf{r}-\mathbf{r}_0|}\right)\\
&\quad= \frac{1}{\varrho^4}\{[(\mathbf{r}-\mathbf{r}_0)\cdot\boldsymbol{\varphi}(\mathbf{r}_0)](\mathbf{r}_0-\mathbf{r}) - (\mathbf{r}-\mathbf{r}_0)[\boldsymbol{\varphi}(\mathbf{r}_0)\cdot(\mathbf{r}_0-\mathbf{r})] - \boldsymbol{\varphi}(\mathbf{r}_0)|\mathbf{r}-\mathbf{r}_0|^2\}\\
&\quad= -\frac{1}{\varrho^2}\boldsymbol{\varphi}(\mathbf{r}_0)\ . \qquad (5.2.19)
\end{aligned}$$

Substituting (5.2.19) into (5.2.17) we write

$$\lim_{\varrho\to 0} \mathbf{C}^{\sigma_\varrho}[\mathbf{r}_0;\ \boldsymbol{\varphi}(\mathbf{r}_0)] = \frac{1}{4\pi}\boldsymbol{\varphi}(\mathbf{r}_0)\lim_{\varrho\to 0}\frac{|\sigma_\varrho|}{\varrho^2}\ , \qquad (5.2.20)$$

where $|\sigma_\varrho|$ is the surface area of the part of the sphere σ_ϱ (see Fig. 34). But the limit of the ratio of the area of the part of the sphere $|\sigma_\varrho|$ to ϱ^2 for $\varrho\to 0$ is equal to the value of a solid angle in steradians:

$$\lim_{\varrho\to 0}\frac{|\sigma_\varrho|}{\varrho^2} = \theta(\mathbf{r}_0)\ .$$

Hence:

$$\lim_{\varrho\to 0} \mathbf{C}^{\sigma_\varrho}[\mathbf{r}_0;\ \boldsymbol{\varphi}(\mathbf{r}_0)] = \frac{1}{4\pi}\,\theta(\mathbf{r}_0)\boldsymbol{\varphi}(\mathbf{r}_0)\ . \qquad (5.2.21)$$

Proceeding now to the limit for $\varrho\to 0$ in (5.2.15) and taking account of (5.2.16) and (5.2.21), we obtain in the final form

$$\mathbf{C}^S[\mathbf{r}_0;\ \boldsymbol{\varphi}(\mathbf{r})] = \mathbf{C}^S[\mathbf{r}_0;\ \boldsymbol{\varphi}(\mathbf{r})-\boldsymbol{\varphi}(\mathbf{r}_0)]+\frac{1}{4\pi}\,\theta(\mathbf{r}_0)\boldsymbol{\varphi}(\mathbf{r}_0)\ . \tag{5.2.22}$$

Formula (5.2.22) permits evaluation of the Cauchy-type singular integral formula both at ordinary (regular) and at singular points of the surface S. In a particular case, if $\mathbf{r}_0$ is an ordinary point, $\theta(\mathbf{r}_0) = 2\pi$ and (5.2.22) takes the form

$$\mathbf{C}^S[\mathbf{r}_0;\ \boldsymbol{\varphi}(\mathbf{r})] = \mathbf{C}^S[\mathbf{r}_0;\ \boldsymbol{\varphi}(\mathbf{r})-\boldsymbol{\varphi}(\mathbf{r}_0)]+\tfrac{1}{2}\boldsymbol{\varphi}(\mathbf{r}_0)\ . \tag{5.2.23}$$

5.2.3 Three-Dimensional Analogs of the Sokhotsky-Plemelj Formulas

In the case of a Laplace vector field defined by the Cauchy-type integral, the very closed integration surface S is a singular surface. Just as in a two-dimensional case, of great value for potential theory is the behavior of Cauchy integral analogs on approaching the surface S from different sides. To solve this problem, we will designate an arbitrary fixed point at the surface S by $\mathbf{r}_0$ and represent integral (5.2.3) as follows:

$$\mathbf{C}^S[\mathbf{r}';\ \boldsymbol{\varphi}(\mathbf{r})] = \mathbf{A}(\mathbf{r}')+\mathbf{C}^S[\mathbf{r}';\ \boldsymbol{\varphi}(\mathbf{r}_0)]\ , \tag{5.2.24}$$

where $\mathbf{A}(\mathbf{r})$ is an auxiliary function defined by the expression

$$\mathbf{A}(\mathbf{r}') = \mathbf{C}^S[\mathbf{r}';\ \boldsymbol{\varphi}(\mathbf{r})-\boldsymbol{\varphi}(\mathbf{r}_0)]\ . \tag{5.2.25}$$

As is seen, the validity of representation (5.2.24) follows directly from the linearity of the Cauchy integral analog with respect to its density.

Just as in Sect. 1.2, we will denote the limit values of the functions $\mathbf{C}^S(\mathbf{r}';\ \boldsymbol{\varphi})$ and $\mathbf{A}(\mathbf{r}')$ by $\mathbf{C}^{S+}(\mathbf{r}_0;\ \boldsymbol{\varphi})$ and $\mathbf{A}^+(\mathbf{r}_0)$ for the point $\mathbf{r}'$ tending to $\mathbf{r}_0 \in S$ from the inside of a finite domain D bounded by S, while those for the point $\mathbf{r}'$ tending from the outside of this domain will be represented by $\mathbf{C}^{S-}(\mathbf{r}_0;\ \boldsymbol{\varphi})$ and $\mathbf{A}^-(\mathbf{r}_0)$. To stress the direction of passage to the limit, we will write $\mathbf{r}'\to\mathbf{r}_0^+$ and $\mathbf{r}'\to\mathbf{r}_0^-$, respectively. Using the three-dimensional analog of the Cauchy integral formula (5.1.16) [or (5.1.26)] and relation (5.2.22), we can write

$$\mathbf{C}^S(\mathbf{r}';\ \boldsymbol{\varphi}(\mathbf{r}_0)) = \begin{cases} \boldsymbol{\varphi}(\mathbf{r}_0)\ , & \mathbf{r}' \in D\ , \\ \mathbf{0}\ , & \mathbf{r}' \in CD\ , \\ \dfrac{1}{4\pi}\,\theta(\mathbf{r}_0)\boldsymbol{\varphi}(\mathbf{r}_0)\ , & \mathbf{r}' = \mathbf{r}_0 \in S\ . \end{cases} \tag{5.2.26}$$

Hence, it is evident that the integral $\mathbf{C}^S(\mathbf{r}';\ \boldsymbol{\varphi}(\mathbf{r}_0))$ has different limit values for $\mathbf{r}'\to\mathbf{r}_0$ from the outside and inside of the domain:

$$\lim_{\mathbf{r}'\to\mathbf{r}_0^+} \mathbf{C}^S[\mathbf{r}';\ \boldsymbol{\varphi}(\mathbf{r}_0)] = \boldsymbol{\varphi}(\mathbf{r}_0)\ , \quad \lim_{\mathbf{r}'\to\mathbf{r}_0^-} \mathbf{C}^S[\mathbf{r}';\ \boldsymbol{\varphi}(\mathbf{r}_0)] = \mathbf{0}\ . \tag{5.2.27}$$

Just as in a two-dimensional case, one can readily see that if the density of the Cauchy-type integral $\boldsymbol{\varphi}(\mathbf{r})$ satisfies the Hölder condition (5.2.14) on S, the vector function $\mathbf{A}(\mathbf{r})$ defined by formula (5.2.25) is continuous everywhere in space; in particular, it is continuous upon transition through the surface S, i.e., with $\mathbf{r}'$ approaching $\mathbf{r}_0$ on any side of the surface S, the function $\mathbf{A}(\mathbf{r}')$ has the same limit value equal, according to (5.2.22), to

$$\lim_{\mathbf{r}'\to\mathbf{r}_0^{\pm}} \mathbf{A}(\mathbf{r}') = \mathbf{A}(\mathbf{r}_0) = \mathbf{C}^S[\mathbf{r}_0;\ \boldsymbol{\varphi}(\mathbf{r})] - \frac{1}{4\pi}\,\theta(\mathbf{r}_0)\boldsymbol{\varphi}(\mathbf{r}_0)\ , \tag{5.2.28}$$

where $\mathbf{C}^S(\mathbf{r}_0;\ \boldsymbol{\varphi}(\mathbf{r}))$ is the singular integral in terms of the Cauchy principal value.

Considering the limits of expression (5.2.24) for $\mathbf{r}'$ approaching $\mathbf{r}_0$ along a path lying inside the domain D or along one outside the domain D, and allowing for (5.2.27) and (5.2.28), we obtain

$$\begin{aligned}\mathbf{C}^{S_+}[\mathbf{r}_0;\ \boldsymbol{\varphi}(\mathbf{r})] &= \mathbf{C}^S[\mathbf{r}_0;\ \boldsymbol{\varphi}(\mathbf{r})] + \left(1 - \frac{\theta(\mathbf{r}_0)}{4\pi}\right)\boldsymbol{\varphi}(\mathbf{r}_0)\ ,\\ \mathbf{C}^{S_-}[\mathbf{r}_0;\ \boldsymbol{\varphi}(\mathbf{r})] &= \mathbf{C}^S[\mathbf{r}_0;\ \boldsymbol{\varphi}(\mathbf{r})] - \frac{\theta(\mathbf{r}_0)}{4\pi}\boldsymbol{\varphi}(\mathbf{r}_0)\ .\end{aligned} \tag{5.2.29}$$

Thus, the three-dimensional analog of the Cauchy-type integral, with densities satisfying the Hölder condition, has limit values on approaching the surface S from each of its sides. But these limit values are different, so that there is a discontinuity on transition through the surface S. The magnitude of the discontinuity as follows from formulas (5.2.29), is equal fo the density $\boldsymbol{\varphi}(\mathbf{r}_0)$:

$$\mathbf{C}^{S_+}(\mathbf{r}_0;\ \boldsymbol{\varphi}) - \mathbf{C}^{S_-}(\mathbf{r}_0;\ \boldsymbol{\varphi}) = \boldsymbol{\varphi}(\mathbf{r}_0)\ . \tag{5.2.30}$$

Relation (5.2.30) holds good both at ordinary points at the surface S and at corners and cusps. Note also that (5.2.29) suggests the following representation for a singular integral in terms of the Cauchy principal value:

$$\mathbf{C}^S[\mathbf{r}_0;\ \boldsymbol{\varphi}(\mathbf{r})] = \tfrac{1}{2}[\mathbf{C}^{S_+}[\mathbf{r}_0;\ \boldsymbol{\varphi}(\mathbf{r})] + \mathbf{C}^{S_-}[(\mathbf{r}_0;\ \boldsymbol{\varphi}(\mathbf{r})]] + [\theta(\mathbf{r}_0)/4\pi - \tfrac{1}{2}]\boldsymbol{\varphi}(\mathbf{r}_0)\ . \tag{5.2.31}$$

Particularly, when $\mathbf{r}_0$ is an ordinary point of the surface S we have $\theta(\mathbf{r}_0) = 2\pi$, and (5.2.29) and (5.2.30) are written in the form

$$\mathbf{C}^{S_+}[\mathbf{r}_0;\ \boldsymbol{\varphi}(\mathbf{r})] = \mathbf{C}^S[\mathbf{r}_0;\ \boldsymbol{\varphi}(\mathbf{r})] + \tfrac{1}{2}\boldsymbol{\varphi}(\mathbf{r}_0)\ , \tag{5.2.32a}$$

$$\mathbf{C}^{S_-}[\mathbf{r}_0;\ \boldsymbol{\varphi}(\mathbf{r})] = \mathbf{C}^S[\mathbf{r}_0;\ \boldsymbol{\varphi}(\mathbf{r})] - \tfrac{1}{2}\boldsymbol{\varphi}(\mathbf{r}_0)\ ; \tag{5.2.32b}$$

$$\mathbf{C}^S[\mathbf{r}_0;\ \boldsymbol{\varphi}(\mathbf{r})] = \tfrac{1}{2}\{\mathbf{C}^S[\mathbf{r}_0;\ \boldsymbol{\varphi}(\mathbf{r})] + \mathbf{C}^{S_-}[\mathbf{r}_0;\ \boldsymbol{\varphi}(\mathbf{r}]\}\ . \tag{5.2.33}$$

Formulas (5.2.29) through (5.2.33) are three-dimensional analogs of the Sokhotsky-Plemelj formulas (1.2.21 a), (1.2.22 b), (1.2.28). They are as important in the theory of three-dimensional Cauchy-type integrals as are the classical formulas in the theory of functions of a complex variable.

5.3 Integral Transforms of the Laplace Vector Fields

The three-dimensional analogs of the Sokhotsky-Plemelj formulas make it possible to extend a variety of theorems of complex analysis relating to the properties of boundary values of analytical functions to a three-dimensional space (see Sect. 1.3).

5.3.1 Integral Boundary Conditions for the Laplace Field

Consider a domain D in space bounded by a smooth surface S, and specify an arbitrary continuous vector field $\boldsymbol{\varphi}$ on S. Is there a potential field $\mathbf{F}(\mathbf{r})$ in the domain $D(C\bar{D})$ whose sources are outside $\bar{D}$ (outside CD) and for which the specified field $\boldsymbol{\varphi}$ is a limit value at the surface S? In other words, is there a Laplace field $\mathbf{F}$ in $D(C\bar{D})$ whose boundary values are the function $\boldsymbol{\varphi}(\mathbf{r})$? This problem is similar to that of reconstructing a complex analytical function inside a plane domain Γ from the values of a continuous function on the boundary of the domain and it has, just as the latter, generally no solution.

Indeed, by the statement of the problem, the sought field in the domain D should satisfy the following equations:

$$\Delta\mathbf{F}(\mathbf{r}) = 0\ , \qquad \mathbf{r}\in D\ , \tag{5.3.1}$$

$$\operatorname{rot}\mathbf{F}(\mathbf{r}) = 0\ , \qquad \mathbf{r}\in D\ , \tag{5.3.2}$$

$$\mathbf{F}|_S = \boldsymbol{\varphi}\ . \tag{5.3.3}$$

We will introduce a Cartesian coordinate system (X, Y, Z) in space and write conditions (5.3.1) and (5.3.3) for the Cartesian components of the field:

$$\Delta F_x(\mathbf{r}) = 0 \quad \text{for} \quad \mathbf{r}\in D\ , \quad F_x|_S = \varphi_x\ , \tag{5.3.4a}$$

$$\Delta F_y(\mathbf{r}) = 0 \quad \text{for} \quad \mathbf{r}\in D\ , \quad F_y|_S = \varphi_y\ , \tag{5.3.4b}$$

$$\Delta F_z(\mathbf{r}) = 0 \quad \text{for} \quad \mathbf{r}\in D\ , \quad F_z|_S = \varphi_z\ . \tag{5.3.4c}$$

Evidently, relations (5.3.4a) through (5.3.4c) describe the Dirichlet problems for scalar components of the field and each of them can be solved independently of the others. However, with arbitrarily specified components φ_x, φ_y, and φ_z, the condition of field potentiality [relation (5.3.2)] is not to be satisfied with certainty. Indeed, define, for example, the function $\boldsymbol{\varphi}$ as follows:

$$\varphi_x = \frac{1}{|\mathbf{r}-\mathbf{r}_1|}\ , \quad \varphi_y = \frac{1}{|\mathbf{r}-\mathbf{r}_2|}\ , \quad \varphi_z = \frac{1}{|\mathbf{r}-\mathbf{r}_3|}\ ,$$

where $\mathbf{r}\in S$, and the points $\mathbf{r}_1$, $\mathbf{r}_2$, $\mathbf{r}_3$ are outside the domain $\bar{D}$. Then it is evident that the solution to a system of (5.3.4a) through (5.3.4c) is the following vector field:

$$\mathbf{F} = \left(\frac{1}{|\mathbf{r}-\mathbf{r}_1|}\ , \ \frac{1}{|\mathbf{r}-\mathbf{r}_2|}\ , \ \frac{1}{|\mathbf{r}-\mathbf{r}_3|}\right)\ , \tag{5.3.5}$$

which obviously does not satisfy Eq. (5.3.2). Thus, for the function $\boldsymbol{\varphi}(\mathbf{r})$ to be a boundary value of some field, the Laplace field in $D(C\bar{D})$ at the surface S, it should meet certain specific conditions at S. These conditions will be called boundary conditions for the Laplace field. Let us establish them. First of all, note the obvious fact that one of the necessary boundary conditions for the Laplace vector field is satisfaction of the relation (5.2.2), $\operatorname{rot}_S\boldsymbol{\varphi}_\tau \equiv \mathbf{0}$ on S. This prerequisite is, however, definitely insufficient, since the choice of the normal component of the field, φ_n, at S remains indeterminate.

Consider the Cauchy-type integral analog (5.2.3) with a Hölder continuous vector density $\boldsymbol{\varphi}(\mathbf{r})$. If $\boldsymbol{\varphi}(\mathbf{r})$ is the boundary value of the Laplace field $\mathbf{F}(\mathbf{r})$ in D, then, according to formula (5.2.12), we have

$$\mathbf{C}^S(\mathbf{r}';\ \boldsymbol{\varphi}) \equiv \mathbf{0} \quad \text{for all} \quad \mathbf{r}' \in C\bar{D} \ . \tag{5.3.6}$$

Now we will prove the sufficiency of condition (5.3.6) in conjunction with (5.2.2) for $\boldsymbol{\varphi}(\mathbf{r})$ to be a boundary value of the field $\mathbf{F}(\mathbf{r})$ or S, defined by

$$\mathbf{F}(\mathbf{r}') = \mathbf{C}^S(\mathbf{r}';\ \boldsymbol{\varphi}) \ , \quad \mathbf{r}' \in D \ , \tag{5.3.7}$$

which is the Laplace field in D and continuously extendible at S from D.

Indeed, in accordance with (5.3.6), the function $\mathbf{C}^S(\mathbf{r}',\ \boldsymbol{\varphi})$ is extendible continuously at S from the outside of the domain D:

$$\mathbf{C}^{S-}(\mathbf{r}_0;\ \boldsymbol{\varphi}) = \lim_{\mathbf{r}' \to \mathbf{r}_0^-} \mathbf{C}^S(\mathbf{r}';\ \boldsymbol{\varphi}) = 0 \ . \tag{5.3.8}$$

On the other hand, according to the analog of the Sokhotsky-Plemelj formulas (5.2.30), the function $\mathbf{F}(\mathbf{r}')$ is extendible continuously at S from the inside of the domain D, with

$$\lim_{\mathbf{r}' \to \mathbf{r}_0^+} \mathbf{F}(\mathbf{r}') = \mathbf{C}^{S+}(\mathbf{r}';\ \boldsymbol{\varphi}) = \boldsymbol{\varphi}(\mathbf{r}_0) \ . \tag{5.3.9}$$

Thus, we have proved the following theorem.

Theorem 5.3.1. For the Hölder continuous vector function, $\boldsymbol{\varphi}(\mathbf{r})$ specified at a closed smooth surface and having thereon differentiable tangential components, to be a boundary value of a vector field which is the Laplace field in D, it is necessary and sufficient that $\operatorname{rot}_S \boldsymbol{\varphi}_\tau \equiv 0$ and

$$\mathbf{C}^S(\mathbf{r}';\ \boldsymbol{\varphi}) = -\frac{1}{4\pi} \iint_S \left[(\mathbf{n}\cdot\boldsymbol{\varphi}) \operatorname{grad} \frac{1}{|\mathbf{r}-\mathbf{r}'|} + (\mathbf{n}\times\boldsymbol{\varphi}) \times \operatorname{grad} \frac{1}{|\mathbf{r}-\mathbf{r}'|} \right] ds = \mathbf{0} \tag{5.3.10}$$

for all $\mathbf{r}' \in C\bar{D}$.

Theorem 5.3.1 can be somewhat sharpened by passing to the limit for $\mathbf{r}' \to \mathbf{r}_0^-$ in formula (5.3.10) and by employing Sokhotsky-Plemelj formulas (5.2.32a) and (5.2.32b). Then the following theorem is valid.

Theorem 5.3.2. The necessary and sufficient condition that the vector function $\boldsymbol{\varphi}(\mathbf{r})$, which is Hölder continuous on a closed smooth surface S and has thereon differentiable tangential components, should serve as a boundary value of the vector field which is of the Laplace kind in D involves the following relations:

$$\operatorname{rot}_S \boldsymbol{\varphi}_\tau \equiv 0 \ ,$$

$$\begin{aligned} \boldsymbol{\varphi}(\mathbf{r}_0) &= -\frac{1}{2\pi} \iint_S \left[(\mathbf{n}\cdot\boldsymbol{\varphi}) \operatorname{grad} \frac{1}{|\mathbf{r}-\mathbf{r}_0|} + [\mathbf{n}\times\boldsymbol{\varphi}] \times \operatorname{grad} \frac{1}{|\mathbf{r}-\mathbf{r}_0|} \right] ds \\ &= 2\mathbf{C}^S(\mathbf{r}_0;\ \boldsymbol{\varphi}) \quad \text{for all} \quad \mathbf{r}_0 \in S. \end{aligned} \tag{5.3.11}$$

Proof. The necessity of condition (5.3.11) follows directly from relation (5.3.11) upon the limiting process for $\mathbf{r}'\to\mathbf{r}_0^-$ with allowance for formulas (5.2.32a) and (5.2.32b).

Let us prove the sufficiency. Let conditions (5.3.11) be the case. Then for the Cauchy integral analog $\mathbf{C}(\mathbf{r}', \varphi)$, according to Sokhotsky-Plemelj formulas (5.2.32a) and (5.2.32b), we have

$$\mathbf{C}^{S_+}(\mathbf{r}_0;\ \varphi) = \varphi(\mathbf{r}_0)\ . \tag{5.3.12}$$

Consequently, $\varphi(\mathbf{r})$ is a boundary value of the Laplace in D field $\mathbf{F}(\mathbf{r}') = \mathbf{C}^S(\mathbf{r}';\ \varphi)$, which was to be proved.

As can be seen, the proof of theorems 5.3.1 and 5.3.2 repeats completely the proof of the corresponding two-dimensional theorems 1.3.1 and 1.3.3 for analytical functions, which became possible due to the above-established analogy between the properties of the classical and three-dimensional Cauchy-type integrals.

By similar reasoning, theorems 1.3.2 and 1.3.4 for boundary values of functions specified in an infinite domain $C\bar{D}$ can be readily extended to a three-dimensional case.

Theorem 5.3.3. For the Hölder continuous vector function $\varphi(\mathbf{r})$ specified on a closed smooth surface S and having differentiable tangential components thereon, to be a boundary value of the vector field $\mathbf{F}(\mathbf{r})$ which is of the Laplace kind in $C\bar{D}$ and tends to a finite limit $\mathbf{F}(\infty)$ at infinity, it is necessary and sufficient that

$$\mathrm{rot}_S\varphi_\tau = \mathbf{0}$$

and

$$\mathbf{C}^S(\mathbf{r}';\ \varphi) = -\frac{1}{4\pi}\iint_S \left[(\mathbf{n}\cdot\varphi)\,\mathrm{grad}\,\frac{1}{|\mathbf{r}-\mathbf{r}'|}+(\mathbf{n}\times\varphi)\times\mathrm{grad}\,\frac{1}{|\mathbf{r}-\mathbf{r}'|}\right] ds = \mathbf{F}(\infty) \tag{5.3.13}$$

for all $\mathbf{r}'\in D$ (where $\mathbf{n}$ is the an outward pointing normal, with respect to the domain D).

Theorem 5.3.4. The necessary and sufficient condition for the Hölder continuous vector function $\varphi(\mathbf{r})$, specified on S and having differentiable tangential components thereon, to be a boundary value on S of the field $\mathbf{F}(\mathbf{r})$ which is of the Laplace kind in the outer domain $C\bar{D}$, are the following relations:

$$\mathrm{rot}_S\varphi_\tau = \mathbf{0}\ ,$$

$$\begin{aligned}\varphi(\mathbf{r}_0) &= -2\mathbf{C}^S(\mathbf{r}_0;\ \varphi)+2\mathbf{F}(\infty)\\ &= \frac{1}{2\pi}\iint_S \left[(\mathbf{n}\cdot\varphi)\,\mathrm{grad}\,\frac{1}{|\mathbf{r}-\mathbf{r}'|}+(\mathbf{n}\times\varphi)\times\mathrm{grad}\,\frac{1}{|\mathbf{r}-\mathbf{r}'|}\right] ds+2\mathbf{F}(\infty)\ ,\end{aligned} \tag{5.3.14}$$

where $\mathbf{n}$ is the same as in (5.3.13).

Just as in a two-dimensional case, formulas (5.3.11) and (5.3.14) permit direct evaluation of three-dimensional singular Cauchy-type integrals where their density is a boundary value of Laplace vector fields.

5.3.2 Piecewise Laplace Vector Fields. Determination of a Piecewise Laplace Field from a Specified Discontinuity

Let $D_1, D_2, \dots D_N$ be several connected domains in space which have no inner points in common but mutually osculate each other along smooth surfaces S_1, $S_2, \dots S_M$, these being the common part of their boundaries. We also assume that in the domains $D_1, D_2 \dots D_N$ there are specified, accordingly, Laplace vector fields $\mathbf{F}^{(1)}(\mathbf{r})$, $\mathbf{F}^{(2)}(\mathbf{r})$, $\dots \mathbf{F}^{(N)}(\mathbf{r})$ which are continuously extendible to the corresponding interfaces $S_1, S_2, \dots, S_M$.

We will denote by $\mathbf{F}(\mathbf{r})$ the following vector field:

$$\mathbf{F}(\mathbf{r}) = \begin{cases} \mathbf{F}^{(1)}(\mathbf{r}) , & \mathbf{r} \in D_1 , \\ \mathbf{F}^{(2)}\mathbf{r}) , & \mathbf{r} \in D_2 , \\ \dots\dots\dots & \\ \mathbf{F}^{(N)}(\mathbf{r}) , & \mathbf{r} \in D_N . \end{cases}$$

This field will be termed a *piecewise Laplace vector field with a discontinuity surface* $S = S_1 \cup S_2 \cup \dots \cup S_M$.

A typical example of a piecewise Laplace field is the three-dimensional Cauchy integral analog $\mathbf{C}^S(\mathbf{r}'; \varphi)$. Here *the integration surface S is the discontinuity surface of the three-dimensional Cauchy-type integral.*

Of great value for applications is the problem of determining a piecewise Laplace vector field $\mathbf{F}(\mathbf{r})$ vanishing at infinity from a specified discontinuity at a closed smooth surface S:

$$\mathbf{F}^+(\mathbf{r}) - \mathbf{F}^-(\mathbf{r}_0) = \varphi(\mathbf{r}_0) , \quad \mathbf{r}_0 \in S , \tag{5.3.15}$$

where φ is the Hölder continuous function at S.

It is evident that, by virtue of the Sokhotsky-Plemelj formulas (5.2.30), a solution to this problem is just the three-dimensional Cauchy integral analog:

$$\mathbf{F}(\mathbf{r}') = \mathbf{C}^S(\mathbf{r}'; \varphi) = -\frac{1}{4\pi} \iint_S \left[(\mathbf{n} \cdot \varphi) \operatorname{grad} \frac{1}{|\mathbf{r}-\mathbf{r}'|} + (\mathbf{n} \times \varphi) \times \operatorname{grad} \frac{1}{|\mathbf{r}-\mathbf{r}'|} \right] ds . \tag{5.3.16}$$

Just as in a two-dimensional case, this solution is unique. Indeed, if there were two solutions $\mathbf{F}(\mathbf{r}')$ and $\tilde{\mathbf{F}}(\mathbf{r}')$, their difference $\Psi(\mathbf{r}') = \mathbf{F}(\mathbf{r}') - \tilde{\mathbf{F}}(\mathbf{r}')$ would be the Laplace field in the whole infinite space tending to zero at infinity. According to the Liouville theorem for harmonic functions, this field is known to be identically zero (Timan, Trofimov 1968).

It is remarkable that the above result is also valid for an arbitrary piecewise smooth closed surface S (finite or tending to infinity).

5.3.3 Inversion Formulas for the Three-Dimensional Cauchy Integral Analog

Of particular interest for applications is the extension of the Cauchy integral transforms (Sect. 1.3.3) to a three-dimensional case, since they are a most elegant way for solving singular integral equations. It will be shown that this extension is readily achievable on the basis of the body of three-dimensional Cauchy integral analogs.

Let S be a closed smooth surface and $\boldsymbol{\varphi}(\mathbf{r})$ be an arbitrary Hölder continuous vector function specified on S and having thereon differential tangential components which satisfy Eq. (5.2.2).

We will denote by $\boldsymbol{\psi}(\mathbf{r}_0)$ the following singular integral:

$$\boldsymbol{\psi}(\mathbf{r}_0) = -\frac{1}{2\pi}\iint_S \left[(\mathbf{n}\cdot\boldsymbol{\varphi})\operatorname{grad}\frac{1}{|\mathbf{r}-\mathbf{r}_0|}+(\mathbf{n}\times\boldsymbol{\varphi})\times\operatorname{grad}\frac{1}{|\mathbf{r}-\mathbf{r}_0|}\right]ds$$

$$= 2\mathbf{C}^S(\mathbf{r}_0;\ \boldsymbol{\varphi}) \ . \tag{5.3.17}$$

It is noteworthy that, similarly to the two-dimensional case, it can be proved that the function $\boldsymbol{\psi}(\mathbf{r})$ is also Hölder continuous on S and meets condition (5.2.2). In particular, if $\boldsymbol{\varphi}(\mathbf{r})$ is a boundary value of some Laplace in the D vector field, then, according to theorem (5.3.2), we have

$$\boldsymbol{\psi}(\mathbf{r}_0)\equiv\boldsymbol{\varphi}(\mathbf{r}_0) \ , \quad \mathbf{r}_0\in S \ . \tag{5.3.18}$$

In any other case $\boldsymbol{\psi}(\mathbf{r}_0)\not\equiv\boldsymbol{\varphi}(\mathbf{r}_0)$.

If $\boldsymbol{\psi}(\mathbf{r}_0)$ is specified, while $\boldsymbol{\varphi}(\mathbf{r}_0)$ is unknown, relation (5.3.17), similarly to (1.3.11) in a two-dimensional case, can be regarded as a *singular integral equation in the function* $\boldsymbol{\varphi}(\mathbf{r}_0)$.

The procedure for solving this equation is as follows. Employing Cauchy integral analog (5.2.3) and Sokhotsky-Plemelj formulas (5.2.33), we can rewrite Eq. (5.3.17) as follows:

$$\mathbf{C}^{S+}(\mathbf{r}_0;\ \boldsymbol{\varphi})+\mathbf{C}^{S-}(\mathbf{r}_0;\ \boldsymbol{\varphi}) = \boldsymbol{\psi}(\mathbf{r}_0) \ , \quad \mathbf{r}_0\in S \ . \tag{5.3.19}$$

Just as in a two-dimensional case (Sect. 1.3.3), the auxiliary piecewise Laplace field $\boldsymbol{\Phi}(\mathbf{r}')$ can be defined as

$$\boldsymbol{\Phi}(\mathbf{r}') = \begin{cases} \mathbf{C}^S(\mathbf{r}';\ \boldsymbol{\varphi}) & \mathbf{r}'\in D \ , \\ -\mathbf{C}^S(\mathbf{r}';\ \boldsymbol{\varphi}) \ , & \mathbf{r}'\in C\bar{D} \ . \end{cases} \tag{5.3.20}$$

Substitution of (5.3.20) into (5.3.19) yields

$$\boldsymbol{\Phi}^+(\mathbf{r}_0)-\boldsymbol{\Phi}^-(\mathbf{r}_0) = \boldsymbol{\psi}(\mathbf{r}_0) \ , \quad \mathbf{r}_0\in S \ . \tag{5.3.21}$$

Thus, $\boldsymbol{\Phi}(\mathbf{r}')$ is a piecewise Laplace field with a specified discontinuity $\boldsymbol{\psi}$. Hence, according to the results of the previous subsection and formula (5.3.16), we have

$$\boldsymbol{\Phi}(\mathbf{r}') = \mathbf{C}^S(\mathbf{r}_0;\ \boldsymbol{\psi}) \ . \tag{5.3.22}$$

Applying again Sokhotsky-Plemelj formula (5.2.33) to the Cauchy-type integral $\mathbf{C}^S(\mathbf{r}_0;\ \boldsymbol{\psi})$, we write

$$\mathbf{\Phi}^{+}(\mathbf{r}_0)+\mathbf{\Phi}^{-}(\mathbf{r}_0)=2\mathbf{C}^{S}(\mathbf{r}_0;\ \boldsymbol{\psi}) \ . \tag{5.3.23}$$

On the other hand, by virtue of (5.3.20) and Sokhotsky-Plemelj formulas (5.2.30) we have

$$\mathbf{\Phi}^{+}(\mathbf{r}_0)+\mathbf{\Phi}^{-}(\mathbf{r}_0)=\mathbf{C}^{S_+}(\mathbf{r}_0;\ \boldsymbol{\varphi})-\mathbf{C}^{S_-}(\mathbf{r}_0;\ \boldsymbol{\varphi})=\boldsymbol{\varphi}(\mathbf{r}_0) \ . \tag{5.3.24}$$

Equating the right-hand sides of (5.3.23) and (5.3.24) we derive in the final form

$$\boldsymbol{\varphi}(\mathbf{r}_0)=2\mathbf{C}^{S}(\mathbf{r}_0;\ \boldsymbol{\psi})=-\frac{1}{2\pi}\iint_S\left[(\mathbf{n}\cdot\boldsymbol{\psi})\,\mathrm{grad}\,\frac{1}{|\mathbf{r}-\mathbf{r}_0|}\right.$$

$$\left.+(\mathbf{n}\times\boldsymbol{\psi})\times\mathrm{grad}\,\frac{1}{|\mathbf{r}-\mathbf{r}_0|}\right]ds \ . \tag{5.3.25}$$

Precisely formula (5.3.25) provides a solution to the singular integral Eq. (5.3.17).

Evidently, similar reasoning may lead us again from (5.3.25) to (5.3.17). Thus, in a three-dimensional case we also have a pair of mutually invertible integral transforms which will be referred to as *three-dimensional Cauchy integral transforms:*

$$\begin{aligned}\boldsymbol{\psi}(\mathbf{r}_0)&=-\frac{1}{2\pi}\iint_S\left[(\mathbf{n}\cdot\boldsymbol{\varphi})\,\mathrm{grad}\,\frac{1}{|\mathbf{r}-\mathbf{r}_0|}+(\mathbf{n}\times\boldsymbol{\varphi})\times\mathrm{grad}\,\frac{1}{|\mathbf{r}-\mathbf{r}_0|}\right]ds \ ,\\ \boldsymbol{\varphi}(\mathbf{r}_0)&=-\frac{1}{2\pi}\iint_S\left[(\mathbf{n}\cdot\boldsymbol{\psi})\,\mathrm{grad}\,\frac{1}{|\mathbf{r}-\mathbf{r}_0|}+(\mathbf{n}\times\boldsymbol{\psi})\times\mathrm{grad}\,\frac{1}{|\mathbf{r}-\mathbf{r}_0|}\right]ds \ .\end{aligned} \tag{5.3.26}$$

Note that, by construction, the unit vector of the normal in (5.3.26) points outward from the domain D, but transforms (5.3.26) remain valid if the direction of $\mathbf{n}$ in the two formulas (5.3.26) is reversed.

5.3.4 Three-Dimensional Hilbert Transforms

Let us introduce a Cartesian coordinate system XYZ in space and write three-dimensional Cauchy integral transforms for the case where S is a coordinate plane XY. In doing so, we assume that $\boldsymbol{\varphi}(\mathbf{r})$ satisfies the Hölder condition and the condition $\partial\varphi_x/\partial y-\partial\varphi_y/\partial x=0$ for any finite values of x and y and it vanishes at infinity. To evaluate the Cauchy-type integral over an infinite plane, it is necessary, just as in a two-dimensional case (Sect. 1.2.5), to define *the principal value of a three-dimensional Cauchy integral analog within infinite limits.* This concept is introduced as follows. Let us fix on a point $\mathbf{r}_0(x_0,\ y_0,\ 0)$ of the plane XY and draw therefrom, as a center, a circle L_R of a radius R in the plane XY. The part of the plane XY bounded by this circle will be denoted by P_R. Write the Cauchy-type integral over the part of the plane P_R:

$$\mathbf{C}^{P_R}(\mathbf{r}_0;\ \boldsymbol{\varphi}(\mathbf{r}))=\frac{1}{4\pi}\iint_{P_R}\{(x-x_0)\varphi_z\mathbf{d}_x+(y-y_0)\varphi_z\mathbf{d}_y$$

$$-[(x-x_0)\varphi_x+(y-y_0)\varphi_y]\,\mathbf{d}_z\}[(x-x_0)^2+(y-y_0)^2]^{-3/2}\,ds \ .$$

The limit of the latter expression for $R \to \infty$ (if any) is called *the principal value of an infinite Cauchy-type integral.* One can readily see, using the polar coordinates, that if $\boldsymbol{\varphi}(x, y, 0)$ vanishes at infinity, the principal value of a three-dimensional Cauchy integral analog over an infinite plane XY does exist. Then relations (5.3.26) take the form (in the assumption of an upward pointing normal $\mathbf{n}$):

$$\boldsymbol{\psi}(x_0, y_0, 0) = \frac{1}{2\pi} \iint_{-\infty}^{+\infty} \{(x-x_0)\varphi_z \mathbf{d}_x + (y-y_0)\varphi_z \mathbf{d}_y - [(x-x_0)\varphi_x + (y-y_0)\varphi_y]\,\mathbf{d}_z\} \times [(x-x_0)^2 + (y-y_0)^2]^{-3/2}\, dx dy \,, \tag{5.3.27a}$$

$$\boldsymbol{\varphi}(x_0, y_0, 0) = \frac{1}{2\pi} \iint_{-\infty}^{+\infty} \{(x-x_0)\psi_z \mathbf{d}_x + (y-y_0)\psi_z \mathbf{d}_y - [(x-x_0)\psi_x + (y-y_0)\psi_y]\,\mathbf{d}_z\} \times [(x-x_0)^2 + (y-y_0)^2]^{-3/2}\, dx dy \,. \tag{5.3.27b}$$

Let the symbols $\hat{G}_x$ and $\hat{G}_y$ stand for the following integral operators:

$$\hat{G}_x f(x, y) = \frac{-1}{2\pi} \iint_{-\infty}^{+\infty} \frac{\partial}{\partial x} \frac{1}{[(x-x_0)^2 + (y-y_0)^2]^{1/2}} f(x, y)\, dx dy = \frac{1}{2\pi} \iint_{-\infty}^{+\infty} \frac{(x-x_0) f(x, y)}{[(x-x_0)^2 + (y-y_0)^2]^{3/2}}\, dx dy \,, \tag{5.3.28a}$$

$$\hat{G}_y f(x, y) = \frac{-1}{2\pi} \iint_{-\infty}^{+\infty} \frac{\partial}{\partial y} \frac{1}{[(x-x_0)^2 + (y-y_0)^2]^{1/2}} f(x, y)\, dx dy = \frac{1}{2\pi} \iint_{-\infty}^{+\infty} \frac{(y-y_0) f(x, y)}{[(x-x_0)^2 + (y-y_0)^2]^{3/2}}\, dx dy \,. \tag{5.3.28b}$$

The operators $\hat{G}_x$ and $\hat{G}_y$ will be termed *three-dimensional Hilbert operators.* Then relations (5.3.27a) and (5.3.27b) are written as follows:

$$\begin{aligned} \psi_x(x_0, y_0, 0) &= \hat{G}_x \varphi_z(x, y, 0) \,, \\ \psi_y(x_0, y_0, 0) &= \hat{G}_y \varphi_z(x, y, 0) \,, \\ \psi_z(x_0, y_0, 0) &= -\hat{G}_x \varphi_x(x, y, 0) - \hat{G}_y \varphi_y(x, y, 0) \,; \end{aligned} \tag{5.3.29a}$$

$$\begin{aligned} \varphi_x(x_0, y_0, 0) &= \hat{G}_x \psi_z(x, y, 0) \,, \\ \varphi_y(x_0, y_0, 0) &= \hat{G}_y \psi_z(x, y, 0) \,, \\ \varphi_z(x_0, y_0, 0) &= -\hat{G}_x \psi_x(x, y, 0) - \hat{G}_y \psi_y(x, y, 0) \,. \end{aligned} \tag{5.3.29b}$$

Relations (5.3.29a) and (5.3.29b) will be called *three-dimensional Hilbert integral transforms.* These transforms can be rearranged as follows:

$$\psi_x = \hat{G}_x \varphi_z \,, \quad \varphi_y = \hat{G}_y \varphi_z \,, \quad \varphi_z = -\hat{G}_x \psi_x - \hat{G}_y \psi_y \,; \tag{5.3.30a}$$

$$\varphi_x = \hat{G}_x \psi_z \,, \quad \varphi_y = \hat{G}_y \psi_z \,, \quad \psi_z = -\hat{G}_x \varphi_x - \hat{G}_y \varphi_y \,. \tag{5.3.30b}$$

Thus, the three-dimensional Hilbert transforms relate the vertical component of the vector field φ to the horizontal component of the vector field ψ and vice versa.

Of great importance for further discussion is the case where $\varphi(x_0, y_0, 0)$ is a boundary value in the horizontal plane $z = 0$ of the field which is of the Laplace kind everywhere in the lower half-space and vanishes at infinity. Then, in accordance with (5.3.18), $\psi(x_0, y_0, 0) \equiv \varphi(x_0, y_0, 0)$. Consequently, relations (5.3.30a) and (5.3.30b) are written in the form

$$\varphi_x = \hat{G}_x \varphi_z \ , \quad \varphi_y = \hat{G}_y \varphi_z \ , \quad \varphi_z = -\hat{G}_x \varphi_x - \hat{G}_y \varphi_y \ . \tag{5.3.31}$$

Similarly, in the case where $\varphi(x_0, y_0, 0)$ is a boundary value in the coordinate plane $z = 0$ of the vector field, which is of the Laplace kind everywhere in the upper half-space and vanishes at infinity, the pertinent relations between the vertical and the horizontal components of the field φ are derived from (5.3.31) by reversing the sign in the right-hand side:

$$\varphi_x = -\hat{G}_x \varphi_z \ , \quad \varphi_y = -\hat{G}_y \varphi_z \ , \quad \varphi_z = \hat{G}_x \varphi_x + \hat{G}_y \varphi_y \ . \tag{5.3.32}$$

Below, the reader will see that the three-dimensional Hilbert operators are as important in the theory of three-dimensional geopotential fields as is the ordinary Hilbert operator in a two-dimensional case.

5.4 Cauchy Integral Analogs in Matrix Notation

Sometimes it is useful to express three-dimensional Cauchy integral analogs in matrix notation. It is worth recalling how the main operations of vector algebra are represented in matrix notation.

5.4.1 Matrix Representation of the Differentiation Operators of Scalar and Vector Fields

Let **A** and **B** be two arbitrary vectors whose decompositions on a certain orthonormalized basis $\{\mathbf{d}_x, \mathbf{d}_y, \mathbf{d}_z\}$ have the form

$$\mathbf{A} = \sum_{\alpha=x,y,z} A_\alpha \mathbf{d}_\alpha \ , \quad \mathbf{B} = \sum_{\beta=x,y,z} B_\beta \mathbf{d}_\beta \ , \tag{5.4.1}$$

where A_x, A_y, A_z and B_x, B_y, B_z are the scalar components of the vectors **A** and **B**.

Let us express the scalar product of the vector **A** and **B** in terms of the scalar products of unit vectors. This operation is linear in each of the arguments and, hence,

$$\mathbf{A} \cdot \mathbf{B} = \sum_{\alpha=x,y,z} \sum_{\beta=x,y,z} A_\alpha B_\beta \mathbf{d}_\alpha \cdot \mathbf{d}_\beta \ .$$

Due to the orthogonality of the vectors $\mathbf{d}_x$, $\mathbf{d}_y$, $\mathbf{d}_z$ we have that $\mathbf{d}_\alpha \cdot \mathbf{d}_\beta = \delta_{\alpha\beta}$, where

$$\delta_{\alpha\beta} = \begin{cases} 1 \ , & \alpha = \beta \ , \\ 0 \ , & \alpha \neq \beta \ , \end{cases}$$

is *the symmetric Kronecker symbol,* or the δ-symbol. Thus, we have

$$\mathbf{A} \cdot \mathbf{B} = \sum_{\alpha=x,y,z} \sum_{\beta=x,y,z} A_\alpha B_\beta \delta_{\alpha\beta} \ . \tag{5.4.2}$$

The procedure of summation over a twice repeating index is often referred to as *convolution.* Thus, the scalar product of the vectors **A** and **B** is expressed as a convolution of the components of these vectors with the δ-symbol.

Similarly, the vector product of vectors A and B can be expressed. Since this operation is linear in each of the arguments, we have

$$\mathbf{A} \times \mathbf{B} = \sum_{\alpha=x,y,z} \sum_{\beta=x,y,z} A_\alpha B_\beta \mathbf{d}_\alpha \times \mathbf{d}_\beta \ .$$

Due to the orthogonality of the vectors $\mathbf{d}_x$, $\mathbf{d}_y$, $\mathbf{d}_z$ we have that $\mathbf{d}_\alpha \times \mathbf{d}_\beta = \sum_{\gamma=x,y,z} \varepsilon_{\alpha\beta\gamma} \mathbf{d}_\gamma$, where $\varepsilon_{\alpha\beta\gamma}$ is *the skew-symmetric Kronecker symbol,* or the ε-symbol. For noncoincident indices we have

$$\varepsilon_{xyz} = \varepsilon_{zxy} = \varepsilon_{yzx} = 1 \ , \quad \varepsilon_{yxz} = \varepsilon_{zyx} = \varepsilon_{xzy} = -1 \ .$$

The values of $\varepsilon_{\alpha\beta\gamma}$ are zero if any two indices happen to coincide (say, $\varepsilon_{xxy} = 0$). We should like to note here that the Kronecker δ- and ε-symbols are related by

$$\sum_{\zeta=x,y,z} \varepsilon_{\alpha\zeta\eta} \varepsilon_{\beta\gamma\zeta} = \delta_{\alpha\gamma} \delta_{\beta\eta} - \delta_{\alpha\beta} \delta_{\gamma\eta} \ . \tag{5.4.3}$$

Using the ε-symbol, we write the vector product of vectors as follows:

$$\mathbf{A} \cdot \mathbf{B} = \sum_{\alpha=x,y,z} \sum_{\beta=x,y,z} \sum_{\gamma=x,y,z} A_\alpha B_\beta \varepsilon_{\alpha\beta\gamma} \mathbf{d}_\gamma \ . \tag{5.4.4}$$

The vector product of the vectors **A** and **B** is a convolution of the components of these vectors with the Kronecker ε-symbol and with the unit vector $\mathbf{d}_\gamma$ of the Cartesian basis.

Now consider the parallelepipedal product of the three vectors: **A**, **B**, and **C**:

$$\mathbf{A} \cdot \mathbf{B} \cdot \mathbf{C} = \mathbf{A} \cdot [\mathbf{B} \times \mathbf{C}] = [\mathbf{A} \times \mathbf{B}] \cdot \mathbf{C}$$
$$= \sum_{\alpha=x,y,z} \sum_{\beta=x,y,z} \sum_{\gamma=x,y,z} A_\alpha B_\beta C_\gamma \mathbf{d}_\alpha \cdot [\mathbf{d}_\beta \times \mathbf{d}_\gamma] \ ,$$

where

$$\mathbf{d}_\alpha \cdot [\mathbf{d}_\beta \times \mathbf{d}_\gamma] = \sum_{\xi=x,y,z} \delta_{\alpha\xi} \varepsilon_{\xi\beta\gamma} = \varepsilon_{\alpha\beta\gamma} \ .$$

Hence,

$$\mathbf{A} \cdot \mathbf{B} \cdot \mathbf{C} = \sum_{\alpha=x,y,z} \sum_{\beta=x,y,z} \sum_{\gamma=x,y,z} A_\alpha B_\beta C_\gamma \varepsilon_{\alpha\beta\gamma} \ . \tag{5.4.5}$$

A double vector product of the vectors **A**, **B**, and **C** can be written, with due references to (5.4.3) and (5.4.4), as follows

$$\mathbf{A}\times[\mathbf{B}\times\mathbf{C}] = \sum_{\alpha=x,y,z}\ \sum_{\beta=x,y,z}\ \sum_{\gamma=x,y,z}\ \sum_{\zeta=x,y,z}\ \sum_{\eta=x,y,z} A_\alpha B_\beta C_\gamma \varepsilon_{\alpha\zeta\eta}\varepsilon_{\beta\gamma\zeta}\mathbf{d}_\eta$$

$$= \sum_{\alpha=x,y,z}\ \sum_{\beta=x,y,z}\ \sum_{\gamma=x,y,z} A_\alpha B_\beta C_\gamma(\delta_{\alpha\gamma}\mathbf{d}_\beta - \delta_{\alpha\beta}\mathbf{d}_\gamma) \ . \tag{5.4.6}$$

In all the above formulas, summation is carried out over twice recurring indices. The summation sign can be omitted upon agreement that the twice recurring index indicates the necessity of summation over this index. This rule is known as *an agreement on summation.* Hereinafter we will use this agreement, assuming that the indices denoted by the Greek letters $\alpha, \beta, \gamma, \ldots$ run through values x, y, z. Employing the agreement on summation, we can rewrite formulas (5.4.2), (5.4.4), (5.4.5), and (5.4.6) in the form

$$\mathbf{A}\cdot\mathbf{B} = A_\alpha B_\beta \delta_{\alpha\beta} \ , \quad \mathbf{A}\times\mathbf{B} = A_\alpha B_\beta \varepsilon_{\alpha\beta\gamma}\mathbf{d}_\gamma \ , \quad \mathbf{A}\cdot\mathbf{B}\cdot\mathbf{C} = A_\alpha B_\beta C_\gamma \varepsilon_{\alpha\beta\gamma} \ ,$$

$$\mathbf{A}\times[\mathbf{B}\times\mathbf{C}] = A_\alpha B_\beta C_\gamma \varepsilon_{\alpha\zeta\eta}\varepsilon_{\beta\gamma\zeta}\mathbf{d}_\eta = A_\alpha B_\beta C_\gamma(\delta_{\alpha\gamma}\mathbf{d}_\beta - \delta_{\alpha\beta}\mathbf{d}_\gamma) \ , \tag{5.4.7}$$

$$\alpha,\ \beta,\ \gamma,\ \zeta,\ \eta = x,\ y,\ z \ .$$

Now we will employ matrix notation for calculating spatial derivatives of the scalar field W and of the vector field $\mathbf{A}$:

$$\operatorname{grad} W = W_{,\alpha}\mathbf{d}_\alpha \ , \quad \operatorname{div}\mathbf{A} = A_{\alpha,\beta}\delta_{\alpha,\beta} \ , \quad \operatorname{rot}\mathbf{A} = A_{\alpha,\beta}\varepsilon_{\gamma\beta\alpha}\mathbf{d}_\gamma \ , \tag{5.4.8}$$

$$\alpha,\ \beta,\ \gamma = x,\ y,\ z \ .$$

Here summation is taken with respect to all twice recurring indices and the index after the comma designates differentiation with respect to the derivative denoted by this index:

$$W_{,\alpha} = \partial W/\partial r_\alpha \ , \quad A_{\alpha,\beta} = \partial A_\alpha/\partial r_\beta \ , \quad \alpha,\ \beta = x,\ y,\ z$$

where

$$r_x = x \ , \quad r_y = y \ , \quad r_z = z \ .$$

The introduced designations permit representation of the three-dimensional Cauchy integral analogs in matrix form.

5.4.2 Matrix Representation of Three-Dimensional Cauchy Integral Analogs

Cauchy-type integral (5.2.3) can be represented in matrix notation. To this end, we will decompose the vectors $\mathbf{C}^S$, $\boldsymbol{\varphi}$, $\mathbf{n}$, and $\operatorname{grad} 1/|\mathbf{r}-\mathbf{r}'|$ on an arbitrary Cartesian basis $\{\mathbf{d}_x, \mathbf{d}_y, \mathbf{d}_z\}$:

$$\mathbf{C}^S = \mathbf{C}^S_\alpha\mathbf{d}_\alpha \ , \quad \boldsymbol{\varphi} = \varphi_\beta\mathbf{d}_\beta \ , \quad \mathbf{n} = n_\gamma\mathbf{d}_\gamma \ , \quad \operatorname{grad}\frac{1}{|\mathbf{r}-\mathbf{r}'|} = \frac{\mathbf{r}'-\mathbf{r}}{|\mathbf{r}-\mathbf{r}'|^3} = \frac{r'_\eta - r_\eta}{|\mathbf{r}-\mathbf{r}'|^3}\,\mathbf{d}_\eta$$

$$\alpha,\ \beta,\ \gamma,\ \eta = x,\ y,\ z \ , \tag{5.4.9}$$

where

$$r_x = x \ , \quad r_y = y \ , \quad r_z = z \ .$$

According to (5.4.7) and (5.4.8), we have

$$(\mathbf{n}\cdot\boldsymbol{\varphi})\,\mathrm{grad}\,\frac{1}{|\mathbf{r}-\mathbf{r}'|}=\varphi_\beta n_\gamma\,\frac{r'_\alpha-r_\alpha}{|\mathbf{r}-\mathbf{r}'|^3}\,\delta_{\beta\gamma}\mathbf{d}_\alpha=\varphi_\beta n_\gamma\,\frac{r'_\eta-r_\eta}{|\mathbf{r}-\mathbf{r}'|^3}\,\delta_{\alpha\eta}\delta_{\beta\gamma}\mathbf{d}_\alpha\ ,$$

$$[\mathbf{n}\times\boldsymbol{\varphi}]\times\mathrm{grad}\,\frac{1}{|\mathbf{r}-\mathbf{r}'|}=\varphi_\beta n_\gamma\,\frac{r'_\eta-r_\eta}{|\mathbf{r}-\mathbf{r}'|^3}\,\varepsilon_{\alpha\eta\zeta}\varepsilon_{\beta\gamma\zeta}\mathbf{d}_\alpha\ ,\quad \alpha,\ \beta,\ \gamma,\ \zeta,\ \eta=x,\ y,\ z\ .$$

Simple manipulation yields

$$\begin{aligned}&(\mathbf{n}\cdot\boldsymbol{\varphi})\,\mathrm{grad}\,\frac{1}{|\mathbf{r}-\mathbf{r}'|}+[\mathbf{n}\times\boldsymbol{\varphi}]\times\mathrm{grad}\,\frac{1}{|\mathbf{r}-\mathbf{r}'|}\\&\quad=\varphi_\beta\,\frac{r'_\eta-r_\eta}{|\mathbf{r}-\mathbf{r}'|^3}\,(\delta_{\alpha\eta}\delta_{\beta\gamma}+\varepsilon_{\alpha\eta\zeta}\varepsilon_{\beta\gamma\zeta})n_\gamma\mathbf{d}_\alpha=\Delta_{\alpha\beta\gamma\eta}\varphi_\beta\,\frac{r'_\eta-r_\eta}{|\mathbf{r}-\mathbf{r}'|^3}\,n_\gamma\mathbf{d}_\alpha\ ,\\&\quad\alpha,\ \beta,\ \gamma,\ \eta=x,\ y,\ z\ ,\end{aligned}\tag{5.4.10}$$

where the four-index Δ-symbol is expressed in terms of the δ- and ε-symbols or, with due account of (5.4.3), in terms of the δ-symbols:

$$\Delta_{\alpha\beta\gamma\eta}=\delta_{\alpha\eta}\delta_{\beta\gamma}+\varepsilon_{\alpha\eta\zeta}\varepsilon_{\beta\gamma\zeta}=\delta_{\alpha\eta}\delta_{\beta\gamma}+\delta_{\alpha\beta}\delta_{\gamma\eta}-\delta_{\alpha\gamma}\delta_{\beta\eta}\ .\tag{5.4.11}$$

Substitution of (5.4.11) into (5.2.3) gives

$$\mathbf{C}_\alpha^S(\mathbf{r}';\ \boldsymbol{\varphi})=\frac{-1}{4\pi}\iint_S\Delta_{\alpha\beta\gamma\eta}\varphi_\beta\,\frac{r'_\eta-r_\eta}{|\mathbf{r}-\mathbf{r}'|^3}\,n_\gamma ds\ ,\quad \alpha,\ \beta,\ \gamma,\ \eta=x,\ y,\ z\ .\tag{5.4.12}$$

Using matrix notation, formula (5.2.12), which is a three-dimensional analog of the Cauchy integral formula, can be cast in the form:

$$-\frac{1}{4\pi}\iint_S\Delta_{\alpha\beta\gamma\eta}\varphi_\beta\,\frac{r'_\eta-r_\eta}{|\mathbf{r}-\mathbf{r}'|^3}\,n_\gamma ds=\begin{cases}\varphi_\alpha(\mathbf{r}')\ , & \mathbf{r}'\in D\ ,\\ 0\ , & \mathbf{r}'\in C\bar{D}\ ,\end{cases}\tag{5.4.13}$$

where $\varphi_\alpha(\alpha=x,\ y,\ z)$ are the scalar components of the Laplace in D vector field $\boldsymbol{\varphi}$. Matrix formulas (5.4.12) and (5.4.13) permit evaluation of the Cauchy-type integrals from components φ_β of their density.

Concluding this section, we should like to note that matrix representation of the three-dimensional Cauchy-integral formula (5.4.13) and of the three-dimensional Cauchy integral analog (5.4.12) provides the basis for extending these formulas to the case of a finite dimensional Euclidean space of an arbitrary number of dimensions. For this purpose, it is sufficient to extend appropriately the concept of integral, taken through a surface, over a surface area, and over a normal thereto to the case of spaces with more than three dimensions (an elementary approach to this extension can be found, for instance, in the book of A.F. Timan and V.N. Trofimov 1968 or in the book of M. Spivak 1965). Thus, formulas (5.4.12) and (5.4.13) point the way to extending the theory of the Cauchy-type integrals to the n-dimensional case. Precisely this circumstance makes matrix formulas important for the theory of Cauchy integral analogs. This extension, however, goes beyond the scope of this book which is devoted primarily to the geophysical aspect of the theory of Cauchy integral analogs. Therefore, it is omitted here.

6 Application of Cauchy Integral Analogs to the Theory of a Three-Dimensional Geopotential Field

The first part of this book has revealed the efficiency of theoretical analysis of plane gravimetric and magnetometric problems when the gravitational and the magnetic fields are represented in terms of the Cauchy-type integrals. In this connection, it seems tempting to derive similar expressions for three-dimensional geopotential fields. Solution to this problem, as well as some important applications of Cauchy integral analogs to the geopotential field theory, are taken up in the present chapter.

6.1 Newton Potential and the Three-Dimensional Cauchy Integral Analog

In Sect. 2.2 we established a simple relationship between the Cauchy-type integral and logarithmic potentials. This relationship is found to hold true also in a three-dimensional case, between the Cauchy integral analogs and the Newton potentials.

6.1.1 Newton Potential

Consider an irrotational vector field $\mathbf{F}(\mathbf{r})$ satisfying the equations

$$\operatorname{div} \mathbf{F}(\mathbf{r}) = q(\mathbf{r}) \ , \quad \operatorname{rot} \mathbf{F}(\mathbf{r}) = \mathbf{0} \ , \tag{6.1.1}$$

where $q(\mathbf{r})$ is *the volume density of field sources.*

As indicated in Sect. 2.1, an irrotational field is potential in the whole space:

$$\mathbf{F}(\mathbf{r}') = \operatorname{grad} U(\mathbf{r}') \ , \tag{6.1.2}$$

where

$$U(\mathbf{r}') = -\frac{1}{4\pi} \iiint_V \frac{q(\mathbf{r})}{|\mathbf{r}-\mathbf{r}'|} \, dv \tag{6.1.3}$$

is the field scalar potential; V is the domain of distribution of sources $q(\mathbf{r})$.

Potential (6.1.3), as distinguished from the logarithmic potential, will be referred to as *the Newton potential,* since it corresponds (under a certain specified density of field sources) to the gravitational attraction potential fitted by Newton's law. It is common knowledge that the Newton potential generally satisfies the Poisson Eq. (2.1.8) while outside sources – it fits the Laplace Eq. (2.1.8a).

Let's take, for example, potentials of simple sources of an irrotational field: point, dipole, linear, surface, and volume sources.

6.1.2 Newton Potential of Simple Field Sources

Consider a point source intensity Q located at a point $\mathbf{r}_0$ of space. Its density can be described by the singular Dirac δ-function $\delta(\mathbf{r}-\mathbf{r}')$:

$$q(\mathbf{r}) = Q\delta(\mathbf{r}-\mathbf{r}_0) , \tag{6.1.4}$$

where the δ-function is defined by the relation

$$\iiint_D \delta(\mathbf{r}-\mathbf{r}_0)f(\mathbf{r})\,dv = \begin{cases} f(\mathbf{r}_0) , & \mathbf{r}_0 \in D , \\ 0 , & \mathbf{r}_0 \in C\bar{D} . \end{cases} \tag{6.1.5}$$

Here $f(\mathbf{r})$ is an arbitrary function continuous in D. Substituting (6.1.4) into (6.1.3) we find

$$U(\mathbf{r}') = -\frac{1}{4\pi}\frac{Q}{|\mathbf{r}_0-\mathbf{r}'|} . \tag{6.1.6}$$

Formula (6.1.6) fits the Newton potential of a point source.

Now let L be a smooth line in space along which point sources are continuously distributed with a linear density $q_L(\mathbf{r})$ which is a continuous function of $\mathbf{r} \in L$. This source is known as *a linear source of an irrotational field.* We will calculate its Newton potential $U^L(\mathbf{r}')$:

$$U^L(\mathbf{r}') = -\frac{1}{4\pi}\int_L \frac{q_L(\mathbf{r})}{|\mathbf{r}-\mathbf{r}'|}\,dl . \tag{6.1.7}$$

Obviously, formula (6.1.7) is derived by integrating (6.1.6) along the line L (upon substitution of $Q = Q(\mathbf{r}) = q_L(\mathbf{r})dl$).

Now assume that point sources of a continuous surface density $q_S(\mathbf{r})$ fill a smooth part of a surface S in space and that $q_S(\mathbf{r})$ is a continuous function of points of the surface S. This source is called *a surface source of an irrotational field,* or *a simple layer.*

The Newton potential of a simple layer is defined by the relation

$$U^S(\mathbf{r}') = -\frac{1}{4\pi}\iint_S \frac{q_S(\mathbf{r})}{|\mathbf{r}-\mathbf{r}'|}\,ds . \tag{6.1.8}$$

Finally, the Newton potential of a volume V filled with point sources of a volume density $q(\mathbf{r})$ is given by formula (6.1.3).

The above models of irrotational field sources will be termed hereinafter *simple sources.*

6.1.3 Newton Potential of Polarized Field Sources

A simple example of a polarized field source is the dipole source made of two closely spaced point sources (located at the ends of the vector $\Delta\mathbf{l}$) with intensities $(+Q)$ and $(-Q)$ that have been drawn together. The product limit

$$\lim_{|\Delta l|\to 0:,\ Q\to\infty} Q\Delta\mathbf{l} = \mathbf{M} \tag{6.1.9}$$

is called *a dipole moment.*

The volume density of a dipole source is defined by the formula

$$q(\mathbf{r}) = \mathbf{M}\cdot \mathrm{grad}^{r_0}\delta(\mathbf{r}-\mathbf{r}_0) \ , \tag{6.1.10}$$

where $\mathbf{r}_0$ is the radius vector of the dipole location point.

Substituting (6.1.10) into (6.1.3) we find *the Newton potential of a dipole source:*

$$U_D(\mathbf{r}') = -\frac{1}{4\pi}M\frac{\partial}{\partial \mathbf{v}}\frac{1}{|\mathbf{r}_0-\mathbf{r}'|} = \frac{1}{4\pi}\mathbf{M}\cdot \mathrm{grad}^{r_0}\frac{1}{|\mathbf{r}_0-\mathbf{r}'|} \ , \tag{6.1.11}$$

where $\mathbf{v}$ is the unit vector oriented along the dipole moment $\mathbf{M}$ ($\mathbf{M} = M\cdot\mathbf{v}$).

Analogously to the simple linear and surface sources, we can define *polarized linear and surface sources.* Indeed, assume that some smooth line L or surface S are continuously filled with dipoles (with their moments oriented orthogonally to L or S, respectively). Let $\mathbf{m}_L$ and $\mathbf{m}_S$ stand for linear and surface densities of dipole moments, respectively, and assume that they are continuous functions of the coordinates of the points on L or S. Then, by reasoning similar to that in Sect. 2.2.1 for logarithmic potentials, we derive

$$U_D^L(\mathbf{r}') = -\frac{1}{4\pi}\int_L \mathbf{m}_L(\mathbf{r})\cdot \mathrm{grad}\frac{1}{|\mathbf{r}-\mathbf{r}'|}\,dl \ , \tag{6.1.12a}$$

$$U_D^S(\mathbf{r}') = -\frac{1}{4\pi}\iint_S \mathbf{m}_S(\mathbf{r})\cdot \mathrm{grad}\frac{1}{|\mathbf{r}-\mathbf{r}'|}\,ds \ . \tag{6.1.12b}$$

Here U_D^L and U_D^S are the Newton potentials of the linear and surface polarized sources. Note that the surface polarized source is often called *a double layer.*

Expression (6.1.12b) for the double layer potential can be cast into a form convenient for analysis. Indeed, we will rewrite the integrand of (6.1.12b) as follows:

$$\mathbf{m}_S(\mathbf{r})\cdot \mathrm{grad}\frac{1}{|\mathbf{r}-\mathbf{r}'|}\,ds = -\frac{m_S(\mathbf{r})}{|\mathbf{r}-\mathbf{r}'|^2}\left(\frac{\mathbf{r}-\mathbf{r}'}{|\mathbf{r}-\mathbf{r}'|}\cdot\mathbf{n}\right)ds = -m_S(\mathbf{r})\frac{ds^*}{|\mathbf{r}-\mathbf{r}'|^2} \ ,$$

where $ds^* = \left(\dfrac{\mathbf{r}-\mathbf{r}'}{|\mathbf{r}-\mathbf{r}'|}\cdot\mathbf{n}\right)ds$ is the projection of the surface element ds on a sphere of a radius $|\mathbf{r}-\mathbf{r}'|$ with its center at a point $\mathbf{r}'$.

Hence, $ds^*/|\mathbf{r}-\mathbf{r}'|^2 = d\Omega$ is the measure of a solid angle at which the surface element ds is seen from the point $\mathbf{r}'$ (its sign is chosen depending on whether the normal $\mathbf{n}$ to the surface makes an acute or an obtuse angle with the vector $\mathbf{r}-\mathbf{r}'$). Thus, formula (6.1.12b) can be written as follows

$$U_D^S(\mathbf{r}') = \frac{1}{4\pi}\iint_S m_S(\mathbf{r})\,d\Omega \ . \tag{6.1.13a}$$

In a particular case of a homogeneous double layer, $\mathbf{m}_S = \mathrm{const}$, and (6.1.13a) is cast in the form

$$U_D^S(\mathbf{r}') = m_S \frac{\Omega_{r'}}{4\pi} , \tag{6.1.13b}$$

where $\Omega_{r'}$ is the solid angle of vision of surface S from the point $\mathbf{r}'$.

Now assume that dipole sources fill continuously a domain of space. This source is known as a *volume polarized source* of an irrotational field. Its Newton potential is

$$U_D^V(\mathbf{r}') = -\frac{1}{4\pi} \iiint_V \mathbf{m}(\mathbf{r}) \cdot \mathrm{grad} \frac{1}{|\mathbf{r}-\mathbf{r}'|} \, dv , \tag{6.1.14}$$

where $\mathbf{m}$ is *the volume density of dipole moments (polarization) of a volume polarized source.*

Let us transform the integrand in (6.1.14):

$$\mathbf{m}(\mathbf{r}) \cdot \mathrm{grad} \frac{1}{|\mathbf{r}-\mathbf{r}'|} = \mathrm{div} \frac{\mathbf{m}(\mathbf{r})}{|\mathbf{r}-\mathbf{r}'|} - \frac{\mathrm{div}\, \mathbf{m}(\mathbf{r})}{|\mathbf{r}-\mathbf{r}'|} . \tag{6.1.15}$$

Substituting (6.1.15) into (6.1.14) and employing the Ostrogradsky-Gauss theorem, we write in the final form

$$U_D^V(\mathbf{r}') = -\frac{1}{4\pi} \iint_S \frac{q_S(\mathbf{r})}{|\mathbf{r}-\mathbf{r}'|} \, ds - \frac{1}{4\pi} \iiint_V \frac{q(\mathbf{r})}{|\mathbf{r}-\mathbf{r}'|} \, dv , \tag{6.1.16}$$

where

$$q_S(\mathbf{r}) = \mathbf{m}(\mathbf{r}) \cdot \mathbf{n}(\mathbf{r}) , \quad q(\mathbf{r}) = -\mathrm{div}\, \mathbf{m}(\mathbf{r}) . \tag{6.1.17}$$

Thus, it is evident that *the volume polarized source is equivalent to a sum of a simple volume source and a simple layer.*

6.1.4 Three-Dimensional Cauchy-Type Integral as a Sum of a Simple and a Double Layer Field

It is shown in Sect. 2.2 that the classical Cauchy-type integral can be represented as a sum of logarithmic potentials of a simple and a double layer. A similar situation is also encountered in a three-dimensional case.

Note first of all that the Cauchy integral analog $\mathbf{C}^S(\mathbf{r}'; \boldsymbol{\varphi})$ according to (5.2.11), can be represented everywhere outside S as a gradient of certain harmonic functions $\Psi^{\pm}(\mathbf{r}')$ (different on the two sides of S):

$$\mathbf{C}^S(\mathbf{r}'; \boldsymbol{\varphi}) = \mathrm{grad}\, \Psi^{\pm}(\mathbf{r}') , \tag{6.1.18}$$

where the sign "+" refers to the domain D, while the sign "–" refers to $C\bar{D}$. Simple transformations yield an explicit expression for $\Psi^{\pm}$. Indeed, in view of (5.2.2), we write

$$\boldsymbol{\varphi}_\tau = \mathrm{grad}^S \varphi , \tag{6.1.19}$$

where φ is a function twice continuously differentiable at S, which will be called *a scalar density* of the Cauchy-type integral.

The second integral in (5.2.4) can be transformed, with due account taken of (6.1.19), as follows:

$$-\frac{1}{4\pi}\,\mathrm{rot}'\iint_S \frac{\mathbf{n}\times\boldsymbol{\varphi}_\tau}{|\mathbf{r}-\mathbf{r}'|}\,ds=\frac{1}{4\pi}\,\mathrm{rot}'\iint_S \frac{\mathrm{grad}^S\varphi\times\mathbf{n}}{|\mathbf{r}-\mathbf{r}'|}\,ds=\frac{1}{4\pi}\,\mathrm{rot}'\iint_S \frac{\mathrm{rot}^S(\varphi\mathbf{n})}{|\mathbf{r}-\mathbf{r}'|}\,ds$$

$$=\frac{1}{4\pi}\,\mathrm{rot}'\iint_S\left(\mathrm{rot}^S\frac{\varphi\mathbf{n}}{|\mathbf{r}-\mathbf{r}'|}+\mathrm{grad}^S\frac{1}{|\mathbf{r}-\mathbf{r}'|}\times\varphi\mathbf{n}\right)ds$$

$$=\frac{1}{4\pi}\,\mathrm{rot}'\left[\iint_S\left(\mathrm{grad}\frac{\varphi}{|\mathbf{r}-\mathbf{r}'|}\times\mathbf{n}\right)ds\right.$$

$$\left.+\iint_S\left(\mathrm{grad}\frac{1}{|\mathbf{r}-\mathbf{r}'|}\times\varphi\mathbf{n}\right)ds\right]. \tag{6.1.20}$$

The first integral in the right-hand side of (6.1.20) is zero, by the Stokes theorem (5.1.11), since S is a closed surface.

Transform the second integral:

$$\frac{1}{4\pi}\,\mathrm{rot}'\iint_S\left[\mathrm{grad}\frac{1}{|\mathbf{r}-\mathbf{r}'|}\times\varphi\mathbf{n}\right]ds=-\frac{1}{4\pi}\,\mathrm{rot}'\iint_S\left[\mathrm{grad}'\frac{1}{|\mathbf{r}-\mathbf{r}'|}\times\varphi\mathbf{n}\right]ds$$

$$=-\frac{1}{4\pi}\,\mathrm{rot}'\iint_S\mathrm{rot}'\frac{\varphi\mathbf{n}}{|\mathbf{r}-\mathbf{r}'|}\,ds=-\frac{1}{4\pi}\,\mathrm{grad}'\iint_S\mathrm{div}'\frac{\varphi\mathbf{n}}{|\mathbf{r}-\mathbf{r}'|}\,ds$$

$$=-\frac{1}{4\pi}\,\mathrm{grad}'\iint_S\mathrm{grad}'\frac{1}{|\mathbf{r}-\mathbf{r}'|}\cdot\varphi\mathbf{n}\,ds$$

$$=\frac{1}{4\pi}\,\mathrm{grad}'\iint_S\mathrm{grad}\frac{1}{|\mathbf{r}-\mathbf{r}'|}\cdot\varphi\mathbf{n}\,ds \tag{6.1.21}$$

for $\mathbf{r}'\notin S$.

Note, in particular, that while deriving formula (6.1.21) we proved the following identity useful for further discussion:

$$-\iint_S\left[[\mathbf{n}\times\boldsymbol{\varphi}]\times\mathrm{grad}\frac{1}{|\mathbf{r}-\mathbf{r}'|}\right]ds=\mathrm{grad}'\iint_S\varphi\mathbf{n}\cdot\mathrm{grad}\frac{1}{|\mathbf{r}-\mathbf{r}'|}\,ds\,. \tag{6.1.22}$$

We will introduce the following designations:

$$q^S=-(\mathbf{n}\cdot\boldsymbol{\varphi})\,,\quad \mathbf{m}^S=-\varphi\mathbf{n}\,. \tag{6.1.23}$$

Then, substituting (6.1.21) into (5.2.4) and allowing for (6.1.23), we obtain

$$\mathbf{C}^S(\mathbf{r}';\boldsymbol{\varphi})=\mathrm{grad}'\left(-\frac{1}{4\pi}\iint_S\frac{q^S}{|\mathbf{r}-\mathbf{r}'|}\,ds-\frac{1}{4\pi}\iint_S\mathbf{m}^S\cdot\mathrm{grad}\frac{1}{|\mathbf{r}-\mathbf{r}'|}\,ds\right). \tag{6.1.24}$$

As a result, we have

$$\Psi^{\pm}(\mathbf{r}') = -\frac{1}{4\pi}\iint_S \frac{q^S}{|\mathbf{r}-\mathbf{r}'|}\,ds - \frac{1}{4\pi}\iint_S \mathbf{m}^S\cdot\operatorname{grad}\frac{1}{|\mathbf{r}-\mathbf{r}'|}\,ds\ . \tag{6.1.25}$$

Thus, one can see that the three-dimensional Cauchy integral analog can be represented as a gradient of the sum of the Newton potentials of a simple layer with a surface density $q^S = -(\mathbf{n}\cdot\boldsymbol{\varphi})$ and of a double layer with a dipole moment density $\mathbf{m}^S = -\varphi\mathbf{n}$:

$$\mathbf{C}^S(\mathbf{r}';\ \boldsymbol{\varphi}) = \operatorname{grad}'\ [U^S(\mathbf{r}') + U^S_D(\mathbf{r}')]\ . \tag{6.1.26}$$

This representation is similar to the relevant representation of the real and imaginary parts of the classical Cauchy-type integral as a superposition of the fields of a simple and a double layer [formula (2.2.30)]. Just as in a two-dimensional case, representation (6.1.26) makes it possible to establish virtually all properties of the Cauchy integral analogs by analyzing the properties of the Newton potentials of simple and double layers. But, as already shown in the first part of the book, the body of the Cauchy-type integral provides means more effective for solving the problems of geopotential field theory than those of a logarithmic potential. The same situation is observed in a three-dimensional case.

6.2 Representation of the Gravitational Field in Terms of the Cauchy Integral Analog

The remarkable advance in the theoretical analysis of plane gravimetric and magnetometric problems has been possible, as indicated earlier, due to the representation of the gravitational and magnetic fields in terms of the Cauchy-type integrals. Hence, it seems essential to derive similar expressions for three-dimensional fields.

6.2.1 Gravitational Field Equations

The gravitational field $\mathbf{g}(\mathbf{r})$ of a three-dimensional distribution of masses with a density $\varrho(\mathbf{r})$ satisfies the equations

$$\operatorname{div}\mathbf{g} = -4\pi\gamma\rho\ , \quad \operatorname{rot}\mathbf{g} = \mathbf{0}\ , \tag{6.2.1}$$

where γ is the universal gravitational constant.

Thus, the density of gravitational field sources is

$$q = -4\pi\gamma\varrho\ . \tag{6.2.2}$$

It follows from (6.2.1) that outside masses the gravitational field is of a Laplace form. It is expressed in terms of the gravitational potential $U(\mathbf{g} = \operatorname{grad} U)$ satisfying the Poisson equation

$$\Delta U = -4\pi\gamma\varrho\ . \tag{6.2.3}$$

Substituting (6.2.2) into (6.1.3) we obtain the known expressions for the gravitational potential and the field of a volume mass distribution

$$U(\mathbf{r}') = \gamma \iiint_D \frac{\varrho(\mathbf{r})}{|\mathbf{r}-\mathbf{r}'|}\, dv \ , \tag{6.2.4}$$

$$g(\mathbf{r}') = \gamma \iiint_D \varrho(\mathbf{r}) \operatorname{grad}' \frac{1}{|\mathbf{r}-\mathbf{r}'|}\, dv \ , \tag{6.2.5}$$

where D is the domain filled with masses.

Formulas (6.2.4) and (6.2.5) represent the gravitational field in terms of volume integrals. At the same time, in a three-dimensional case, just as in a two-dimensional case, it proves extremely helpful in the solution of many theoretical and practical problems to have the gravitational fields expressed in terms of integrals taken over the surface of mass-filled volumes.

6.2.2 Representation of the Gravitational Field of a Three-Dimensional Homogeneous Body in Terms of the Cauchy-Type Integral

We will employ the body of Cauchy integral analogs to calculate the gravitational field of a homogeneous body D filled with masses of a density ϱ_0 and bounded by a piecewise smooth surface S. To this end, take the three-dimensional Pompei formula (5.1.40a) rewritten in the form

$$\mathbf{C}^S[\mathbf{r}';\ \mathbf{F}(\mathbf{r})] + \frac{1}{4\pi} \iiint_D \operatorname{div}\mathbf{F} \operatorname{grad} \frac{1}{|\mathbf{r}-\mathbf{r}'|}\, dv = \begin{cases} 0 \ , & \mathbf{r}' \in C\bar{D} \ , \\ \mathbf{F}(\mathbf{r}') \ , & \mathbf{r}' \in D \ . \end{cases} \tag{6.2.6}$$

Taking in (6.2.6) that

$$\mathbf{F} = (4\pi/3)\gamma\varrho_0\mathbf{r} \ , \tag{6.2.7}$$

we have then that

$$\operatorname{div}\mathbf{F} = 4\pi\gamma\varrho_0 \tag{6.2.8}$$

and (6.2.6) can be cast as follows:

$$\mathbf{C}^S\left(\mathbf{r}';\ \frac{4\pi}{3}\gamma\varrho_0\mathbf{r}\right) + \gamma \iiint_D \varrho_0 \operatorname{grad} \frac{1}{|\mathbf{r}-\mathbf{r}'|}\, dv = \begin{cases} 0 \ , & \mathbf{r}' \in C\bar{D} \ , \\ \dfrac{4\pi}{3}\gamma\varrho_0\mathbf{r}' \ , & \mathbf{r}' \in D \ . \end{cases} \tag{6.2.9}$$

It is evident that, in accordance with (6.2.5), the volume integral in the left-hand side of (6.2.9) is (with the negative sign) the gravitational field $\mathbf{g}(\mathbf{r})$ of a homogeneous body D. Hence,

$$\mathbf{g}(\mathbf{r}') = \begin{cases} \mathbf{C}^S\left(\mathbf{r}';\ \dfrac{4\pi}{3}\gamma\varrho_0\mathbf{r}\right) \ , & \mathbf{r}' \in C\bar{D} \ , \\ -\dfrac{4\pi}{3}\gamma\varrho_0\mathbf{r}' + \mathbf{C}^S\left(\mathbf{r}';\ \dfrac{4\pi}{3}\gamma\varrho_0\mathbf{r}\right) \ , & \mathbf{r}' \in D \ . \end{cases} \tag{6.2.10}$$

Thus, we have arrived at a formula for the gravitational field $\mathbf{g}(\mathbf{r})$ both outside and inside a homogeneous body expressed in terms of the Cauchy-type integral over the body surface. It is noteworthy that, by virtue of Sokhotsky-Plemelj formulas (5.2.29), the right-hand sides of (6.2.10) have identical limits for a point $\mathbf{r}'$ tending to the surface S both from the inside and the outside of the body D. Hence, the gravitational field $\mathbf{g}(\mathbf{r}')$ is a function continuous in the whole space.

Let us examine expression (6.2.10). With this purpose in mind, we will represent the gravitational field first outside masses:

$$\mathbf{g}(\mathbf{r}') = \mathbf{C}^S\left(\mathbf{r}'; \frac{4\pi}{3}\gamma\varrho_0\mathbf{r}\right) . \tag{6.2.11}$$

As can be seen, the problem of determining the gravitational field outside masses boils down to the evaluation of the Cauchy-type integral over the surface S bounding masses, with a density

$$\boldsymbol{\varphi}(\mathbf{r}) = (4\pi/3)\gamma\varrho_0\mathbf{r} . \tag{6.2.12}$$

Representation (6.2.11) is an extension of the Tsirulsky formula (2.5.7) for the complex intensity of a plane field to a three-dimensional case:

$$g(\zeta') = \frac{1}{2\pi i}\int_L \frac{(-2\pi\gamma\varrho_0\bar{\zeta})}{\zeta-\zeta'}\,d\zeta , \tag{6.2.13}$$

where L is the cross-sectional contour of a two-dimensional body extended along the y-axis; $g(\zeta')$ is the complex intensity of the gravitational field related to $\mathbf{g}(\mathbf{r}') = (g_x, 0, g_z)$ by the formula $g(\zeta') = -g_x(x', z')+ig_z(x', z')$; $\bar{\zeta}$ is the complex conjugate to $\zeta = x+iz$.

In view of the results obtained in Sect. 5.1.4, it is evident that in a two-dimensional situation (6.2.11) switches automatically into (6.2.13), upon substitution of the coefficient 2π for $4\pi/3$ (which is related to transition from a three-dimensional finite domain to a two-dimensional one, i.e., the latter is extended infinitely along the y-axis).

Now represent the gravitational field inside masses:

$$\mathbf{g}(\mathbf{r}') = -\frac{4\pi}{3}\gamma\varrho_0\mathbf{r}'+\mathbf{C}^S\left(\mathbf{r}'; \frac{4\pi}{3}\gamma\varrho_0\mathbf{r}\right) . \tag{6.2.14}$$

The items in the right-hand side of (6.2.14) have a simple physical meaning. Indeed, designate

$$\mathbf{g}_0(\mathbf{r}') = -\frac{4\pi}{3}\gamma\varrho_0\mathbf{r}' , \tag{6.2.15}$$

$$\mathbf{g}_{CD}(\mathbf{r}') = -\mathbf{C}^S\left(\mathbf{r}'; \frac{4\pi}{3}\gamma\varrho_0\mathbf{r}\right) . \tag{6.2.16}$$

Obviously, the field $\mathbf{g}_0$ everywhere in space meets the equations

$$\operatorname{div}\mathbf{g}_0 = -4\pi\gamma\varrho_0 , \quad \operatorname{rot}\mathbf{g}_0 = \mathbf{0} . \tag{6.2.17}$$

A comparison of (6.2.17) and (6.2.1) reveals that $\mathbf{g}_0$ is the gravitational field of masses filling the whole space uniformly with a density ϱ_0.

On the other hand, according to (6.2.14) through (6.2.16), we have

$$\mathbf{g}_{CD}(\mathbf{r}') = \mathbf{g}_0(\mathbf{r}') - \mathbf{g}(\mathbf{r}') \ . \tag{6.2.18}$$

At the same time, $\mathbf{g}(\mathbf{r}')$ is the field of masses concentrated at a density ϱ_0 inside the domain D. As a result, according to the principle of superposition of fields, $\mathbf{g}_{CD}(\mathbf{r}')$ is the gravitational field of masses distributed uniformly at a density ϱ_0 in an infinite domain $C\bar{D}$ complementing the domain D with respect to the whole space. Thus, outside D the three-dimensional Cauchy integral analog with a density of $\frac{4\pi}{3}\gamma\varrho_0\mathbf{r}$ implies the gravitational field of the domain D, while inside D it refers to the gravitational field of the domain $C\bar{D}$ (taken with a negative sign).

Note that the right-hand sides of (6.2.10) can be unified. To this end, it is sufficient to take into consideration that the function $(4\pi/3)\,\gamma\varrho_0\mathbf{r}'$ is constant under $\mathbf{r}$, i.e., it satisfies the Laplace equation in the domain D with respect to $\mathbf{r}$. Hence, according to the three-dimensional Cauchy formula (5.1.26), we have

$$\mathbf{C}^S\left(\mathbf{r}';\frac{4\pi}{3}\gamma\varrho_0\mathbf{r}'\right) = \begin{cases} \frac{4\pi}{3}\gamma\varrho_0\mathbf{r}' \ , & \mathbf{r}' \in D \ , \\ 0 \ , & \mathbf{r}' \in C\bar{D} \ . \end{cases} \tag{6.2.19}$$

Allowing for (6.2.19) and for the linearity of the Cauchy-type integral (Sect. 5.2.1), we can cast formula (6.2.10) as follows:

$$\mathbf{g}(\mathbf{r}') = \mathbf{C}^S\left[\mathbf{r}';\frac{4\pi}{3}\gamma\varrho_0\,(\mathbf{r}-\mathbf{r}')\right] \ . \tag{6.2.20}$$

Thus, we have derived a unified analytical expression for the gravitational field of a homogeneous body both inside and outside masses. Expression (6.2.20) is an extension of formula (2.5.13) to a three-dimensional case.

Expand formula (6.2.20) in an explicit form:

$$\mathbf{g}(\mathbf{r}') = \frac{1}{3}\gamma\varrho_0 \iint_S \left[2\mathbf{n}\cdot(\mathbf{r}-\mathbf{r}')\frac{\mathbf{r}-\mathbf{r}'}{|\mathbf{r}-\mathbf{r}'|^3} + \frac{\mathbf{n}}{|\mathbf{r}-\mathbf{r}'|}\right] ds \ . \tag{6.2.21}$$

Expression (6.2.21) proves to be helpful in the practical solution of direct problems of gravimetry.

It should be borne in mind, however, that since the density $(4\pi/3)\gamma\varrho_0(\mathbf{r}-\mathbf{r}')$ depends upon $\mathbf{r}'$ (the point at which the Cauchy-type integral is evaluated), the properties of this integral (studied in Sect. 5.2) are generally satisfied no longer. In particular,

$$\operatorname{div}'\mathbf{C}^S\left[\mathbf{r}';\frac{4\pi}{3}\gamma\varrho_0(\mathbf{r}-\mathbf{r}')\right] = -4\pi\gamma\varrho_0 \neq 0 \ .$$

6.2.3 Gravitational Field of a Body with an Arbitrary Density Distribution

Now consider the case where the density of a body D is an arbitrary continuous function of coordinates $\varrho(\mathbf{r})$. Let us redefine the function $\varrho(\mathbf{r})$ in a certain domain

D^* enclosing completely the domain D and surface $S(D^* \supset D \cup S)$, so that $\varrho(\mathbf{r})$ might be a continuous function in D^*[1]. Let $\boldsymbol{\sigma}(\mathbf{r})$ be an arbitrary particular solution to the equations

$$\operatorname{div}\boldsymbol{\sigma}(\mathbf{r}) = 4\pi\gamma\varrho(\mathbf{r}) \ , \quad \operatorname{rot}\boldsymbol{\sigma}(\mathbf{r}) = 0 \ , \quad \mathbf{r} \in D^* \ . \tag{6.2.22}$$

Note that Eq. (6.2.22) has inside D a general solution fitted by the formula

$$\boldsymbol{\sigma}(\mathbf{r}') = \boldsymbol{\varphi}(\mathbf{r}') + \frac{1}{4\pi} \iiint_D q(\mathbf{r}) \cdot \operatorname{grad} \frac{1}{|\mathbf{r}-\mathbf{r}'|} \, dv \ , \tag{6.2.23}$$

where $\boldsymbol{\varphi}(\mathbf{r})$ is an arbitrary Laplace vector field in D. Formula (6.2.23) follows from the three-dimensional Pompei formula (6.2.6), provided it is claimed in the latter that $\boldsymbol{\varphi}(\mathbf{r}') = \mathbf{C}^S(\mathbf{r}'; \boldsymbol{\sigma}(\mathbf{r}))$. Particular solutions to Eq. (6.2.22) in domain D can be derived by taking $\boldsymbol{\varphi}(\mathbf{r}) = \text{const}$ in (6.2.23).

Substitution of any particular solution $\boldsymbol{\sigma}(\mathbf{r})$ of (6.2.23) into Pompei formula (6.2.6) yields

$$\mathbf{C}^S[\mathbf{r}'; \boldsymbol{\sigma}(\mathbf{r})] + \gamma \iiint_D \varrho(\mathbf{r}) \operatorname{grad} \frac{1}{|\mathbf{r}-\mathbf{r}'|} \, dv = \begin{cases} 0 \ , & \mathbf{r}' \in C\bar{D} \ , \\ \boldsymbol{\sigma}(\mathbf{r}') \ , & \mathbf{r}' \in D \ . \end{cases}$$

The volume integral in the left-hand side of the latter equation is equal (with a negative sign) to the gravitational field $\mathbf{g}(\mathbf{r}')$ of the domain D filled with masses of a density $\varrho(\mathbf{r})$:

$$\mathbf{g}(\mathbf{r}) = -\gamma \iiint_D \varrho(\mathbf{r}) \operatorname{grad} \frac{1}{|\mathbf{r}-\mathbf{r}'|} \, dv = \gamma \iiint_D \varrho(\mathbf{r}) \operatorname{grad}' \frac{1}{|\mathbf{r}-\mathbf{r}'|} \, dv \ .$$

Consequently, we have

$$\mathbf{g}(\mathbf{r}') = \begin{cases} \mathbf{C}^S[\mathbf{r}'; \boldsymbol{\sigma}(\mathbf{r})] \ , & \mathbf{r}' \in C\bar{D} \ , \\ -\boldsymbol{\sigma}(\mathbf{r}') + \mathbf{C}^S[\mathbf{r}'; \boldsymbol{\sigma}(\mathbf{r})] \ , & \mathbf{r}' \in D \ . \end{cases} \tag{6.2.24}$$

Thus, we have derived an expression for the gravitational field of an arbitrary continuous mass distribution in terms of the Cauchy-type integral over the surface of the domain occupied by masses. Note that in this case, by virtue of Sokhotsky-Plemelj formulas (5.2.29), $\mathbf{g}(\mathbf{r}')$ is a continuous function in the whole space. [The Hölder conditions for the function $\boldsymbol{\sigma}(\mathbf{r})$ are satisfied automatically due to its differentiability.]

Just as in the previous subsection, the right-hand sides of formula (6.2.24) can be cast as a single analytical expression, allowing for the fact that

$$\mathbf{C}^S[\mathbf{r}'; \boldsymbol{\sigma}(\mathbf{r}')] = \begin{cases} \boldsymbol{\sigma}(\mathbf{r}') \ , & \mathbf{r}' \in D \ , \\ 0 \ , & \mathbf{r}' \in C\bar{D} \ . \end{cases} \tag{6.2.25}$$

Combining (6.2.25) with (6.2.24) we obtain the following representation for the gravitational field in terms of the Cauchy-type integral over the surface S of the domain filled with masses:

[1] Note that the domain D^* may coincide with the whole three-dimensional space.

$$\mathbf{g}(\mathbf{r}') = \mathbf{C}^S[\mathbf{r}';\ \boldsymbol{\sigma}(\mathbf{r}) - \boldsymbol{\sigma}(\mathbf{r}')] \ , \tag{6.2.26}$$

where $\boldsymbol{\sigma}(\mathbf{r})$ is an arbitrary particular solution to Eqs. (6.2.22).

We should like to stress, however, that the integral in the right-hand side of (6.2.26) is only formally, by its analytical structure, a Cauchy-type integral. The thing is that its density $\boldsymbol{\sigma}(\mathbf{r}) - \boldsymbol{\sigma}(\mathbf{r}')$ is determined by the coordinates of the point at which the integral itself is evaluated. Therefore, the properties of the Cauchy-type integral (studied in Sect. 5.2) are generally satisfied no longer. For instance,

$$\operatorname{div}' \mathbf{C}^S[\mathbf{r}';\ \boldsymbol{\sigma}(\mathbf{r}) - \boldsymbol{\sigma}(\mathbf{r}')] = -4\pi\gamma\varrho \neq 0 \ .$$

6.2.4 Case of Vertical or One-Dimensional Horizontal Variations in the Density

The above-developed general theory will be illustrated using, as example, direct problems of gravimetry for the case of vertical or one-dimensional horizontal variations in the density. As is known (see Sect. 2.5), to solve the stated problem in a two-dimensional case, there are available very simple and convenient formulas. We will derive similar formulas for a three-dimensional field. Take that

$$\varrho(\mathbf{r}) = \varrho(z) \ , \tag{6.2.27}$$

i.e., the density of masses distributed over the domain D varies only with depth. Then, we can take as a particular solution to Eqs. (6.2.22) the following vector function

$$\boldsymbol{\sigma}(\mathbf{r}) = 4\pi\gamma R(z)\,\mathbf{d}_z \ , \tag{6.2.28}$$

where

$$R(z) = \int \varrho(z)\,dz \ , \tag{6.2.29}$$

and $\mathbf{d}_z$ is the unit vector of the z-axis. Substituting (6.2.28) into (6.2.26) we write

$$\mathbf{g}(\mathbf{r}') = \mathbf{C}^S\{\mathbf{r}';\ 4\pi\gamma[R(z) - R(z')]\,\mathbf{d}_z\} \ . \tag{6.2.30}$$

In particular, outside the body D formula (6.2.30) is simplified:

$$\mathbf{g}(\mathbf{r}') = \mathbf{C}^S(\mathbf{r}';\ 4\pi\gamma R(z)\,\mathbf{d}_z) \ . \tag{6.2.31}$$

Explicit expansion of the expression for the Cauchy-type integral yields

$$\begin{aligned}\mathbf{g}(\mathbf{r}') &= -\gamma \iint_S \left\{[\mathbf{n}\cdot R(z)\,\mathbf{d}_z]\operatorname{grad}\frac{1}{|\mathbf{r}-\mathbf{r}'|} + [\mathbf{n}\times R(z)\,\mathbf{d}_z]\times\operatorname{grad}\frac{1}{|\mathbf{r}-\mathbf{r}'|}\right\} ds \\ &= -\gamma \iint_S \left[\left(\mathbf{n}\cdot\operatorname{grad}\frac{1}{|\mathbf{r}-\mathbf{r}'|}\right)\mathbf{d}_z + \left(\mathbf{n}\times\operatorname{grad}\frac{1}{|\mathbf{r}-\mathbf{r}'|}\right)\times\mathbf{d}_z\right] R(z)\,ds \ .\end{aligned} \tag{6.2.32}$$

Formula (6.2.32) suggests, in particular, a simple relation for determining the vertical component of the gravitational field of a body D bounded by the surface S, whose density is a function of only the vertical coordinate:

$$g_z(\mathbf{r}') = \gamma \iint_S R(z) \left(\frac{\mathbf{r}-\mathbf{r}'}{|\mathbf{r}-\mathbf{r}'|^3} \cdot \mathbf{n} \right) ds \ . \tag{6.2.33}$$

Note that, according to the findings of Sect. 6.1.3, we have

$$\frac{\mathbf{r}-\mathbf{r}'}{|\mathbf{r}-\mathbf{r}'|^3} \cdot \mathbf{n}\, ds = d\Omega \ , \tag{6.2.34}$$

where $d\Omega$ is the solid angle of vision of a surface element ds from the point $\mathbf{r}'$. Then (6.2.33) is written in the form

$$g_z(\mathbf{r}') = \gamma \iint_S R(z) d\Omega \ . \tag{6.2.35}$$

A comparison of (6.2.35) and (6.1.13a) reveals that the vertical component of the gravitational field of a body with a vertically variable density coincides with the potential of a double layer S of a dipole moment density $\mathbf{m}_S = 4\pi\gamma R(z)$.

Formulas (6.2.35) are very convenient for evaluating the gravitational field of bodies whose densities are solely vertical functions of coordinates.

Take, for example, the explicit expression for formula (6.2.35) in several simple instances of vertical distributions.

1) A homogeneous body: $\varrho(z) = \varrho_0 = \text{const}$. Then we have

$$R(z) = \varrho_0 z \ . \tag{6.2.36}$$

Substituting (6.2.36) into (6.2.35) we obtain

$$g_z(\mathbf{r}') = \gamma\varrho_0 \iint_S z d\Omega \ . \tag{6.2.37}$$

2) A body with a linear vertical variation of the density: $\varrho(z) = \varrho_0 + az$. Then, we have

$$R(z) = \varrho_0 z + \tfrac{1}{2} a z^2 \ . \tag{6.2.38}$$

Substituting (6.2.38) into (6.2.35) we obtain

$$g_z(\mathbf{r}') = \gamma\varrho_0 \iint_S z d\Omega + \tfrac{1}{2}\gamma a \iint_S z^2 d\Omega \ . \tag{6.2.39}$$

3) A body with an exponential vertical variation of the density: $\varrho(z) = \varrho_0 + ae^{\lambda e}$. In this case we have

$$R(z) = \varrho_0 z \text{-} + (a/\lambda) e^{\lambda z} \ . \tag{6.2.40}$$

Hence, for the vertical component of the gravitational field we obtain

$$g_z(\mathbf{r}') = \gamma\varrho_0 \iint_S z d\Omega + \gamma \frac{a}{\lambda} \iint_S e^{\lambda z} d\Omega \ . \tag{6.2.41}$$

4) An infinitely extended horizontal homogeneous layer bounded by planes $z = z_1$ and $z = z_2$ $(z_1 < z_2)$ with an arbitrary density variation along the z-axis, $\varrho = \varrho(z)$.

Evidently, formula (6.2.35) in this case can be rewritten as follows:

$$g_z(\mathbf{r}') = \gamma R(z_1) \iint\limits_{z=z_1} d\Omega - \gamma R(z_2) \iint\limits_{z=z_2} d\Omega \ , \quad z' > z_2 \ . \tag{6.2.42}$$

The solid angles of vision of the planes $z = z_1$ and $z = z_2$ are equal to 2π. Hence,

$$g_z(\mathbf{r}') = 2\pi\gamma [R(z_1 - R(z_2)] \ . \tag{6.2.43}$$

For instance, in the case of a homogeneous layer we have, by virtue of (6.2.36),

$$g_z(\mathbf{r}') = -2\pi\gamma\varrho_0(z_2 - z_1) \ , \quad z' > z_2 \ . \tag{6.2.44}$$ [2]

For a layer with a linear density variation, we have, according to (6.2.38),

$$g_z(\mathbf{r}') = -2\pi\gamma\varrho_0(z_2 - z_1) - \pi\gamma a(z_2^2 - z_1^2) \ , \quad z' > z_2 \ . \tag{6.2.45}$$

And finally, for a layer with an exponential density variation we have

$$g_z(\mathbf{r}') = -2\pi\gamma\varrho_0(z_1 - z_2) - 2\pi\gamma \frac{a}{\lambda} (e^{\lambda z_2} - e^{\lambda z_1}) \ . \tag{6.2.46}$$

The cited examples are sufficient to demonstrate the simplicity and convenience of formula (6.2.35) as applied to practical calculations.

Now examine the situation where the density ϱ varies only in the horizontal. For instance,

$$\varrho = \varrho(x) \ . \tag{6.2.47}$$

Then, a particular solution to Eq. (6.2.22) is provided by the function

$$\boldsymbol{\sigma}(\mathbf{r}) = 4\pi\gamma P(x)\mathbf{d}_x \ , \tag{6.2.48}$$

where

$$P(x) = \int \varrho(x)\, dx \tag{6.2.49}$$

and $\mathbf{d}_x$ is the unit vector of the x-axis. Substituting (6.2.48) into (6.2.26) we obtain

$$\mathbf{g}(\mathbf{r}') = \mathbf{C}^S[\mathbf{r}';\ 4\pi\gamma(P(x) - P(x'))\ \mathbf{d}_x] \ , \tag{6.2.50}$$

or outside a body:

$$\mathbf{g}(\mathbf{r}') = \mathbf{C}^S[\mathbf{r}';\ 4\pi\gamma P(x)\ \mathbf{d}_x] \ . \tag{6.2.51}$$

Similarly to (6.2.32), the right-hand side of formula (6.2.51) will be written in greater detail:

$$\mathbf{g}(\mathbf{r}') = -\gamma \iint\limits_S \left[\left(\mathbf{n} \cdot \operatorname{grad} \frac{1}{|\mathbf{r} - \mathbf{r}'|} \right) \mathbf{d}_x + \left(\mathbf{n} \times \operatorname{grad} \frac{1}{|\mathbf{r} - \mathbf{r}'|} \right) \times \mathbf{d}_x \right] P(x)\, ds \ . \tag{6.2.52}$$

[2] Note that in this book the z-axis points upward. Therefore, the sign in the right-hand side of (6.2.44) is negative, unlike the similar formula presented in conventional treatises on gravimetry, where the z-axis faces, as a rule, downward.

In the projections on the coordinate axis

$$g_x(\mathbf{r}') = -\gamma \iint\limits_S \left(\mathbf{n}\cdot\text{grad}\,\frac{1}{|\mathbf{r}-\mathbf{r}'|}\right) P(x)\,ds = \gamma \iint\limits_S P(x)\,d\Omega\ ,$$

$$g_y(\mathbf{r}') = -\gamma \iint\limits_S \frac{n_x(y'-y)-n_y(x'-x)}{|\mathbf{r}-\mathbf{r}'|^3}\, P(x)\,ds\ , \tag{6.2.53}$$

$$g_z(\mathbf{r}') = \gamma \iint\limits_S \frac{n_x(z'-z)-n_z(x'-x)}{|\mathbf{r}-\mathbf{r}'|^3}\, P(x)\,ds\ .$$

Formulas (6.2.52) and (6.2.53) enable us to reduce the problem of calculating the gravitational field of a three-dimensional body D with an arbitrary density behavior along the x-axis to surface integration. Similar formulas are obtainable for the case where $\varrho = \varrho(y)$.

6.2.5 Some Special Cases of Density Distribution

Now consider more complicated density distributions inside a body. Let the density of a body $\varrho(\mathbf{r})$ be represented as a sum of three arbitrary integrable functions dependent on the x-, y-, and z-coordinates separately:

$$\varrho(\mathbf{r}) = \varrho_x(x)+\varrho_y(y)+\varrho_z(z)\ . \tag{6.2.54}$$

We will specify the function $\boldsymbol{\sigma}(\mathbf{r})$ in the form

$$\boldsymbol{\sigma}(\mathbf{r}) = 4\pi\gamma\,[P(x)\mathbf{d}_x+Q(y)\mathbf{d}_y+R(z)\mathbf{d}_z]\ , \tag{6.2.55}$$

where

$$P(x) = \int \varrho_x(x)\,dx\ ,\quad Q(y) = \int \varrho_y(y)\,dy\ ,\quad R(z) = \int \varrho_z(z)\,dz\ . \tag{6.2.56}$$

Evidently, the function $\boldsymbol{\sigma}(\mathbf{r})$ defined by expression (6.2.55) satisfies Eq. (6.2.22). As a result, substituting (6.2.55) into (6.2.26) we determine the field of a body with a given density (6.2.54):

$$\mathbf{g}(\mathbf{r}') = 4\pi\gamma\,\mathbf{C}^S\{\mathbf{r}', [(P(x)-P(x'))\mathbf{d}_x+(Q(y)+Q(y'))\mathbf{d}_y+(R(z)-R(z'))\mathbf{d}_z]\}\ . \tag{6.2.57}$$

The density distribution, as represented in formula (6.2.54), can be exemplified by *linear density variation.* Then we have

$$\varrho(\mathbf{r}) = a_x x+a_y y+a_z z+\varrho_0\ ,\quad P(x) = \tfrac{1}{2}a_x x^2+\tfrac{1}{3}\varrho_0 x\ ,$$

$$Q(y) = \tfrac{1}{2}a_y y^2+\tfrac{1}{3}\varrho_0 y\ ,\qquad R(z) = \tfrac{1}{2}a_z z^2+\tfrac{1}{3}\varrho_0 z\ .$$

Hence,

$$\boldsymbol{\sigma}(\mathbf{r}) = 2\pi\gamma(a_x x^2\mathbf{d}_x+a_y y^2\mathbf{d}_y+a_z z^2\mathbf{d}_z)+\frac{4\pi}{3}\,\gamma\varrho_0\mathbf{r}\ . \tag{6.2.58}$$

Substituting (6.2.58) into (6.2.57) we obtain

$$\mathbf{g}(\mathbf{r}') = 2\pi\gamma\mathbf{C}^S\{\mathbf{r}', [a_x(x^2-x'^2)\mathbf{d}_x + a_y(y^2-y'^2)\mathbf{d}_y + a_z(z^2-z'^2)\mathbf{d}_z + \tfrac{2}{3}(\mathbf{r}-\mathbf{r}')]\} \ . \tag{6.2.59}$$

Formula (6.2.59) extends relation (2.5.37) to a three-dimensional case.

6.2.6 Calculation of the Gravitational Field of a Three-Dimensional Infinitely Extended Homogeneous Domain

Formulas (6.2.10) and (6.2.20) have been derived for fields of masses concentrated in finite domains of space. But these formulas can be extended, just as it was done in Sect. 2.5.7 for two-dimensional fields, to the case of particular interest for structural gravimetric prospecting where a disturbing body extends at infinity.

Consider a domain D bounded by the surface $\Gamma: z = \Delta z(z, y) - h$ and the plane $P: z = -h$; here $h \geqslant \Delta z(x, y) \geqslant 0$ and $(\Delta z(x, y) - a) \to 0$ for $\sqrt{x^2+y^2} \to \infty$, where $a = \text{const}$. We will calculate the gravitational effect of the domain D filled with disturbing masses of a uniform density ϱ_0. To this end, we shall cut from the domain D a domain D_R bounded by a spherical surface $O_R(x^2+y^2+z^2 = R^2)$ and by parts Γ_R and P_R of the surfaces Γ and P cut therefrom by the sphere O_R. The lateral surface of the sphere O_R enclosed between Γ and P will be denoted by O_R^* (Fig. 35). Designating the external gravitational field of the domain D_R by $\mathbf{g}^R(\mathbf{r}')$, we express the gravitational effect of the domain D as follows:

$$\mathbf{g}(\mathbf{r}') = \lim_{R\to\infty} \mathbf{g}^R(\mathbf{r}') \ . \tag{6.2.60}$$

The function $\mathbf{g}^R(\mathbf{r}')$ is defined, according to (6.2.11), in the form

$$\mathbf{g}^R(\mathbf{r}') = \frac{4\pi}{3}\gamma\varrho_0\mathbf{C}^{S_R}(\mathbf{r}', \mathbf{r}) \ , \tag{6.2.61}$$

where $S_R = O_R^* \cup \Gamma_R \cup P_R$ – is the boundary of the domain D_R.

Evaluate the last integral employing the property (a) of Sect. 5.2.1:

$$\mathbf{C}^{S_R}(\mathbf{r}', \mathbf{r}) = \mathbf{C}^{\Gamma_P \cup O_R^*}(\mathbf{r}'; \mathbf{r}) + \mathbf{C}^{P_R}(\mathbf{r}'; \mathbf{r}) \ . \tag{6.2.62}$$

The Cauchy-type integral over the part of the plane P_R can be evaluated in the following way:

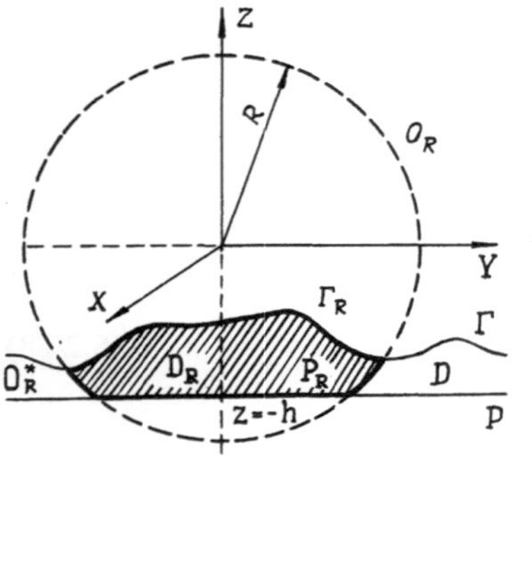

Fig. 35. Scheme explaining the procedure of calculating the gravitational field of a three-dimensional infinitely extended domain

$$\mathbf{C}^{P_R}(\mathbf{r}'; \mathbf{r}) = \mathbf{C}^{P_R}[\mathbf{r}'; \boldsymbol{\varphi}^P(\mathbf{r})] , \tag{6.2.63}$$

where $\boldsymbol{\varphi}^P(\mathbf{r})$ is a function equal to $\mathbf{r}$ in the plane P_R:

$$\boldsymbol{\varphi}^P(\mathbf{r}) = x\mathbf{d}_x + y\mathbf{d}_y - (2z+3h)\mathbf{d}_z . \tag{6.2.64}$$

Indeed, for $\mathbf{r} \in P_R$

$$\boldsymbol{\varphi}^P(\mathbf{r}) = \boldsymbol{\varphi}^P(x, y, -h) = x\mathbf{d}_x + y\mathbf{d}_y - h\mathbf{d}_z = \mathbf{r}|_{P_R} . \tag{6.2.65}$$ [3]

On the other hand, the function $\boldsymbol{\varphi}^P(\mathbf{r})$ in any finite domain of space satisfies the following equations:

$$\operatorname{div} \boldsymbol{\varphi}^P(\mathbf{r}) = 0 , \quad \operatorname{rot} \boldsymbol{\varphi}^P(\mathbf{r}) = 0 . \tag{6.2.66}$$

The above-mentioned property permits transformation of the integral in the right-hand side of (6.2.63). Indeed, by virtue of (6.2.66) and thanks to the property of the Cauchy-type integral (5.2.12), we have

$$\mathbf{C}^{S_R}[\mathbf{r}'; \boldsymbol{\varphi}^P(r)] = 0 , \tag{6.2.67}$$

if the points $\mathbf{r}'$ are outside D ($\mathbf{r}' \in C\bar{D}$). Separating the integral in (6.2.67) into two integrals – over the surface P_R and $\Gamma_R \cup O_R^*$, respectively – we find

$$\mathbf{C}^{P_R}[\mathbf{r}'; \boldsymbol{\varphi}^P(\mathbf{r})] = -\mathbf{C}^{\Gamma_P \cup O_R^*}[\mathbf{r}'; \boldsymbol{\varphi}^P(\mathbf{r})] . \tag{6.2.68}$$

Substituting (6.2.68) into (6.2.62) we write

$$\begin{aligned}\mathbf{C}^{S_R}(\mathbf{r}'; \mathbf{r}) &= \mathbf{C}^{\Gamma_P \cup O_R^*}(\mathbf{r}'; \mathbf{r}) - \mathbf{C}^{\Gamma_P \cup O_R^*}[\mathbf{r}'; \boldsymbol{\varphi}^P(\mathbf{r})] \\ &= \mathbf{C}^{\Gamma_R}[\mathbf{r}'; \mathbf{r} - \boldsymbol{\varphi}^P(\mathbf{r})] + \mathbf{C}^{O_R^*}[\mathbf{r}'; \mathbf{r} - \boldsymbol{\varphi}^P(\mathbf{r})] .\end{aligned} \tag{6.2.69}$$

Hence, by virtue of (6.2.60) and (6.2.61), we obtain

$$\mathbf{g}(\mathbf{r}') = \frac{4\pi}{3} \gamma \varrho_0 \{\lim_{R\to\infty} \mathbf{C}^{\Gamma_R}[\mathbf{r}'; \mathbf{r} - \boldsymbol{\varphi}^P(\mathbf{r})] + \lim_{R\to\infty} \mathbf{C}^{O_R^*}[\mathbf{r}'; \mathbf{r} - \boldsymbol{\varphi}^P(\mathbf{r})]\} , \tag{6.2.70}$$

where

$$\mathbf{r} - \boldsymbol{\varphi}^P(\mathbf{r}) = 3(z+h)\mathbf{d}_z . \tag{6.2.71}$$

One can readily see that the second limit in (6.2.70) is zero. Indeed, estimate the integral $\mathbf{C}^{O_R^*}$ in modulus:

$$\begin{aligned}|\mathbf{C}^{O_R^*}[\mathbf{r}'; \mathbf{r} - \boldsymbol{\varphi}^P(\mathbf{r})]| &= \frac{3}{4\pi} \left| \iint_{O_R^*} \left\{ [\mathbf{n} \cdot (z+h)\mathbf{d}_z] \operatorname{grad} \frac{1}{|\mathbf{r}-\mathbf{r}'|} \right.\right. \\ &\quad \left.\left. + [\mathbf{n} \times (z+h)\mathbf{d}_z] \times \operatorname{grad} \frac{1}{|\mathbf{r}-\mathbf{r}'|} \right\} ds \right| \\ &\leqslant \frac{6}{4\pi} \iint_{O_R^*} |\mathbf{n}|\,|z+h| \left| \operatorname{grad} \frac{1}{|\mathbf{r}-\mathbf{r}'|} \right| ds .\end{aligned}$$

[3] The relation of type (6.2.65) will be referred to in the later discussion as an equation for the surface P_R in a harmonic form (see Sect. 7.2.1).

But for large R, the condition

$$\left|\operatorname{grad}\frac{1}{|\mathbf{r}-\mathbf{r}'|}\right| < \frac{C_1}{R^2}\,, \quad |\mathbf{n}| = 1\,, \quad |z+h| = |\Delta z| < 2a$$

is definitely fulfilled. Hence,

$$|\mathbf{C}^{O_R^*}[\mathbf{r}';\, \mathbf{r} - \boldsymbol{\varphi}^P(\mathbf{r})]| \leqslant \frac{3C_1 a}{\pi R^2} \iint\limits_{O_R^*} ds\,. \tag{6.2.72}$$

For large R, the area of the surface O_R^* seems to be proportional to the area of a circular ring having a height a and radius R: $\iint\limits_{O_R^*} ds \leqslant C_2 2\pi R a$. Hence we find in the final form

$$|\mathbf{C}^{O_R^*}[\mathbf{r}';\, \mathbf{r} - \boldsymbol{\varphi}(\mathbf{r})]| \leqslant \frac{6C_1 C_2 a^2}{R} \to 0 \quad \text{for} \quad R \to \infty\,, \tag{6.2.73}$$

which was to be proved.

Thus, according to (6.2.70) and (6.2.71), we have

$$\begin{aligned}
\mathbf{g}(\mathbf{r}') &= \frac{4\pi}{3}\,\gamma\varrho_0 \mathbf{C}^{\Gamma}[\mathbf{r}';\, \mathbf{r} - \boldsymbol{\varphi}^P(\mathbf{r})] = 4\pi\gamma\varrho_0 \mathbf{C}^{\Gamma}[\mathbf{r}';\, (z+h)\mathbf{d}_z] \\
&= -\gamma\varrho_0 \iint\limits_{\Gamma} \Big\{ [\mathbf{n}\cdot(z+h)\,\mathbf{d}_z] \operatorname{grad}\frac{1}{|\mathbf{r}-\mathbf{r}'|} \\
&\quad + [\mathbf{n}\times(z+h)\,\mathbf{d}_z] \times \operatorname{grad}\frac{1}{|\mathbf{r}-\mathbf{r}'|} \Big\}\, ds\,,
\end{aligned} \tag{6.2.74}$$

where $\mathbf{n}$ points upward.

By means of a simple vector identity, formula (6.2.74) can be rewritten in the following equivalent form:

$$\begin{aligned}
\mathbf{g}(\mathbf{r}') = -\gamma\varrho_0 \iint\limits_{\Gamma} \Big[&\left(\mathbf{n}\cdot\operatorname{grad}\frac{1}{|\mathbf{r}-\mathbf{r}'|}\right)\cdot(z+h)\,\mathbf{d}_z \\
&+ \left(\mathbf{n}\times\operatorname{grad}\frac{1}{|\mathbf{r}-\mathbf{r}'|}\right)\times(z+h)\,\mathbf{d}_z \Big]\, ds\,,
\end{aligned} \tag{6.2.75}$$

which directly implies simple expressions for the scalar components of the gravitational field:

$$g_x(\mathbf{r}') = -\gamma\varrho_0 \iint\limits_{\Gamma} (z+h)\, \frac{n_z(x'-x) - n_x(z'-z)}{|\mathbf{r}-\mathbf{r}'|^3}\, ds\,, \tag{6.2.76a}$$

$$g_y(\mathbf{r}') = -\gamma\varrho_0 \iint\limits_{\Gamma} (z+h)\, \frac{n_z(y'-y) - n_y(z'-z)}{|\mathbf{r}-\mathbf{r}'|^3}\, ds\,, \tag{6.2.76b}$$

$$g_z(\mathbf{r}') = -\gamma\varrho_0 \iint\limits_{\Gamma} (z+h)\, \frac{\mathbf{n}\cdot(\mathbf{r}'-\mathbf{r})}{|\mathbf{r}-\mathbf{r}'|^3}\, ds\,. \tag{6.2.76c}$$

Note that formula (6.2.76c) for the vertical component of the gravitational field can be cast, with due reference to (6.2.34), as follows:

$$g_z(\mathbf{r}') = \gamma \varrho_0 \iint_\Gamma (z+h)\, d\Omega \;, \tag{6.2.77}$$

where $d\Omega$ is the solid angle at which the surface element ds is seen from the point $\mathbf{r}'$. It is evident that the latter formula is just the expression for the Newton potential of a double layer (coincident with the surface Γ) with a density of dipole moments

$$m_S(\mathbf{r}) = 4\pi\gamma\varrho_0(z+h) = 4\pi\gamma_0 \Delta z(x, y) \;. \tag{6.2.78}$$

We will expand explicitly the expressions for the components of the gravitational field of an infinite domain, allowing for the fact that on Γ the following relations

$$n_x ds = -\frac{\partial \Delta z(x, y)}{\partial x}\, dx dy \;, \quad n_y ds = -\frac{\partial \Delta z(x, y)}{\partial y}\, dx dy \;,$$

$$n_z ds = dx dy \;, \qquad (z+h) = \Delta z(x, y)$$

are fulfilled. Then, relations (6.2.76a) through (6.2.76c) can be rewritten as follows:

$$g_x(x', y', z') = -\gamma\varrho_0 \iint_{-\infty}^{+\infty} z(x, y)$$

$$\cdot \frac{(x'-x)+(z'-\Delta z(x, y)+h)\,\partial \Delta z(x, y)/\partial x}{[(x'-x)^2+(y'-y)^2+(z'-\Delta z(x, y)+h)^2]^{3/2}}\, dx dy \tag{6.2.79a}$$

$$g_y(x', y', z') = -\gamma\varrho_0 \iint_{-\infty}^{+\infty} \Delta z(x, y) \left[(y'-y)+(z'-\Delta z(x, y)+h)\, \frac{\partial \Delta z(x, y)}{\partial y} \right]$$

$$\cdot [(x'-x)^2+(y'-y)^2+(z-\Delta z(x, y)+h)^2]^{-3/2} dx dy \;, \tag{6.2.79b}$$

$$g_z(x', y', z') = \gamma\varrho_0 \iint_{-\infty}^{+\infty} \Delta z(x, y)$$

$$\cdot \left[(x'-x)\, \frac{\partial \Delta z(x, y)}{\partial x} + (y'-y)\, \frac{\partial \Delta z(x, y)}{\partial y} - (z'-\Delta z(x, y)+h) \right]$$

$$\cdot [(x'-x)^2+(y'-y)^2+(z'-\Delta z(x, y)+h)^2]^{-3/2} dx dy \;. \tag{6.2.79c}$$

Formulas (6.2.79a) through (6.2.79c) do provide a solution to the problem of the gravitational field of an infinitely extended domain.

Just as in a two-dimensional case, formulas (6.2.79a) through (6.2.79c) are basic to *three-dimensional linearized representations of the gravitational field.* Indeed, assume that the surface Γ is a gradually varying slightly deviating from the plane $P(z=-h)$, i.e., assume that

$$\left|\frac{\partial \Delta z(x, y)}{\partial x}\right| \ll 1 , \quad \left|\frac{\partial \Delta z(x, y)}{\partial y}\right| \ll 1 , \quad \left|\frac{\Delta z(x, y)}{h}\right| \ll 1 .$$

Then formulas (6.2.79a)–(6.2.79c) turn into the known expressions for the derivatives of the Newton potential of a double layer with a surface density equal to $q_s(x, y)=\varrho_0 \Delta z(x, y)$:

$$\begin{aligned}\mathbf{g}(\mathbf{r}') &= -\gamma\varrho_0 \iint_{-\infty}^{+\infty} \left\{[\mathbf{d}_z\cdot(z+h)\mathbf{d}_z]\operatorname{grad}\frac{1}{|\mathbf{r}-\mathbf{r}'|}\right.\\ &\quad \left.+[\mathbf{d}_z\times(z+h)\mathbf{d}_z]\times\operatorname{grad}\frac{1}{|\mathbf{r}-\mathbf{r}'|}\right\} dx\,dy\\ &= -\gamma\varrho_0 \iint_{-\infty}^{+\infty} \Delta z(x, y)\operatorname{grad}\frac{1}{|\mathbf{r}-\mathbf{r}'|}\,dx\,dy\\ &= \gamma \iint_{-\infty}^{+\infty} \varrho_0\Delta z(x, y)\operatorname{grad}'\frac{1}{|\mathbf{r}-\mathbf{r}'|}\,dx\,dy .\end{aligned} \tag{6.2.80}$$

Formula (6.2.80) is a three-dimensional analog of formulas (2.5.55).

6.2.7 Field of an Infinitely Extended Domain Filled with Masses of a Z-Variable Density

All the findings of the previous subsection can be extended to an exceptionally important practical case where the density inside an infinite domain D varies along Z. Indeed, assume that everywhere inside D $\varrho=\varrho(z)$. Following Sect. 6.2.4, we will designate $R(z)=\int \varrho(z)\,dz$. Then, according to (6.2.31), expression (6.2.61) for the gravitational field $\mathbf{g}^R$ of the part D_R of the domain D can be written in the form

$$\mathbf{g}^R(\mathbf{r}')=4\pi\gamma\mathbf{C}^{S_R}[\mathbf{r}';\, R(z)\mathbf{d}_z] . \tag{6.2.81}$$

The latter integral can be represented in the form similar to (6.2.62):

$$\mathbf{C}^{S_R}[\mathbf{r}';\, R(z)\mathbf{d}_z]=\mathbf{C}^{\Gamma_R\cup O_R^*}[\mathbf{r}';\, R(z)\mathbf{d}_z]+\mathbf{C}^{P_R}[\mathbf{r}';\, R(z)\mathbf{d}_z] ,$$

where the second term is calculated as follows:

$$\mathbf{C}^{P_R}[\mathbf{r}';\, R(z)\mathbf{d}_z]=R(-h)\mathbf{C}^{P_R}(\mathbf{r}';\, \mathbf{d}_z)=R(-h)\mathbf{C}^{P_R}[\mathbf{r}';\, \boldsymbol{\psi}(\mathbf{r})] .$$

Here, the density $\boldsymbol{\psi}(\mathbf{r})\equiv\mathbf{d}_z$. It is evident that

$$\operatorname{div}\boldsymbol{\psi}=0 , \quad \operatorname{rot}\boldsymbol{\psi}=\mathbf{0} .$$

Consequently, on the basis of the reasoning in Sect. 6.2.6 $\mathbf{C}^{P_R}(\mathbf{r}';\, \mathbf{d}_z)$ can be switched to the form

$$\mathbf{C}^{P_R}(\mathbf{r}';\ \mathbf{d}_z) = -\mathbf{C}^{\Gamma_R \cup O_R^*}(\mathbf{r}';\ \mathbf{d}_z)\ .$$

Hence

$$\mathbf{C}^{S_R}[\mathbf{r}';\ R(z)\mathbf{d}_z] = \mathbf{C}^{\Gamma_R}\{\mathbf{r}';\ [R(z)-R(-h)]\,\mathbf{d}_z\} + \mathbf{C}^{O_R^*}\{\mathbf{r}';\ [R(z)-R(-h)]\,\mathbf{d}_z\}\ . \tag{6.2.82}$$

For the sake of simplicity, take that $\Delta z(x, y) \to 0$ for $\sqrt{x^2+y^2} \to \infty$, i.e., assume in this case that $a = 0$. Then repeating the reasoning in Sect. 6.2.6 one can readily see that

$$\lim_{R\to\infty} \mathbf{C}^{O_R^*}\{\mathbf{r}';\ [R(z)-R(-h)]\,\mathbf{d}_z\} = 0\ .$$

Hence, substituting (6.2.82) into (6.2.81) and proceeding to the limit for $R \to \infty$, we derive finally the following expression for the field of the whole domain D:

$$\begin{aligned}\mathbf{g}(\mathbf{r}') &= 4\pi\gamma\mathbf{C}^{\Gamma}\{\mathbf{r}';\ [R(z)-R(-h)]\,\mathbf{d}_z\}\\ &= -\gamma \iint_{\Gamma} \left\{(\mathbf{n}\cdot[R(z)-R(-h)]\,\mathbf{d}_z)\,\mathrm{grad}\,\frac{1}{|\mathbf{r}-\mathbf{r}'|}\right.\\ &\quad \left. +\{\mathbf{n}\times[R(z)-R(-h)]\,\mathbf{d}_z\}\times\mathrm{grad}\,\frac{1}{|\mathbf{r}-\mathbf{r}'|}\right\} ds\ .\end{aligned} \tag{6.2.83}$$

As is evident, formula (6.2.83) is obtained from formula (6.2.74) by a simple substitution of $R(z)-R(-h)$ for $\varrho_0(z+h)$. In a similar way, formulas (6.2.75) through (6.2.79c) of Sect. 6.2.2 can be extended to the case of an arbitrary distribution of the density in the vertical. In particular, formula (6.2.77) for the vertical component of the gravitational field takes the form:

$$g_z(\mathbf{r}') = \gamma \iint_{\Gamma} [R(z)-R(-h)]\,d\Omega\ . \tag{6.2.84}$$

For instance, once the density of a body increases linearly with depth, then, according to (6.2.38), $R(z) = \varrho_0 z + \frac{1}{2}az^2$ and (6.2.84) is written as

$$g_z(\mathbf{r}') = \gamma \iint_{\Gamma} \varrho_0(z+h)\left(1+\frac{a}{2\varrho_0}\,z-\frac{ah}{2\varrho_0}\right) d\Omega\ . \tag{6.2.85}$$

6.3 Representation of a Fixed Magnetic Field in Terms of the Cauchy Integral Analog

Now we will represent the magnetic field in terms of three-dimensional Cauchy integral analogs over the boundary of a domain filled with magnetized masses.

6.3.1 Intensity and Potential of a Fixed Magnetic Field

First of all, we should like to recall the calculation of the magnetic field $\mathbf{H}(\mathbf{r}')$ and magnetic potential $W(\mathbf{r}')$ of a domain filled with magnetized masses with a magnetization intensity (magnetization vector) $\mathbf{I}(\mathbf{r}) = (I_x(\mathbf{r}),\ I_y(\mathbf{r}),\ I_z(\mathbf{r}))$.

Assume that $\mathbf{I}(\mathbf{r})$ is a continuously differentiable function everywhere in D right to its boundary S. The density of dipole moments of the corresponding volume polarized source is known here to be

$$\mathbf{m}(\mathbf{r}) = 4p\mathbf{I}(\mathbf{r}) \ . \tag{6.3.1}$$

Hence, according to (6.1.14), we have

$$W(\mathbf{r}') = \iiint_D \mathbf{I}(\mathbf{r})\cdot \text{grad}' \frac{1}{|\mathbf{r}-\mathbf{r}'|}\, dv \ . \tag{6.3.2}$$

As a result, the magnetic field $\mathbf{H}(\mathbf{r}')$ can be represented as follows:

$$\mathbf{H}(\mathbf{r}') = \text{grad}' \iiint_D \mathbf{I}(\mathbf{r})\cdot \text{grad}' \frac{1}{|\mathbf{r}-\mathbf{r}'|}\, dv \ . \tag{6.3.3}$$

Expression (6.3.2) for the magnetic potential can be transformed, allowing for the equivalence of a volume polarized source to a sum of simple volume source and a simple layer [formula (6.1.16)], to

$$W(\mathbf{r}') = -\iint_S \frac{\mathbf{I}(\mathbf{r})\cdot\mathbf{n}}{|\mathbf{r}-\mathbf{r}'|}\, ds + \iiint_D \frac{\text{div}\,\mathbf{I}(\mathbf{r})}{|\mathbf{r}-\mathbf{r}'|}\, dv \ , \tag{6.3.4}$$

where S is the boundary of D. Thus, the magnetic field is represented as

$$\mathbf{H}(\mathbf{r}') = -\iint_D [\mathbf{I}(\mathbf{r})\cdot\mathbf{n}]\, \text{grad}' \frac{1}{|\mathbf{r}-\mathbf{r}'|}\, ds + \iiint_D \text{div}\,\mathbf{I}(\mathbf{r})\, \text{grad}' \frac{1}{|\mathbf{r}-\mathbf{r}'|}\, dv \ . \tag{6.3.5}$$

We will express the right-hand side of formula (6.3.5) in terms of the analog of the Cauchy-type integral taken over the surface S bounding magnetized masses.

6.3.2 Representation of a Magnetic Field with an Arbitrary Distribution of Magnetized Masses

Note, first of all, that the surface integral in the right-hand side of (6.3.5) is an analog of the Cauchy-type integral with a density

$$\varphi(\mathbf{r}) = -4\pi \mathbf{I}_n(\mathbf{r})|_S \ , \tag{6.3.6}$$

where $\mathbf{I}_n(\mathbf{r})|_S$ is the normal component of the magnetization vector on the surface S:

$$-\iint_S (\mathbf{I}(\mathbf{r})\cdot\mathbf{n})\, \text{grad}' \frac{1}{|\mathbf{r}-\mathbf{r}'|}\, ds = \mathbf{C}^S[\mathbf{r}';\ -4\pi\mathbf{I}_n(\mathbf{r})] \ . \tag{6.3.7}$$

Now transform the volume integral in (6.3.5) by employing, just as in the case of the gravitational field, the three-dimensional Pompei formula (5.1.40a) cast in form (6.2.6). To this end, redefine the function $\mathbf{I}(\mathbf{r})$ to a certain domain D^* enclosing completely the domain D and the surface S, so that $\mathbf{I}(\mathbf{r})$ might be a continuously differentiable function in D^*.

Let $\mathbf{h}(\mathbf{r})$ be an arbitrary particular solution in D^* to the equations

$$\operatorname{div}\mathbf{h} = 4\pi \operatorname{div}\mathbf{I} \ , \quad \operatorname{rot}\mathbf{h} = \mathbf{0} \ . \tag{6.3.8}$$

It is noteworthy that general solution (6.3.8) in D can be immediately written in terms of the three-dimensional Pompei formula (6.2.6)

$$\mathbf{h}(\mathbf{r}') = \boldsymbol{\Psi}(\mathbf{r}') + \iiint_D \operatorname{div}\mathbf{I}\operatorname{grad}\frac{1}{|\mathbf{r}-\mathbf{r}'|}\,dv \ , \tag{6.3.9}$$

where $\boldsymbol{\Psi}(\mathbf{r}')$ is an arbitrary Laplace in D vector field.

Substituting any particular solution (6.3.8) into Pompei formula (6.2.6), we obtain

$$\mathbf{C}^S[\mathbf{r}';\, \mathbf{h}(\mathbf{r})] + \iiint_D \operatorname{div}\mathbf{I}\operatorname{grad}\frac{1}{|\mathbf{r}-\mathbf{r}'|}\,dv = \begin{cases} 0 \ , & \mathbf{r}' \in C\bar{D} \ , \\ \mathbf{h}(\mathbf{r}') \ ; & \mathbf{r}' \in D \ . \end{cases} \tag{6.3.10}$$

Hence, we derive an expression for the volume integral in (6.3.5)

$$\iiint_D \operatorname{div}\mathbf{I}\operatorname{grad}'\frac{1}{|\mathbf{r}-\mathbf{r}'|}\,dv = \begin{cases} \mathbf{C}^S[\mathbf{r}';\, \mathbf{h}(\mathbf{r})] \ , & \mathbf{r}' \in C\bar{D} \ , \\ -\mathbf{h}(\mathbf{r}') + \mathbf{C}^S[\mathbf{r}',\, \mathbf{h}(\mathbf{r})] \ , & \mathbf{r}' \in D \ . \end{cases} \tag{6.3.11}$$

Substituting (6.3.11) and (6.3.7) into (6.3.5) we determine finally

$$\mathbf{H}(\mathbf{r}') = \begin{cases} \mathbf{C}^S[\mathbf{r}';\, \mathbf{h}(\mathbf{r}) - 4\pi\mathbf{I}_n(\mathbf{r})] \ , & \mathbf{r}' \in C\bar{D} \ , \\ -\mathbf{h}(\mathbf{r}') + \mathbf{C}^S[\mathbf{r}';\, \mathbf{h}(\mathbf{r}) - 4\pi\mathbf{I}_n(\mathbf{r})] \ , & \mathbf{r}' \in D \ . \end{cases} \tag{6.3.12}$$

Thus, we have represented the magnetic field intensity $\mathbf{H}(\mathbf{r}')$ inside and outside a domain filled with magnetized masses in terms of the Cauchy-type integral. It should be noted that, by virtue of Sokhotsky-Plemelj formulas (5.2.30), on the boundary S of the domain occupied by magnetized masses, the following relations

$$\begin{aligned} \mathbf{H}^+(\mathbf{r}_0) - \mathbf{H}^-(\mathbf{r}_0) &= -\mathbf{h}(\mathbf{r}_0) + \mathbf{C}^{S_+}(\mathbf{r}';\, \mathbf{h} - 4\pi\mathbf{I}_n) - \mathbf{C}^{S_-}(\mathbf{r}';\, \mathbf{h} - 4\pi\mathbf{I}_n) \\ &= -\mathbf{h}(\mathbf{r}_0) + \mathbf{h}(\mathbf{r}_0) - 4\pi\mathbf{I}_n(\mathbf{r}_0) = -4\pi\mathbf{I}_n(\mathbf{r}_0) \ , \end{aligned} \tag{6.3.13}$$

[$\mathbf{r}_0 \in S$ and $\mathbf{I}_n(\mathbf{r}_0) = (\mathbf{I}(\mathbf{r}_0)\cdot\mathbf{n})\mathbf{n}$ being the normal component of the magnetization vector at the surface S] are satisfied. As a result, the tangential components of the magnetic field are continuous on transition through the body boundary S, while the normal components undergo a discontinuity equal to $-4\pi\mathbf{I}_n(\mathbf{r}_0)$.

The right-hand side of formula (6.3.12) can be represented, just as in the case of the gravitational field, as a unified analytical expression:

$$\mathbf{H}(\mathbf{r}') = \mathbf{C}^S(\mathbf{r}',\, \mathbf{h}(\mathbf{r}) - \mathbf{h}(\mathbf{r}') - 4\pi\mathbf{I}_n(\mathbf{r})) \ . \tag{6.3.14}$$

Expression (6.3.14) provides one of the most convenient representations of the magnetic field with arbitrarily distributed magnetized masses.

6.3.3 Potential Distribution of Magnetization

Now assume that the magnetization **I** is potential in D:

$$\mathbf{I} = \operatorname{grad} U \ , \tag{6.3.15a}$$

$$\operatorname{rot} \mathbf{I} = \mathbf{0} \ . \tag{6.3.15b}$$

Then, formulas (6.3.12) and (6.3.14) can be markedly simplified. Indeed, in this case, a particular solution to Eqs. (6.3.8) is provided by the magnetization vector itself multiplied by 4π:

$$\mathbf{h}(\mathbf{r}) = 4\pi \mathbf{I}(\mathbf{r}) \ . \tag{6.3.16}$$

Substitution of (6.3.16) into (6.3.14) yields

$$\mathbf{H}(\mathbf{r}') = 4\pi \mathbf{C}^S[\mathbf{r}';\ \mathbf{I}(\mathbf{r}) - \mathbf{I}_n(\mathbf{r}) - \mathbf{I}(\mathbf{r}')] \ . \tag{6.3.17}$$

Or, representing $\mathbf{I}(\mathbf{r})$ on S as a sum of the tangential $\mathbf{I}_\tau$ and of the normal $\mathbf{I}_n$ components:

$$\mathbf{I}(\mathbf{r}) = \mathbf{I}_\tau(\mathbf{r}) + \mathbf{I}_n(\mathbf{r}) \ , \quad \mathbf{r} \in S \ ,$$

we write

$$\mathbf{H}(\mathbf{r}') = 4\pi \mathbf{C}^S[\mathbf{r}';\ \mathbf{I}_\tau(\mathbf{r}) - \mathbf{I}(\mathbf{r}')] \ . \tag{6.3.18}$$

In particular, for the magnetic field outside masses we obtain, by virtue of the identity $\mathbf{C}^S(\mathbf{r}';\ \mathbf{I}(\mathbf{r}')) \equiv \mathbf{0}$,

$$\mathbf{H}(\mathbf{r}') = 4\pi \mathbf{C}^S[\mathbf{r}';\ \mathbf{I}_\tau(\mathbf{r})] \ , \tag{6.3.19}$$

or in greater detail

$$\mathbf{H}(\mathbf{r}') = -\iint_S \left\{ [\mathbf{n} \times \mathbf{I}_\tau(\mathbf{r})] \times \operatorname{grad} \frac{1}{|\mathbf{r}-\mathbf{r}'|} \right\} ds \ . \tag{6.3.20}$$

Using identity (6.1.22) we can cast expression (6.3.20) in the form

$$\mathbf{H}(\mathbf{r}) = \operatorname{grad}' \left\{ \iint_S U \left(\mathbf{n} \cdot \operatorname{grad} \frac{1}{|\mathbf{r}-\mathbf{r}'|} \right) ds \right\} \ , \tag{6.3.21}$$

whence the magnetic potential is

$$W(\mathbf{r}') = \iint_S U \left(\mathbf{n} \cdot \operatorname{grad} \frac{1}{|\mathbf{r}-\mathbf{r}'|} \right) ds \ . \tag{6.3.22}$$

Thus, *the magnetic potential of a body with an arbitrary potential distribution of magnetization is equal to the potential of a double layer which coincides with the boundary S of the body and has a dipole moment density equal to*

$$\mathbf{m}_S = -4\pi U \mathbf{n} \ , \tag{6.3.23}$$

where U is the potential of the magnetization density vector **I**.

With allowance for formula (6.1.13 a), expression (6.3.22) can be cast as

$$W(\mathbf{r}') = -\iint_S U(\mathbf{r})\,d\Omega \ , \tag{6.3.24}$$

where $d\Omega$ is the solid angle at which the surface element ds is seen from the point $\mathbf{r}'$.

6.3.4 Laplace Distribution of Magnetization

Now assume that the magnetization $\mathbf{I}(\mathbf{r})$ is the Laplace field everywhere in D, i.e.,

$$\operatorname{div}\mathbf{I}(\mathbf{r}) = 0 \ , \quad \operatorname{rot}\mathbf{I}(\mathbf{r}) = \mathbf{0} \ , \quad \mathbf{r} \in D \ . \tag{6.3.25}$$

We should like to elucidate how formula (6.3.14) is simplified in this case. Note that under condition (6.3.25) we can again posit that $\mathbf{h} = 4\pi\mathbf{I}$ in (6.3.14), which leads again to formula (6.3.17). But by virtue of the Laplace nature of $\mathbf{I}(\mathbf{r})$ in D, we have, according to (5.2.12),

$$\mathbf{C}^S[\mathbf{r}'; \mathbf{I}(\mathbf{r})] = \begin{cases} 0 \ , & \mathbf{r}' \in C\bar{D} \ , \\ \mathbf{I}(\mathbf{r}') \ , & \mathbf{r}' \in D \ . \end{cases} \tag{6.3.26}$$

On the other hand, it is clear that the relation

$$\mathbf{C}^S[\mathbf{r}'; \mathbf{I}(\mathbf{r}')] = \begin{cases} 0 \ , & \mathbf{r}' \in C\bar{D} \ , \\ \mathbf{I}(\mathbf{r}') \ , & \mathbf{r}' \in D \end{cases} \tag{6.3.27}$$

is valid. Comparing (6.3.27) and (6.3.26) one can notice the following important identity:

$$\mathbf{C}^S[\mathbf{r}'; \mathbf{I}(\mathbf{r})] \equiv \mathbf{C}^S[\mathbf{r}'; \mathbf{I}(\mathbf{r}')] \ . \tag{6.3.28}$$

Substitution of (6.3.28) into (6.3.17) yields

$$\mathbf{H}(\mathbf{r}') = -4\pi\mathbf{C}^S[\mathbf{r}'; \mathbf{I}_n(\mathbf{r})] \ . \tag{6.3.29}$$

Thus, we have two equivalent representations for the magnetic field of masses with a Laplace distribution of the magnetization vector $\mathbf{I}(\mathbf{r})$:

$$\mathbf{H}(\mathbf{r}') = -4\pi\mathbf{C}^S[\mathbf{r}'; \mathbf{I}_\tau(\mathbf{r}) - \mathbf{I}(\mathbf{r}')] \ , \tag{6.3.30}$$

$$\mathbf{H}(\mathbf{r}') = -4\pi\mathbf{C}^S[\mathbf{r}'; \mathbf{I}_n(\mathbf{r})] \ . \tag{6.3.31}$$

It is remarkable that formulas (6.3.30) and (6.3.31) are valid for observation points lying both outside ($\mathbf{r}' \in C\bar{D}$) and inside ($\mathbf{r}' \in D$) masses.

Expanding expression (6.3.31), according to (5.2.3), we derive in the final form

$$\mathbf{H}(\mathbf{r}') = \iint_S [\mathbf{I}(\mathbf{r})\cdot\mathbf{n}]\operatorname{grad}\frac{1}{|\mathbf{r}-\mathbf{r}'|}\,ds \ , \tag{6.3.32}$$

whence the magnetic potential $W(\mathbf{r}')$ is

$$W(\mathbf{r}') = -\iint\limits_S \frac{\mathbf{I}(\mathbf{r})\cdot\mathbf{n}}{|\mathbf{r}-\mathbf{r}'|}\,ds\ . \tag{6.3.33}$$

As is seen, the magnetic potential of a body with an arbitrary Laplace distribution of magnetization is equal to the potential of a simple layer coinciding with the boundary S of the body and having a surface density equal to

$$q_s = 4\pi\,[\mathbf{I}(\mathbf{r})\cdot\mathbf{n}]\,|_S\ . \tag{6.3.34}$$

We should like to stress that formulas (6.3.31) and (6.3.33) hold good for the whole space. Thus, *the magnetic field of a Laplace distribution of magnetization is a Laplace field both outside and inside masses.* However, the Laplace fields describing the magnetic field in D and $C\bar{D}$ are *different,* since the Cauchy-type integral (6.3.31) experiences a discontinuity on transition through the boundary of a body. Note finally that outside masses the Laplace distribution of magnetization is also defined by formulas (6.3.19), (6.3.22), and (6.3.24).

6.3.5 Magnetic Field of a Uniformly Magnetized Body

A particular but extremely important example of the Laplace magnetization distribution is the case of uniform magnetization:

$$\mathbf{I}(\mathbf{r}) = \mathbf{I}^0 = \text{const.} \tag{6.3.35}$$

In accordance with formulas (6.3.30) and (6.3.31), the magnetic field of a uniformly magnetized body can be represented as follows:

$$\mathbf{H}(\mathbf{r}') = 4\pi\,\mathbf{C}^S(\mathbf{r}';\ \mathbf{I}^0_\tau - \mathbf{I}^0) = -4\pi\,\mathbf{C}^S(\mathbf{r}';\ \mathbf{I}^0_n)\ . \tag{6.3.36}$$

It is remarkable that *representation* (6.3.36) *describes the field in the whole space: both outside and inside magnetized masses.*

If we confine ourselves only to the external field (outside masses), then, due to the identity

$$\mathbf{C}^S(\mathbf{r}';\ \mathbf{I}^0) \equiv 0\ ,\quad \mathbf{r}' \in C\bar{D}\ ,$$

the following representation of the magnetic field, equivalent to (6.3.36),

$$\mathbf{H}(\mathbf{r}') = 4\pi\,\mathbf{C}^S(\mathbf{r}';\ \mathbf{I}^0_\tau) \tag{6.3.37}$$

is valid.

Expanding the expression for the three-dimensional Cauchy integral [see Sect. (5.2.3)] we derive explicitly the following analytical representations for the magnetic field:

a) at a point of space

$$\mathbf{H}(\mathbf{r}') = \iint\limits_S (\mathbf{I}^0\cdot\mathbf{n})\,\text{grad}\,\frac{1}{|\mathbf{r}-\mathbf{r}'|}\,ds\ ; \tag{6.3.38}$$

b) outside masses

$$\mathbf{H}(\mathbf{r}') = -\iint_S \left[\text{grad} \frac{1}{|\mathbf{r}-\mathbf{r}'|} \times (\mathbf{n} \times \mathbf{I}^0) \right] ds \ . \tag{6.3.39}$$

Expressions (6.3.38) and (6.3.39) imply, in accordance with formulas (6.3.33) and (6.3.22), the following representations of the magnetic potential $W(\mathbf{r}')$:

a) at any point of space

$$W(\mathbf{r}') = -\mathbf{I}^0 \cdot \iint_S \frac{\mathbf{n}}{|\mathbf{r}-\mathbf{r}'|} ds \ ; \tag{6.3.40}$$

b) outside masses

$$W(\mathbf{r}') = \iint_S U \left(\mathbf{n} \cdot \text{grad} \frac{1}{|\mathbf{r}-\mathbf{r}'|} \right) ds \ , \tag{6.3.41}$$

where U, according to (6.3.15a), is

$$U = \mathbf{I}_0 \cdot \mathbf{r} \ . \tag{6.3.42}$$

The two formulas (6.3.40) and (6.3.41) are useful for practical solution of direct problems in magnetometry.

6.4 Generalized Kertz-Siebert Technique for Separation of Three-Dimensional Geopotential Fields

In Sect. 3.2 consideration was given to the Kertz-Siebert technique used to separate a plane field into an external and an internal part. The body of three-dimensional Cauchy integral analogs permits extension of this method to the case of three variables.

6.4.1 Statement of the Problem of Separation of a Three-Dimensional Field

Take a geopotential field $\mathbf{F}(\mathbf{r})$ produced by two sources: an external source distributed in a domain Q_e with a density $q_e(\mathbf{r})$ and an internal source concentrated with a density $q_i(\mathbf{r})$ in a domain Q_i (Fig. 36). It is evident that

$$\mathbf{F}(\mathbf{r}) = \mathbf{F}^e(\mathbf{r}) + \mathbf{F}^i(\mathbf{r}) \ , \tag{6.4.1}$$

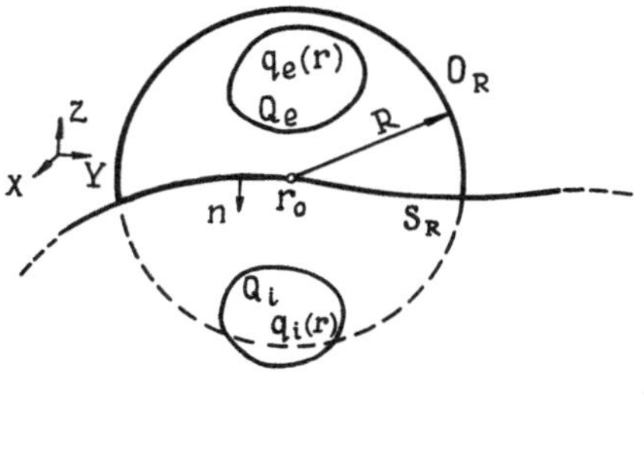

Fig. 36. Substantiation of the generalized Kertz-Siebert technique for separation of three-dimensional geopotential fields: q_e stands for external sources, q_i represents internal sources

where $\mathbf{F}^e$ and $\mathbf{F}^i$ are the fields produced by each of the sources q_e and q_i separately. These fields are called, just as in a two-dimensional case, an *external* and an *internal* field. Let S designate an arbitrary piecewise smooth surface separating the domains Q_e and Q_i, bearing in mind that S may be either closed or go through a point at infinity. The problem is to find, knowing the field $\mathbf{F}$ specified on S, the fields $\mathbf{F}^e$ and $\mathbf{F}^i$ on S. Let us solve this problem.

6.4.2 Separation of Fields at Ordinary Points of the Surface

Consider an arbitrary point $\mathbf{r}_0$ belonging to the surface S and draw therefrom a sphere of such a large radius R that the domain Q_e might be completely enclosed in this sphere. Part of the sphere bounded by the surface S and lying in the same half-space as Q_e will be denoted by O_R. Part of the surface S bounded by the sphere will be represented by S_R.

We will determine a three-dimensional analog of the Cauchy-type integral over a closed surface $\Gamma_R = O_R \cup S_R$: $\mathbf{C}^{\Gamma_R}(\mathbf{r}'; \mathbf{F}(\mathbf{r}))$. By virtue of formula (6.4.1) and thanks to the property of linearity of the Cauchy-type integral, we have

$$\mathbf{C}^{\Gamma_R}[\mathbf{r}'; \mathbf{F}(\mathbf{r})] = \mathbf{C}^{\Gamma_R}[\mathbf{r}'; \mathbf{F}^e(\mathbf{r})] + \mathbf{C}^{\Gamma_R}[\mathbf{r}'; \mathbf{F}^i(\mathbf{r})] \ . \tag{6.4.2}$$

Let D_R designate the domain bounded by Γ_R. Since the field $\mathbf{F}^i$ is the Laplace field everywhere inside D_R and continuous in $\bar{D}_R$, we have, according to the three-dimensional analog of the Cauchy integral formula (5.2.12),

$$\mathbf{C}^{\Gamma_R}[\mathbf{r}'; \mathbf{F}^i(\mathbf{r})] = \begin{cases} \mathbf{F}^i(\mathbf{r}') \ , & \mathbf{r}' \in D_R \ , \\ \mathbf{0} \ , & \mathbf{r}' \in C\bar{D}_R \ . \end{cases} \tag{6.4.3}$$

Similarly, as $\mathbf{F}^e$ is the Laplace field everywhere outside $\bar{D}_R$ (in the domain $C\bar{D}_R$), it is continuous in CD_R, and vanishes at infinity, we have, in accordance with (5.1.37),

$$\mathbf{C}^{\Gamma_R}[\mathbf{r}'; \mathbf{F}^e(\mathbf{r})] = \begin{cases} \mathbf{0} \ , & \mathbf{r}' \in D_R \ , \\ -\mathbf{F}(\mathbf{r}') \ , & \mathbf{r}' \in C\bar{D}_R \ . \end{cases} \tag{6.4.4}$$

Substituting formulas (6.4.3) and (6.4.4) into (6.4.2) we obtain

$$\mathbf{C}^{\Gamma_R}[\mathbf{r}'; \mathbf{F}(\mathbf{r})] = \begin{cases} \mathbf{F}^i(\mathbf{r}') \ , & \mathbf{r}' \in D_R \ , \\ -\mathbf{F}^e(\mathbf{r}') \ , & \mathbf{r}' \in C\bar{D}_R \ . \end{cases} \tag{6.4.5}$$

Now let the point $\mathbf{r}'$ in formula (6.4.5) tend to the surface S, remaining either inside the domain $D_R(\mathbf{r}' \to \mathbf{r}_0^+)$ or outside it $(\mathbf{r}' \to \mathbf{r}_0^-)$. Then, by virtue of the continuity of the external and the internal field in the neighborhood of the point $\mathbf{r}_0 \in S$ we find

$$\begin{aligned} \mathbf{F}^i(\mathbf{r}_0) &= \lim_{r' \to r_0^+} \mathbf{C}^{\Gamma_R}[\mathbf{r}'; \mathbf{F}(\mathbf{r})] = \mathbf{C}^{\Gamma_{R^+}}[\mathbf{r}_0; \mathbf{F}(\mathbf{r})] \ , \\ \mathbf{F}^e(\mathbf{r}_0) &= -\lim_{r' \to r_0^-} \mathbf{C}^{\Gamma_R}[\mathbf{r}'; \mathbf{F}(\mathbf{r})] = -\mathbf{C}^{\Gamma_{R^-}}[\mathbf{r}_0; \mathbf{F}(\mathbf{r})] \ . \end{aligned} \tag{6.4.6}$$

Assume that $\mathbf{r}_0$ is an ordinary point of the surface S (i.e., at this point there is a tangential plane to S). Then, substituting Sokhotsky-Plemelj formulas (5.2.32a) and (5.2.32b) into (6.4.6) we write

$$\begin{aligned} \mathbf{F}^i(\mathbf{r}_0) &= \tfrac{1}{2}\mathbf{F}(\mathbf{r}_0)+\mathbf{C}^{\Gamma_R}[\mathbf{r}_0;\ \mathbf{F}(\mathbf{r})]\ , \\ \mathbf{F}^e(\mathbf{r}_0) &= \tfrac{1}{2}\mathbf{F}(\mathbf{r}_0)-\mathbf{C}^{\Gamma_R}[\mathbf{r}_0;\ \mathbf{F}(\mathbf{r})]\ , \end{aligned} \tag{6.4.7}$$

where $\mathbf{C}^{\Gamma_R}(\mathbf{r}_0;\ \mathbf{F}(\mathbf{r}))$ is the singular integral in terms of the Cauchy principal value.

Let us evaluate the limit of the singular integral for $R\to\infty$. To this end, we will represent it in the form

$$\mathbf{C}^{\Gamma_R}[\mathbf{r}_0;\ \mathbf{F}(\mathbf{r})] = \mathbf{C}^{O_R}[\mathbf{r}_0;\ \mathbf{F}(\mathbf{r})]+\mathbf{C}^{S_R}[\mathbf{r}_0;\ \mathbf{F}(\mathbf{r})]\ . \tag{6.4.8}$$

Transformations similar to those in Sect. 5.1.5 render the first integral in the right-hand side of Eq. (6.4.8) to the form

$$\mathbf{C}^{O_R}[\mathbf{r}_0;\ \mathbf{F}(\mathbf{r})] = \frac{1}{4\pi R^2}\iint_{O_R} \mathbf{F}(\mathbf{r})\,ds\ . \tag{6.4.9}$$

Since $\lim\limits_{|\mathbf{r}|\to\infty} \mathbf{F}(\mathbf{r}) = 0$, the right-hand side of (6.4.9) also tends to zero at $R\to\infty$. Hence, we have

$$\lim_{R\to\infty} \mathbf{C}^{\Gamma_R}[\mathbf{r}_0;\ \mathbf{F}(\mathbf{r})] = \lim_{R\to\infty} \mathbf{C}^{S_R}[\mathbf{r}_0;\ \mathbf{F}(\mathbf{r})] = \mathbf{C}^{S}[\mathbf{r}_0;\ \mathbf{F}(\mathbf{r})]\ . \tag{6.4.10}$$

Thus, proceeding to the limit for $R\to\infty$ in (6.4.7) and allowing for (6.4.10) we derive

$$\begin{aligned} \mathbf{F}^i(\mathbf{r}_0) &= \tfrac{1}{2}\mathbf{F}(\mathbf{r}_0)+\mathbf{C}^{S}[\mathbf{r}_0;\ \mathbf{F}(\mathbf{r})]\ , \\ \mathbf{F}^e(\mathbf{r}_0) &= \tfrac{1}{2}\mathbf{F}(\mathbf{r}_0)-\mathbf{C}^{S}[\mathbf{r}_0;\ \mathbf{F}(\mathbf{r})]\ . \end{aligned} \tag{6.4.11}$$

Expanding the expression for the singular integral in an explicit form we write finally

$$\begin{aligned} \mathbf{F}^i(\mathbf{r}_0) &= \tfrac{1}{2}\mathbf{F}(\mathbf{r}_0)-\frac{1}{4\pi}\iint_S \left[(\mathbf{n}\cdot\mathbf{F})\,\mathrm{grad}\,\frac{1}{|\mathbf{r}-\mathbf{r}_0|}+(\mathbf{n}\times\mathbf{F})\times\mathrm{grad}\,\frac{1}{|\mathbf{r}-\mathbf{r}_0|}\right] ds\ ; \\ \mathbf{F}^e(\mathbf{r}_0) &= \tfrac{1}{2}\mathbf{F}(\mathbf{r}_0)+\frac{1}{4\pi}\iint_S \left[(\mathbf{n}\cdot\mathbf{F})\,\mathrm{grad}\,\frac{1}{|\mathbf{r}-\mathbf{r}_0|}+(\mathbf{n}\times\mathbf{F})\times\mathrm{grad}\,\frac{1}{|\mathbf{r}-\mathbf{r}_0|}\right] ds\ , \end{aligned} \tag{6.4.12}$$

where the normal $\mathbf{n}$ points downward, as is shown in Fig. 36.

Formulas (6.4.12) *solve the problem of separating an arbitrary potential field* $\mathbf{F}(\mathbf{r})$ *specified on a piecewise smooth surface S into an external and an internal part at ordinary points of this surface.*

6.4.3 Separation of Fields at Singular Points of the Surface

Let $\mathbf{r}_0 \in S$ be a singular point of the surface S and $\theta(\mathbf{r}_0)$ be the value of the solid angle (in steradians) of a conic surface formed by one-side tangents to S at the point $\mathbf{r}_0$ (see Sect. 5.2.2). In this case, the limit values of the Cauchy-type integrals in formula (6.4.6) are defined by relations (5.2.29)

$$\mathbf{F}^i(\mathbf{r}_0) = \left(1-\frac{\theta(\mathbf{r}_0)}{4\pi}\right)\mathbf{F}(\mathbf{r}_0)+\mathbf{C}^{\Gamma_R}(\mathbf{r}_0;\, \mathbf{F}(\mathbf{r}))\ ,$$
$$\mathbf{F}^e(\mathbf{r}_0) = \frac{\theta(\mathbf{r}_0)}{4\pi}\mathbf{F}(\mathbf{r}_0)-\mathbf{C}^{\Gamma_R}(\mathbf{r}_0;\, \mathbf{F}(\mathbf{r}))\ . \tag{6.4.13}$$

Passing, just as previously, to the limit for $R\to\infty$ in formulas (6.4.13) and expanding explicitly the values of the Cauchy-type singular integral, we find in the final form

$$\mathbf{F}^i(\mathbf{r}_0) = \left(1-\frac{\theta(\mathbf{r}_0)}{4\pi}\right)\mathbf{F}(\mathbf{r}_0)-\frac{1}{4\pi}\iint_S\left\{(\mathbf{n}\cdot\mathbf{F})\,\mathrm{grad}\,\frac{1}{|\mathbf{r}-\mathbf{r}_0|}\right.$$
$$\left.+[\mathbf{n}\times\mathbf{F}]\times\mathrm{grad}\,\frac{1}{|\mathbf{r}-\mathbf{r}_0|}\right\}ds\ ;$$
$$\mathbf{F}^e(\mathbf{r}_0) = \frac{\theta(\mathbf{r}_0)}{4\pi}\mathbf{F}(\mathbf{r}_0)+\frac{1}{4\pi}\iint_S\left\{(\mathbf{n}\cdot\mathbf{F})\,\mathrm{grad}\,\frac{1}{|\mathbf{r}-\mathbf{r}_0|}\right.$$
$$\left.+[\mathbf{n}\times\mathbf{F}]\times\mathrm{grad}\,\frac{1}{|\mathbf{r}-\mathbf{r}_0|}\right\}ds\ . \tag{6.4.14}$$

The latter relations enable us to solve the problem of separating potential fields at singular points of the surface S.

6.4.4 Generalized Kertz-Siebert Formulas

Now assume that the surface S coincides with the coordinate plane XY. Then, formulas (6.4.14) take the following form (allowing for the fact that the normal $\mathbf{n}$ points in the opposite direction with respect to the z-axis):

$$\mathbf{F}^i(x_0,\, y_0,\, 0) = \frac{1}{2}\mathbf{F}(\mathbf{r}_0)+\frac{1}{4\pi}\iint_{-\infty}^{+\infty}$$
$$\cdot\frac{(x-x_0)F_z\mathbf{d}_x+(y-y_0)F_z\mathbf{d}_y-[(x-x_0)F_x+(y-y_0)F_y]\mathbf{d}_z}{[(x-x_0)^2+(y-y_0)^2]^{3/2}}\,dx\,dy\ ; \tag{6.4.15 a}$$

$$\mathbf{F}^e(x_0, y_0, 0) = \frac{1}{2}\,\mathbf{F}(\mathbf{r}_0) - \frac{1}{4\pi} \iint\limits_{-\infty}^{+\infty} \cdot \frac{(x-x_0)F_z\mathbf{d}_x + (y-y_0)F_z\mathbf{d}_y - [(x-x_0)F_x + (y-y_0)F_y]\,\mathbf{d}_z}{[(x-x_0)^2 + (y-y_0)^2]^{3/2}}\, dx\,dy \; . \tag{6.4.15b}$$

If three-dimensional Hilbert operators (5.3.28a) and (5.3.28b) are employed, the latter relations take the form

$$F_x^i = \tfrac{1}{2}(F_x - \hat{G}_x F_z) \; , \quad F_y^i = \tfrac{1}{2}(F_y - \hat{G}_y F_z) \; , \quad F_z^i = \tfrac{1}{2}(F_z + \hat{G}_x F_x + \hat{G}_y F_y) \; ; \tag{6.4.16a}$$

$$F_x^e = \tfrac{1}{2}(F_x + \hat{G}_x F_z) \; , \quad F_y^e = \tfrac{1}{2}(F_y + \hat{G}_y F_z) \; , \quad F_z^e = \tfrac{1}{2}(F_z - \hat{G}_x F_x - \hat{G}_y F_y) \; . \tag{6.4.16b}$$

Formulas (6.4.16a) and (6.4.16b) are *extensions of the Kertz-Siebert formulas* (3.2.19) *and* (3.2.20) *to a three-dimensional case.* Just as in a two-dimensional case, three-dimensional Kertz-Siebert formulas make it possible to write the integral relations between the components of an external or an internal field. Indeed, taking a zero external field in (6.4.16a), we have

$$F_x^i = -\hat{G}_x F_z^i \; , \quad F_y^i = -\hat{G}_y F_z^i \; , \quad F_z^i = \hat{G}_x F_x^i + \hat{G}_y F_y^i \; . \tag{6.4.17a}$$

Similarly, assuming that the internal field in (6.4.16b) is zero, we find

$$F_x^e = \hat{G}_x F_z^e \; , \quad F_y^e = \hat{G}_y F_z^e \; , \quad F_z^e = -\hat{G}_x F_x^e - \hat{G}_y F_y^e \; . \tag{6.4.17b}$$

Thus, knowing the vertical component of an external or an internal field we can establish the field itself.

7 Analytical Continuation of a Three-Dimensional Geopotential Field

7.1 Fundamentals of Analytical Continuation of the Laplace Field

In Chap. 4 we recalled the fundamentals underlying the theory of analytical continuation of functions of a complex variable. It turns out that a greater part of the concepts of this theory also apply to three-dimensional Laplace vector fields described by real vector functions of a real variable. The basis for this relationship is provided by the analytical nature of Laplace vector fields.

7.1.1 Analytical Nature of Laplace Vector Fields

It is worth remembering that the neighborhood $\varepsilon(\mathbf{r}_0)$ of the point $\mathbf{r}_0$ in space implies the domain bounded by a sphere of a radius ε circumscribed from the point $\mathbf{r}_0$ as a center. *The function of three variables $U(\mathbf{r}) = U(x, y, z)$ is called analytical in the domain D,* provided one can specify the neighborhood of any inner point $\mathbf{r}_0(x_0, y_0, z_0)$ from D where $U(\mathbf{r})$ is the sum of an absolutely convergent power series (Taylor series):

$$U(\mathbf{r}) = U(x, y, z) = \sum_{k=0}^{\infty} \sum_{l=0}^{\infty} \sum_{m=0}^{\infty} U_{klm}(x-x_0)^k (y-y_0)^l (z-z_0)^m , \tag{7.1.1}$$

where U_{klm} stands for the Taylor series coefficients.

The above definition is readily extended to the case of vector fields. *The vector field $\mathbf{F}(\mathbf{r})$ is called analytical in the domain D,* provided there is a neighborhood for any inner point $\mathbf{r}' \in D$ where $\mathbf{F}(\mathbf{r})$ can be represented as an absolutely convergent power series:

$$\mathbf{F}(\mathbf{r}) = \sum_{k=0}^{\infty} \sum_{l=0}^{\infty} \sum_{m=0}^{\infty} \mathbf{f}_{klm}(x-x_0)^k (y-y_0)^l (z-z_0)^m , \tag{7.1.2}$$

where $\mathbf{f}_{klm}$ is the power series vector coefficient.

Now we will show that any vector field which is the Laplace field in D is analytical in D. For this purpose, we will fix on a point $\mathbf{r}_0$ from D and circumscribe therefrom a sphere S_a of a radius a lying completely inside D (Fig. 37). Using the three-dimensional analog of the Cauchy integral formula (5.1.26) we can represent $\mathbf{F}(\mathbf{r}')$ at any point $\mathbf{r}'$ lying inside the sphere S_a in terms of the values of this field on the sphere itself:

$$\mathbf{F}(\mathbf{r}') = -\frac{1}{4\pi} \iint_{S_a} \left[(\mathbf{n}\cdot\mathbf{F})\,\mathrm{grad}\,\frac{1}{|\mathbf{r}-\mathbf{r}'|} + (\mathbf{n}\times\mathbf{F})\times\mathrm{grad}\,\frac{1}{|\mathbf{r}-\mathbf{r}'|} \right] ds . \tag{7.1.3}$$

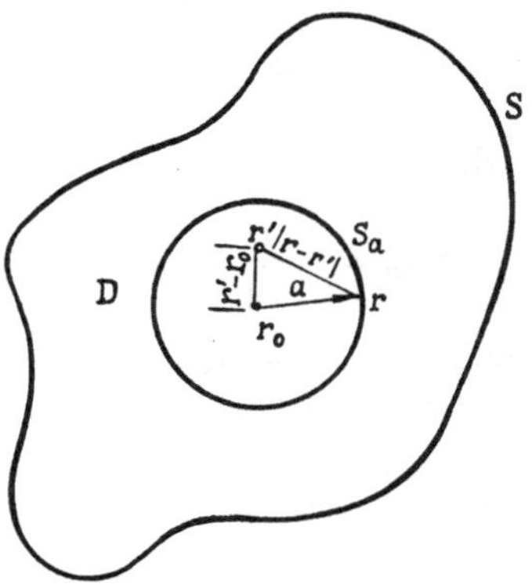

Fig. 37. Proof of the analyticity of the Laplace field: $\mathbf{r}_0$ is the center of the circle of convergence of a power series

Transform the expression $|\mathbf{r}-\mathbf{r}'|$ by the theorem of Carnot (see Fig. 37)

$$|\mathbf{r}-\mathbf{r}'| = [a^2+|\mathbf{r}'-\mathbf{r}_0|^2+2(\mathbf{r}-\mathbf{r}_0)\cdot(\mathbf{r}'-\mathbf{r}_0)]^{1/2} = a\sqrt{1+t} \ , \tag{7.1.4}$$

where

$$t = \frac{1}{a^2}[|\mathbf{r}'-\mathbf{r}_0|^2-2(\mathbf{r}-\mathbf{r}_0)\cdot(\mathbf{r}'-\mathbf{r}_0)] = \frac{1}{a^2}[(x'-x_0)^2+(y'-y_0)^2$$

$$+(z'-z_0)^2-2(x-x_0)(x'-x_0)-2(y-y_0)(y'-y_0)-2(z-z_0)(z'-z_0)] \ , \tag{7.1.5}$$

and $-1<t<3$. Hence, we have

$$\frac{1}{|\mathbf{r}-\mathbf{r}'|} = a^{-1}(1+t)^{-1/2} \ . \tag{7.1.6}$$

The expression $(1+t)^{-1/2}$ for $|t|<1$ can be expanded into an absolutely convergent series involving the powers of t:

$$(1+t)^{-1/2} = \sum_{n=0}^{\infty} c_n t^n \ . \tag{7.1.7}$$

It follows from (7.1.5) that t is expressed through the first and second powers of $(x'-x_0)$, $(y'-y_0)$, and $(z'-z_0)$. Hence, substituting (7.1.5) into (7.1.7) and the latter relation into (7.1.6), we write in the final form

$$\frac{1}{|\mathbf{r}-\mathbf{r}'|} = \sum_{k=0}^{\infty}\sum_{l=0}^{\infty}\sum_{m=0}^{\infty} b_{klm}(x'-x_0)^k(y'-y_0)^l(z'-z_0)^m \ , \tag{7.1.8}$$

where the coefficients $b_{klm} = b_{klm}(\mathbf{r}|\mathbf{r}_0)$ are the functions dependent upon the locations of the points $\mathbf{r}$ and $\mathbf{r}_0$. This expansion holds good under the condition that $|t|<1$. According to (7.1.5), this condition is fulfilled if

$$|\mathbf{r}_0-\mathbf{r}'|<\delta \ , \tag{7.1.9}$$

where $\delta = (\sqrt{2}-1)a$.

Substituting power series (7.1.8) into (7.1.3) and allowing for the fact that this series admits termwise differentiation and integration, we obtain

$$\mathbf{F}(\mathbf{r}') = \sum_{k=0}^{\infty} \sum_{l=0}^{\infty} \sum_{m=0}^{\infty} \mathbf{f}_{klm}(x'-x_0)^k(y'-y_0)^l(z'-z_0)^m \; , \tag{7.1.10}$$

where

$$\mathbf{f}_{klm} = -\frac{1}{4\pi} \iint_{S_a} [(\mathbf{n}\cdot\mathbf{F})\,\mathrm{grad}\, b_{klm} + (n\times\mathbf{F})\times\mathrm{grad}\, b_{klm}]\; ds \; . \tag{7.1.11}$$

The latter expansion, just as (7.1.8) is valid in the neighborhood of $|\mathbf{r}_0 - \mathbf{r}'| < \delta$.

Thus, we have established the neighborhood of the point $\mathbf{r}_0$ within which the Laplace field $\mathbf{F}$ is representable in the form of an absolutely and uniformly convergent power series. Similarly, one can show that $\mathbf{F}(\mathbf{r})$ is expanded in power series in the neighborhood of any inner point of the domain D. Consequently, everywhere in D the Laplace field $\mathbf{F}(\mathbf{r})$ is an analytical function.

7.1.2 Uniqueness of Laplace Vector Fields and Harmonic Functions

The analytical nature of Laplace vector fields ensures the property of *uniqueness* which is formulated as follows. *If two fields* $\mathbf{F}_1(\mathbf{r})$ *and* $\mathbf{F}_2(\mathbf{r})$, *which are Laplace fields in the domain D, take identical values in some neighborhood* $\varepsilon(\mathbf{r}_0)$ *of a point* $\mathbf{r}_0$ *of this domain, they coincide everywhere inside D.*

The proof of this important claim will be presented following the work of Timan and Trofimov (1968). Let $\mathbf{F}(\mathbf{r})$ denote the difference $\mathbf{F}(\mathbf{r}) = \mathbf{F}_1(\mathbf{r}) - \mathbf{F}_2(\mathbf{r})$. It seems evident that $\mathbf{F}(\mathbf{r}) \equiv 0$ for $\mathbf{r} \in \varepsilon(\mathbf{r}_0)$.

Let $\mathbf{r}$ be an arbitrary point out of D. Connect the point $\mathbf{r}$ with $\mathbf{r}_0$ by an arbitrary piecewise smooth line l lying inside D. Take δ_1 to be smaller than the distance[1] between l and S (S is the boundary of D). Assume also that $\delta_1 < \varepsilon$.

Let $\delta\,(\delta < \delta_1)$ be a radius of the neighborhood where expansion (7.1.10) holds. Consider a sphere of a radius equal to $\frac{3}{4}\delta$ with its center at the point $\mathbf{r}_0$. Everywhere in this sphere $\mathbf{F}(\mathbf{r}) \equiv 0$. As a result, once the point $\mathbf{r}$ is inside this sphere, the vector field at this point is zero, which was to be proved.

In the opposite case we deal with a sphere $O(\mathbf{r}_1)$ of a radius equal to $\frac{3}{4}\delta$ with its center at the point $\mathbf{r}_1$ common for the curve l and for the boundary of the former sphere with its center at the point $\mathbf{r}_0$ (Fig. 38). By virtue of the analyticity of the field everywhere in the neighborhood of the point $\mathbf{r}_1(x, y, z)$ of the radius δ, we find the expansion

$$\mathbf{F}(\mathbf{r}') = \sum_{k=0}^{\infty} \sum_{l=0}^{\infty} \sum_{m=0}^{\infty} \mathbf{f}_{klm}(x'-x_1)^k(y'-y_1)^l(z'-z_1)^m \; . \tag{7.1.12}$$

On the other hand, since $\mathbf{r}_1 \in \varepsilon(\mathbf{r}_0)$ (see Fig. 38), in a certain specific neighborhood of $\mathbf{r}_1$ $\mathbf{F}(\mathbf{r}) \equiv 0$ and, as a consequence, all the coefficients of expansion (7.1.12) vanish in the sphere $O(\mathbf{r}_1)$. If $\mathbf{r} \in O(\mathbf{r}_1)$, the proof has been already presented.

[1] The distance between l and S is defined as the lower limit between all possible pairs of points, one of which lies on l, while the other belongs to S.

Fig. 38. Illustration of the proof of the uniqueness property of Laplace vector fields; $\mathbf{r}_0$, $\mathbf{r}_1$, $\mathbf{r}_2$, and $\mathbf{r}_3$ are the centers of the circles of convergence of power series

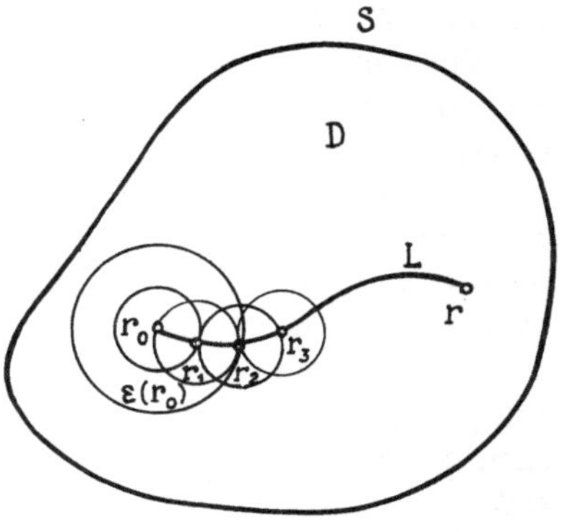

Otherwise, we consider a sphere $O(\mathbf{r}_2)$ of a $\frac{3}{4}\delta$-radius with its center at a point $\mathbf{r}_2$ common for the boundary of the latter sphere and the arc of the curve l connecting $\mathbf{r}_0$ and $\mathbf{r}$, etc. Since the curve l is of a finite length, we get to the point $\mathbf{r}$ in a finite number of steps. Thus, we have proved that $\mathbf{F}(\mathbf{r})\equiv 0$ everywhere in D and, hence, $\mathbf{F}_1(\mathbf{r})\equiv\mathbf{F}_2(\mathbf{r})$ everywhere in D, too.

Note that the property of uniqueness of Laplace vector fields in the scalar case implies uniqueness of harmonic functions. The property of uniqueness of functions permitting expansion in a power series suggests, in turn, that *if a function analytical in the domain D is harmonic in the neighborhood* $O(\mathbf{r}_0)$ *of some point* $\mathbf{r}_0$ *out of D, it is harmonic in the whole domain.*

Indeed, considering the function $\Delta U(\mathbf{r})$ [Laplace operator of $U(\mathbf{r})$] and its corresponding expansion in a power series, we see that it is analytical everywhere in the domain D [because $U(\mathbf{r})$ is analytical], and by virtue of the harmonicity of $U(\mathbf{r})$ it is identically zero in $O(\mathbf{r}_0)$. Hence, by virtue of the uniqueness property, $\Delta U(\mathbf{r})$ vanishes everywhere inside D, i.e., $U(\mathbf{r})$ is a function harmonic in D.

7.1.3 Concept of Analytical Continuation of a Vector Field and Its Riemann Space

Just as in a two-dimensional case, we can introduce the concept of analytical continuation of three-dimensional vector fields.

Definition 7.1.1. Analytical continuation of a vector field $\mathbf{\Phi}_0(\mathbf{r})$ specified in some domain D_0 of space implies its redefinition to the field $\mathbf{\Phi}(\mathbf{r})$ specified in a larger domain $D \supset D_0$ so that $\mathbf{\Phi}(\mathbf{r})$ is analytical in D, while its striction in the domain D_0 coincides with $\mathbf{\Phi}_0(\mathbf{r})$: $\mathbf{\Phi}(\mathbf{r})|_{D_0}\equiv\mathbf{\Phi}_0(\mathbf{r})$.

By virtue of the uniqueness property formulated in the foregoing it is evident that if analytical continuation is existent it is unique indeed.

Analogous to the two-dimensional case (Sect. 4.1.3), the analytical field $\mathbf{\Phi}(\mathbf{r})$ together with the domain D of its definition will be called an *element* and denoted by $(\mathbf{\Phi}, D)$. The element $(\mathbf{\Phi}_2, D_2)$ is termed *direct analytical continuation* of the element ($\mathbf{\Phi}_1$, D_1), provided $D_1 \cap D_2 = \Delta$ is a domain and $\mathbf{\Phi}_1 \equiv \mathbf{\Phi}_2$ in Δ. A set of a finite number of elements ($\mathbf{\Phi}_0$, D_0), ($\mathbf{\Phi}_1$, D_1), ... ($\mathbf{\Phi}_M$, D_M) forms a *chain* if each subsequent element in this set is the direct analytical continuation of the preceding element.

Just as in a two-dimensional case, we can define *the concept of a general analytical function of a vector field as a set $\mathscr{H}$ of elements* ($\mathbf{\Phi}$, D) *where one element of any two is derived from the other by means of a chain all elements of which belong to $\mathscr{H}$.* Vector fields $\mathbf{\Phi}_j$ belonging to different elements will be referred to as *branches* of a general analytical function of a vector field. The general analytical function of a vector field containing all analytical continuations of each of its elements will be termed a *complete analytical function of a vector field.*

Obviously, the situation depicted in Fig. 16 may be encountered in a three-dimensional case, too, i.e., analytical continuation of a vector field $\mathbf{\Phi}_0$ may bring us back to the range of definition of the initial element ($\mathbf{\Phi}_0$, D_0). Thus, general and complete analytical functions of a vector field may be *multiple-valued*, as is the case with the theory of functions of a complex variable.

To study multiple-valued analytical functions and multiple-valued analytical vector fields in a three-dimensional space, it is necessary, just as in a two-dimensional case, to deal with multiple-sheeted spaces composed of several copies (sheets) of three-dimensional domains rather than with ordinary domains of these functions. These multiple-sheeted spaces can be thought of as located above ordinary three-dimensional domains, with so many sheets (copies of three-dimensional domains) above the point $\mathbf{r}'$ as is the number of values ascribed to this point by an analytical function or by an analytical vector field. Thus, in these spaces, (general and complete) analytical functions of a vector field can be treated as *single-valued* functions, which simplifies their study very much.

The above-introduced multiple-sheeted space is called Riemann space of a given analytical function of three real variables or of a given analytical vector field. Thus, a *complete analytical function of a vector field is a function of a point of its Riemann space (rather than of a point* $\mathbf{r}'$ *of the ordinary Euclidean space).* This circumstance should be invariably borne in mind while analyzing analytical continuation of vector fields.

We should like to stress here that it is not our purpose to provide the reader with a comprehensive theory of Riemann spaces, which can be found in the relevant mathematical courses. We just introduce the basic ideas underlying the analytical continuation of vector fields that are needed for understanding the subsequent discussion.

Analogously to the analytical continuation of an arbitrary analytical vector field, we can also define the concept of analytical continuation of the Laplace field.

Definition 7.1.2. Analytical continuation of the Laplace field $\mathbf{F}_0(\mathbf{r})$ specified in some domain of space implies its redefinition to the function $\mathbf{F}(\mathbf{r})$ specified in a larger domain $D \supset D_0$ so that $\mathbf{F}(\mathbf{r})$ is the Laplace field in D and its striction in the domain D_0 coincides with $\mathbf{F}_0(\mathbf{r})$: $\mathbf{F}(\mathbf{r})|_{D_0} \equiv \mathbf{F}_0(\mathbf{r})$.

By virtue of the uniqueness property of Laplace vector fields which was proved in the foregoing, it is evident that if analytical continuation is existent it is unique indeed.

All the concepts introduced in the above discussion for describing analytical continuation of an arbitrary vector field seem to apply to Laplace fields as well. In particular, we can talk about an *element, direct analytical continuation, chain, general and complete analytical functions of the Laplace field and about its Riemann space.* All these concepts will be employed later.

In concluding this subsection, we wish to mention that the property of uniqueness of Laplace vector fields has been virtually proved on the basis of their analyticity. Therefore, the property of uniqueness can be formulated for any analytical vector field: *if two fields* $\mathbf{\Phi}_1(\mathbf{r})$ *and* $\mathbf{\Phi}_2(\mathbf{r})$ *analytical in the domain D take the same values in the neighborhood* $\varepsilon(\mathbf{r}_0)$ *of a point* $\mathbf{r}_0$ *of this domain, they coincide everywhere inside D.*

7.1.4 Continuation of the Laplace Field Using the Taylor Series

Owing to the analytical nature of Laplace vector fields, we can talk about their analytical continuation. Indeed, let the Laplace field $\mathbf{F}(\mathbf{r})$ be specified in a domain D of space. Expand the Laplace field into power series in the neighborhoods of all inner points of the domain D. These power series converge within the corresponding spherical neighborhoods of the points $\varepsilon(\mathbf{r})$, where ε stands for the radii of convergence of the power series. Consider the domain $D^{(1)}$ formed by combining all the spheres $\varepsilon(\mathbf{r})$. This domain may turn out to be larger than the initial domain, $D^{(1)} \supset D$. The field $\mathbf{F}(\mathbf{r})$ has been specified only in D, but since power series of type (7.1.10) permit calculation of the Laplace field at any point within the sphere of convergence, $\mathbf{F}(\mathbf{r})$ can be also evaluated in the domain $D^{(1)}$. Now expand again $\mathbf{F}(\mathbf{r})$ in power series in the neighborhood of any point $\mathbf{r}$ out of $D^{(1)}$. Construct the corresponding spheres of convergence and, having combined them, form a new domain $D^{(2)}$. If the domain $D^{(2)}$ exceeds $D^{(1)}$, $D^{(2)} \supset D^{(1)}$, this implies that we succeeded in continuing the Laplace field further beyond the domain D. Reiterating this operation, we will continue the function $\mathbf{F}(\mathbf{r})$ into all possible domains of space. This continuation is known as *Weierstrass analytical continuation,* while the resultant function is a *complete analytical function of the Laplace vector field.*

Points to which the Laplace field cannot be continued in the above way are termed its *singular points.*

Thus, *a singular point of the Laplace field (or of any analytical vector field) is a point in whose neighborhood the field cannot be represented by a convergent power series.* Sets of singular points may form *singular lines, surfaces and even domains.* It is remarkable that along with *isolated singular points* one can deal with *isolated singular lines.* The latter fall, just as singular points in the two-dimensional case, into single- and multiple-valued singular lines, depending on whether the function in their sufficiently close neighborhood is single- or multiple-valued. Multiple-valued singular lines are also known as *branch lines* of a field. Branch lines of Laplace vector fields will be exemplified below, in the analytical continuation of geopotential fields inside and outside masses (see Sects. 7.3 and 7.4).

And finally, it is noteworthy that the function derived by analytical continuation of a harmonic function is harmonic itself. That is why we deal with *harmonic continuation* of harmonic functions and with their singular points. These terms will be employed subsequently. Similarly, it is shown that the *complete analytical function of the Laplace vector field satisfies the Laplace equation in the whole domain of its existence.*

7.1.5 Stal Theorem (Principle of Continuity for the Laplace Field)

Along with the above-given definition of the Weierstrass analytical continuation of a function of three variables, we can also define, just as in a two-dimensional case, the procedure of Penleve continuation. This definition relies on the following Stal theorem (Sretensky 1946).

Theorem 7.1.1. Let D_1 and D_2 be two three-dimensional domains whose intersection is a void set and which have a common segment of the boundaries, composed of a smooth surface S. If the vector fields $\mathbf{F}_1(\mathbf{r})$ and $\mathbf{F}_2(\mathbf{r})$ are the Laplace fields in the domains D_1 and D_2, respectively, continuous up to S and, besides,

$$\mathbf{F}_1(\mathbf{r}) \equiv \mathbf{F}_2(\mathbf{r}) \ , \quad \mathbf{r} \in S \ , \tag{7.1.13}$$

the field $\mathbf{F}_2(\mathbf{r})$ is the analytical continuation of the field $\mathbf{F}_1(\mathbf{r})$ from the domain D_1 to the domain D_2 through the surface S.

We will prove the theorem. To this end, let us define the field $\mathbf{F}(\mathbf{r})$ as follows:

$$\mathbf{F}(\mathbf{r}) = \begin{cases} \mathbf{F}_1(\mathbf{r}) \ , & \mathbf{r} \in D_1 \ , \\ \mathbf{F}_2(\mathbf{r}) \ , & \mathbf{r} \in D_2 \ , \\ \mathbf{F}_1(\mathbf{r}) = \mathbf{F}_2(\mathbf{r}) \ , & \mathbf{r} \in S \ . \end{cases} \tag{7.1.14}$$

Obviously, we have to establish the analyticity of the field $\mathbf{F}(\mathbf{r})$ in the neighborhood of an arbitrary point $\mathbf{r}_0$ of the surface S (maybe except for its boundary-contour l). Draw a sphere O_δ from the point $\mathbf{r}_0$ with such a small radius δ that it might cut only one connected part δS from S (Fig. 39). This seems

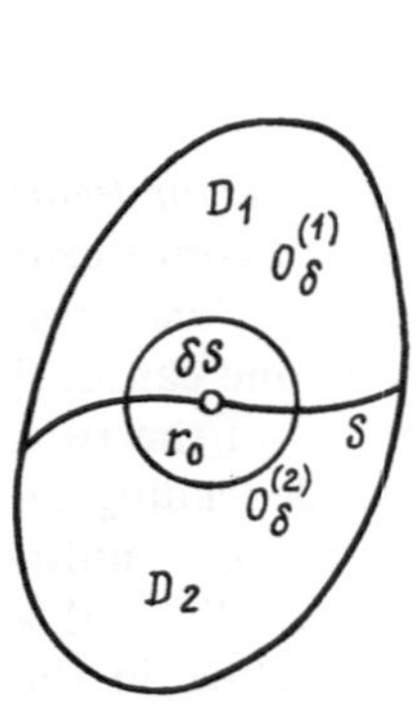

Fig. 39. Substantiation of the principle of continuity for the Laplace field

to be quite possible, because S is a smooth surface. Let $O_\delta^{(1)}$ designate the part of the sphere O_δ lying in D_1, and $O_\delta^{(2)}$ the part of the sphere inside D_2. Then, $V_\delta^{(1)}$ and $V_\delta^{(2)}$ are the parts of the sphere bounded by O_δ and lying inside D_1 and D_2, respectively.

By virtue of the three-dimensional Cauchy integral formula (5.2.12), we have

$$\mathbf{C}^{O_\delta^{(1)}+\delta S}(\mathbf{r}'; \mathbf{F}_1(\mathbf{r})) = \begin{cases} \mathbf{F}_1(\mathbf{r}) , & \mathbf{r}' \in V_\delta^{(1)} , \\ \mathbf{0} , & \mathbf{r}' \in V_\delta^{(2)} , \end{cases} \tag{7.1.15a}$$

$$\mathbf{C}^{O_\delta^{(2)}+\delta S}(\mathbf{r}'; \mathbf{F}_1(\mathbf{r})) = \begin{cases} \mathbf{0} , & \mathbf{r}' \in V_\delta^{(1)} , \\ \mathbf{F}_2(\mathbf{r}) , & \mathbf{r}' \in V_\delta^{(2)} . \end{cases} \tag{7.1.15b}$$

Adding up the left- and the right-hand sides of (7.1.15a) and (7.1.15b) and allowing for (7.1.14) we obtain

$$\mathbf{C}^{O_\delta^{(1)}+\delta S}(\mathbf{r}'; \mathbf{F}_1(\mathbf{r})) + \mathbf{C}^{O_\delta^{(2)}+\delta S}(\mathbf{r}'; \mathbf{F}_2(\mathbf{r})) = \mathbf{F}(\mathbf{r}') . \tag{7.1.16}$$

At the same time, thanks to the coincident boundary values of the fields $\mathbf{F}_1(\mathbf{r})$ and $\mathbf{F}_2(\mathbf{r})$ on S, the Cauchy-type integrals taken along δS (with the normals to δS pointing in different directions) are contracted and we obtain

$$\mathbf{F}(\mathbf{r}') = \mathbf{C}^{O_\delta^{(1)}}(\mathbf{r}'; \mathbf{F}_1(\mathbf{r})) + \mathbf{C}^{O_\delta^{(2)}}(\mathbf{r}'; \mathbf{F}_2(\mathbf{r})) = \mathbf{C}^{O_\delta}(\mathbf{r}'; \mathbf{F}(\mathbf{r})) . \tag{7.1.17}$$

The Cauchy-type integral included in the right-hand side of (7.1.17) defines, in turn, everywhere inside O_δ the Laplace vector field, i.e., an analytical function, which was to be proved.

Theorem 7.1.1 can be regarded as a *principle of continuity* for three-dimensional Laplace fields. The fact that the function $\mathbf{F}(\mathbf{r})$ fitted by relation (7.1.14) is analytical everywhere in $D = D_1 \cup D_2$ implies precisely that $\mathbf{F}_2(\mathbf{r})$ is the analytical continuation of the field $\mathbf{F}_1(\mathbf{r})$ from D_1 to a larger domain D.

7.2 Analytical Continuation of the Three-Dimensional Cauchy Integral Analog Through the Integration Surface

As indicated in Sect. 5.2, the integration surface is a singular surface for functions fitted by the Cauchy-type integrals. Meanwhile, it is sometimes possible to continue the Cauchy-type integrals through the integration surface. This problem is of great value in geophysical applications. Let us establish the class of integration surfaces for which the above property of three-dimensional Cauchy-type integrals is valid.

7.2.1 Concept of an Analytical Part of the Surface; Surface Equations in a Harmonic Form

Consider a simple smooth part of surface whose equation is parametrically represented as

$$x=x(s,\ t)\ ,\quad y=y(s,\ t)\ ,\quad z=z(s,\ t),\quad s_a\leqslant s\leqslant s_b\ ,\quad t_a\leqslant t\leqslant t_b\ , \tag{7.2.1}$$

where s_a, t_a and s_δ, t_b are some constant quantities.

The surface Γ is called analytical, provided $x(s, t)$, $y(s, t)$ and $z(s, t)$ are analytical functions of real variables s and t in a rectangle: $\left\{\begin{matrix} s_a\leqslant s\leqslant s_b \\ t_a\leqslant t\leqslant t_b \end{matrix}\right\}$.

The analytical part of surface Γ is called *regular* if the rank of the matrix

$$\left\|\begin{matrix} \dfrac{\partial x}{\partial s} & \dfrac{\partial y}{\partial s} & \dfrac{\partial z}{\partial s} \\ \dfrac{\partial x}{\partial t} & \dfrac{\partial y}{\partial t} & \dfrac{\partial z}{\partial t} \end{matrix}\right\|$$

is 2. This is equivalent to the condition

$$\frac{\partial \mathbf{r}}{\partial s}\times\frac{\partial \mathbf{r}}{\partial t}\neq 0\ , \tag{7.2.2}$$

where

$$\mathbf{r}=\mathbf{r}(s,\ t)=(x(s,\ t)\ ,\quad y(s,\ t)\ ,\quad z(s,\ t))\ .$$

Evidently, any smooth analytical surface is regular and vice versa.

Note that along with representation (7.2.1), there is an equivalent expression for an analytical part of surface

$$\Phi(x,\ y,\ z)=0\ , \tag{7.2.3}$$

where $\Phi(x, y, z)$ is the analytical function of three real variables in some three-dimensional neighborhood of the surface Γ.

Subsequently, of special concern is a particular class of surfaces referred to as harmonic surfaces.

Definition 7.2.1. Some part Γ of the surface S bounding the domain D is said to have a harmonic inner side [to be one-side (inside) harmonic], provided there exists a harmonic function $\mathbf{\Phi}^{+\Gamma}(\mathbf{r})$ in the domain $D^+\subset D$ adjacent to the surface Γ (Fig. 40a), so that

$$\operatorname{grad}\Phi^{+\Gamma}(\mathbf{r})=\mathbf{r}\quad \text{for}\quad \mathbf{r}\in\Gamma\ . \tag{7.2.4}$$

Definition 7.2.2. Similarly, the surface Γ is said to have a harmonic outer side [to be one-side (outside) harmonic], provided there exists a harmonic function

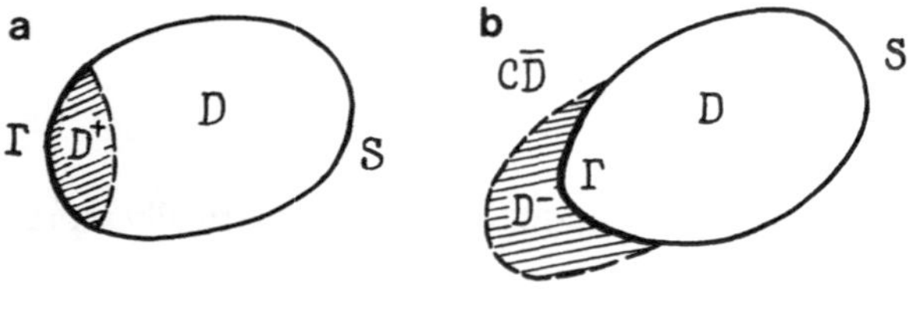

Fig. 40 a, b. Harmonic surfaces

$\boldsymbol{\Phi}^{-\Gamma}(\mathbf{r})$ in the domain $D^- \subset C\bar{D}$ adjacent to the surface Γ (Fig. 40b), so that

$$\operatorname{grad} \Phi^{-\Gamma}(\mathbf{r}) = \mathbf{r} \quad \text{for} \quad \mathbf{r} \in \Gamma . \tag{7.2.5}$$

Definition 7.2.3. Once the surface Γ has harmonic inner and outer sides, Γ is said to be a harmonic surface. In this case, the equation for this surface takes the form

$$\operatorname{grad} \Phi^{\Gamma}(\mathbf{r}) = \mathbf{r} \quad \text{for} \quad \mathbf{r} \in \Gamma , \tag{7.2.6}$$

where

$$\Phi^{\Gamma}(\mathbf{r}) = \Phi^{+\Gamma}(\mathbf{r}) \equiv \Phi^{-\Gamma}(\mathbf{r}) .$$

One is readily convinced that any harmonic surface is a regular-analytical surface, i.e., the equation for this surface is defined by analytical functions and condition (7.2.2) sets in. Indeed, the function $\boldsymbol{\Phi}^{\Gamma}(\mathbf{r})$ specifying Eq. (7.2.6) of a harmonic surface is harmonic and any harmonic function is, as shown earlier, analytical.

The regularity of the surface Γ is proved as follows. Assume the converse statement, i.e., $\mathbf{r}_0 \in \Gamma$ is a differentially-geometrically singular point of the surface Γ, while Γ is a harmonic surface. Then, according to (7.2.6),

$$\operatorname{grad} \Phi^{\Gamma}(\mathbf{r}) - \mathbf{r} = \mathbf{P}(\mathbf{r}) = 0 .$$

Since $\mathbf{r}_0$ is a geometrically singular point of the surface Γ, we have (Rashevsky 1956)

$$\frac{\partial \mathbf{P}(\mathbf{r})}{\partial x} = 0 , \quad \frac{\partial \mathbf{P}(\mathbf{r})}{\partial y} = 0 , \quad \frac{\partial \mathbf{P}(\mathbf{r})}{\partial z} = 0 , \quad \text{for} \quad \mathbf{r} = \mathbf{r}_0 .$$

The last expression implies, in particular,

$$\left.\frac{\partial^2 \Phi}{\partial x^2}\right|_{\Gamma = \Gamma_0} - 1 = 0 ; \quad \left.\frac{\partial^2 \Phi}{\partial y^2}\right|_{\Gamma = \Gamma_0} - 1 = 0 ; \quad \left.\frac{\partial^2 \Phi}{\partial z^2}\right|_{\Gamma = \Gamma_0} - 1 = 0 .$$

Summing up, we obtain

$$\Delta \Phi|_{\Gamma = \Gamma_0} = -3 ,$$

which is inconsistent with the property of harmonicity of the function $\Phi(\mathbf{r})$.

Now we will prove the converse. Any regular-analytical surface Γ is harmonic, i.e., in the neighborhood of Γ there exists such a harmonic function $\boldsymbol{\Phi}^{\Gamma}(\mathbf{r})$ that $\mathbf{r} = \operatorname{grad} \boldsymbol{\Phi}^{\Gamma}(\mathbf{r})$ on Γ.

Proof. Let Γ be a regular-analytical surface. Fix on Γ a point $\mathbf{r}_0 \in \Gamma$ and consider the equation

$$\Delta \varphi(\mathbf{r}) = -3 . \tag{7.2.7}$$

According to the Cauchy-Kovalevskaya theorem (Sretensky 1946, p. 208), one can find an analytical integral $\varphi(\mathbf{r})$ of Eq. (7.2.7) which, together with its normal derivative would vanish at the surface Γ in a certain neighborhood Γ_0 of the point $\mathbf{r}_0$ (Fig. 41). As a result,

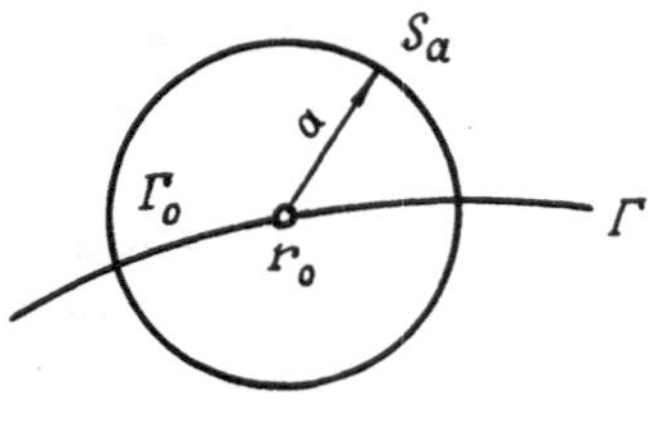

Fig. 41. According to the Cauchy-Kovalevskaya theorem, there exists an analytical integral $\varphi(\mathbf{r})$ of Eq. (7.2.7) which, together with its normal derivative, vanishes at the surface Γ in a certain neighborhood Γ_0 of the point $\mathbf{r}_0$

$$\operatorname{grad}\varphi(\mathbf{r}) = 0 \quad \text{for} \quad \mathbf{r} \in \Gamma_0 \ . \tag{7.2.8}$$

Construct a sphere S_a with its center at the point $\mathbf{r}_0$ having such a small radius a that the function $\varphi(\mathbf{r})$ inside this sphere might be analytical, and define the function $\Phi^\Gamma(\mathbf{r})$ inside the sphere S_a:

$$\Phi^\Gamma(\mathbf{r}) = \varphi(\mathbf{r}) + \tfrac{1}{2}\,\mathbf{r}^2 \ . \tag{7.2.9}$$

It is evident that

$$\Delta\,\Phi^\Gamma(\mathbf{r}) = \Delta\,\varphi(\mathbf{r}) + 3 = 0 \ .$$

This means that $\Phi^\Gamma(\mathbf{r})$ is a harmonic function inside S_a. On the other hand, according to (7.2.8) and (7.2.9), we have

$$\operatorname{grad}\Phi^\Gamma(\mathbf{r}) = \operatorname{grad}\varphi(\mathbf{r}) + \mathbf{r} = \mathbf{r} \quad \text{for} \quad \mathbf{r} \in \Gamma_0 \ . \tag{7.2.10}$$

Consequently, by definition, Γ_0 is a harmonic surface. Reiterating this reasoning for each point $\mathbf{r}_0 \in \Gamma$, we find out that Γ is a harmonic surface.

Thus, the concept of a harmonic surface is consistent with that of a regular-analytical surface. Later on, Eqs. (7.2.4) through (7.2.6) will be referred to as *equations of the surface* (or of its inner or outer sides) *in a harmonic form.* Take, as an example of a harmonic surface, a sphere C_{R_0} of a radius R_0 with its center at the origin of coordinates $x^2+y^2+z^2 = R_0^2$. The harmonic function $\Phi^\Gamma(\mathbf{r})$ is defined by

$$\Phi^\Gamma(\mathbf{r}) = \frac{-R_0^3}{(x^2+y^2+z^2)^{1/2}} = -\frac{R_0^3}{r} \ ,$$

where $r = (x^2+y^2+z^2)^{1/2}$. Then, we have

$$\operatorname{grad}\Phi^\Gamma(\mathbf{r}) = \frac{R_0^2}{r^3}\,\mathbf{r} \ . \tag{7.2.11}$$

By definition, the function $\Phi^\Gamma(\mathbf{r})$ meets the Laplace equation $\Delta\,\Phi^\Gamma(\mathbf{r}) = 0$ in the whole space and the condition

$$\operatorname{grad}\Phi^\Gamma(\mathbf{r}) = \frac{R_0^3}{R_0^3}\,\mathbf{r} = \mathbf{r} \quad \text{for} \quad r \in O_{R_0} \ . \tag{7.2.12}$$

Equation (7.2.12) is just the equation for the sphere O_{R_0} in a harmonic form.

Another simple example of a harmonic surface is a plane P: $c_x x + c_y y + c_z z = c_0$, i.e., $\mathbf{c}\cdot\mathbf{r} = c_0$, where $\mathbf{c} = (c_x, c_y, c_z)$.

Indeed, the corresponding harmonic function can be specified in the form

$$\Phi^{\Gamma}(\mathbf{r}) = -\frac{3}{2}\frac{(\mathbf{c}\cdot\mathbf{r})^2}{\mathbf{c}\cdot\mathbf{c}}+\frac{1}{2}r^2+\frac{3c_0(\mathbf{c}\cdot\mathbf{r})}{\mathbf{c}\cdot\mathbf{c}} \; . \tag{7.2.13}$$

One can readily see that

$$\operatorname{grad}\Phi^{\Gamma}(\mathbf{r}) = -3\frac{(\mathbf{c}\cdot\mathbf{r})}{\mathbf{c}\cdot\mathbf{c}}\mathbf{c}+\mathbf{r}+\frac{3c_0\mathbf{c}}{\mathbf{c}\cdot\mathbf{c}} \; .$$

Hence,

$$\Delta\Phi^{\Gamma}(\mathbf{r}) = \operatorname{div}\operatorname{grad}\Phi^{\Gamma}(\mathbf{r}) = -3+3 = 0 \; ,$$

i.e., $\Phi^{\Gamma}(\mathbf{r})$ is a harmonic function.

On the other hand, since we have $\mathbf{c}\cdot\mathbf{r} = c_0$ on the plane P, $\operatorname{grad}\Phi^{\Gamma}(\mathbf{r}) = \mathbf{r}$ for $\mathbf{r}\in P$. As a result, the function $\Phi^{\Gamma}(\mathbf{r})$ fitted by formula (7.2.13) specifies the equation for the plane P in a harmonic form.

7.2.2 Relationship Between the Surface Equation in a Harmonic Form and the Plane Curve Equation in the Herglotz-Tsirulsky Form

In a particular case where the regular-analytical surface Γ is cylindrical, with its generator parallel to the y-axis, we can derive from (7.2.6) an equation for the projection L of the surface Γ onto the plane XZ:

$$\frac{\partial\Phi^{\Gamma}(x,\,0,\,z)}{\partial x} = x \; , \quad \frac{\partial\Phi^{\Gamma}(x,\,0,\,z)}{\partial z} = z \; . \tag{7.2.14}$$

Introduction of a complex variable $\zeta = x+iz$ in the plane XZ and designations $\partial\Phi^{\Gamma}(x,\,0,\,z)/\partial x = a(x,\,z)$, $\partial\Phi^{\Gamma}(x,\,0,\,z)/\partial z = -b(x,\,z)$ will render relation (7.2.14) to a complex form

$$x-iz = a(x,\,z)+ib(x,\,z) \; . \tag{7.2.15}$$

One can readily see that by virtue of the harmonicity of the function $\Phi^{\Gamma}(\mathbf{r})$, the functions a and b satisfy the Cauchy-Riemann conditions:

$$\partial a/\partial z = -\partial b/\partial x \; , \quad \partial a/\partial x = \partial b/\partial z \; .$$

Hence, the function $\psi(\zeta) = a(x,\,z)+ib(x,\,z)$ included in the right-hand side of (7.2.15) is complex analytical and Eq. (7.2.15) is written in the final form

$$\bar{\zeta} = \psi(\zeta) \; , \tag{7.2.16}$$

where the bar over ζ indicates the taking of a complex conjugate. Expression (7.2.16) is an equation for a plane curve in the Herglotz-Tsirulsky form (4.2.11). Thus, in the case of a two-dimensional disturbing bodies bounded by cylindrical surfaces, the equations for the latter in the harmonic form (7.2.6) go over into the equations of contours of the disturbing body cross section in the Herglotz-Tsirulsky form (7.2.16). The established relationship between Eqs. (7.2.6) and (7.2.16) makes it possible to extend the results obtained for three-dimensional fields to a two-dimensional case and vice versa.

7.2.3 Continuation of the Cauchy-Type Integral Through the Integration Surface

Consider a regular analytical part Γ of the surface S on which a Cauchy integral analog with a vector density $\boldsymbol{\varphi}(\mathbf{r})$ is specified. It is assumed that the components of $\boldsymbol{\varphi}(\mathbf{r})$ are described on Γ by analytical functions of coordinates. Note that they are definitely differentiable on Γ and, hence, satisfy the Hölder condition.

Fix on a certain specific point $\mathbf{r}_0 \in \Gamma$. Then, according to the Cauchy-Kovalevskaya theorem (Sretensky 1946), in the neighborhood of the point $\mathbf{r}_0$ there is a harmonic function $\Phi^{\Gamma}(\mathbf{r})$ meeting on the surface Γ the condition

$$\left.\frac{\partial \Phi^{\Gamma}}{\partial n}\right|_{\Gamma} = \varphi_n \, , \quad \Phi^{\Gamma}|_{\Gamma} = \varphi \, , \tag{7.2.17}$$

where φ is the scalar density of the Cauchy-type integral related to the vector density by (6.1.19). Clearly,

$$\operatorname{grad} \Phi^{\Gamma}|_{\Gamma} = \boldsymbol{\varphi} \, . \tag{7.2.18}$$

Now we will employ the Sokhotsky-Plemelj formulas to examine the Cauchy integral analogs. In accordance with (5.2.30) and (7.2.18), we have

$$\mathbf{C}^{-}(\mathbf{r}_0; \boldsymbol{\varphi}) = \mathbf{C}^{+}(\mathbf{r}_0; \boldsymbol{\varphi}) - \operatorname{grad} \Phi^{\Gamma}(\mathbf{r}_0) \, . \tag{7.2.19}$$

Evidently, the right-hand side of (7.2.19) represents boundary values of vector fields on Γ, which are Laplace fields in some domain D^+ adjacent to Γ and lying completely inside D. The left-hand side describes boundary values of vector fields which are Laplace fields everywhere outside D. Consequently, according to the Stal theorem (Sect. 7.1.5), the right-hand side is the analytical continuation of the left-hand side through the surface Γ. In a similar way, one can prove the possibility of outward continuation of the Cauchy-type integrals through an analytical part of the surface Γ.

Formula (7.2.19) suggests that the values of the Cauchy integral analogs continued through the surface, $\mathbf{C}^{\text{cont}}(\mathbf{r}; \boldsymbol{\varphi})$ differ from those of the integrals themselves $\mathbf{C}(\mathbf{r}; \boldsymbol{\varphi})$ by a value equal to $-\operatorname{grad} \Phi^{\Gamma}(\mathbf{r})$:

$$\mathbf{C}^{\text{cont}}(\mathbf{r}; \boldsymbol{\varphi}) - \mathbf{C}(\mathbf{r}; \boldsymbol{\varphi}) = -\operatorname{grad} \Phi^{\Gamma}(\mathbf{r}) \, . \tag{7.2.20}$$

This implies two corollaries. First, singular points of $\mathbf{C}^{\text{cont}}$ coincide with those of $\operatorname{grad} \Phi^{\Gamma}(\mathbf{r})$. Second, the lines joining different analytical parts of the surface S are generally branch lines for Cauchy integral analogs. These properties of Cauchy integral analogs enable us to examine the issue of analytical continuation of external geopotential fields into disturbing masses.

7.3 Analytical Continuation of a Three-Dimensional Gravitational Field into a Homogeneous Material Body

The representations of external gravitational and magnetic fields in terms of Cauchy integral analogs, derived in Sects. 6.2 and 6.3, provide the basis for the theory of field continuation into and beyond masses. The fundamentals of this theory will be outlined, starting with the general properties of the gravitational field of a homogeneous material body bounded by an analytical surface.

7.3.1 Properties of the Gravitational Field of a Body Bounded by an Analytical Surface

Consider an arbitrary body of a finite volume D filled with masses of a constant positive density ϱ_0 and bounded by a smooth surface S. The external potential of this body is denoted by $U(\mathbf{r})$, $\mathbf{r} \in C\bar{D}$. Let us determine also the external potential $U_{CD}(\mathbf{r})$, $\mathbf{r} \in D$ of the domain $C\bar{D}$ filled with masses of the same density (Fig. 42). On the basis of formulas (6.2.11) and (6.2.16), we have

$$\begin{aligned} \mathbf{g}(\mathbf{r}') &= \operatorname{grad}' U(\mathbf{r}') = \mathbf{C}^S\left(\mathbf{r}'; \frac{4\pi}{3}\gamma\varrho_0\mathbf{r}\right), \quad \mathbf{r}' \in C\bar{D}, \\ \mathbf{g}_{CD}(\mathbf{r}') &= \operatorname{grad}' U_{CD}(\mathbf{r}') = -\mathbf{C}^S\left(\mathbf{r}'; \frac{4\pi}{3}\gamma\varrho_0\mathbf{r}\right), \quad \mathbf{r}' \in D. \end{aligned} \tag{7.3.1}$$

The difference between the limit values of the integral contained in the right-hand side of (7.3.1) on the surface S approached from the inside and outside of the domain D is defined by the Sokhotsky-Plemelj formula (5.2.30):

$$\lim_{\mathbf{r}\to\mathbf{r}_0^+} \mathbf{C}^S\left(\mathbf{r}'; \frac{4\pi}{3}\gamma\varrho_0\mathbf{r}\right) - \lim_{\mathbf{r}\to\mathbf{r}_0^-} \mathbf{C}^S\left(\mathbf{r}'; \frac{4\pi}{3}\gamma\varrho_0\mathbf{r}\right) = \frac{4\pi}{3}\gamma\varrho_0\mathbf{r}_0 .$$

With due reference to (7.3.1), we write

$$\lim_{\mathbf{r}\to\mathbf{r}_0^+} [-\mathbf{g}_{CD}(\mathbf{r})] - \lim_{\mathbf{r}\to\mathbf{r}_0^-} [\mathbf{g}(\mathbf{r})] = \frac{4\pi}{3}\gamma\varrho_0\mathbf{r}_0 .$$

Hence, we obtain in the final form

$$\mathbf{g}(\mathbf{r}_0) + \mathbf{g}_{CD}(\mathbf{r}_0) = -(4\pi/3)\gamma\varrho_0\mathbf{r}_0 . \tag{7.3.2}$$

Relation (7.3.2) suggests the following conclusions, extending the results obtained by Tsirulsky for plane fields (Sect. 4.4.1) to a three-dimensional case.

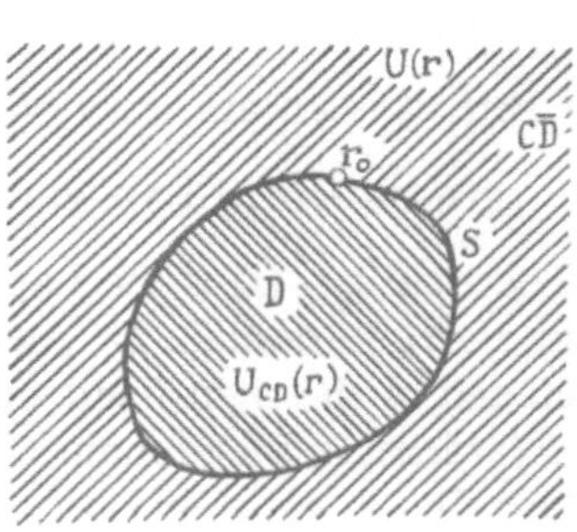

Fig. 42. System of two domains (internal D and external CD) filled with masses of a constant density

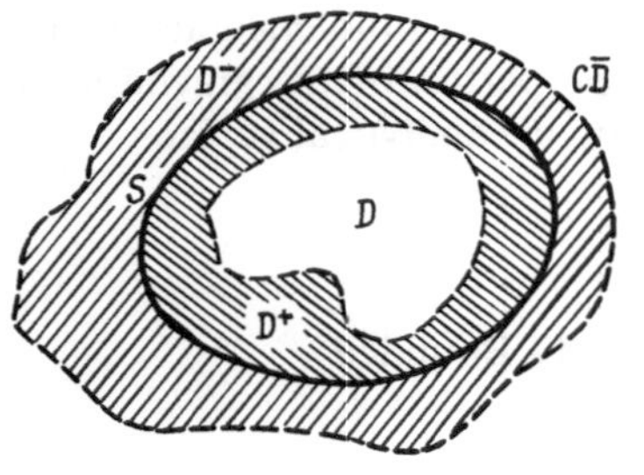

Fig. 43. The problem of analytical continuation of the gravitational potential through the boundary of a homogeneous material body D

Theorem 7.3.1. For a closed smooth surface S to be harmonic, it is necessary and sufficient that the function $U(\mathbf{r})$ might be analytically (harmonically) continued into a domain, while the function $U_{CD}(\mathbf{r})$ beyond a domain bounded by S through any point of the surface S.

Proof. Let us prove the sufficiency of the theorem condition. Assume that the function $U(\mathbf{r})$ can be continued into some part D^+ of the domain D adjacent to S, and the function $U_{CD}(\mathbf{r})$ can be continued outwards, i.e., into some part D^- of the domain adjacent to S (Fig. 43). Let D^S denote the domain formed upon combination of the domains D^+ and D^- (shaded in Fig. 43). Then, the functions $U(\mathbf{r})$ and $U_{CD}(\mathbf{r})$ are harmonic everywhere in D^S. Define the function $\Phi^S(\mathbf{r})$ as

$$\Phi^S(\mathbf{r}) = -\frac{3}{4\pi\gamma\varrho_0}[U(\mathbf{r})+U_{CD}(\mathbf{r})] \ , \quad \mathbf{r}\in D^S \ . \tag{7.3.3}$$

By construction, this function is harmonic in D^S and, according to (7.3.2), we have

$$\operatorname{grad}\Phi^S(\mathbf{r}) = -\frac{3}{4\pi\gamma\varrho_0}[\mathbf{g}(\mathbf{r})+\mathbf{g}_{CD}(\mathbf{r})] = \mathbf{r} \ , \quad \mathbf{r}\in S \ . \tag{7.3.4}$$

Expression (7.3.4) is the equation for the surface S in a harmonic form. Hence, S is a harmonic surface, which was to be proved.

Now we will prove the necessity of the theorem condition. Let the surface S be harmonic and its equation in a harmonic form be

$$\operatorname{grad}\Phi^S(\mathbf{r}) = \mathbf{r} \ . \tag{7.3.5}$$

Then, substituting (7.3.5) into (7.3.2) we derive

$$\mathbf{g}(\mathbf{r}_0)+\mathbf{g}_{CD}(\mathbf{r}_0) = -(4\pi/3)\gamma\varrho_0\operatorname{grad}\Phi^S(\mathbf{r}_0) \ , \tag{7.3.6}$$

where $\mathbf{r}_0$ is an arbitrary point of the surface S. It follows from (7.3.6) that

$$\mathbf{g}(\mathbf{r}_0) = -(4\pi/3)\gamma\varrho_0\operatorname{grad}\Phi^S(\mathbf{r}_0)-\mathbf{g}_{CD}(\mathbf{r}_0) \ . \tag{7.3.7}$$

The field $\mathbf{g}_{CD}(\mathbf{r})$ is the Laplace field in D, while $\operatorname{grad}\Phi^S$ is of the Laplace kind in a certain domain D^+ belonging to D (Fig. 43). Therefore, these functions included in the right-hand side of Eq. (7.3.7) permit analytical continuations into D (at least in the domain D^+). As a result $\mathbf{g}(\mathbf{r})$ can be also continued analytically into D:

$$\mathbf{g}^{\text{cont}}(\mathbf{r}) = -(4\pi/3)\gamma\varrho_0 \operatorname{grad}\Phi^S(\mathbf{r}) - \mathbf{g}_{CD}(\mathbf{r}) \ . \tag{7.3.8}$$

Integrating (7.3.8) and allowing for the fact that $\mathbf{g} = \operatorname{grad} U$ and $\mathbf{g}_{CD} = \operatorname{grad} U_{CD}$, we obtain

$$U^{\text{cont}}(\mathbf{r}) = -(4\pi/3)\gamma\varrho_0 \Phi^S(\mathbf{r}) - U_{CD}(\mathbf{r}) \ , \quad \mathbf{r} \in D^+ \ , \tag{7.3.9}$$

where the arbitrary constant arising in the integration is introduced into the function $\Phi^S(\mathbf{r})$.

In a similar way, one can prove the possibility of continuing $U_{CD}(\mathbf{r})$ beyond the domain D, as well as the validity of the formula

$$U_{CD}^{\text{cont}}(\mathbf{r}) = -\left(\frac{4\pi}{3}\right)\gamma\varrho_0 \Phi^S(\mathbf{r}) - U(\mathbf{r}) \ , \quad \mathbf{r} \in D^- \ . \tag{7.3.10}$$

Relations (7.3.9) and (7.3.10) imply that the set of singular points of fields continued analytically through the surface S coincides with the set of singularities of the function $\Phi^S(\mathbf{r})$ inside and outside the domain D, respectively.

Now turn to formula (6.2.14) defining the gravitational field $\mathbf{g}(\mathbf{r}')$ inside a body ($\mathbf{r}' \in D$). With due account taken of (6.2.16), this formula can be cast in the form

$$\mathbf{g}(\mathbf{r}) = -(4\pi/3)\gamma\varrho_0 \mathbf{r} - \mathbf{g}_{CD}(\mathbf{r}) \ , \quad \mathbf{r} \in D \ . \tag{7.3.11}$$

A comparison of formulas (7.3.11) and (7.3.8) reveals that the analytical continuation of the external gravitational field into masses differs from the true field by a value of

$$\mathbf{g}^{\text{cont}}(\mathbf{r}) - \mathbf{g}(\mathbf{r}) = \frac{4\pi}{3}\gamma\varrho_0[\mathbf{r} - \operatorname{grad}\Phi^S(\mathbf{r})] \ . \tag{7.3.12}$$

Now consider a particular case where the function $\Phi^S(\mathbf{r})$ describing the harmonic surface S has no singularities outside S. It is evident from the proof of theorem 7.3.1 that the function $U_{CD}(\mathbf{r})$ continued analytically to the whole space has no singularities at all, i.e., it is harmonic in the whole space. By the Liouville theorem (Sretensky 1946), the function which is harmonic in the whole space is identically equal to a constant:

$$U_{CD}^{\text{cont}}(\mathbf{r}) \equiv \text{const} \ . \tag{7.3.13}$$

Substitution of (7.3.13) into (7.3.10) yields

$$U(\mathbf{r}) = -(4\pi/3)\gamma\varrho_0 \Phi^S(\mathbf{r}) + \text{const.} \tag{7.3.14}$$

We will pass to the limit for $\mathbf{r} \to \infty$ in (7.3.14):

$$\lim_{r\to\infty} U(\mathbf{r}) = -(4\pi/3)\gamma\varrho_0 \lim_{r\to\infty} \Phi^S(\mathbf{r}) + \text{const.}$$

But $\lim_{r\to\infty} U(\mathbf{r}) = 0$, hence,

$$\text{const} = (4\pi/3)\gamma\varrho_0 \lim_{r\to\infty} \Phi^S(\mathbf{r}) \ . \tag{7.3.15}$$

Substituting (7.3.15) into (7.3.14) we obtain

$$U(\mathbf{r}) = -(4\pi/3)\gamma\varrho_0\Phi^S(\mathbf{r}) + (4\pi/3)\gamma\varrho_0\lim_{r\to\infty}\Phi^S(\mathbf{r}) \ . \tag{7.3.16}$$

Formula (7.3.16) allows us to solve, without integration, the direct problem of the three-dimensional potential theory for a body bounded by a harmonic surface S fitted by the function $\Phi^S(\mathbf{r})$ harmonic everywhere outside S. For instance, for a sphere of a radius R_0 with its center at the origin of coordinates, we have, according to (7.2.11)

$$\Phi(\mathbf{r}) = -R_0^3/r \ . \tag{7.3.17}$$

Substituting (7.3.17) into (7.3.16) we find

$$U(\mathbf{r}) = \frac{4\pi}{3}\gamma\varrho_0\frac{R_0^3}{r} = \gamma\frac{M}{r} \ , \tag{7.3.18}$$

where $M = \varrho_0(4\pi/3)R_0^3$ is the mass of the sphere.

7.3.2 Relationship Between the Shape of the Surface of a Three-Dimensional Homogeneous Material Body and the Location of Singularities of the Gravitational Field Continued Analytically into the Body

Let us extend the findings of Sect. 7.3.1 to the case of a body bounded by a piecewise smooth surface S. Of primary concern here is the relationship between the singular points of an analytically (harmonically) continued gravitational field and the shape of the surface of a disturbing body.

This issue was first studied by Schmiedt and Wavre (Wavre 1934), who showed that the gravitational potential could be continued into a body through its surface, provided the latter is analytical. Further development of these investigations is found in the works of A. V. Tsirulsky (1963), A. V. Tsirulsky and M. I. Sirotin (1964), G. Ya. Golizdra (1966), V. N. Strakhov (1970d), suggesting a complete and at the same time rather simple solution to the stated problem for the case of two-dimensional (plane) fields, a solution relying on the body of theory of functions of a complex variable and mainly of the Cauchy integral. The results of these investigations were considered in Chap. 4. In this subsection we will show that this problem is solvable comprehensively and readily in a three-dimensional case as well, thanks to the above elaborated body of three-dimensional Cauchy integral analogs.

According to formula (7.3.1), the Cauchy-type integral

$$\mathbf{C}^S\left(\mathbf{r}'; \frac{4\pi}{3}\gamma\varrho_0\mathbf{r}\right) \tag{7.3.19}$$

outside D implies the gravitational field of the domain D, while inside D it means the gravitational field (with a negative sign) of the domain $C\bar{D}$, which are filled with masses of a constant density ϱ_0. The difference between the limit values of integral (7.3.19) over a piecewise smooth surface S bounding the domain D, when this surface is approached from the inside and the outside of this domain, is defined, analogously to (7.3.2), by the Sokhotsky-Plemelj formula (5.2.30):

$$\mathbf{g}(\mathbf{r}_0)+\mathbf{g}_{CD}(\mathbf{r}) = -\frac{4\pi}{3}\gamma\varrho_0\mathbf{r}_0 \ , \quad \mathbf{r}_0 \in S \ . \tag{7.3.20}$$

Now assume that the gravitational field $\mathbf{g}(\mathbf{r})$ of the domain D is specified in some domain Ω lying outside $D(\Omega \subset C\bar{D})$. The function $\mathbf{g}(\mathbf{r})$ can be continued analytically to all the points of the domain $C\bar{D}$, since it is harmonic there. We will elucidate the conditions under which the external gravitational field can be continued into the domain D, i.e., *into masses*, and locate the singularities of this continuation.

Theorem 7.3.2. The function $\mathbf{g}(\mathbf{r})$ ($\mathbf{g}_{CD}(\mathbf{r})$) can be continued analytically into (beyond) the domain D through any point of a surface Γ belonging to S, maybe except for its bounding contour l, in the only case where the surface Γ has a harmonic inner (outer) side.

Proof. We assume at first that the surface S has a harmonic inner side and the equation for its inner side in a harmonic form is

$$\operatorname{grad}\Phi^{+\Gamma} = \mathbf{r} \ . \tag{7.3.21}$$

Then, substituting (7.3.21) into (7.3.20) we have

$$\mathbf{g}(\mathbf{r}_0)+\mathbf{g}_{CD}(\mathbf{r}_0) = -(4\pi/3)\gamma\varrho_0 \operatorname{grad}\Phi^{+\Gamma}(\mathbf{r}_0) \ , \tag{7.3.22}$$

whence

$$\mathbf{g}(\mathbf{r}_0) = -(4\pi/3)\gamma\varrho_0 \operatorname{grad}\Phi^{+\Gamma}(\mathbf{r}_0)-\mathbf{g}_{CD}(\mathbf{r}_0) \ , \tag{7.3.23}$$

where $\mathbf{r}_0$ is an arbitrary inner point of the surface Γ.

Since the functions contained in the right-hand side of (7.2.23) admit analytical continuation into D, $\mathbf{g}(\mathbf{r})$ can be continued analytically into D, into some domain D^+ (see Fig. 43):

$$\mathbf{g}^{\text{cont}}(\mathbf{r}) = -\frac{4\pi}{3}\gamma\varrho_0 \operatorname{grad}\Phi^{+\Gamma}(\mathbf{r})-\mathbf{g}_{CD}(\mathbf{r}) \ . \tag{7.3.24}$$

Taking into consideration that $\mathbf{g}^{\text{cont}} = \operatorname{grad} U^{\text{cont}}$ and integrating (7.3.24), we obtain

$$U^{\text{cont}}(\mathbf{r}) = -(4\pi/3)\gamma\varrho_0 \Phi^{+\Gamma}(\mathbf{r})-U_{CD}(\mathbf{r}) \ , \tag{7.3.25}$$

where the arbitrary constant arising in the integration is introduced into the function $\Phi^{+\Gamma}(\mathbf{r})$.

Assuming now, on the contrary, that the function $\mathbf{g}(\mathbf{r})$ can be continued analytically into some domain $D^+ \subset D$ (see Fig. 43) adjacent to Γ (maybe except for the contour l) and introducing the designations

$$\operatorname{grad}\Phi^{+\Gamma}(\mathbf{r}) = -\frac{3}{4\pi\gamma\varrho_0}[\mathbf{g}(\mathbf{r})+\mathbf{g}_{CD}(\mathbf{r})] \ , \quad \mathbf{r} \in D^+ \ , \tag{7.3.26}$$

we derive on the basis of (7.3.20)

$$\operatorname{grad}\Phi^{+\Gamma}(\mathbf{r}_0) = \mathbf{r}_0 \ , \quad \mathbf{r}_0 \in \Gamma \ , \tag{7.3.27}$$

suggesting that the surface Γ has a harmonic inner side.

Similarly, the proof is provided for the function $\mathbf{g}_{CD}$ with its harmonic continuation in $C\bar{D}$ fitted by the formula

$$\mathbf{g}_{CD}^{\text{cont}}(\mathbf{r}) = -(4\pi/3)\gamma\varrho_0 \operatorname{grad} \Phi^{-\Gamma}(\mathbf{r}) - \mathbf{g}(\mathbf{r}) \ , \tag{7.3.28}$$

whence

$$U_{CD}^{\text{cont}}(\mathbf{r}) = -(4\pi/3)\gamma\varrho_0 \Phi^{-\Gamma}(\mathbf{r}) - U(\mathbf{r}) \ . \tag{7.3.29}$$

Relations (7.3.25) and (7.3.29) imply that *the set of singular points of fields continued analytically through the surface Γ coincides with the set of singularities of the function $\Phi^{+\Gamma}(\mathbf{r})$ inside D and of the function $\Phi^{-\Gamma}(\mathbf{r})$ outside D, respectively.*

Now we will clarify which points *at the surface* of a disturbing body are singular for analytically continued potential fields. Let the smooth parts Γ_i making up the surface S have harmonic inner sides and let the equations for these sides in a harmonic form be

$$\operatorname{grad} \Phi^{+\Gamma_i}(\mathbf{r}) = \mathbf{r} \ , \quad i = 1, 2, \ldots, N,$$

with

$$\Phi^{+\Gamma_i}(\mathbf{r}) \not\equiv \Phi^{+\Gamma_j}(\mathbf{r}) \quad \text{for} \quad i \neq j \ . \tag{7.3.30}$$

Assume that the contours l_i and l_{i+1} bounding the surfaces Γ_i and Γ_{i+1} are coincident over some segment $l_{i,i+1}$. This segment will be referred to as a *joining line* of two surfaces Γ_i and Γ_{i+1}. We will show that the lines joining the surfaces are branch lines for the potential and its derivatives.

Indeed, continuing $U(\mathbf{r})$ harmonically through the surface Γ_i we find, by virtue of (7.3.25), that

$$U^{(i)}(\mathbf{r}) = -(4\pi/3)\gamma\varrho_0 \Phi^{+\Gamma_i}(\mathbf{r}) - U_{CD}(\mathbf{r}) \ , \tag{7.3.31}$$

while continuing $U(\mathbf{r})$ through the surface Γ_{i+1} we define

$$U^{(i+1)}(\mathbf{r}) = -(4\pi/3)\gamma\varrho_0 \Phi^{+\Gamma_{i+1}}(\mathbf{r}) - U_{CD}(\mathbf{r}) \ . \tag{7.3.32}$$

By virtue of condition (7.3.30):

$$U^{(i)}(\mathbf{r}) \not\equiv U^{(i+1)}(\mathbf{r}) \ ; \tag{7.3.33}$$

as a result, $l_{i,\,i+1}$ is a branch line of the potential. Thus, we have proved the following theorem.

Theorem 7.3.3. Lines joining different harmonic (analytical) parts of the surface S bounding the domain D are branch lines for the values of the external gravitational field of the domain D harmonically continued into D filled with masses of a constant density ϱ_0.

It is remarkable that not only singular lines but also singular points of potential fields may lie at the surface of a disturbing body.

Theorem 7.3.4. Any point of the surface S bounding the domain D filled with disturbing masses of a constant density ϱ_0, which is singular in the differential geometrical sense (a corner or a cusp) is also singular for the corresponding potential fields (either for the field $\mathbf{g}$, or for the field $\mathbf{g}_{CD}$, or for the two fields $\mathbf{g}$ and $\mathbf{g}_{CD}$).

Proof. Assume the contrary of the theorem assertion, i.e., assume that the point $\mathbf{r}_0 \in S$ is a singular point of the surface in the differential geometric sense and at the same time the fields $\mathbf{g}$ and $\mathbf{g}_{CD}$ can be continued harmonically through any point of the surface S belonging to a certain neighborhood of the point $\mathbf{r}_0$, including $\mathbf{r}_0$ itself. Designate the part of the surface S belonging to this neighborhood of the point $\mathbf{r}_0$ by Γ_0. Then, by theorem 7.3.2, Γ_0 is a harmonic (regular-analytical) surface. This means that Γ_0 is a smooth surface, which is inconsistent with the conditions of the theorem. Thus, the theorem has been proved.

7.3.3 Definition of the Shape of the Surface of Three-Dimensional Material Bodies by Analytical Continuation of the Gravitational Field

The results obtained in the previous subsection bring us to the solution of the problem of defining the shape of the surface of disturbing masses of a known uniform density ϱ_0.

It is evident, first of all, that this problem has a unique solution, once the gravitational field $\mathbf{g}$ in the domains $\Omega^- \subset C\bar{D}$ and $\Omega^+ \subset D$ is known. Indeed, in this case, using the operation of harmonic (analytical) continuation we can find, by theorem 7.3.3, all the lines joining different harmonic parts of the surface S, i.e., we can determine all the contours l_i bounding the surfaces Γ_i, respectively. Thus, it is possible to construct a "spatial frame" over which the sought surface S is "stretched." Then, employing formula (7.3.11) and knowing the field $\mathbf{g}(\mathbf{r})$ inside D, we can define the function $\mathbf{g}_{CD}(\mathbf{r})$ fitting the gravitational field of the domain $C\bar{D}$ filled with masses of a density ϱ_0. After that, the functions $\Phi^{(i)}(\mathbf{r})$ specifying the equations for harmonic parts Γ_i are defined in principle unequivocally, since, by theorem 7.3.2 and formulas (7.3.25), (7.3.27), the sets of singular points of the fields $\mathbf{g}$ and $\mathbf{g}_{CD}$ harmonically continued through the contours l_i coincide with the sets of singularities of the functions $\Phi^i(\mathbf{r})$ inside and outside D, respectively.

Denoting the results of harmonic continuation of the functions $\mathbf{g}$ and $\mathbf{g}_{CD}$ through the contours l_i by $\mathbf{g}^{(i)}(\mathbf{r})$ and $\mathbf{g}_{CD}^{(i)}(\mathbf{r})$, we can calculate the functions $\operatorname{grad} \Phi^{(i)}(\mathbf{r})$ on the basis of (7.3.24) and (7.3.28) directly by the formula

$$\operatorname{grad} \Phi^{(i)}(\mathbf{r}) = -\frac{\mathbf{g}^{(i)}(\mathbf{r}) + \mathbf{g}_{CD}^{(i)}(\mathbf{r})}{(4\pi/3)\gamma\varrho_0} . \tag{7.3.34}$$

Hence, the equations for surfaces Γ_i in a harmonic form are

$$\mathbf{r} = -[\mathbf{g}^{(i)}(\mathbf{r}) + \mathbf{g}_{CD}^{(i)}(\mathbf{r})] \Big/ \frac{4\pi}{3} \gamma\varrho_0 . \tag{7.3.35}$$

It is worthwile noting that in common practice the field $\mathbf{g}(\mathbf{r})$ inside masses is not known, so the problem of defining the shape of the surface S from the value of $\mathbf{g}$, measured outside masses, is generally ambiguous. But if we know at least one harmonic part of the surface S, the problem of establishing the overall surface S from the external potential becomes unequivocal again.

Indeed, let the function $\Phi^{(i)}(\mathbf{r})$ fitting the harmonic part Γ_i be known; then, in accordance with (7.3.34), we have

$$\mathbf{g}_{CD}(\mathbf{r}) = -(4\pi/3)\gamma\varrho_0 \operatorname{grad} \Phi^{(1)}(\mathbf{r}) - \mathbf{g}^{(1)}(\mathbf{r}) \tag{7.3.36}$$

and by (7.3.11)

$$\mathbf{g}(\mathbf{r}) = \mathbf{g}^{(1)}(\mathbf{r}) + (4\pi/3)\gamma\varrho_0[\operatorname{grad} \Phi^{(1)}(\mathbf{r}) - \mathbf{r}] \ , \quad \mathbf{r} \in D \ . \tag{7.3.37}$$

Since we have evaluated $\mathbf{g}$ and $\mathbf{g}_{CD}$ in some domain inside D, we deal already with the conditions of the problem considered earlier, which has a unique solution.

Continuing $\mathbf{g}$ into D through the contours l_i, we define the functions $\mathbf{g}^{(i)}(\mathbf{r})$ bearing in mind that, according to (7.3.24)

$$\operatorname{grad} \Phi^{(i)}(\mathbf{r}) = -\frac{\mathbf{g}^{(i)}(\mathbf{r}) + \mathbf{g}_{CD}(\mathbf{r})}{(4\pi/3)\gamma\varrho_0} \ . \tag{7.3.38}$$

Substituting (7.3.36) into (7.3.38) we obtain

$$\operatorname{grad} \Phi^{(i)}(\mathbf{r}) = \frac{\mathbf{g}^{(1)}(\mathbf{r}) - \mathbf{g}^{(i)}(\mathbf{r})}{(4\pi/3)\gamma\varrho_0} + \operatorname{grad} \Phi^{(1)}(\mathbf{r}) \ . \tag{7.3.39}$$

Accordingly, the equations for surfaces Γ_i in a harmonic form are cast, with due reference to (7.3.39), as follows

$$\mathbf{r} = [\mathbf{g}^{(1)}(\mathbf{r}) - \mathbf{g}^{(i)}(\mathbf{r})]/(4\pi/3)\gamma\varrho_0 + \operatorname{grad} \Phi^{(1)}(\mathbf{r}) \ , \quad i = 2, 3, \ldots, N \ . \tag{7.3.40}$$

Formulas (7.3.34) and (7.3.40) admit a solution for the problem of defining the shape of the surface S composed of a finite number of harmonic (analytical) parts by means of analytical continuation of the gravitational potential.

The above method for seeking the surface S is an extension of the relevant methods used to determine the contours of the cross section of two-dimensional disturbing bodies, and it has been developed by A. A. Zamorev (1942) and V. N. Strakhov (1970d) to a three-dimensional case.

7.4 Continuation of the Gravitational and Magnetic Fields into a Domain with an Arbitrary Analytical Distribution of Field Sources

The findings of the previous subsection are applicable to the magnetic field and readily extendible to the case of gravitational and magnetic fields produced by bodies with an arbitrary distribution of density or magnetization.

7.4.1 Analytical Representations of Fields Continued into Masses

Consider a three-dimensional domain D bounded by a surface S and filled with gravitational or magnetic masses with a density $\varrho(\mathbf{r})$ or magnetization $\mathbf{I}(\mathbf{r})$, respectively. Let Γ be an analytical part of the surface S, and let the density $\varrho(\mathbf{r})$ and magnetization $\mathbf{I}(\mathbf{r})$ be everywhere in D^* (a domain including D together with the boundary S) analytical functions of coordinates. Then it is obvious that the functions $\boldsymbol{\sigma}|_{\Gamma}$, $\mathbf{h}|_{\Gamma}$, and $\mathbf{I}_n|_{\Gamma}$ are analytical on Γ [$\boldsymbol{\sigma}(\mathbf{r})$ being the function satisfying Eq. (6.2.22) and $\mathbf{h}(\mathbf{r})$ meeting Eq. (6.3.8)].

Hence, by the Cauchy-Kovalevskaya theorem, in the neighborhood of Γ there exist such harmonic functions $\Phi^g(\mathbf{r})$ and $\Phi^H(\mathbf{r})$ that we have on Γ

$$\operatorname{grad}\Phi^g|_{\Gamma} = \boldsymbol{\sigma}|_{\Gamma} , \tag{7.4.1 a}$$

$$\operatorname{grad}\Phi^H|_{\Gamma} = (\mathbf{h}-4\pi\mathbf{I}_n)|_{\Gamma} . \tag{7.4.1 b}$$

For instance, once the domain D is filled with masses of uniform density ϱ_0, $\boldsymbol{\sigma}(\mathbf{r}) = (4\pi/3)\gamma\varrho_0\mathbf{r}$ and Eq. (7.4.1 a) turns into the equation for an analytical part of the surface Γ in a harmonic form:

$$\operatorname{grad}\Phi^g(\mathbf{r})|_{\Gamma} = (4\pi/3)\gamma\varrho_0\mathbf{r} .$$

Formulas (7.4.1 a) and (7.4.1 b) specifying the equation for the analytical part of the surface S make it possible to employ the results of Sects. 6.2, 6.3, and 7.2.3, according to which external gravitational and magnetic fields fitted by the three-dimensional Cauchy-type integrals can be continued into masses. Indeed, by virtue of (6.2.24) and (6.3.12), the gravitational $\mathbf{g}(\mathbf{r}')$ and the magnetic $\mathbf{H}(\mathbf{r}')$ fields outside masses are defined by the expressions

$$\mathbf{g}(\mathbf{r}') = \mathbf{C}^S[\mathbf{r}'; \boldsymbol{\sigma}(\mathbf{r})] , \quad \mathbf{r}' \in C\bar{D} , \tag{7.4.2 a}$$

$$\mathbf{H}(\mathbf{r}') = \mathbf{C}^S[\mathbf{r}'; \mathbf{h}(\mathbf{r})-4\pi\mathbf{I}_n(\mathbf{r})] , \quad \mathbf{r} \in C\bar{D} . \tag{7.4.2 b}$$

Consequently, in view of (7.2.20), the continued fields are defined by the following formulas:

$$\mathbf{g}^{\text{cont}}(\mathbf{r}) = -\operatorname{grad}\Phi^g(\mathbf{r}) + \mathscr{F}^g(\mathbf{r}) , \tag{7.4.3 a}$$

$$\mathbf{H}^{\text{cont}}(\mathbf{r}) = -\operatorname{grad}\Phi^H(\mathbf{r}) + \mathscr{F}^H(\mathbf{r}) , \tag{7.4.3 b}$$

where

$$\begin{aligned} \mathscr{F}^g(\mathbf{r}) &= \mathbf{C}^S(\mathbf{r}; \boldsymbol{\sigma}) = \mathbf{C}^S(\mathbf{r}; \operatorname{grad}\Phi^g) , \\ \mathscr{F}^H(\mathbf{r}) &= \mathbf{C}^S(\mathbf{r}; \mathbf{h}-4\pi\mathbf{I}_n) = \mathbf{C}^S(\mathbf{r}; \operatorname{grad}\Phi^H) . \end{aligned} \tag{7.4.4}$$

Formulas (7.4.3 a) and (7.3.4 b) enable us to study the issue of analytical continuation of the external field into the domain D in a most comprehensive manner.

7.4.2 Case of a Domain Bounded by an Analytical Surface

If the surface S is analytical and the distributions $\varrho(\mathbf{r})$ and $\mathbf{I}(\mathbf{r})$ are fitted by analytical functions, the following theorems extending the Strakhov theorems (V.N. Strakhov 1970c) to a three-dimensional case are valid[2].

Theorem 7.4.1. If S is a closed regular-analytical surface described by Eqs. (7.4.1), the functions $\Phi^g(\mathbf{r})$ and $\Phi^H(\mathbf{r})$ are sure to have singularities inside S and the set of their singularities coincides with those of functions $\mathbf{g}(\mathbf{r})$ or $\mathbf{H}(\mathbf{r})$ continued analytically into S.

Theorem 7.4.2. For two different domains D_1 and D_2 filled with masses of different analytical distributions of density $\varrho_1(\mathbf{r})$ and $\varrho_2(\mathbf{r})$ or magnetization $\mathbf{I}_1(\mathbf{r})$ and $\mathbf{I}_2(\mathbf{r})$, and bounded by regular-analytical surfaces S_1 and S_2, to produce identically equal fields

$$\mathbf{g}^{(1)}(\mathbf{r}) \equiv \mathbf{g}^{(2)}(\mathbf{r}) \; , \tag{7.4.5a}$$

$$\mathbf{H}^{(1)}(\mathbf{r}) \equiv \mathbf{H}^{(2)}(\mathbf{r}) \; , \tag{7.4.5b}$$

it is necessary and sufficient that the domains D_1 and D_2 intersect and the functions $\Phi^{g(1)}$ and $\Phi^{g(2)}$ or $\Phi^{H(1)}$ and $\Phi^{H(2)}$ specifying the surfaces S_1 and S_2 should have singularities only in their common part $D_0 = D_1 \cap D_2$ and the differences $\Phi^{g(1)} - \Phi^{g(2)} = \delta\Phi^g$ or $\Phi^{H(1)} - \Phi^{H(2)} = \delta\Phi^H$ should be functions harmonic in D_0.

The validity of theorem 7.4.1 follows directly from relations (7.4.3a) through (7.4.4). We will prove theorem 7.4.2 confining ourselves, for the sake of brevity, only to magnetized masses.

Necessity. By virtue of theorem 7.4.1, the functions $\Phi^{H(1)}$ and $\Phi^{H(2)}$ must obviously have some common singularities to be found simultaneously inside D_1 and D_2. As a result, the domains D_1 and D_2 either intersect or enclose one another, $D_1 \cap D_2 = D_0 \neq \varnothing$. For the sake of concreteness, it will be assumed below that $D_2 \backslash D_0$ is not void, i.e., D_1 is either completely in D_2 (Fig. 44a) or it intersects the latter (Fig. 44b).

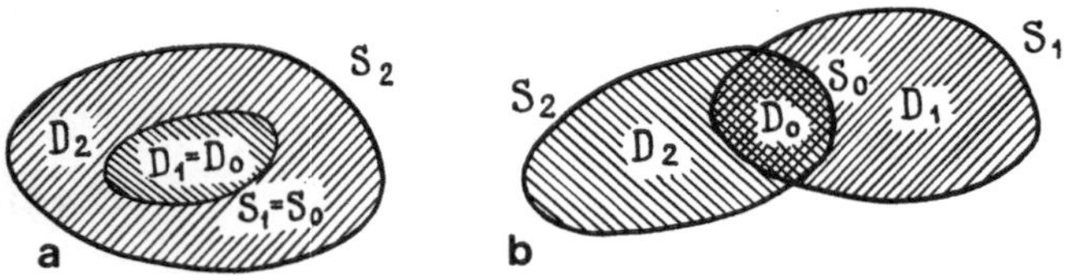

Fig. 44. **a** Case where D_1 is completely within D_2; **b** Case where D_1 intersects D_2

[2] For a domain of a uniform density or magnetization, these theorems were first formulated for a two-dimensional case and proved by A.V. Tsirulsky (1963, 1969), while for a three-dimensional case this was done in our study (Zhdanov 1973b).

Further, we have from formulas (7.4.3 a) and (7.4.3 b) that

$$\mathbf{H}^{(1)}(\mathbf{r}) = -\operatorname{grad}\Phi^{H(1)}(\mathbf{r}) + \mathscr{F}^{H(1)}(\mathbf{r}) \;,$$
$$\mathbf{H}^{(2)}(\mathbf{r}) = -\operatorname{grad}\Phi^{H(2)}(\mathbf{r}) + \mathscr{F}^{H(2)}(\mathbf{r}) \;, \tag{7.4.6}$$

where the functions $\mathscr{F}^{H(1)}$ and $\mathscr{F}^{H(2)}$ harmonic in D_1 and D_2 are defined by the formulas

$$\mathscr{F}^{H(1)} = \mathbf{C}^{S_1}(\mathbf{r};\, \operatorname{grad}\Phi^{H(1)}) \;, \quad \mathscr{F}^{H(2)} = \mathbf{C}^{S_2}(\mathbf{r};\, \operatorname{grad}\Phi^{H(2)}) \;,$$

respectively. Evidently, the function $\mathbf{H}^{(1)}(\mathbf{r})$ is harmonic in $D_2 \backslash D_1$. Consequently, by the condition of the theorem, the function $\mathbf{H}^{(2)}(\mathbf{r})$ (or its analytical continuation) is also harmonic in $D_2 \backslash D_0$. Thus, according to Eqs. (7.4.6), $\operatorname{grad}\Phi^{H(2)}$, as a difference between two harmonic functions, is itself a harmonic function in $D_2 \backslash D_1$. Therefore, employing property (5.2.12) of the Cauchy integral analogs, we can write

$$\mathbf{H}^{(2)}(\mathbf{r}') = \mathbf{C}^{S_2}(\mathbf{r}';\, \mathbf{h} - 4\pi \mathbf{I}_n^{(2)}) = \mathbf{C}^{S_2}(\mathbf{r}';\, \operatorname{grad}\Phi^{H(2)}) = \mathbf{C}^{S_0}(\mathbf{r}';\, \operatorname{grad}\Phi^{H(2)}) \;, \tag{7.4.7}$$

where S_0 is the boundary of D_0; $\mathbf{r}' \in C\bar{D}_0$.

Similarly, it is shown that

$$\mathbf{H}^{(1)}(\mathbf{r}') = \mathbf{C}^{S_0}(\mathbf{r}';\, \operatorname{grad}\Phi^{H(1)}) \;. \tag{7.4.8}$$

Hence, we derive

$$\mathbf{C}^{S_0}(\mathbf{r}';\, \operatorname{grad}\Phi^{H(1)} - \operatorname{grad}\Omega^{H(2)}) = \mathbf{C}^{S_0}(\mathbf{r}';\, \operatorname{grad}\delta\Phi^{H}) \equiv 0 \;, \quad \mathbf{r}' \in C\bar{D}_0 \;.$$

As a result, by theorem 5.3.1, the vector field $\operatorname{grad}\delta\Phi^H$ is the Laplace field everywhere in D: $\operatorname{div}\operatorname{grad}\delta\Phi^H(\mathbf{r}) \equiv 0$, $\mathbf{r} \in D_0$, i.e., $\Delta\,\delta\Phi^H(\mathbf{r}) = 0$, $\mathbf{r} \in D_0$, and $\delta\Phi^H$ is a function harmonic everywhere in D_0, which was to be proved.

Sufficiency. Let us calculate the difference between the fields $\mathbf{H}^{(2)}(\mathbf{r})$ and $\mathbf{H}^{(1)}(\mathbf{r})$, using formulas (7.4.7) and (7.4.8):

$$\mathbf{H}^{(2)}(\mathbf{r}) - \mathbf{H}^{(1)}(\mathbf{r}) = \mathbf{C}^{S_0}(\mathbf{r};\, \operatorname{grad}\delta\Phi^{H}) \;, \quad \mathbf{r} \in C\bar{D}_0 \;. \tag{7.4.9}$$

But once $\delta\Phi^H$ is a function harmonic everywhere in D_0, the Cauchy-type integral in the right-hand side of (7.4.9) is, by theorem 5.3.1, identically zero. Hence,

$$\mathbf{H}^{(1)}(\mathbf{r}) \equiv \mathbf{H}^{(2)}(\mathbf{r}) \;, \quad \mathbf{r} \in C\bar{D}_0 \;.$$

Thus, theorem 7.4.2 has been proved conclusively.

7.4.3 Case of a Domain Bounded by a Piecewise Analytical Surface

Generally, if the boundary S of the domain D is a piecewise analytical surface, the following claims extending the Strakhov theorem developed for plane fields (Sects. 4.3 and 4.4) to a three-dimensional case are valid.

Theorem 7.4.3. An external gravitational or magnetic field produced by analytical distributions of density or magnetization inside the domain D (so that the functions $\boldsymbol{\sigma}|_S$, $\mathbf{h}|_S$, and $\mathbf{I}_n|_S$ on the boundary S of the domain D satisfy the Hölder condition) admits analytical continuation through any regular-analytical part Γ of the surface, except maybe for its boundary. The singular points of the continued field coincide with those of the functions describing the equations for surfaces Γ in form (7.4.1 a) and (7.4.1 b).

Theorem 7.4.4. If the boundary S of the domain D consists of a finite number of different regular-analytical parts Γ_i ($i = 1, 2, \ldots N$) [different in the sense that different analytical functions $\Phi^{(i)}(\mathbf{r})$, $i = 1, 2, \ldots N$, appear in the equations for these surfaces (7.4.1 a) and (7.4.1 b)], the lines joining the different analytical parts Γ_i and Γ_j are branch lines of the external field for any analytical distribution of density or magnetization inside D (ensuring that $\boldsymbol{\sigma}|_S$, $\mathbf{h}|_S$, and $\mathbf{I}_n|_S$ satisfy the Hölder conditions).

Proof the theorems 7.4.3 and 7.4.4 follows directly from formulas (7.4.3 a), (7.4.3 b), and (7.4.4). These formulas can be also used to solve some inverse problems of gravimetry and magnetometry on the basis of analytical continuation. An example of this problem for a homogeneous material body was considered in Sect. 7.3.3. It is pertinent to mention that the method employed therein to solve the inverse problem relying on analytical continuation is of theoretical rather than of practical concern, since it establishes the conditions in which the solution to the inverse problem is unambiguous. We will extend this result to three-dimensional bodies with an arbitrary distribution of density (or magnetization).

Let the surface S be a sum of several regular-analytical parts Γ_i, $i = 0, 1, 2, \ldots N$ and assume that the analytical distribution of density inside S is known [in which case the corresponding function $\boldsymbol{\sigma}(\mathbf{r})$ – a particular solution to Eq. (6.2.22) – satisfies the Hölder condition on S]. We also assume that we know the analytical function $\Phi^{g(1)}$ defining the equation for surface Γ_1 in the form of (7.4.1 a) and that this function is analytical everywhere inside D (except maybe for a finite number of isolated singular points). The problem is to find the whole surface S from the gravitational field $\mathbf{g}(\mathbf{r})$ known outside masses, using the procedure of analytical continuation.

A solution for this problem follows from formula (7.4.3 a). Indeed, let $\mathbf{g}^{(i)}(\mathbf{r})$ stand for the result of analytical continuation of the external gravitational field through the surface Γ_i. Then we have

$$\mathscr{F}^g(\mathbf{r}) = \mathbf{g}^{(i)}(\mathbf{r}) + \operatorname{grad} \Phi^{g^{(i)}}(\mathbf{r}) \ , \tag{7.4.10}$$

and, in particular, for the known part of the surface

$$\mathscr{F}^g(\mathbf{r}) = \mathbf{g}^{(1)}(\mathbf{r}) + \operatorname{grad} \Phi^{g^{(1)}}(\mathbf{r}) \ . \tag{7.4.11}$$

By equating the right-hand sides of (7.4.10) and (7.4.11) we find

$$\operatorname{grad} \Phi^{g^{(i)}}(\mathbf{r}) = \operatorname{grad} \Phi^{g^{(1)}}(\mathbf{r}) - \mathbf{g}^{(i)}(\mathbf{r}) + \mathbf{g}^{(1)}(\mathbf{r}) \ . \tag{7.4.12}$$

Consequently, the equation for the surface S, allowing for (7.4.1 a) and (7.4.1 b), takes the form

$$\operatorname{grad} \Phi^{g^{(1)}}(\mathbf{r}) - \mathbf{g}^{(i)}(\mathbf{r}) + \mathbf{g}^{(1)}(\mathbf{r}) = \boldsymbol{\sigma}(\mathbf{r}) \ , \tag{7.4.13}$$

where $\boldsymbol{\sigma}(\mathbf{r})$ is a particular solution of Eq. (6.2.22).

Formula (7.4.13) admits a solution for the problem of defining the shape of the surface S composed of a finite number of regular-analytical parts on the basis of analytical continuation. As can be seen, the problem has a unique solution, provided not only the density distribution inside S but also some (in principle, arbitrarily small) part of this surface is known. In the opposite case, as theorem 7.4.2 asserts, the inverse problem is ambiguous. A similar result is observed with the magnetic field.

7.5 Integral Techniques for Analytical Continuation of Three-Dimensional Laplace Fields

The methods for analytical continuation of Laplace vector fields outlined in Sect. 7.1 are useful for theoretical considerations but are impracticable in numerical calculations. Meanwhile, in the interpretation of gravitational and magnetic anomalies one has to tackle practical problems of harmonic continuation of functions measured at the Earth's surface (at the surface of observation). Below, we will consider an approach to this problem which relies on three-dimensional Cauchy integral analogs.

7.5.1 Analytical Continuation of the Laplace Field into the Upper Half-Space

The surface of observation will be specified in the form of an arbitrary piecewise smooth surface S passing through a point at infinity (Fig. 45), S: $[z = z(x, y), -\infty < x < +\infty, -\infty < y < +\infty]$, with $0 \geqslant z(x, y) \geqslant -h$, $h > 0$.

The upper half-space bounded by the surface S will be denoted by Ω^+. Assume that the field $\mathbf{F}(\mathbf{r})$ is the Laplace field in Ω^+ and that the values of this field on S are known. The problem is to continue $\mathbf{F}(\mathbf{r})$ analytically into the upper half-space Ω^+. To solve this problem, take an arbitrary point $\mathbf{r}'$ out of Ω^+ and seek to reconstruct the field $\mathbf{g}(\mathbf{r})$ at this point. Circumscribe a sphere $O_R(\mathbf{r}')$ of a radius R from the point $\mathbf{r}'$. The part of the surface of this sphere belonging to Ω^+ will be designated by O_R^+. The sphere $O_R(\mathbf{r}')$ cuts a certain part S_R from the

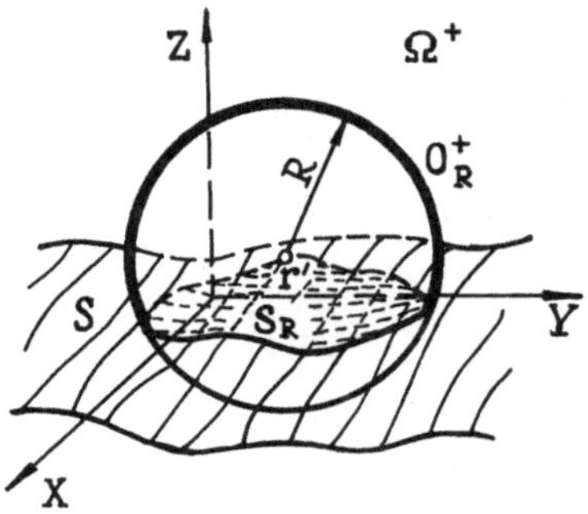

Fig. 45. Analytical continuation of the Laplace field into the upper half-space

surface S. Construct a closed surface γ_R composed of the surfaces S_R and O_R^+; then we have, by Cauchy formula (5.2.12)

$$\mathbf{F}(\mathbf{r}') = \mathbf{C}^{\gamma_R}[\mathbf{r}';\ \mathbf{F}(\mathbf{r})] \ . \tag{7.5.1}$$

The Cauchy-type integral taken over the surface γ_R can be divided into two integrals

$$\mathbf{C}^{\gamma_R}[\mathbf{r}';\ \mathbf{F}(\mathbf{r})] = \mathbf{C}^{O_R^+}[\mathbf{r}';\ \mathbf{F}(\mathbf{r})] + \mathbf{C}^{S_R}[\mathbf{r}';\ \mathbf{F}(\mathbf{r})] \ . \tag{7.5.2}$$

The first integral extended along the part of the sphere O_R^+ is, according to (6.4.9),

$$\mathbf{C}^{O_R^+}[\mathbf{r}';\ \mathbf{F}(\mathbf{r})] = \frac{1}{4\pi R^2} \iint_{O_R^+} \mathbf{F}(\mathbf{r})\, ds \ . \tag{7.5.3}$$

Employing the law of averages, we find

$$\mathbf{C}^{O_R^+}[\mathbf{r}';\ \mathbf{F}(\mathbf{r})] = \frac{1}{4\pi R^2} \iint_{O_R^+} ds\, \mathbf{F}(\tilde{\mathbf{r}}) \ , \tag{7.5.4}$$

where $\tilde{\mathbf{r}} \in O_R^+$.

It is clear that

$$\frac{1}{4\pi R^2} \iint_{O_R^+} ds \leqslant 1 \ . \tag{7.5.5}$$

Hence, estimating integral (7.5.4) by modulus and allowing for (7.5.5) we find

$$|\mathbf{C}^{O_R^+}[\mathbf{r}';\ \mathbf{F}(\mathbf{r})]| \leqslant |\mathbf{F}(\tilde{\mathbf{r}})| \ . \tag{7.5.6}$$

Let the following condition

$$\max_{\tilde{\mathbf{r}} \in O_R^+} |\mathbf{F}(\tilde{\mathbf{r}})| \to 0 \quad \text{for} \quad R \to \infty \tag{7.5.7}$$

set in.

Then

$$|\mathbf{C}^{O_R^+}[\mathbf{r}';\ \mathbf{F}(\mathbf{r})]| \to 0 \quad \text{for} \quad R \to \infty \ . \tag{7.5.8}$$

It is noteworthy that the Laplace vector fields, whose sources are concentrated in finite domains of space, satisfy invariably condition (7.5.7).

The second integral in (7.5.2) for $R \to \infty$ tends to the integral taken over the whole surface. Thus, proceeding to the limit for $R \to \infty$ in (7.5.1) and allowing for (7.5.2) and (7.5.8), we obtain

$$\mathbf{F}(\mathbf{r}') = \mathbf{C}^{S}[\mathbf{r}';\ \mathbf{F}(\mathbf{r})]$$

$$= -\frac{1}{4\pi} \iint_S \left[(\mathbf{n} \cdot \mathbf{F}) \operatorname{grad} \frac{1}{|\mathbf{r}-\mathbf{r}'|} + (\mathbf{n} \times \mathbf{F}) \times \operatorname{grad} \frac{1}{|\mathbf{r}-\mathbf{r}'|} \right] ds \ , \tag{7.5.9}$$

where $\mathbf{n}$ is the unit vector of the normal to the surface pointing to the lower half-space.

Formula (7.5.9) permits reconstruction of values of the Laplace field $\mathbf{F}(\mathbf{r})$ everywhere in the upper half-space from its values at the surface of observation, i.e., it resolves the problem of an upward analytical continuation.

7.5.2 Analytical Continuation of the Laplace Field into the Lower Half-Space

Now we will proceed to the problem of continuing Laplace fields into the lower half-space. Assume that the conditions of Sect. 7.5.1 are fulfilled and the field $\mathbf{F}(\mathbf{r})$ is the Laplace field in the whole upper half-space $z > -A$ and it decreases rather fast at infinity. [The required rate at which $\mathbf{F}(\mathbf{r})$ decreases at infinity will be evaluated accurately later.] The values of the function $\mathbf{F}(\mathbf{r})$ at the surface S are known. The problem is to find $\mathbf{F}(\mathbf{r})$ everywhere for $z > -A$.

Assume that $\mathbf{r}'(x', y', z')$ is a point of space, lying below S and $z' > -A$ (Fig. 46). Take ε to be so small that $z' > -A+\varepsilon$ and draw a horizontal plane Σ at a depth $z = -A+\varepsilon$. Try to seek the value of $\mathbf{F}(\mathbf{r}')$ at the point $\mathbf{r}'$. Note, first of all, that if the field $\mathbf{F}(\mathbf{r})$ at the surface Σ were known, we could evaluate it at the point $\mathbf{r}'$ by formula (7.5.9) of an upward analytical continuation, since the point $\mathbf{r}'$ is above Σ:

$$\mathbf{F}(\mathbf{r}') = \mathbf{C}^{\Sigma}(\mathbf{r}'; \mathbf{F}(\mathbf{r}))$$

$$= -\frac{1}{4\pi}\iint_{\Sigma}\left[(\mathbf{n}\cdot\mathbf{F})\,\mathrm{grad}\,\frac{1}{|\mathbf{r}-\mathbf{r}'|}+(\mathbf{n}\times\mathbf{F})\times\mathrm{grad}\,\frac{1}{|\mathbf{r}-\mathbf{r}'|}\right]ds\ . \qquad (7.5.10)$$

Transform the integral over the surface Σ into the integral over the surface S. To this end, we will employ the known spectral decomposition

$$\frac{1}{|\mathbf{r}-\mathbf{r}'|} = \frac{1}{2\pi}\iint_{-\infty}^{+\infty}\frac{f(\mathbf{r}')g(\mathbf{r})}{\eta_0}\,dk_x dk_y\ , \qquad (7.5.11)$$

where the spectral harmonics f and g are equal:

$$f(\mathbf{r}') = \exp[i(k_x x' + k_y y']\exp(-\eta_0 z')\ ,$$

$$\mathbf{g}(\mathbf{r}) = \exp[-i(k_x x + k_y y)]\exp(\eta_0 z)\,|_{z'>z}\ , \qquad (7.5.12)$$

$$\eta_0 = \sqrt{k_x^2+k_y^2}\ .$$

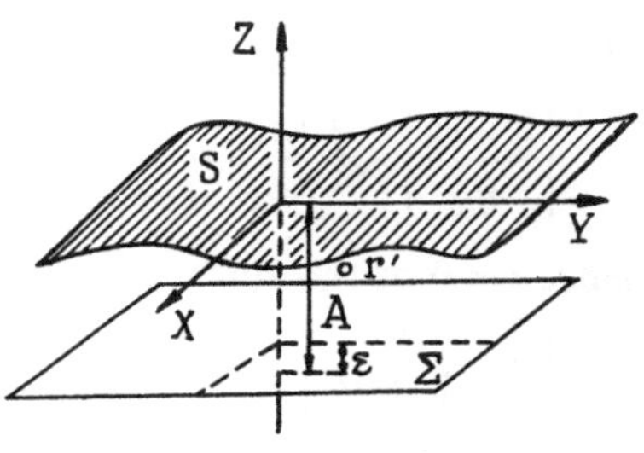

Fig. 46. Analytical continuation of the Laplace field into the lower half-space

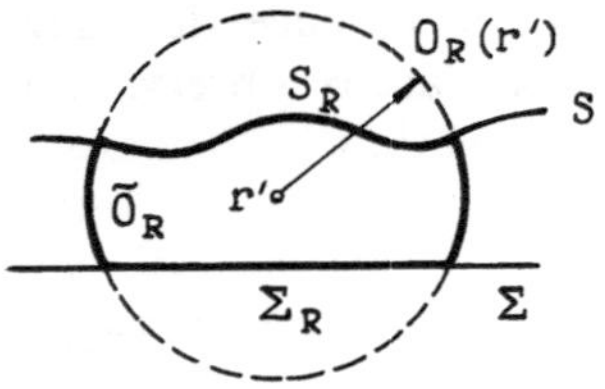

Fig. 47. Scheme illustrating the possibility for transforming integral (7.5.10) taken over the surface S into an integral over the surface Γ

Substitute (7.5.11) into (7.5.10) and exchange the order of integration, since all the integrals converge uniformly in corresponding parameters:

$$\mathbf{F}(\mathbf{r}') = -\frac{1}{8\pi^2} \iint_{-\infty}^{+\infty} \frac{f(\mathbf{r}')}{\eta_0} \iint_{\Sigma} [(\mathbf{n}\cdot\mathbf{F})\operatorname{grad} g + (\mathbf{n}\times\mathbf{F})\times\operatorname{grad} g]\, ds\, dk_x dk_y \ . \quad (7.5.13)$$

Just as in the previous subsection, draw a sphere $O_R(\mathbf{r}')$ of a radius R with its center at the point $\mathbf{r}'$ (Fig. 47). The part of the sphere found between S and Σ will be denoted by $\tilde{O}_R$. The sphere $O_R(\mathbf{r}')$ cuts some part S_R from the surface S and some part Σ_R from the surface Σ. Consider the closed surface $\Gamma_R = S_R \cup \tilde{O}_R \cup \Sigma_R$.

One can readily see that the function $g(\mathbf{r})$ is harmonic inside the domain D_R bounded by Γ_R, $\Delta g(\mathbf{r}) = 0$. Hence, we obtain by formula (5.1.14) that

$$\frac{1}{\eta_0} \iint_{\Gamma_R} [(\mathbf{n}\cdot\mathbf{F})\operatorname{grad} g + (\mathbf{n}\times\mathbf{F})\times\operatorname{grad} g]\, ds = \frac{1}{\eta_0} \iiint_{D_R} (\operatorname{div}\mathbf{F}\operatorname{grad} g + \Delta g\mathbf{F})\, dv = 0 \ , \quad (7.5.14)$$

since, by virtue of the Laplace nature of $\mathbf{F}$: $\operatorname{div}\mathbf{F} \equiv 0$ in D_R.

The surface integral in (7.5.14) can be represented as a sum of three integrals taken over the corresponding parts of the surface Γ_R:

$$\frac{1}{\eta_0}\left(\iint_{S_R} \cdots ds + \iint_{\Sigma_R} \cdots ds + \iint_{\tilde{O}_R} \cdots ds\right) = 0 \ . \quad (7.5.15)$$

Let us estimate the modulus of the integral taken over $\tilde{O}_R$:

$$\left|\frac{1}{\eta_0} \iint_{\tilde{O}_R} \cdots ds\right| = \left|\frac{1}{\eta_0} \iint_{\tilde{O}_R} \{(\mathbf{n}\cdot\mathbf{F})\operatorname{grad} g + [\mathbf{n}\times\mathbf{F}]\times\operatorname{grad} g\} ds\right|$$

$$\leqslant 2 \iint_{\tilde{O}_R} |\mathbf{n}|\,|\mathbf{F}| \left|\frac{\operatorname{grad} g}{\eta_0}\right| ds \ . \quad (7.5.16)$$

But

$$\left|\frac{\operatorname{grad} g}{\eta_0}\right| \leqslant \left(\frac{|k_x|}{\eta_0} + \frac{|k_y|}{\eta_0} + 1\right) \exp(\eta_0 z) \leqslant 3 \ , \quad (7.5.17)$$

since $z<0$ at the surface $\tilde{O}_R$. Hence,

$$\left|\frac{1}{\eta_0} \iint_{\tilde{O}_R} \cdots ds\right| \leqslant 6 \iint_{\tilde{O}_R} |\mathbf{F}|\, ds \leqslant 6 \max_{\tilde{\mathbf{r}}\in\tilde{O}_R} |\mathbf{F}(\tilde{\mathbf{r}})| \iint_{\tilde{O}_R} ds \ . \quad (7.5.18)$$

By construction, the surface $\tilde{O}_R$ is always between the planes $z=0$ and $z=-A$ for any R; it is remarkable that the larger the radius R, the closer is the form of the ring $\tilde{O}_R$ to that of a direct circular ring of the radius R; therefore

$$\iint\limits_{\tilde{O}_R} ds \leqslant M 2\pi RA \ , \tag{7.5.19}$$

where M is a constant. Substituting (7.5.19) into (7.5.18) we find that

$$\left| \frac{1}{\eta_0} \iint\limits_{\tilde{O}_R} \cdots ds \right| \leqslant 6 \max_{\tilde{r} \in \tilde{O}_R} |\mathbf{F}(\tilde{\mathbf{r}})| M 2\pi RA \ . \tag{7.5.20}$$

Let us make the function $\mathbf{F}(\mathbf{r})$ decrease at infinity faster than the reciprocal power of R does:

$$\lim_{R\to\infty} R \max_{\mathbf{r}\in\tilde{O}_R} |\mathbf{F}(\mathbf{r})| = 0 \ .$$

Then, for $R\to\infty$, i.e., for an ever expanding sphere $O_R(\mathbf{r}')$ the integral over $\tilde{O}_R$ in (7.5.15) tends to zero. Thus, passing to the limit for $R\to\infty$ in (7.5.15), we obtain

$$\frac{1}{\eta_0}\left(\iint\limits_{S} \cdots ds + \iint\limits_{\Sigma} \cdots ds \right) = 0 \ ,$$

or

$$\frac{1}{\eta_0} \iint\limits_{\Sigma} \cdots ds = -\frac{1}{\eta_0} \iint\limits_{S} \cdots ds \ . \tag{7.5.21}$$

Thus, we have transformed the integral over the surface Σ into the integral over the surface S, where the field $\mathbf{F}(\mathbf{r})$ is known. Since this passage to the limit is uniform with respect to η_0, it can be accomplished also inside the external integral sign in (7.5.13):

$$\mathbf{F}(\mathbf{r}') = -\frac{1}{8\pi^2} \iint\limits_{-\infty}^{+\infty} \frac{f(\mathbf{r}')}{\eta_0} \iint\limits_{S} \{(\mathbf{n}\cdot\mathbf{F}) \operatorname{grad} g + [\mathbf{n}\times\mathbf{F}] \times \operatorname{grad} g\} ds\, dk_x dk_y \ , \tag{7.5.22}$$

where $\mathbf{n}$ is the unit vector of the normal to the surface S pointing to the lower half-space.

Relation (7.5.22) yields a solution to the problem of continuing the Laplace field analytically into the lower half-space down to a horizontal plane going through the nearest singular point of the function $\mathbf{F}(\mathbf{r})$.

We should like to note in conclusion of this part that the procedure of downward analytical continuation of the Laplace vector field is an ill-posed problem, just as is the case with plane fields (Sect. 4.5). That is why its numerical solution calls for application of appropriate regularizing algorithms (Tikhonov and Arsenin 1974; Strakhov 1969a, b), which can be elaborated on the basis of the above integral formulas (Zhdanov 1976b). Some examples of construction of such algorithms for the case of electromagnetic and seismic fields will be given below.

Part III
Stratton-Chu Type Integrals in the Theory of Electromagnetic Fields

8 Stratton-Chu Type Integrals

8.1 Electromagnetic Field Equations

As indicated earlier, the body of three-dimensional Cauchy integral analogs developed in the previous part permits the extension of many ideas and methods established in the theory of functions of a complex variable to three-dimensional situations important in the study of geopotential fields. It seems tempting to apply these methods to other geophysical fields, particularly to the electromagnetic field. A natural extension of the concept of the Cauchy-type integral to the electromagnetic field theory involves, as will be shown later, the Stratton-Chu type integrals introduced on the basis of the known integral formulas of Stratton-Chu (Stratton 1941).

For the sake of a thorough exposition, this part starts with a description of the basic equations of the electromagnetic field theory.

8.1.1 Maxwell Equations

The basis of the theory of electromagnetic fields studied by geophysicists is provided by the macro-electrodynamic equations, i.e., the Maxwell equations:

$$\operatorname{rot}\mathbf{H} = \mathbf{j}+\mathbf{j}^e+\frac{\partial \mathbf{D}}{\partial t} , \tag{8.1.1 a}$$

$$\operatorname{rot}\mathbf{E} = -\frac{\partial \mathbf{B}}{\partial t} , \tag{8.1.1 b}$$

$$\operatorname{div}\mathbf{B} = 0 , \tag{8.1.1 c}$$

$$\operatorname{div}\mathbf{D} = q+q^e , \tag{8.1.1 d}$$

where $\mathbf{E}$ and $\mathbf{D}$ are the vectors of the electric field; $\mathbf{H}$ and $\mathbf{B}$ are the vectors of the magnetic field; $\mathbf{j}$ is the conduction current density; q is the spatial density of free electric charges; $\mathbf{j}^e$ and q^e are the densities of extraneous electric currents and charges interrelated by the continuity equations

$$\operatorname{div}\mathbf{j}^e = -\partial q^e/\partial t . \tag{8.1.2}$$

Equations (8.1.1 a) through (8.1.1 d) should be supplemented by the constraint equations reflecting the electromagnetic properties of a medium, which are for linear and isotropic media as follows:

$$\mathbf{D} = \varepsilon \mathbf{E} \ , \quad \mathbf{B} = \mu \mathbf{H} \ , \tag{8.1.3}$$

where ε and μ are the dielectric constant (permittivity) and the magnetic permeability, respectively.

The conduction current density is expressed in terms of the electric field by means of Ohm's law (in a differential form):

$$\mathbf{j} = \sigma \mathbf{E} \ , \tag{8.1.4}$$

where σ is the electric conductivity of a medium which is the reciprocal of its resistivity ϱ, $\sigma = 1/\varrho$.

8.1.2 Field in Homogeneous Domains of a Medium

In domains where the values of ε, μ and σ are constant, Eqs. (8.1.1 a)–(8.1.1 d) take the form

$$\operatorname{rot} \mathbf{H} = \sigma \mathbf{E} + \mathbf{j}^e + \varepsilon \partial \mathbf{E}/\partial t \ , \tag{8.1.5 a}$$

$$\operatorname{rot} \mathbf{E} = -\mu \partial \mathbf{H}/\partial t \ , \tag{8.1.5 b}$$

$$\operatorname{div} \mathbf{H} = 0 \ , \tag{8.1.5 c}$$

$$\operatorname{div} \mathbf{E} = \frac{q+q^e}{\varepsilon} \ . \tag{8.1.5 d}$$

In many problems, it is useful *to separate the Maxwell Eqs.* (8.1.5 a)–(8.1.5 d), i.e., to write the equations for the electric and for the magnetic fields separately. To do this, we will apply the operator rot to the left- and right-hand sides of (8.1.5 b) and substitute the result into (8.1.5 a):

$$\operatorname{rot}\operatorname{rot} \mathbf{E} + \mu\varepsilon \frac{\partial^2 \mathbf{E}}{\partial t^2} + \mu\sigma \frac{\partial \mathbf{E}}{\partial t} = -\mu \frac{\partial \mathbf{j}^e}{\partial t} \ . \tag{8.1.6}$$

Similarly, using the operator rot with (8.1.5 a) and allowing for (8.1.5 b) we arrive at

$$\operatorname{rot}\operatorname{rot} \mathbf{H} + \mu\varepsilon \frac{\partial^2 \mathbf{H}}{\partial t^2} + \mu\sigma \frac{\partial \mathbf{H}}{\partial t} = \operatorname{rot} \mathbf{j}^e \ . \tag{8.1.7}$$

In view of the vector identity

$$\operatorname{rot}\operatorname{rot} \mathbf{E} = \operatorname{grad}\operatorname{div} \mathbf{E} - \Delta \mathbf{E}$$

and of Maxwell Eq. (8.1.5 d), relation (8.1.6) can be cast in the form

$$\Delta \mathbf{E} - \mu\varepsilon \frac{\partial^2 \mathbf{E}}{\partial t^2} - \mu\sigma \frac{\partial \mathbf{E}}{\partial t} = \mu \frac{\partial \mathbf{j}^e}{\partial t} + \frac{1}{\varepsilon} \operatorname{grad}(q+q^e) \ . \tag{8.1.8 a}$$

It is common knowledge (Stratton 1948) that in the absence of extraneous electric currents and charges, free electric charges disappear (relax) very rapidly in a homogeneous medium; hence, we can write

$$\Delta\mathbf{E}-\mu\varepsilon\frac{\partial^2\mathbf{E}}{\partial t^2}-\mu\sigma\frac{\partial\mathbf{E}}{\partial t}=0\ . \tag{8.1.8b}$$

Similar transforms [using Maxwell Eq. (8.1.5c)] will bring (8.1.7) to the form

$$\Delta\mathbf{H}-\mu\varepsilon\frac{\partial^2\mathbf{H}}{\partial t^2}-\mu\sigma\frac{\partial\mathbf{H}}{\partial t}=\operatorname{rot}\mathbf{j}^e\ , \tag{8.1.9a}$$

or for $\mathbf{j}^e=0$

$$\Delta\mathbf{H}-\mu\varepsilon\frac{\partial^2\mathbf{H}}{\partial t^2}-\mu\sigma\frac{\partial\mathbf{H}}{\partial t}=0\ . \tag{8.1.9b}$$

Equations (8.1.8b) and (8.1.9b) fitting the vectors of the electric and magnetic fields in a homogeneous domain of a medium, in the absence of extraneous currents and charges, are called *homogeneous telegraph equations.*

8.1.3 Boundary Conditions

At the interface of different homogeneous domains, the values of the parameters ε, μ, and σ may undergo steplike variations. In this case, according to formulas (8.1.3) and (8.1.4) some field vectors are also bound to change suddenly. To solve the problems of electrodynamics, it is necessary therefore to know the boundary conditions, that is the relations between the vectors of the field at two adjacent points on the different sides of the interface of media exhibiting different electromagnetic parameters.

We will formulate the boundary conditions when they are of critical importance for practical applications, namely where a smooth surface separates two media, 1 and 2, whose parameters are either constant or vary from point to point very slowly, so that in a close neighborhood of any point at the interface the latter can be regarded as a plane one and the medium parameters are thought to be constant. Let $\mathbf{n}$ denote the vector of the unit normal to the surface S at a given point. We also assume that no extraneous current or charges are present at the interface S. Then the following relations hold good:

1) For normal components of the field:

$$(\mathbf{B}^{(2)}-\mathbf{B}^{(1)})\cdot\mathbf{n}=0\ ,\quad (\mathbf{D}^{(2)}-\mathbf{D}^{(1)})\cdot\mathbf{n}=\eta\ , \tag{8.1.10}$$

where η is the surface density of the electric charge on S. Hence, the normal component of the magnetic field $\mathbf{B}$ is continuous on transition through the media interface, while the normal component of the electric field $\mathbf{D}$ experiences a discontinuity equal to the surface density of the electric charge.

2) For the tangential components of the field:

$$\mathbf{n}\times(\mathbf{H}^{(2)}-\mathbf{H}^{(1)})=\mathbf{j}^S\ , \tag{8.1.11}$$

where $\mathbf{j}^S$ is the density of the surface electric current,

$$\mathbf{n}\times(\mathbf{E}^{(2)}-\mathbf{E}^{(1)})=0 \ . \tag{8.1.12}$$

The surface current density $\mathbf{j}^S$ is nonzero only at the surface of a perfect conductor; hence for real media, (8.1.11) and (8.1.12) can be cast in the form

$$\mathbf{H}_\tau^{(2)}=\mathbf{H}_\tau^{(1)}=0 \ , \quad \mathbf{E}_\tau^{(2)}-\mathbf{E}_\tau^{(1)}=0 \ , \tag{8.1.13}$$

where the subscript τ refers to the field component tangential to S. Thus, the tangential components of the vector of the magnetic field $\mathbf{H}$ are continuous at interfaces of real media, while those of the vector of the electric field $\mathbf{E}$ are continuous at interfaces of any media.

An important particular case is related to the boundary conditions for transient fields at the surface of a perfect conductor. As is known in electrostatics, there is no electric field in a material whatever its conductivity might be. Yet, transient fields may penetrate a material of a finite conductivity. But if the conductor is perfect, the electromagnetic field is absent therein too.

Indeed, the charges inside a perfect conductor are so mobile that they respond instantaneously to arbitrarily rapid variations in the field. In doing so, they give rise to such a high surface density of the charge η at the surface of a perfect conductor that the electric field inside this conductor becomes zero. In addition, as the magnetic field varies with time, the surface electric charges move, thereby producing such a surface current $\mathbf{j}^S$ that inside the conductor there is no magnetic field. Thus, in a perfect conductor, the electromagnetic field is zero: $\mathbf{D}\equiv\mathbf{B}\equiv\mathbf{E}\equiv\mathbf{H}\equiv 0$. If the superscript 2 in the above formulas is ascribed to a perfect conductor the boundary conditions will be the following

$$\mathbf{D}^{(1)}\cdot\mathbf{n}=-\eta \ , \tag{8.1.14a}$$

$$\mathbf{H}^{(1)}\times\mathbf{n}=\mathbf{j}^S \ , \tag{8.1.14b}$$

$$\mathbf{B}^{(1)}\cdot\mathbf{n}=0 \ , \tag{8.1.14c}$$

$$\mathbf{E}^{(1)}\times\mathbf{n}=0 \ . \tag{8.1.14d}$$

So the normal component of the magnetic field $\mathbf{B}$ and the tangential component of the electric field $\mathbf{E}$ at the surface of a perfect conductor are zero. The normal components $\mathbf{D}$ is equal (with a negative sign) to the surface density of the electric charge, while the density of the surface electric current is equal in magnitude, and perpendicular to the direction of the tangential component of the magnetic field $\mathbf{H}$ at the conductor surface.

8.1.4 Monochromatic Field Equations

By virtue of the linearity of the Maxwell equations, the field varying arbitrarily in time can be represented as a sum of harmonic fields whose time dependence is expressed by means of the factor $\exp(-i\omega t)$. For a monochromatic field, Eqs. (8.1.1a) through (8.1.1d) take the form

$$\operatorname{rot}\mathbf{H} = \mathbf{j}+\mathbf{j}^e - i\omega\mathbf{D} \ , \quad \operatorname{rot}\mathbf{E} = i\omega\mathbf{B} \ ,$$
$$\operatorname{div}\mathbf{B} = 0 \ , \quad \operatorname{div}\mathbf{D} = q+q^e \ . \tag{8.1.15}$$

In a homogeneous domain of a medium, where the parameters ε, μ, and σ are maintained at a constant level, Eqs. (8.1.5a) through (8.1.5d) are cast in the form

$$\operatorname{rot}\mathbf{H} = \sigma^*\mathbf{E}+\mathbf{j}^e \ , \quad \operatorname{rot}\mathbf{E} = i\omega\mu\mathbf{H} \ ,$$
$$\operatorname{div}\mathbf{H} = 0 \ , \quad \operatorname{div}\mathbf{E} = q^e/\varepsilon^* \ , \tag{8.1.16}$$

where $\varepsilon^* = \varepsilon + i\sigma/\omega$ stands for the complex permittivity of a medium and $\sigma^* = \sigma - i\omega\varepsilon$ is its complex electric conductivity.

Accordingly, Eqs. (8.1.8a) and (8.1.9a), which are valid for the case of ε, μ, $\sigma = \text{const}$, take the form

$$\varDelta\mathbf{E}+k^{*2}\mathbf{E} = -i\omega\mu\mathbf{j}^e+(1/\varepsilon^*)\operatorname{grad} q^e \ ,$$
$$\varDelta\mathbf{H}+k^{*2}\mathbf{H} = -\operatorname{rot}\mathbf{j}^e \ , \tag{8.1.17}$$

where the coefficient k^* is the wave number of a medium:

$$k^{*2} = i\omega\mu\sigma+\omega^2\mu\varepsilon = i\omega\mu\sigma^* \ . \tag{8.1.18}$$

Equations (8.1.17) are known as *the Helmholtz equations.* The value of k^* is chosen to have a positive real part:

$$\operatorname{Re} k^* > 0 \ . \tag{8.1.19}$$

If the homogeneous domain in question is devoid of extraneous currents or extraneous charges, then we have

$$\operatorname{rot}\mathbf{H} = \sigma^*\mathbf{E} \ , \quad \operatorname{rot}\mathbf{E} = i\omega\mu\mathbf{H} \ ,$$
$$\operatorname{div}\mathbf{H} = 0 \ , \quad \operatorname{div}\mathbf{E} = 0 \ . \tag{8.1.20}$$

In this case, Eqs. (8.1.17) are reduced to the homogeneous Helmholtz equations

$$\varDelta\mathbf{E}+k^{*2}\mathbf{E} = 0 \ , \quad \varDelta\mathbf{H}+k^{*2}\mathbf{H} = 0 \ . \tag{8.1.21}$$

8.1.5 Quasi-Stationary Electromagnetic Field

In many problems, geophysicists deal with electromagnetic fields varying in time very slowly. In the equations for this field the second derivative with respect to time $\partial^2/\partial t^2$ can be discarded. This model is called *quasi-stationary.* A quasi-stationary electromagnetic field is fitted by the following equations

$$\operatorname{rot}\mathbf{H} = \sigma\mathbf{E}+\mathbf{j}^e \ , \quad \operatorname{rot} = -\mu\partial\mathbf{H}/\partial t \ ,$$
$$\operatorname{div}\mathbf{H} = 0 \ , \quad \operatorname{div}\mathbf{E} = -\frac{1}{\sigma}(\mathbf{E}\cdot\operatorname{grad}\sigma+\operatorname{div}\mathbf{j}^e) \ . \tag{8.1.22}$$

In domains of a constant electric conductivity and magnetic permeability we have

$$\operatorname{rot}\operatorname{rot}\mathbf{E}+\mu\sigma\frac{\partial\mathbf{E}}{\partial t}=-\mu\frac{\partial\mathbf{j}^e}{\partial t}\ ,\quad \operatorname{rot}\operatorname{rot}\mathbf{H}+\mu\sigma\frac{\partial\mathbf{H}}{\partial t}=\operatorname{rot}\mathbf{j}^e\ . \tag{8.1.23}$$

As a consequence, the telegraph equations go over into the *equations of diffusion*

$$\Delta\mathbf{E}-\mu\sigma\frac{\partial\mathbf{E}}{\partial t}=\mu\frac{\partial\mathbf{j}^e}{\partial t}-\frac{1}{\sigma}\operatorname{grad}\operatorname{div}\mathbf{j}^e\ ,$$
$$\Delta\mathbf{H}-\mu\sigma\frac{\partial\mathbf{H}}{\partial t}=-\operatorname{rot}\mathbf{j}^e\ . \tag{8.1.24}$$

Just as in a general case, the quasi-stationary field can be represented by Fourier transforms as a sum of harmonic fields whose time dependence is reflected by the factor $\exp(-i\omega t)$. The quasi-stationary field harmonics are fitted by the equations

$$\operatorname{rot}\mathbf{H}=\sigma\mathbf{E}+\mathbf{j}^e\ ,\quad \operatorname{rot}\mathbf{E}=i\omega\mu\mathbf{H}\ ,\quad \operatorname{div}\mathbf{H}=0\ ,$$
$$\operatorname{div}\mathbf{E}=-\frac{1}{\sigma}(\mathbf{E}\cdot\operatorname{grad}\sigma+\operatorname{div}\mathbf{j}^e)\ . \tag{8.1.25}$$

Separating the electric and magnetic fields in (8.1.25) for $\sigma=\text{const}$, $\mu=\text{const}$, we write

$$\Delta\mathbf{E}+k^2\mathbf{E}=-i\omega\mu\mathbf{j}^e-\frac{1}{\sigma}\operatorname{grad}\operatorname{div}\mathbf{j}^e\ ,\quad \Delta\mathbf{H}+k^2\mathbf{H}=-\operatorname{rot}\mathbf{j}^e\ . \tag{8.1.26}$$

Here the wave number of a medium for a quasi-stationary model of the field is defined by the equation

$$k^2=i\omega\mu\sigma\ . \tag{8.1.27}$$

The k values are also chosen to have a positive real part:

$$k=\sqrt{\frac{\omega\mu\sigma}{2}}(1+i)=\frac{2\pi}{\lambda}(1+i)\ ,$$

where λ is the quasi-stationary electromagnetic wave length calculate by the formula $\lambda=2\pi\sqrt{2}/\sqrt{\omega\mu\sigma}$.

8.1.6 Field Wave Equations

In an insulator (for $\sigma=0$), Eq. (8.1.1a) takes the form

$$\operatorname{rot}\mathbf{H}=\mathbf{j}^e+\frac{\partial\mathbf{D}}{\partial t}\ . \tag{8.1.28}$$

If here $\varepsilon=\text{const}$ and $\mu=\text{const}$, field Eqs. (8.1.8a) and (8.1.9a) go over into the *wave equations*:

$$\Delta\mathbf{E}-\mu\varepsilon\frac{\partial^2\mathbf{E}}{\partial t^2}=\mu\frac{\partial\mathbf{j}^e}{\partial t}+\frac{1}{\varepsilon}\operatorname{grad}q^e\ ,$$

$$\Delta\mathbf{H}-\mu\varepsilon\frac{\partial^2\mathbf{H}}{\partial t^2}=-\operatorname{rot}\mathbf{j}^e\ . \tag{8.1.29}$$

8.1.7 Field Equations Allowing for Magnetic Currents and Charges

Refer again to Eqs. (8.1.5a) through (8.1.5d) for an arbitrary electromagnetic field in a homogeneous and isotropic domain of a medium. The first striking feature is asymmetry of the right-hand sides of equations for the electric and magnetic fields. This asymmetry arises from the absence of extraneous magnetic currents and charges in the equations concerned. This situation can be improved by introducing, after Stratton (1948), the required additional terms

$$\operatorname{rot}\mathbf{H}=\sigma\mathbf{E}+\mathbf{j}^e+\varepsilon\frac{\partial\mathbf{E}}{\partial t}\ . \tag{8.1.30a}$$

$$\operatorname{rot}\mathbf{E}=-\mu\frac{\partial\mathbf{H}}{\partial t}-\mathbf{j}^m\ , \tag{8.1.30b}$$

$$\operatorname{div}\mathbf{H}=\frac{q^m}{\mu}\ , \tag{8.1.30c}$$

$$\operatorname{div}\mathbf{E}=\frac{\check{q}^e}{\varepsilon}\ . \tag{8.1.30d}$$

Here $\check{q}^e=q+q^e$ is the sum of free and extraneous electric charges; $\mathbf{j}^m$ and q^m are the densities of extraneous "magnetic currents" and "magnetic charges", respectively, which are interrelated by the continuity equation

$$\operatorname{div}\mathbf{j}^m=-\frac{\partial q^m}{\partial t}\ . \tag{8.1.31}$$

Evidently, the quantities $\mathbf{j}^m$ and q^m are fictitious, since to the best of our knowledge, no magnetic charges exist in nature. Yet, these fictitious magnetic currents and charges may prove extremely useful for description of electromagnetic field discontinuities in certain theoretical problems (see below).

Employing the operator rot with the left- and right-hand sides of Eqs. (8.1.30a) and (8.1.30b) we obtain

$$\operatorname{rot}\operatorname{rot}\mathbf{E}+\mu\varepsilon\frac{\partial^2\mathbf{E}}{\partial t^2}+\mu\sigma\frac{\partial\mathbf{E}}{\partial t}=-\mu\frac{\partial\mathbf{j}^e}{\partial t}-\operatorname{rot}\mathbf{j}^m\ ,$$

$$\operatorname{rot}\operatorname{rot}\mathbf{H}+\mu\varepsilon\frac{\partial^2\mathbf{H}}{\partial t^2}+\mu\sigma\frac{\partial\mathbf{H}}{\partial t}=-\sigma\mathbf{j}^m-\varepsilon\frac{\partial\mathbf{j}^m}{\partial t}+\operatorname{rot}\mathbf{j}^e\ . \tag{8.1.32}$$

Equations (8.1.30a)–(8.1.30d) become particularly elegant and symmetric on transition to a monochromatic field

$$\operatorname{rot}\mathbf{H} = \sigma^*\mathbf{E}+\mathbf{j}^e \ , \tag{8.1.33a}$$

$$\operatorname{rot}\mathbf{E} = i\omega\mu\mathbf{H}-\mathbf{j}^m \ , \tag{8.1.33b}$$

$$\operatorname{div}\mathbf{H} = \frac{q^m}{\mu} \ , \tag{8.1.33c}$$

$$\operatorname{div}\mathbf{E} = \frac{q^e}{\varepsilon^*} \ . \tag{8.1.33d}$$

In particular, Helmholtz Eqs. (8.1.17) can be written as follows

$$\Delta\mathbf{E}+k^{*2}\mathbf{E} = -i\omega\mu\mathbf{j}^e+\frac{1}{\varepsilon^*}\operatorname{grad}q^e+\operatorname{rot}\mathbf{j}^m \ , \tag{8.1.34}$$

$$\Delta\mathbf{H}+k^{*2}\mathbf{H} = \sigma^*\mathbf{j}^m+\frac{1}{\mu}\operatorname{grad}q^m-\operatorname{rot}\mathbf{j}^e \ . \tag{8.1.35}$$

8.1.8 Stationary Electromagnetic Field

A stationary field is known to be independent of time. Its equations for a homogeneous and isotropic domain of a medium are derived from (8.1.30a)–(8.1.30d) by setting all the time derivatives equal to zero and by considering that $\check{q}^e = q+q^e = 0$:

$$\operatorname{rot}\mathbf{H} = \sigma\mathbf{E}+\mathbf{j}^e \ , \quad \operatorname{rot}\mathbf{E} = -\mathbf{j}^m \ , \quad \operatorname{div}\mathbf{H} = q^m/\mu \ , \quad \operatorname{div}\mathbf{E} = 0 \ . \tag{8.1.36}$$

Here the continuity equations for extraneous electric and magnetic currents take the form

$$\operatorname{div}\mathbf{j}^e = 0 \ , \quad \operatorname{div}\mathbf{j}^m = 0 \ . \tag{8.1.37}$$

Accordingly, Eqs. (8.1.32) are cast as follows

$$\operatorname{rot}\operatorname{rot}\mathbf{E} = -\operatorname{rot}\mathbf{j}^m \ , \quad \operatorname{rot}\operatorname{rot}\mathbf{H} = -\sigma\mathbf{j}^m+\operatorname{rot}\mathbf{j}^e \ . \tag{8.1.38}$$

As a result, stationary electric and magnetic fields satisfy the Poisson equations

$$\Delta\mathbf{E} = \operatorname{rot}\mathbf{j}^m \ , \quad \Delta\mathbf{H} = \frac{1}{\mu}\operatorname{grad}q^m+\sigma\mathbf{j}^m-\operatorname{rot}\mathbf{j}^e \ . \tag{8.1.39}$$

To conclude, note that it is monochromatic electromagnetic fields fitting Eqs. (8.1.33a) through (8.1.33d), and (8.1.34), (8.1.35) that will be studied in most detailed fashion. Yet, certain sections of this book will be concerned with transient electromagnetic fields as well as with quasi-stationary and wave models of the field.

8.2 Integration of Equations for an Arbitrary Vector Field

Prior to taking up the direct integration of equations for the electromagnetic field, consider a similar problem for an arbitrary vector field satisfying the equations

$$\operatorname{div}\mathbf{F} = q \ , \quad \operatorname{rot}\mathbf{F} = \mathbf{j} \ , \tag{8.2.1}$$

where q and $\mathbf{j}$ are the volume densities of sources and rotations of the field $\mathbf{F}$, respectively.

8.2.1 Auxiliary Integral Identities

Derive several useful relations for the vector field $\mathbf{F}$ and its spatial derivatives, following from the Ostrogradsky-Gauss formulas. For this purpose, we will specify an auxiliary vector field $\boldsymbol{\Phi}(\mathbf{r})$:

$$\boldsymbol{\Phi}(\mathbf{r}) = (\mathbf{C}\cdot\mathbf{F})\operatorname{grad}P + (\mathbf{C}\times\mathbf{F})\times\operatorname{grad}P - (\mathbf{C}\times\operatorname{rot}\mathbf{F})P \ , \tag{8.2.2}$$

where $\mathbf{F}(\mathbf{r})$ and $P(\mathbf{r})$ are arbitrary functions twice continuously differentiable in D (up to its boundary S); $\mathbf{C}$ is an arbitrary constant vector.

Let us calculate $\operatorname{div}\boldsymbol{\Phi}$. This operation is performed best of all using the Hamiltonian operator:

$$\operatorname{div}\boldsymbol{\Phi} = \boldsymbol{\nabla}\cdot\boldsymbol{\Phi} = \boldsymbol{\nabla}\cdot[(\mathbf{C}\cdot\mathbf{F})\boldsymbol{\nabla}P] + \boldsymbol{\nabla}\cdot[(\mathbf{C}\times\mathbf{F})\times\boldsymbol{\nabla}P] - \boldsymbol{\nabla}\cdot\{[\mathbf{C}\times(\boldsymbol{\nabla}\times\mathbf{F})]P\} \ . \tag{8.2.3}$$

Calculate all three terms in the right-hand side of (8.2.3) successively:

$$\boldsymbol{\nabla}\cdot[(\mathbf{C}\cdot\mathbf{F})\boldsymbol{\nabla}P] = \boldsymbol{\nabla}P\cdot\boldsymbol{\nabla}(\mathbf{C}\cdot\mathbf{F}) + (\mathbf{C}\cdot\mathbf{F})\Delta P \ , \tag{8.2.4}$$

$$\boldsymbol{\nabla}\cdot[(\mathbf{C}\times\mathbf{F})\times\boldsymbol{\nabla}P] = \boldsymbol{\nabla}P\cdot[\boldsymbol{\nabla}\times(\mathbf{C}\times\mathbf{F})] = \boldsymbol{\nabla}P\cdot[\mathbf{C}(\boldsymbol{\nabla}\cdot\mathbf{F}) - (\mathbf{C}\cdot\boldsymbol{\nabla})\mathbf{F}] \ , \tag{8.2.5}$$

$$\begin{aligned} -\boldsymbol{\nabla}\cdot\{[\mathbf{C}\times(\boldsymbol{\nabla}\times\mathbf{F})]P\} &= -P\boldsymbol{\nabla}\cdot(\mathbf{C}\times[\boldsymbol{\nabla}\times\mathbf{F}]) - (\mathbf{C}\times[\boldsymbol{\nabla}\times\mathbf{F}])\cdot\boldsymbol{\nabla}P \\ &= P\mathbf{C}\cdot(\boldsymbol{\nabla}\times[\boldsymbol{\nabla}\times\mathbf{F}]) - \boldsymbol{\nabla}P\cdot[\boldsymbol{\nabla}(\mathbf{C}\cdot\mathbf{F}) - (\mathbf{C}\cdot\boldsymbol{\nabla})\mathbf{F}] \ . \end{aligned} \tag{8.2.6}$$

Substituting (8.2.4)–(8.2.6) into (8.2.3) and canceling the like terms we write finally

$$\begin{aligned} \operatorname{div}\boldsymbol{\Phi} &= \mathbf{C}\cdot\Delta P\mathbf{F} + \mathbf{C}\cdot\boldsymbol{\nabla}P(\boldsymbol{\nabla}\cdot\mathbf{F}) + \mathbf{C}\cdot P[\boldsymbol{\nabla}\times(\boldsymbol{\nabla}\times\mathbf{F})] \\ &= \mathbf{C}\cdot(\Delta P\cdot\mathbf{F} + \operatorname{grad}P\cdot\operatorname{div}\mathbf{F} + P\operatorname{rot}\operatorname{rot}\mathbf{F}) \ . \end{aligned} \tag{8.2.7}$$

On the other hand, the formula

$$\begin{aligned} \boldsymbol{\Phi}\cdot\mathbf{n} &= \mathbf{C}\cdot[(\mathbf{n}\cdot\mathbf{F})]\operatorname{grad}P + (\mathbf{n}\times\mathbf{F})\times\operatorname{grad}P + (\mathbf{n}\times\operatorname{rot}\mathbf{F})P] \\ &= \mathbf{C}\cdot[(\mathbf{n}\cdot\operatorname{grad}P)\mathbf{F} + (\mathbf{n}\times\operatorname{grad}P)\times\mathbf{F} + (\mathbf{n}\times\operatorname{rot}\mathbf{F})P] \ , \end{aligned} \tag{8.2.8}$$

($\mathbf{n}$ being the unit vector of an outward pointing normal to S) is valid. Applying Ostrogradsky-Gauss formula (5.1.1) to the vector function $\boldsymbol{\Phi}(\mathbf{r})$ and taking into account, just as in Sect. 5.1, that the vector $\mathbf{C}$ is chosen arbitrarily, we derive the following significant integral identity:

$$\iiint_D (\Delta P \cdot \mathbf{F} + \operatorname{grad} P \operatorname{div} \mathbf{F} + P \operatorname{rot}\operatorname{rot}\mathbf{F})\, dv$$

$$= \begin{cases} \iint_S [(\mathbf{n}\cdot\mathbf{F})\operatorname{grad} P + (\mathbf{n}\times\mathbf{F})\times\operatorname{grad} P + (\mathbf{n}\times\operatorname{rot}\mathbf{F})P]\, ds\ , & (8.2.9\text{a}) \\ \iint_S [(\mathbf{n}\cdot\operatorname{grad} P)\mathbf{F} + (\mathbf{n}\times\operatorname{grad} P)\times\mathbf{F} + (\mathbf{n}\times\operatorname{rot}\mathbf{F})P]\, ds\ . & (8.2.9\text{b}) \end{cases}$$

It should be noted that if $\operatorname{rot}\mathbf{F} = 0$, formulas (8.2.9a) and (8.2.9b) yield

$$\iiint_D (\Delta P \cdot \mathbf{F} + \operatorname{grad} P \operatorname{div} \mathbf{F})\, dv$$

$$= \begin{cases} \iint_S [(\mathbf{n}\cdot\mathbf{F})\operatorname{grad} P + (\mathbf{n}\times\mathbf{F})\times\operatorname{grad} P]\, ds\ , \\ \iint_S [(\mathbf{n}\cdot\operatorname{grad} P)\mathbf{F} + (\mathbf{n}\times\operatorname{grad} P)\times\mathbf{F}]\, ds\ . \end{cases} \quad (8.2.10)$$

This directly implies relations (5.1.14a) and (5.1.14b), if $\mathbf{F} = \operatorname{grad} h$.

8.2.2 Vector Analogs of the Pompei Formulas

Relations (8.2.9a) and (8.2.9b) yield integral formulas admitting an analytical solution to vector field Eqs. (8.2.1). Thus, let the twice continuously differentiable vector field $\mathbf{F}$ satisfy Eqs. (8.2.1) everywhere in the domain D (up to its boundary S). Assume that the densities of sources q and rotations $\mathbf{j}$ of the vector field are known in D, while the field $\mathbf{F}$ itself has been specified on the boundary S. It is necessary to find the field $\mathbf{F}$ inside D.

To solve this problem, we will employ formulas (8.2.9a) and (8.2.9b) setting therein that $P = -1/4\pi|\mathbf{r}-\mathbf{r}'|$. Then we derive

$$-\frac{1}{4\pi}\iint_S \left[(\mathbf{n}\cdot\mathbf{F})\operatorname{grad}\frac{1}{|\mathbf{r}-\mathbf{r}'|} + (\mathbf{n}\times\mathbf{F})\times\operatorname{grad}\frac{1}{|\mathbf{r}-\mathbf{r}'|} + (\mathbf{n}\times\operatorname{rot}\mathbf{F})\frac{1}{|\mathbf{r}-\mathbf{r}'|}\right] ds$$

$$+\frac{1}{4\pi}\iiint_D \left(\operatorname{div}\mathbf{F}\cdot\operatorname{grad}\frac{1}{|\mathbf{r}-\mathbf{r}'|} + \operatorname{rot}\operatorname{rot}\mathbf{F}\cdot\frac{1}{|\mathbf{r}-\mathbf{r}'|}\right) dv$$

$$= \begin{cases} \mathbf{F}(\mathbf{r}')\ , & \mathbf{r}'\in D\ , \\ 0\ , & \mathbf{r}'\in C\bar{D}\ , \end{cases} \quad (8.2.11\text{a})$$

or

$$-\frac{1}{4\pi}\iint_S\left\{\left(\mathbf{n}\cdot\operatorname{grad}\frac{1}{|\mathbf{r}-\mathbf{r}'|}\right)\mathbf{F}+\left(\mathbf{n}\times\operatorname{grad}\frac{1}{|\mathbf{r}-\mathbf{r}'|}\right)\times\mathbf{F}\right.$$
$$\left.+(\mathbf{n}\times\operatorname{rot}\mathbf{F})\frac{1}{|\mathbf{r}-\mathbf{r}'|}\right\}ds$$
$$+\frac{1}{4\pi}\iiint_D\left(\operatorname{div}\mathbf{F}\cdot\operatorname{grad}\frac{1}{|\mathbf{r}-\mathbf{r}'|}+\operatorname{rot}\operatorname{rot}\mathbf{F}\cdot\frac{1}{|\mathbf{r}-\mathbf{r}'|}\right)dv$$
$$=\begin{cases}\mathbf{F}(\mathbf{r}'), & \mathbf{r}'\in D,\\ 0, & \mathbf{r}'\in C\bar{D}.\end{cases} \tag{8.2.11 b}$$

Formulas (8.2.11 a) and (8.2.11 b) can be somewhat simplified by using the following identity suggested by the third vector statement of the Ostrogradsky-Gauss formula (5.1.7).

$$\iiint_D \operatorname{rot}\operatorname{rot}\mathbf{F}\cdot\frac{1}{|\mathbf{r}-\mathbf{r}'|}dv$$
$$=\iint_S(\mathbf{n}\times\operatorname{rot}\mathbf{F})\frac{1}{|\mathbf{r}-\mathbf{r}'|}ds+\iiint_D\left(\operatorname{rot}\mathbf{F}\times\operatorname{grad}\frac{1}{|\mathbf{r}-\mathbf{r}'|}\right)dv\,. \tag{8.2.12}$$

Substituting (8.2.12) into (8.2.11) we find

$$\left\{\begin{aligned}&-\frac{1}{4\pi}\iint_S\left[(\mathbf{n}\cdot\mathbf{F})\operatorname{grad}\frac{1}{|\mathbf{r}-\mathbf{r}'|}+(\mathbf{n}\times\mathbf{F})\times\operatorname{grad}\frac{1}{|\mathbf{r}-\mathbf{r}'|}\right]ds\\&-\frac{1}{4\pi}\iint_S\left[\left(\mathbf{n}\cdot\operatorname{grad}\frac{1}{|\mathbf{r}-\mathbf{r}'|}\right)\mathbf{F}+\left(\mathbf{n}\times\operatorname{grad}\frac{1}{|\mathbf{r}-\mathbf{r}'|}\right)\times\mathbf{F}\right]ds\end{aligned}\right\}$$
$$+\frac{1}{4\pi}\iiint_D\left(\operatorname{div}\mathbf{F}\cdot\operatorname{grad}\frac{1}{|\mathbf{r}-\mathbf{r}'|}+\operatorname{rot}\mathbf{F}\times\operatorname{grad}\frac{1}{|\mathbf{r}-\mathbf{r}'|}\right)dv$$
$$=\begin{cases}\mathbf{F}(\mathbf{r}'), & \mathbf{r}'\in D,\\ 0, & \mathbf{r}'\in C\bar{D}.\end{cases} \tag{8.2.13}$$

Formulas (8.2.13) extend formulas (5.1.40a) and (5.1.40b) to the case of an arbitrary vector field, hence they will be termed vector analogs for the Pompei formulas. Precisely these formulas yield a solution to the above-stated problem. Indeed, substituting Eqs. (8.2.1) into (8.2.13) we obtain

$$\mathbf{F}(\mathbf{r}')=\left\{\begin{aligned}&-\frac{1}{4\pi}\iint_S\left[(\mathbf{n}\cdot\mathbf{F})\operatorname{grad}\frac{1}{|\mathbf{r}-\mathbf{r}'|}+[\mathbf{n}\times\mathbf{F}]\times\operatorname{grad}\frac{1}{|\mathbf{r}-\mathbf{r}'|}\right]ds\\&-\frac{1}{4\pi}\iint_S\left[\left(\mathbf{n}\cdot\operatorname{grad}\frac{1}{|\mathbf{r}-\mathbf{r}'|}\right)\mathbf{F}+\left(\mathbf{n}\times\operatorname{grad}\frac{1}{|\mathbf{r}-\mathbf{r}'|}\right)\times\mathbf{F}\right]ds\end{aligned}\right\}$$
$$+\frac{1}{4\pi}\iiint_D\left(q\operatorname{grad}\frac{1}{|\mathbf{r}-\mathbf{r}'|}+\mathbf{j}\times\operatorname{grad}\frac{1}{|\mathbf{r}-\mathbf{r}'|}\right)dv\,. \tag{8.2.14}$$

Once the field **F** is potential in the domain D, formulas (8.2.14) go over into (5.1.41 a) and (5.1.42 b), while in the case of the Laplace vector field, (8.2.14) suggests directly the three-dimensional Cauchy integral formula (5.1.26).

8.3 Stratton-Chu Integral Formulas

We will employ the relations derived in the previous subsection to elaborate a theory of integral representations of the electromagnetic field in a homogeneous and isotropic domain of space. Behind this theory are the Stratton-Chu integral formulas (Stratton 1948) whose derivation will be presented now.

In doing so we will try to provide a sufficiently complete set of integral formulas for various models of the electromagnetic field whose equations were presented in Sect. 8.1. We should like to stress here that this section deals, primarily, with reference information. Therefore, the reader interested in geoelectrical applications of the body of the Stratton-Chu type integrals can proceed directly to Sect. 8.4.

8.3.1 Stratton-Chu Formulas for a Transient Electromagnetic Field (General Case)

Consider the following electrodynamic problem: *express the electromagnetic field* **E**, **H** *in a certain domain D through the values of* **E**, **H** *on the inner side of the surface S bounding this domain.* Assume that the electromagnetic parameters ε, μ, and σ of a medium inside D are constant. Then the electromagnetic field in D satisfies Eqs. (8.1.30 a)–(8.1.30 d) and (8.1.32). Allowing for these equations we will employ formula (8.2.9 a) postulating therein that $\mathbf{F}\equiv\mathbf{E}$. Then we have

$$\iiint_D \left(\Delta P\mathbf{E}-\mu\varepsilon P\frac{\partial^2\mathbf{E}}{\partial t^2}-\mu\sigma P\frac{\partial\mathbf{E}}{\partial t}+\frac{\check{q}^e}{\varepsilon}\operatorname{grad}P-\mu\frac{\partial\mathbf{j}^e}{\partial t}P-\operatorname{rot}\mathbf{j}^m P\right)dv$$
$$=\iint_S\left[(\mathbf{n}\cdot\mathbf{E})\operatorname{grad}P+(\mathbf{n}\times\mathbf{E})\times\operatorname{grad}P-\mu\left(\mathbf{n}\times\frac{\partial\mathbf{H}}{\partial t}\right)P-(\mathbf{n}\times\mathbf{j}^m)P\right]ds\ . \tag{8.3.1}$$

Note that in view of the third vector statement of Ostrogradsky-Gauss formula (5.1.17) the following identity

$$\iiint_D \operatorname{rot}\mathbf{j}^m P\,dv=\iint_S(\mathbf{n}\times\mathbf{j}^m)P\,ds+\iiint_D(\mathbf{j}^m\times\operatorname{grad}P)\,dv\ , \tag{8.3.2}$$

holds good and permits some simplification of (8.3.1):

$$\iiint_D\left\{\Delta P\mathbf{E}-\mu\varepsilon P\frac{\partial^2\mathbf{E}}{\partial t^2}-\mu\sigma P\frac{\partial\mathbf{E}}{\partial t}+\frac{\check{q}^e}{\varepsilon}\operatorname{grad}P-\mu\frac{\partial\mathbf{j}^e}{\partial t}P-\mathbf{j}^{\mathrm{m}}\times\operatorname{grad}P\right\}dv$$
$$=\iint_S\left[(\mathbf{n}\cdot\mathbf{E})\operatorname{grad}P+(\mathbf{n}\times\mathbf{E})\times\operatorname{grad}P-\mu\left(\mathbf{n}\times\frac{\partial\mathbf{H}}{\partial t}\right)P\right]ds\ . \tag{8.3.3}$$

A similar substitution for the magnetic field $\mathbf{F}=\mathbf{H}$ yields

$$\iiint_D \left(\Delta P\mathbf{H} - \mu\varepsilon P \frac{\partial^2 \mathbf{H}}{\partial t^2} - \mu\sigma P \frac{\partial \mathbf{H}}{\partial t} + \frac{q^m}{\mu} \operatorname{grad} P \right.$$

$$\left. - \sigma \mathbf{j}^m P - \varepsilon \frac{\partial \mathbf{j}^m}{\partial t} P + \mathbf{j}^e \times \operatorname{grad} P \right) dv$$

$$= \iint_S \left[(\mathbf{n}\cdot\mathbf{H}) \operatorname{grad} P + (\mathbf{n}\times\mathbf{H}) \times \operatorname{grad} P + \sigma (\mathbf{n}\times\mathbf{E}) P + \varepsilon \left(\mathbf{n} \times \frac{\partial \mathbf{E}}{\partial t} \right) P \right] ds \;. \tag{8.3.4}$$

Integrating (8.3.3) and (8.3.4) infinitely with respect to time and bringing, through integration by parts, the derivative $\partial/\partial t$ under the integral from the fields to the function P, we write

$$\int_{-\infty}^{+\infty} \iiint_D \left[\mathbf{E} \left(\Delta P - \mu\varepsilon \frac{\partial^2 P}{\partial t^2} + \mu\sigma \frac{\partial P}{\partial t} \right) + \frac{q^e}{\varepsilon} \operatorname{grad} P + \mu \mathbf{j}^e \frac{\partial P}{\partial t} - \mathbf{j}^m \times \operatorname{grad} P \right] dv\,dt$$

$$= \int_{-\infty}^{+\infty} \iint_S \left[(\mathbf{n}\cdot\mathbf{E}) \operatorname{grad} P + (\mathbf{n}\times\mathbf{E}) \times \operatorname{grad} P + \mu (\mathbf{n}\times\mathbf{H}) \frac{\partial P}{\partial t} \right] ds\,dt \;, \tag{8.3.5a}$$

$$\int_{-\infty}^{+\infty} \iiint_D \left[H \left(\Delta P - \mu\varepsilon \frac{\partial^2 P}{\partial t^2} + \mu\sigma \frac{\partial P}{\partial t} \right) + \frac{q^m}{\mu} \operatorname{grad} P - \sigma \mathbf{j}^m P \right.$$

$$\left. + \varepsilon \mathbf{j}^m \frac{\partial P}{\partial t} + \mathbf{j}^e \times \operatorname{grad} P \right] dv\,dt$$

$$= \int_{-\infty}^{+\infty} \iint_S \left[(\mathbf{n}\cdot\mathbf{H}) \operatorname{grad} P + (\mathbf{n}\times\mathbf{H}) \times \operatorname{grad} P \right.$$

$$\left. + \sigma (\mathbf{n}\times\mathbf{E}) P - \varepsilon (\mathbf{n}\times\mathbf{E}) \frac{\partial P}{\partial t} \right] ds\,dt \;. \tag{8.3.5b}$$

Take for P the function G^t satisfying the equation

$$\Delta G^t - \mu\varepsilon \frac{\partial^2 G^t}{\partial t^2} - \mu\sigma \frac{\partial G^t}{\partial t} = \delta(\mathbf{r}-\mathbf{r}')\delta(t-t') \;. \tag{8.3.6}$$

Here δ is the delta function of Dirac and the Laplacian operates on the variable $\mathbf{r}$.

As is seen, G^t is the fundamental Green function for the telegraph equation. It is known (Morse and Feshbach 1953) to have the following form:

$$G^t(\mathbf{r}, t|\mathbf{r}', t') = \frac{-\exp\{[-\sigma(t'-t)]/2\varepsilon\}}{4\pi\sqrt{\mu\varepsilon}\,|\mathbf{r}-\mathbf{r}'|}\left\{\delta\left(\frac{t'-t-\sqrt{\mu\varepsilon}\,|\mathbf{r}-\mathbf{r}'|}{\sqrt{\mu\varepsilon}}\right)\right.$$

$$+\frac{1}{2}\sqrt{\frac{\mu\sigma^2}{\varepsilon}}\frac{|\mathbf{r}-\mathbf{r}'|}{\sqrt{|\mathbf{r}-\mathbf{r}'|^2-(t'-t)^2/\mu\varepsilon}}$$

$$\times J_1\left[\frac{1}{2}\sqrt{\frac{\mu\sigma^2}{\varepsilon}}\sqrt{|\mathbf{r}-\mathbf{r}'|^2-(t'-t)^2/\mu\varepsilon}\right]$$

$$\left.\times\varkappa\left(\frac{t'-t-\sqrt{\mu\varepsilon}\,|\mathbf{r}-\mathbf{r}'|}{\sqrt{\mu\varepsilon}}\right)\right\}, \tag{8.3.7}$$

where J_1 is the Bessel function and $\varkappa$ is the Heavyside unit function:

$$\varkappa(t) = \begin{cases} 0\,, & t\leqslant 0\,, \\ 1\,, & t>0\,. \end{cases} \tag{8.3.8}$$

Substituting (8.3.7) for P in (8.3.5) we obtain

$$\iint_S \int_{-\infty}^{t'-\sqrt{\mu\varepsilon}|\mathbf{r}-\mathbf{r}'|}\left[(\mathbf{n}\cdot\mathbf{E})\operatorname{grad}G^t+(\mathbf{n}\times\mathbf{E})\times\operatorname{grad}G^t+\mu(\mathbf{n}\times\mathbf{H})\frac{\partial G^t}{\partial t}\right]dt\,ds$$

$$-\iiint_D \int_{-\infty}^{t'-\sqrt{\mu\varepsilon}|\mathbf{r}-\mathbf{r}'|}\left(\frac{\check{q}^e}{\varepsilon}\operatorname{grad}G^t+\mu\mathbf{j}^e\frac{\partial G^t}{\partial t}-\mathbf{j}^m\times\operatorname{grad}G^t\right)dt\,dv$$

$$=\begin{cases}\mathbf{E}(\mathbf{r}', t')\,, & \mathbf{r}'\in D\,, \\ 0\,, & \mathbf{r}'\in C\bar{D}\,; \end{cases} \tag{8.3.9a}$$

$$\iint_S \int_{-\infty}^{t'-\sqrt{\mu\varepsilon}|\mathbf{r}-\mathbf{r}'|}\left[(\mathbf{n}\cdot\mathbf{H})\operatorname{grad}G^t+(\mathbf{n}\times\mathbf{H})\times\operatorname{grad}G^t+\sigma(\mathbf{n}\times\mathbf{E})G^t\right.$$

$$\left.-\varepsilon(\mathbf{n}\times\mathbf{E})\frac{\partial G^t}{\partial t}\right]dt\,ds$$

$$-\iiint_D \int_{-\infty}^{t'-\sqrt{\mu\varepsilon}|\mathbf{r}-\mathbf{r}'|}\left(\frac{q^m}{\mu}\operatorname{grad}G^t-\sigma\mathbf{j}^m G^t+\varepsilon\mathbf{j}^m\frac{\partial G^t}{\partial t}+\mathbf{j}^e\times\operatorname{grad}G^t\right)dt\,dv$$

$$=\begin{cases}\mathbf{H}(\mathbf{r}', t')\,, & \mathbf{r}'\in D\,; \\ 0\,, & \mathbf{r}'\in C\bar{D}\,. \end{cases} \tag{8.3.9b}$$

Formulas (8.3.9a) and (8.3.9b), yielding a solution to the problem formulated at the beginning of this subsection, will be referred to as the *integral Stratton-Chu formulas for a transient electromagnetic field.*

If there are no extraneous currents or charges inside the domain D, the volume integrals in (8.3.9a) and (8.3.9b) do not exist any longer and the Stratton-Chu formulas take the form

$$\iint_S \int_{-\infty}^{t'-\sqrt{\mu\varepsilon}|\mathbf{r}-\mathbf{r}'|} \left[(\mathbf{n}\cdot\mathbf{E})\,\mathrm{grad}\, G^t + (\mathbf{n}\times\mathbf{E})\times\mathrm{grad}\, G^t + \mu(\mathbf{n}\times\mathbf{H})\frac{\partial G^t}{\partial t} \right] dt\, ds$$

$$= \begin{cases} \mathbf{E}(\mathbf{r}', t')\,, & \mathbf{r}'\in D\,, \\ 0\,, & \mathbf{r}'\in C\bar{D}; \end{cases} \tag{8.3.10a}$$

$$\iint_S \int_{-\infty}^{t'-\sqrt{\mu\varepsilon}|\mathbf{r}-\mathbf{r}'|} \Big\{ (\mathbf{n}\cdot\mathbf{H})\,\mathrm{grad}\, G^t + [\mathbf{n}\times\mathbf{H}]\times\mathrm{grad}\, G^t + \sigma[\mathbf{n}\times\mathbf{E}]\, G^t - \varepsilon[\mathbf{n}\times\mathbf{E}]$$

$$\cdot\frac{\partial G^t}{\partial t} \Big\}\, dt\, ds = \begin{cases} \mathbf{H}(\mathbf{r}', t')\,, & \mathbf{r}'\in D\,, \\ 0\,, & \mathbf{r}'\in C\bar{D}\,. \end{cases} \tag{8.3.10b}$$

Thus, the values of the electromagnetic field $\mathbf{E}$, $\mathbf{H}$ everywhere inside the domain D are defined from its values on the boundary S.

8.3.2 Stratton-Chu Formulas for a Quasi-Stationary Field

Now let us integrate the quasi-stationary field equations over the domain D (assuming, just as in the foregoing discussion, that inside D we have ε, μ, $\sigma = \mathrm{const}$). To this end, set $\mathbf{F}\equiv\mathbf{E}$ in (8.2.9) and use Eqs. (8.1.22) and (8.1.23)

$$\iiint_D \left(\Delta P\mathbf{E} - \mathrm{grad}\, P\frac{1}{\sigma}\,\mathrm{div}\,\mathbf{j}^e - \mu P\frac{\partial\mathbf{j}^e}{\partial t} - \mu\sigma P\frac{\partial\mathbf{E}}{\partial t} \right) dv$$

$$= \iint_S \left[(\mathbf{n}\cdot\mathbf{E})\,\mathrm{grad}\, P + (\mathbf{n}\times\mathbf{E})\times\mathrm{grad}\, P - \mu\left(\mathbf{n}\times\frac{\partial\mathbf{H}}{\partial t}\right) P \right] ds\,. \tag{8.3.11}$$

Accordingly, for a quasi-stationary magnetic field we obtain

$$\iiint_D \left(\Delta P\,\mathbf{H} - \mu\sigma P\frac{\partial\mathbf{H}}{\partial t} + \mathrm{rot}\,\mathbf{j}^e P \right) dv$$

$$= \iint_S [(\mathbf{n}\cdot\mathbf{H})\,\mathrm{grad}\, P + (\mathbf{n}\times\mathbf{H})\times\mathrm{grad}\, P + \sigma(\mathbf{n}\times\mathbf{E})P + (\mathbf{n}\times\mathbf{j}^e)P]\, ds\,. \tag{8.3.12}$$

With due account taken of the identity

$$\iiint_D \mathrm{rot}\,\mathbf{j}^e P dv = \iint_S (\mathbf{n}\times\mathbf{j}^e) P ds + \iiint_D (\mathbf{j}^e\times\mathrm{grad}\, P)\, dv$$

formula (8.3.12) is cast in the form

$$\iiint_D \left(\Delta P\mathbf{H} - \mu\sigma P\frac{\partial\mathbf{H}}{\partial t} + \mathbf{j}^e\times\mathrm{grad}\, P \right) dv$$

$$= \iint_S [(\mathbf{n}\cdot\mathbf{H})\,\mathrm{grad}\, P + (\mathbf{n}\times\mathbf{H})\times\mathrm{grad}\, P + \sigma(\mathbf{n}\times\mathbf{E})P]\, ds\,. \tag{8.3.13}$$

Integrating, just as before, expressions (8.3.11) and (8.3.13) infinitely with respect to t, we obtain

$$\int_{-\infty}^{+\infty}\iiint_D\left[\mathbf{E}\left(\Delta P+\mu\sigma\frac{\partial P}{\partial t}\right)+\frac{1}{\sigma}\operatorname{div}\mathbf{j}^e\operatorname{grad}P+\mu\mathbf{j}^e\frac{\partial P}{\partial t}\right]dv\,dt$$
$$=\int_{-\infty}^{+\infty}\iint_S\left[(\mathbf{n}\cdot\mathbf{E})\operatorname{grad}P+(\mathbf{n}\times\mathbf{E})\times\operatorname{grad}P+\mu(\mathbf{n}\times\mathbf{H})\frac{\partial P}{\partial t}\right]ds\,dt\ , \tag{8.3.14a}$$

$$\int_{-\infty}^{+\infty}\iiint_D\left[\mathbf{H}\left(\Delta P+\mu\sigma\frac{\partial P}{\partial t}\right)+\mathbf{j}^e\times\operatorname{grad}P\right]dv\,dt$$
$$=\int_{-\infty}^{+\infty}\iint_S[(\mathbf{n}\cdot\mathbf{H})\operatorname{grad}P+(\mathbf{n}\times\mathbf{H})\times\operatorname{grad}P+\sigma(\mathbf{n}\times\mathbf{E})\ P]\,ds\,dt\ . \tag{8.3.14b}$$

Let P be the Green function G^d for the equation of diffusion:

$$G^d(\mathbf{r},\ t\,|\,\mathbf{r}',\ t')=-\frac{\sqrt{\mu\sigma}}{8\pi^{3/2}(t'-t)^{3/2}}\exp\left(-\frac{\mu\sigma\ |\mathbf{r}-\mathbf{r}'|^2}{4(t'-t)}\right)\varkappa(t'-t)\ , \tag{8.3.15}$$

satisfying the equation

$$\Delta G^d+\mu\sigma\frac{\partial G^d}{\partial t}=\delta(\mathbf{r}-\mathbf{r}')\delta(t-t')\ . \tag{8.3.16}$$

Then, we obtain from (8.3.14)

$$\int_{-\infty}^{t'}\left\{\iint_S\left[(\mathbf{n}\cdot\mathbf{E})\operatorname{grad}G^d+(\mathbf{n}\times\mathbf{E})\times\operatorname{grad}G^d+\mu(\mathbf{n}\times\mathbf{H})\frac{\partial G^d}{\partial t}\right]ds\right.$$
$$\left.-\iiint_D\left(\frac{1}{\sigma}\operatorname{div}\mathbf{j}^e\operatorname{grad}G^d+\mu\mathbf{j}^e\frac{\partial G^d}{\partial t}\right)dv\right\}dt$$
$$=\begin{cases}\mathbf{E}(\mathbf{r}',\ t')\ , & \mathbf{r}'\in D\ ,\\ 0\ , & \mathbf{r}'\in C\bar{D}\ ;\end{cases} \tag{8.3.17a}$$

$$\int_{-\infty}^{t'}\left\{\iint_S\{(\mathbf{n}\cdot\mathbf{H})\operatorname{grad}G^d+[\mathbf{n}\times\mathbf{H}]\times\operatorname{grad}G^d+\sigma[\mathbf{n}\times\mathbf{E}]\,G^d\}ds\right.$$
$$\left.-\iiint_D[\mathbf{j}^e\times\operatorname{grad}G^d]\,dv\right\}dt$$
$$=\begin{cases}\mathbf{H}(\mathbf{r}',\ t')\ , & \mathbf{r}'\in D\ ,\\ 0\ , & \mathbf{r}'\in C\bar{D}\ .\end{cases} \tag{8.3.17b}$$

The latter relations are *the Stratton-Chu integral formulas for a quasi-stationary electromagnetic field.* In the absence of extraneous currents inside D, these formulas take the form

$$\int_{-\infty}^{t'} \iint_S \left[(\mathbf{n}\cdot\mathbf{E})\,\mathrm{grad}\,G^d + (\mathbf{n}\times\mathbf{E})\times\mathrm{grad}\,G^d + \mu(\mathbf{n}\times\mathbf{H})\,\frac{\partial G^d}{\partial t}\right] ds\,dt$$

$$= \begin{cases} \mathbf{E}(\mathbf{r}', t')\,, & \mathbf{r}'\in D\,, \\ 0\,, & \mathbf{r}'\in C\bar{D}\,; \end{cases} \tag{8.3.18a}$$

$$\int_{-\infty}^{t'} \iint_S \left[(\mathbf{n}\cdot\mathbf{H})\,\mathrm{grad}\,G^d + (\mathbf{n}\times\mathbf{E})\times\mathrm{grad}\,G^d + \sigma(\mathbf{n}\times\mathbf{E})\,G^d\right] ds\,dt$$

$$= \begin{cases} \mathbf{H}(\mathbf{r}', t')\,, & \mathbf{r}'\in D\,, \\ 0\,, & \mathbf{r}'\in C\bar{D}\,. \end{cases} \tag{8.3.18b}$$

8.3.3 Wave Model of the Field

The Stratton-Chu integral formulas for a wave electromagnetic field can be directly derived, in accordance with Sect. 8.1.6, from relations (8.3.5a) and (8.3.5b) by positing therein that $\sigma = 0$ and $\check{q}^e = q^e$ (by setting free electric charges q to zero) and

$$P = G^W = G^W(\mathbf{r}', t'\,|\,\mathbf{r}, t) = -\frac{1}{4\pi|\mathbf{r}-\mathbf{r}'|}\,\delta(t'-t-\sqrt{\mu\varepsilon}\,|\mathbf{r}-\mathbf{r}'|)\,, \tag{8.3.19}$$

where G^W is the Green fundamental function for the wave equation:

$$\Delta G^W - \mu\varepsilon\,\frac{\partial^2 G^W}{\partial t^2} = \delta(\mathbf{r}-\mathbf{r}')\,\delta(t-t')\,. \tag{8.3.20}$$

Making the relevant substitution we write

$$\int_{-\infty}^{+\infty} \left\{ \iint_S \left[(\mathbf{n}\cdot\mathbf{E})\,\mathrm{grad}\,G^W + (\mathbf{n}\times\mathbf{E})\times\mathrm{grad}\,G^W + \mu(\mathbf{n}\times\mathbf{H})\,\frac{\partial G^W}{\partial t}\right] ds \right.$$

$$\left. - \iiint_D \left(\frac{q^e}{\varepsilon}\,\mathrm{grad}\,G^W + \mu\mathbf{j}^e\,\frac{\partial G^W}{\partial t} - \mathbf{j}^m\times\mathrm{grad}\,G^W \right) dv \right\} dt$$

$$= \begin{cases} \mathbf{E}(\mathbf{r}', t')\,, & \mathbf{r}'\in D\,, \\ 0\,, & \mathbf{r}'\in C\bar{D}\,; \end{cases} \tag{8.3.21a}$$

$$\int_{-\infty}^{+\infty} \left\{ \iint_S \left[(\mathbf{n}\cdot\mathbf{H})\,\mathrm{grad}\,G^W + (\mathbf{n}\times\mathbf{H})\times\mathrm{grad}\,G^W - \varepsilon(\mathbf{n}\times\mathbf{E})\,\frac{\partial G^W}{\partial t}\right] ds \right.$$

$$\left. - \iiint_D \left(\frac{q^m}{\mu}\,\mathrm{grad}\,G^W + \varepsilon\mathbf{j}^m\,\frac{\partial G^W}{\partial t} + \mathbf{j}^e\times\mathrm{grad}\,G^W \right) dv \right\} dt$$

$$= \begin{cases} \mathbf{H}(\mathbf{r}', t')\,, & \mathbf{r}'\in D\,, \\ 0\,, & \mathbf{r}'\in C\bar{D}\,. \end{cases} \tag{8.3.21b}$$

We will factor the differential operators in formulas (8.3.21 a) and (8.3.21 b) outside the integral sign. Indeed, since the operator grad operates only on the Green function, we can write

$$\operatorname{grad} G^w = -\operatorname{grad}' G^w \ , \quad (\mathbf{n}\cdot\mathbf{E}) \operatorname{grad} G^w = -\operatorname{grad}'(\mathbf{n}\cdot\mathbf{E}) G^w \ ,$$

$$(\mathbf{n}\times\mathbf{E})\times \operatorname{grad} G^w = -(\mathbf{n}\times\mathbf{E})\times \operatorname{grad}' G^w = \operatorname{rot}'(\mathbf{n}\times\mathbf{E}) G^w \ , \tag{8.3.22a}$$

$$\frac{q^e}{\varepsilon} \operatorname{grad} G^W = -\operatorname{grad}' \frac{q^e}{\varepsilon} G^W \ , \quad \mathbf{j}^n \times \operatorname{grad} G^W = \operatorname{rot}' \mathbf{j}^n G^W \ .$$

Furthermore,

$$\partial G^w/\partial t = -\partial G^w/\partial t' \ . \tag{8.3.22b}$$

Allowing for (8.3.22a) and (8.3.22b), rewrite (8.3.21 a)

$$\int\limits_{-\infty}^{+\infty} \left[-\operatorname{grad}' \iint\limits_S (\mathbf{n}\cdot\mathbf{E}) G^W ds + \operatorname{rot}' \iint\limits_S (\mathbf{n}\times\mathbf{E}) G^W ds - \mu \frac{\partial}{\partial t'} \iint\limits_S (\mathbf{n}\times\mathbf{H}) G^W ds \right.$$

$$\left. + \operatorname{grad}' \iiint\limits_D \frac{q^e}{\varepsilon} G^W dv + \mu \frac{\partial}{\partial t'} \iiint\limits_D \mathbf{j}^e G^W dv + \operatorname{rot}' \iiint\limits_D \mathbf{j}^n G^W dv \right] dt$$

$$= \begin{cases} \mathbf{E}(\mathbf{r}', t') \ , & \mathbf{r}' \in D \ , \\ 0 \ , & \mathbf{r}' \in C\bar{D} \ . \end{cases} \tag{8.3.23}$$

Substituting (8.3.19) into (8.3.23) we find

$$-\frac{1}{4\pi} \left(-\operatorname{grad}' \iint\limits_S \frac{\mathbf{n}\cdot\tilde{\mathbf{E}}}{|\mathbf{r}-\mathbf{r}'|} ds + \operatorname{rot}' \iint\limits_S \frac{\mathbf{n}\times\tilde{\mathbf{E}}}{|\mathbf{r}-\mathbf{r}'|} ds - \mu \frac{\partial}{\partial t'} \iint\limits_S \frac{\mathbf{n}\times\tilde{\mathbf{H}}}{|\mathbf{r}-\mathbf{r}'|} ds \right.$$

$$\left. + \frac{1}{\varepsilon} \operatorname{grad}' \iiint\limits_D \frac{\tilde{q}^e}{|\mathbf{r}-\mathbf{r}'|} dv + \mu \frac{\partial}{\partial t'} \iiint\limits_D \frac{\tilde{\mathbf{j}}^e}{|\mathbf{r}-\mathbf{r}'|} dv + \operatorname{rot}' \iiint\limits_D \frac{\tilde{\mathbf{j}}^m}{|\mathbf{r}-\mathbf{r}'|} dv \right)$$

$$= \begin{cases} \mathbf{E}(\mathbf{r}', t') \ , & \mathbf{r}' \in D \ , \\ 0 \ , & \mathbf{r}' \in C\bar{D} \ , \end{cases} \tag{8.3.24a}$$

where $\tilde{\mathbf{E}}$, $\tilde{\mathbf{H}}$, $\tilde{q}^{e,m}$, and $\tilde{\mathbf{j}}^{e,m}$ are the values of the electric and magnetic fields, of the charges and currents at delayed instants of time:

$$\tilde{\mathbf{E}} = \mathbf{E}(\mathbf{r}', t' - \sqrt{\mu\varepsilon}\ |\mathbf{r}-\mathbf{r}'|) \ , \quad \tilde{\mathbf{H}} = \mathbf{H}(\mathbf{r}', t' - \sqrt{\mu\varepsilon}\ |\mathbf{r}-\mathbf{r}'|) \ ,$$

$$\tilde{q}^{e,m} = \tilde{q}^{e,m}(\mathbf{r}', t' - \sqrt{\mu\varepsilon}\ |\mathbf{r}-\mathbf{r}'|) \ , \quad \tilde{\mathbf{j}}^{e,m}(\mathbf{r}', t' - \sqrt{\mu\varepsilon}\ |\mathbf{r}-\mathbf{r}'|) \ .$$

Similar transformations of expression (8.3.21 b) yield

$$-\frac{1}{4\pi}\left(-\text{grad}'\iint_S \frac{\mathbf{n}\cdot\tilde{\mathbf{H}}}{|\mathbf{r}-\mathbf{r}'|}ds+\text{rot}'\iint_S \frac{\mathbf{n}\times\tilde{\mathbf{H}}}{|\mathbf{r}-\mathbf{r}'|}ds+\varepsilon\frac{\partial}{\partial t'}\iint_S \frac{\mathbf{n}\times\tilde{\mathbf{E}}}{|\mathbf{r}-\mathbf{r}'|}ds\right.$$

$$\left.+\frac{1}{\mu}\text{grad}'\iiint_D \frac{\tilde{q}^m}{|\mathbf{r}-\mathbf{r}'|}dv+\varepsilon\frac{\partial}{\partial t'}\iiint_D \frac{\tilde{\mathbf{j}}^m}{|\mathbf{r}-\mathbf{r}'|}dv-\text{rot}'\iiint_D \frac{\tilde{\mathbf{j}}^e}{|\mathbf{r}-\mathbf{r}'|}dv\right)$$

$$=\begin{cases}\mathbf{H}(\mathbf{r}',t')\ , & \mathbf{r}'\in D\ ,\\ 0\ , & \mathbf{r}'\in C\bar{D}\ .\end{cases} \qquad (8.3.24\text{b})$$

Relations (8.3.24a) and (8.3.24b) are *the Stratton-Chu integral formulas for a wave electromagnetic field.* If no extraneous currents or charges are present inside the domain D, Stratton-Chu formulas (8.3.24a) and (8.3.24b) take the form

$$-\frac{1}{4\pi}\left(-\text{grad}'\iint_S \frac{\mathbf{n}\cdot\tilde{\mathbf{E}}}{|\mathbf{r}-\mathbf{r}'|}ds+\text{rot}'\iint_S \frac{\mathbf{n}\times\tilde{\mathbf{E}}}{|\mathbf{r}-\mathbf{r}'|}ds-\mu\frac{\partial}{\partial t'}\iint_S \frac{\mathbf{n}\times\tilde{\mathbf{H}}}{|\mathbf{r}-\mathbf{r}'|}ds\right)$$

$$=\begin{cases}\mathbf{E}(\mathbf{r}',t')\ , & \mathbf{r}'\in D\ ,\\ 0\ , & \mathbf{r}'\in C\bar{D}\ ;\end{cases} \qquad (8.3.25\text{a})$$

$$-\frac{1}{4\pi}\left(-\text{grad}'\iint_S \frac{\mathbf{n}\cdot\tilde{\mathbf{H}}}{|\mathbf{r}-\mathbf{r}'|}ds+\text{rot}'\iint_S \frac{\mathbf{n}\times\tilde{\mathbf{H}}}{|\mathbf{r}-\mathbf{r}'|}ds+\varepsilon\frac{\partial}{\partial t'}\iint_S \frac{\mathbf{n}\times\tilde{\mathbf{E}}}{|\mathbf{r}-\mathbf{r}'|}ds\right)$$

$$=\begin{cases}\mathbf{H}(\mathbf{r}',t')\ , & \mathbf{r}'\in D\ ,\\ 0\ , & \mathbf{r}'\in C\bar{D}\ .\end{cases} \qquad (8.3.25\text{b})$$

In concluding this subsection, we should like to note that taking the operators grad′ and rot′ in formulas (8.3.25a) and (8.3.25b) again inside the integral sign and allowing for Eqs. (8.1.1a) through (8.1.1d) for a wave field (for $\sigma=0$), we shall obtain the known Kirchhoff integral formulas (Stratton 1948) (see Chap. 11).

8.3.4 Case of a Stationary Field

In the case of a stationary electromagnetic field, substitution $\mathbf{F}\equiv\mathbf{E}$ in (8.2.9a), with due account taken of (8.1.36) and (8.1.38), yields

$$\iiint_D (\mathbf{E}\Delta P-\text{rot}\,\mathbf{j}^m P)\,dv=\iint_S [(\mathbf{n}\cdot\mathbf{E})\,\text{grad}\,P+(\mathbf{n}\times\mathbf{E})\times\text{grad}\,P-(\mathbf{n}\times\mathbf{j}^m)P]\,ds\ . \qquad (8.3.26)$$

Hence, allowing for identity (8.3.2) we derive

$$\iiint_D (\mathbf{E}\Delta P-\mathbf{j}^m\times\text{grad}\,P)\,dv=\iint_S [(\mathbf{n}\cdot\mathbf{E})\,\text{grad}\,P+(\mathbf{n}\times\mathbf{E})\times\text{grad}\,P]\,ds \qquad (8.3.27\text{a})$$

and similarly for the magnetic field

$$\iiint_D \left(\mathbf{H} \Delta P + \frac{q^m}{\mu} \operatorname{grad} P - \sigma \mathbf{j}^m P + \mathbf{j}^e \times \operatorname{grad} P \right) dv$$

$$= \iint_S [(\mathbf{n} \cdot \mathbf{H}) \operatorname{grad} P + (\mathbf{n} \times \mathbf{H}) \times \operatorname{grad} P + \sigma (\mathbf{n} \times \mathbf{E}) P] \, ds \; . \tag{8.3.27b}$$

Replacing P in (8.3.27a) and (8.3.27b) by the Green fundamental function for the Laplace equation

$$P = G = \frac{-1}{4\pi \, |\mathbf{r} - \mathbf{r}'|} \, ,$$

we find finally

$$-\frac{1}{4\pi} \iint_S \left[(\mathbf{n} \cdot \mathbf{E}) \operatorname{grad} \frac{1}{|\mathbf{r} - \mathbf{r}'|} + (\mathbf{n} \times \mathbf{E}) \times \operatorname{grad} \frac{1}{|\mathbf{r} - \mathbf{r}'|} \right] ds$$

$$-\frac{1}{4\pi} \iiint_D \left(\mathbf{j}^m \times \operatorname{grad} \frac{1}{|\mathbf{r} - \mathbf{r}'|} \right) dv = \begin{cases} \mathbf{E}(\mathbf{r}') \, , & \mathbf{r}' \in D \, , \\ 0 \, , & \mathbf{r}' \in C\bar{D} \; ; \end{cases} \tag{8.3.28a}$$

$$-\frac{1}{4\pi} \iint_S \left[(\mathbf{n} \cdot \mathbf{H}) \operatorname{grad} \frac{1}{|\mathbf{r} - \mathbf{r}'|} + (\mathbf{n} \times \mathbf{H}) \times \operatorname{grad} \frac{1}{|\mathbf{r} - \mathbf{r}'|} + \sigma [\mathbf{n} \times \mathbf{E}] \frac{1}{|\mathbf{r} - \mathbf{r}'|} \right] ds$$

$$+\frac{1}{4\pi} \iiint_D \left(\frac{q^m}{\mu} \operatorname{grad} \frac{1}{|\mathbf{r} - \mathbf{r}'|} - \sigma \mathbf{j}^m \frac{1}{|\mathbf{r} - \mathbf{r}'|} + \mathbf{j}^e \times \operatorname{grad} \frac{1}{|\mathbf{r} - \mathbf{r}'|} \right) dv$$

$$= \begin{cases} \mathbf{H}(\mathbf{r}') \, , & \mathbf{r}' \in D \; . \\ 0 \, , & \mathbf{r}' \in C\bar{D} \; . \end{cases} \tag{8.3.28b}$$

Thus, we have derived *the Stratton-Chu integral formulas for a stationary electromagnetic field.* In the absence of extraneous electric and magnetic currents and charges, formulas (8.3.28a) and (8.3.28b) take the form

$$-\frac{1}{4\pi} \iint_S \left[(\mathbf{n} \cdot \mathbf{E}) \operatorname{grad} \frac{1}{|\mathbf{r} - \mathbf{r}'|} + (\mathbf{n} \times \mathbf{E}) \times \operatorname{grad} \frac{1}{|\mathbf{r} - \mathbf{r}'|} \right] ds$$

$$= \begin{cases} \mathbf{E}(\mathbf{r}') \, , & \mathbf{r}' \in D \, , \\ 0 \, , & \mathbf{r}' \in C\bar{D} \; ; \end{cases} \tag{8.3.29a}$$

$$-\frac{1}{4\pi} \iint_S \left[(\mathbf{n} \cdot \mathbf{H}) \operatorname{grad} \frac{1}{|\mathbf{r} - \mathbf{r}'|} + (\mathbf{n} \times \mathbf{H}) \times \operatorname{grad} \frac{1}{|\mathbf{r} - \mathbf{r}'|} + \sigma (\mathbf{n} \times \mathbf{E}) \frac{1}{|\mathbf{r} - \mathbf{r}'|} \right] ds$$

$$= \begin{cases} \mathbf{H}(\mathbf{r}') \, , & \mathbf{r}' \in D \, , \\ 0 \, , & \mathbf{r}' \in C\bar{D} \; . \end{cases} \tag{8.3.29b}$$

It is noteworthy that formula (8.3.29a) coincides completely with the three-dimensional Cauchy integral formula (5.1.26). This result is virtually a simple consequence of the fact that the stationary electric field $\mathbf{E}$ is Laplacian in a domain devoid of extraneous charges or currents.

Thus, the value of the electromagnetic field **E**, **H** everywhere inside the domain D is determined from its values on the domain boundary S.

8.3.5 Stratton-Chu Formulas for a Monochromatic Field (General Case)

Now assume that there is a specified monochromatic electromagnetic field satisfying Eqs. (8.1.33 a) through (8.1.33 d) in the domain D up to its boundary. Then, taking $\mathbf{F} \equiv \mathbf{E}$ in (8.2.9 a) and allowing for (8.1.34) we obtain

$$\iiint_D \left[\mathbf{E}(\Delta P + k^{*2} P) + \frac{q^e}{\varepsilon^*} \operatorname{grad} P + i\omega\mu \mathbf{j}^e P - \operatorname{rot} \mathbf{j}^m P \right] dv$$
$$= \iint_S [(\mathbf{n} \cdot \mathbf{E}) \operatorname{grad} P + (\mathbf{n} \times \mathbf{E}) \times \operatorname{grad} P + i\omega\mu(\mathbf{n} \times \mathbf{H}) P - (\mathbf{n} \times \mathbf{j}^m) P] \, ds \ . \tag{8.3.30}$$

On account of identity (8.3.12), the latter expression can be written as follows

$$\iiint_D \left[\mathbf{E}(\Delta P + k^{*2} P) + \frac{q^e}{\varepsilon^*} \operatorname{grad} P + i\omega\mu \mathbf{j}^e P - \mathbf{j}^m \times \operatorname{grad} P \right] dv$$
$$= \iint_S [(\mathbf{n} \cdot \mathbf{E}) \operatorname{grad} P + (\mathbf{n} \times \mathbf{E}) \times \operatorname{grad} P + i\omega\mu(\mathbf{n} \times \mathbf{H}) P] \, ds \ . \tag{8.3.31}$$

A similar transformation for the magnetic field yields

$$\iiint_D \left[\mathbf{H}(\Delta P + k^{*2} P) + \frac{q^m}{\mu} \operatorname{grad} P - \sigma^* \mathbf{j}^m P + \mathbf{j}^e \times \operatorname{grad} P \right] dv$$
$$= \iint_S [(\mathbf{n} \cdot \mathbf{H}) \operatorname{grad} P + (\mathbf{n} \times \mathbf{H}) \times \operatorname{grad} P + \sigma^*(\mathbf{n} \times \mathbf{E}) P] \, ds \ . \tag{8.3.32}$$

Let P be provided by the Green fundamental function for the Helmholtz equation

$$G^h = G^h(\mathbf{r}' | \mathbf{r}) = -\frac{e^{ik^* |\mathbf{r} - \mathbf{r}'|}}{4\pi |\mathbf{r} - \mathbf{r}'|} \ , \tag{8.3.33}$$

satisfying the equation

$$\Delta G^h + k^{*2} G^h = \delta(\mathbf{r} - \mathbf{r}') \ . \tag{8.3.34}$$

Substituting (8.3.33) into (8.3.32) we derive the *Stratton-Chu integral formulas for a monochromatic field*

$$\iint_S [(\mathbf{n} \cdot \mathbf{E}) \operatorname{grad} G^h + (\mathbf{n} \times \mathbf{E}) \times \operatorname{grad} G^h + i\omega\mu(\mathbf{n} \times \mathbf{H}) G^h] \, ds$$
$$- \iiint_D \left(\frac{q^e}{\varepsilon^*} \operatorname{grad} G^h + i\omega\mu \mathbf{j}^e G^h - \mathbf{j}^m \times \operatorname{grad} G^h \right) dv = \begin{cases} \mathbf{E}(\mathbf{r}') \ , & \mathbf{r}' \in D \ , \\ 0 \ , & \mathbf{r}' \in C\bar{D} \ ; \end{cases} \tag{8.3.35 a}$$

$$\iint\limits_S [(\mathbf{n}\cdot\mathbf{H})\,\mathrm{grad}\,G^h+(\mathbf{n}\times\mathbf{H})\times\mathrm{grad}\,G^h+\sigma^*(\mathbf{n}\times\mathbf{E})\,G^h]\;ds$$

$$-\iiint\limits_D\left(\frac{q^m}{\mu}\,\mathrm{grad}\,G^h-\sigma^*\mathbf{j}^m G^h+\mathbf{j}^e\times\mathrm{grad}\,G^h\right)dv = \;=\begin{cases}\mathbf{H}(\mathbf{r}')\;, & \mathbf{r}'\in D\;,\\ 0\;, & \mathbf{r}'\in C\bar{D}\;.\end{cases} \qquad (8.3.35\,\mathrm{b})$$

And finally, if there are no extraneous currents or charges in the domain D, formulas (8.3.35 a) and (8.3.35 b) take the form

$$\iint\limits_S [(\mathbf{n}\cdot\mathbf{E})\,\mathrm{grad}\,G^h+(\mathbf{n}\times\mathbf{E})\times\mathrm{grad}\,G^h+i\omega\mu(\mathbf{n}\times\mathbf{H})\,G^h]\;ds$$

$$=\begin{cases}\mathbf{E}(\mathbf{r}')\;, & \mathbf{r}'\in D\;,\\ 0\;, & \mathbf{r}'\in C\bar{D}\;;\end{cases} \qquad (8.3.36\,\mathrm{a})$$

$$\iint\limits_S [(\mathbf{n}\cdot\mathbf{H})\,\mathrm{grad}\,G^h+(\mathbf{n}\times\mathbf{H})\times\mathrm{grad}\,G^h+\sigma^*(\mathbf{n}\times\mathbf{E})\,G^h]\;ds$$

$$=\begin{cases}\mathbf{H}(\mathbf{r}')\;, & \mathbf{r}'\in D\;,\\ 0\;, & \mathbf{r}'\in C\bar{D}\;.\end{cases} \qquad (8.3.36\,\mathrm{b})$$

8.3.6 Modified Stratton-Chu Formulas for a Monochromatic Field

There exists another representation of the Stratton-Chu formulas where the differential operators are factored outside the integral sign. Since the operator grad operates only on the Green function, we can make the substitution $\mathrm{grad} = -\mathrm{grad}'$, just as in Sect. 8.3.3, where the operator with a prime operates on $\mathbf{r}'$. Then

$$(\mathbf{n}\cdot\mathbf{H})\,\mathrm{grad}\,G^h = -\mathrm{grad}'(\mathbf{n}\cdot\mathbf{E})\,G^h\;,$$

$$(\mathbf{n}\times\mathbf{E})\times\mathrm{grad}\,G^h = -(\mathbf{n}\times\mathbf{E})\times\mathrm{grad}'\,G^h = \mathrm{rot}'(\mathbf{n}\times\mathbf{E})\,G^h\;.$$

Upon this substitution, formula (8.3.36 a) takes the form

$$\mathrm{rot}'\iint\limits_S(\mathbf{n}\times\mathbf{E})\,G^h ds-\mathrm{grad}'\iint\limits_S(\mathbf{n}\cdot\mathbf{E})\,G^h ds+i\omega\mu\iint\limits_S(\mathbf{n}\times\mathbf{H})\,G^h ds$$

$$=\begin{cases}\mathbf{E}(\mathbf{r}')\;, & \mathbf{r}'\in D\;,\\ 0\;, & \mathbf{r}'\in C\bar{D}\;.\end{cases} \qquad (8.3.37)$$

Similarly, formula (8.3.36 b) is transformed

$$\mathrm{rot}'\iint\limits_S(\mathbf{n}\times\mathbf{H})\,G^h ds-\mathrm{grad}'\iint\limits_S(\mathbf{n}\cdot\mathbf{H})\,G^h ds+\sigma^*\iint\limits_S(\mathbf{n}\times\mathbf{E})\,G^h ds$$

$$=\begin{cases}\mathbf{H}(\mathbf{r}')\;, & \mathbf{r}'\in D\;,\\ 0\;, & \mathbf{r}'\in C\bar{D}\;.\end{cases} \qquad (8.3.38)$$

The Stratton-Chu formulas are notable for a simple physical interpretation. Let electric and magnetic currents and charges be distributed over a closed surface

Γ with surface densities $\mathbf{j}_s^e$, $\mathbf{j}_s^m$ and q_s^e, q_s^m, respectively. They excite an electromagnetic field

$$\mathbf{E}(\mathbf{r}') = -\iint_S \left(\frac{q_s^e}{\varepsilon^*} \operatorname{grad} G^h - \mathbf{j}_s^m \times \operatorname{grad} G^h + i\omega\mu \mathbf{j}^e G^h \right) ds \ ,$$

$$\mathbf{H}(\mathbf{r}') = -\iint_S \left(\frac{q_s^m}{\mu} \operatorname{grad} G^h - \mathbf{j}_s^e \times \operatorname{grad} G^h - \sigma^* \mathbf{j}^m G^h \right) ds \ , \quad \mathbf{r}' \notin S \ . \qquad (8.3.39)$$

A comparison of (8.3.39) with (8.3.36a) and (8.3.36b) shows that the electromagnetic field in D coincides with that excited by currents and charges distributed over S with a density

$$\mathbf{j}_s^e = -\mathbf{n} \times \mathbf{H} \ , \quad \mathbf{j}_s^m = \mathbf{n} \times \mathbf{E} \ ; \quad q_s^e = -\varepsilon^*(\mathbf{n} \cdot \mathbf{E}) \ , \quad q_s^m = -\mu(\mathbf{n} \cdot \mathbf{H}) \ . \qquad (8.3.40)$$

Thus, in the Stratton-Chu formulas, the real exciters distributed outside the domain D are replaced by equivalent (fictitious) exciters distributed over the surface S.

Formulas (8.3.36a) through (8.3.38) include the products $\mathbf{n} \cdot \mathbf{H} = E_n$, $\mathbf{n} \cdot \mathbf{H} = H_n$, $\mathbf{n} \times \mathbf{E} = \mathbf{n} \times \mathbf{E}_\tau$, $\mathbf{n} \times \mathbf{H} = \mathbf{n} \times \mathbf{H}_\tau$ related to the normal E_n, H_n and to the tangential $\mathbf{E}_\tau$, $\mathbf{H}_\tau$ components of the field. But E_n, H_n can always be eliminated from consideration, since they are expressed, according to the Maxwell equations, in terms of $\mathbf{E}_\tau$, $\mathbf{H}_\tau$.

Indeed, the relations between the normal E_n, H_n and tangential $\mathbf{E}_\tau$, $\mathbf{H}_\tau$ components of the electromagnetic field at the surface S are derived using two-dimensional differentiation operators introduced in Sect. 5.2 [formulas (5.2.1)]. Write, for instance, using two-dimensional operators and Eqs. (8.1.25), the normal components of the magnetic field:

$$H_n = \mathbf{n} \cdot \mathbf{H} = \frac{\mathbf{n} \cdot \operatorname{rot} \mathbf{E}}{i\omega\mu} = \frac{\mathbf{n} \cdot \operatorname{rot}_S \mathbf{E}_\tau}{i\omega\mu} = \frac{1}{i\omega\mu} \lim_{\Delta S \to 0} \frac{1}{\Delta S} \oint_L \mathbf{E}_\tau \cdot \mathbf{l}\, dl$$

$$= \frac{1}{i\omega\mu} \lim_{\Delta S \to 0} \frac{1}{\Delta S} \oint_L \mathbf{E}_\tau \cdot [\mathbf{n} \times \mathbf{v}]\, dl = \frac{1}{i\omega\mu} \lim_{\Delta S \to 0} \frac{1}{\Delta S} \oint_L [\mathbf{E}_\tau \times \mathbf{n}] \cdot \mathbf{v}\, dl$$

$$= -\frac{1}{i\omega\mu} \operatorname{div}_S [\mathbf{n} \times \mathbf{E}_\tau] = -\frac{1}{i\omega\mu} \operatorname{div}_S [\mathbf{n} \times \mathbf{E}] \ . \qquad (8.3.41)$$

In a similar way

$$E_n = \mathbf{n} \cdot \mathbf{E} = \frac{\mathbf{n} \cdot \operatorname{rot}_S \mathbf{H}_\tau}{\sigma^*} = -\frac{1}{\sigma^*} \operatorname{div}_S (\mathbf{n} \times \mathbf{H}_\tau) = -\frac{1}{\sigma^*} \operatorname{div}_S (\mathbf{n} \times \mathbf{H}) \ . \qquad (8.3.42)$$

The two-dimensional analog of the Ostrogradsky-Gauss formula has the form

$$\iint_S \operatorname{div}_S \mathbf{h}_\tau ds = \oint_L \mathbf{h}_\tau \cdot \mathbf{v}\, dl \ . \qquad (8.3.43)$$

If the surface S is closed, then we have

$$\iint_S \operatorname{div}_S \mathbf{h}_\tau ds = 0 \ . \qquad (8.3.44)$$

Allowing for formulas (8.3.41) and (8.3.42), eliminate the normal field components from (8.3.37) and (8.3.38). The simplest result is obtained by making the operator rot′ operate on the latter formulas. Then, the integrals containing E_n and H_n disappear and the Stratton-Chu formula, allowing for (8.1.35), take the form

$$\frac{1}{\sigma^*}\,\mathrm{rot}'\mathrm{rot}'\iint_S (\mathbf{n}\times\mathbf{H})G^h ds + \mathrm{rot}'\iint_S (\mathbf{n}\times\mathbf{E})G^h ds = \begin{cases}\mathbf{E}(\mathbf{r}')\,, & \mathbf{r}'\in D\,,\\ 0\,, & \mathbf{r}'\in C\bar{D}\,;\end{cases} \tag{8.3.45}$$

$$\frac{1}{i\omega\mu}\,\mathrm{rot}'\mathrm{rot}'\iint_S (\mathbf{n}\times\mathbf{E})G^h ds + \mathrm{rot}'\iint_S (\mathbf{n}\times\mathbf{H})G^h ds = \begin{cases}\mathbf{H}(\mathbf{r}')\,, & \mathbf{r}'\in D\,.\\ 0\,, & \mathbf{r}'\in C\bar{D}\,.\end{cases} \tag{8.3.46}$$

These formulas suggested by H. Hönl, A.W. Maue and K. Westpfahl (1961) will be called *modified Stratton-Chu integral formulas.*

8.3.7 Two-Dimensional Stratton-Chu Formulas

If the domain D is bounded by a cylindrical surface S whose generator is parallel to the y-axis and the electromagnetic field **E**, **H** is uniform along this axis, the problem is reduced to a two-dimensional one. The two-dimensional Stratton-Chu formulas retain a three-dimensional structure, while integration is carried out along a closed contour Γ (projection of the surface S onto the plane XZ) and the Green fundamental function has the form

$$G^h = G^h(\mathbf{r}'|\mathbf{r}) = -\frac{i}{4}\,H_0^{(1)}(k^*|\mathbf{r}-\mathbf{r}'|)\,, \tag{8.3.47}$$

where $H_0^{(1)}$ is the zero-order Hankel function of the first kind.

Take, for example, two-dimensional analogs of formulas (8.3.36a) (8.3.36b). Allowing for (8.3.47) we obtain

$$\oint_\Gamma [(\mathbf{n}\cdot\mathbf{E})\,\mathrm{grad}\,G^h + (\mathbf{n}\times\mathbf{E})\times\mathrm{grad}\,G^h + i\omega\mu(\mathbf{n}\times\mathbf{H})G^h]\,dl = \begin{cases}\mathbf{E}(\mathbf{r}')\,, & \mathbf{r}'\in D\,,\\ 0\,, & \mathbf{r}'\in C\bar{D}\,;\end{cases} \tag{8.3.48}$$

$$\oint_\Gamma [(\mathbf{n}\cdot\mathbf{H})\,\mathrm{grad}\,G^h + (\mathbf{n}\times\mathbf{H})\times\mathrm{grad}\,G^h + \sigma^*(\mathbf{n}\times\mathbf{E})G^h]\,dl = \begin{cases}\mathbf{H}(\mathbf{r}')\,, & \mathbf{r}'\in D\,,\\ 0\,, & \mathbf{r}'\in C\bar{D}\,;\end{cases} \tag{8.3.49}$$

where **n** is the unit vector of an outward pointing normal; the contour Γ is traversed counter-clockwise.

8.3.8 Stratton-Chu Formulas as a Cauchy Formula Analog

There is a remarkable analogy between the Stratton-Chu integral formulas and the Cauchy integral formula. In all these formulas, the function at inner points of a domain is evaluated from its values on the boundary, while the integrals at outer points are zero. This analogy is not formal, it reflects the common character of the relations in question.

Indeed, for $\sigma^* = 0$ formula (8.3.36b) goes over into a three-dimensional Cauchy integral formula

$$\frac{1}{4\pi}\iint_S \left[(\mathbf{n}\cdot\mathbf{H})\,\mathrm{grad}\,\frac{1}{|\mathbf{r}-\mathbf{r}'|}+(\mathbf{n}\times\mathbf{H})\times\mathrm{grad}\,\frac{1}{|\mathbf{r}-\mathbf{r}'|}\right] ds = \begin{cases}\mathbf{H}(\mathbf{r}') \ , & \mathbf{r}'\in D \ , \\ 0 \ , & \mathbf{r}'\in C\bar{D} \ ,\end{cases} \tag{8.3.50}$$

which is identical in a two-dimensional case with the classical Cauchy integral formula

$$\frac{1}{2\pi i}\int_\Gamma \frac{f(\zeta)}{\zeta-\zeta'}\,d\zeta = \begin{cases} f(\zeta') \ , & \zeta'\in D \ , \\ 0 \ , & \zeta\in C\bar{D} \ ,\end{cases} \tag{8.3.51}$$

where $\zeta = x+iz$; Γ is the contour of intersection of the cylindrical surface S with the plane XZ; $f(\zeta) = -H_x+iH_z$ (see Sect. 5.1.4).

Thus, the Stratton-Chu formulas can be regarded as an extension of the Cauchy formula to the electromagnetic field **E**, **H**. This extension renders many ideas of the theory of functions of a complex variable applicable in electrodynamics.

8.4 Stratton-Chu Type Integrals

By analogy with the Cauchy-type integrals, we can take into consideration the Stratton-Chu type integrals (Zhdanov 1975b, 1980a; Berdichevsky and Zhdanov 1981, 1984). For the sake of simplification, consider the theory of Stratton-Chu type integrals mainly for a monochromatic field. At the same time, it is no problem to make similar constructions for general electrodynamic equations. An example of these extensions will be given at the end of this subsection, where the Stratton-Chu type integrals are employed for a quasi-stationary field.

8.4.1 Concept of the Stratton-Chu Type Integral for a Monochromatic Field

Let some vector fields **E** and **H** be specified on the inner side of a closed smooth surface S bounding the domain D, fields satisfying the following conditions: (a) they have continuously differentiable tangential components $\mathbf{E}_\tau$, $\mathbf{H}_\tau$; (b) their normal components E_n, H_n are related to the tangential components $\mathbf{E}_\tau$, $\mathbf{H}_\tau$ by formulas (8.3.41) and (8.3.42):

$$E_n = -\frac{1}{\sigma^*}\,\mathrm{div}_S(\mathbf{n}\times\mathbf{H}_\tau) \ , \quad H_n = -\frac{1}{i\omega\mu}\,\mathrm{div}_S(\mathbf{n}\times\mathbf{E}_\tau) \ , \tag{8.4.1}$$

where σ^*, μ, and ω are some constants. Since relations (8.3.42) and (8.3.43) are a two-dimensional consequence of the Maxwell equations, the fields **E**, **H** fall into a class of electromagnetic fields.

Consider the expressions

$$\mathbf{C}^E(\mathbf{r}') = \iint_S [(\mathbf{n}\cdot\mathbf{E})\operatorname{grad} G^h + (\mathbf{n}\times\mathbf{E})\times\operatorname{grad} G^h + i\omega\mu(\mathbf{n}\times\mathbf{H})G^h]\, ds$$

$$= \operatorname{rot}' \iint_S (\mathbf{n}\times\mathbf{E})G^h ds - \operatorname{grad}' \iint_S (\mathbf{n}\cdot\mathbf{E})G^h ds + i\omega\mu \iint_S (\mathbf{n}\times\mathbf{H})G^h ds \;; \tag{8.4.2}$$

$$\mathbf{C}^H(\mathbf{r}') = \iint_S [(\mathbf{n}\cdot\mathbf{H})\operatorname{grad} G^h + (\mathbf{n}\times\mathbf{H})\times\operatorname{grad} G^h + \sigma^*(\mathbf{n}\times\mathbf{E})G^h]\, ds$$

$$= \operatorname{rot}' \iint_S (\mathbf{n}\times\mathbf{H})G^h ds - \operatorname{grad}' \iint_S (\mathbf{n}\cdot\mathbf{H})G^h ds + \sigma^* \iint_S (\mathbf{n}\times\mathbf{E})G^h ds \;, \tag{8.4.3}$$

where $\mathbf{n}$ is the unit vector of an inward pointing normal; G^h is the Green fundamental function,

$$G^h = G^h(\mathbf{r}'|\mathbf{r}) = \frac{-\exp ik^* \;|\mathbf{r}-\mathbf{r}'|}{4\pi\;|\mathbf{r}-\mathbf{r}'|} \;; \quad k^* = \sqrt{i\omega\mu\sigma + \omega^2\mu\varepsilon} \;, \quad \operatorname{Re} k^* > 0 \;.$$

These expressions coincide with the integrals from Stratton-Chu formulas (8.3.36a)–(8.3.38). They can be referred to as *Stratton-Chu type integrals.* The functions $\mathbf{E}$ and $\mathbf{H}$ are *"electric" and "magnetic" densities* of the Stratton-Chu type integrals, respectively.

8.4.2 Properties of the Stratton-Chu Type Integrals

The properties of the Stratton-Chu type integrals are very much similar to those of the three-dimensional Cauchy integral analogs, which are closely outlined in Sects. 5.1. and 5.2. Therefore, in the description of these properties some details of the proof will be often omitted.

Let us sketch the main properties of the Stratton-Chu type integrals, following our book (Berdichevsky and Zhdanov 1984).

1) The Stratton-Chu type integrals beyond the surface S satisfy the Maxwell equations. For the sake of proof, determine rot′ $\mathbf{C}^H$. According to (8.4.3), we have

$$\operatorname{rot}' \mathbf{C}^H(\mathbf{r}') = \operatorname{rot}'\operatorname{rot}' \iint_S (\mathbf{n}\times\mathbf{H})G^h ds + \sigma^* \operatorname{rot}' \iint_S (\mathbf{n}\times\mathbf{E})G^h ds \;.$$

The first integrals in the right-hand side of this expression can be transformed, with due account taken of (8.3.42) and (8.3.44), as follows

$$\begin{aligned}\mathrm{rot}'\mathrm{rot}' \iint_S (\mathbf{n}\times\mathbf{H})G^h ds &= \mathrm{grad}'\mathrm{div}' \iint_S (\mathbf{n}\times\mathbf{H})G^h ds - \Delta' \iint_S (\mathbf{n}\times\mathbf{H})G^h ds \\ &= \mathrm{grad}' \iint_S (\mathbf{n}\times\mathbf{H})\cdot\mathrm{grad}' G^h ds - \iint_S (\mathbf{n}\times\mathbf{H})\Delta' G^h ds \\ &= -\mathrm{grad}' \iint_S (\mathbf{n}\times\mathbf{H})\cdot\mathrm{grad}_S G^h ds + k^{*2} \iint_S (\mathbf{n}\times\mathbf{H})G^h ds \\ &= -\mathrm{grad}' \iint_S \mathrm{div}_S(\mathbf{n}\times\mathbf{H})G^h ds \\ &\quad + \mathrm{grad}' \iint_S G^h \mathrm{div}_S(\mathbf{n}\times\mathbf{H}) ds + i\omega\mu\sigma^* \iint_S (\mathbf{n}\times\mathbf{H})G^h ds \\ &= -\sigma^*\mathrm{grad}' \iint_S (\mathbf{n}\cdot\mathbf{E})G^h ds + i\omega\mu\sigma^* \iint_S (\mathbf{n}\times\mathbf{H})G^h ds \\ &= \sigma^* \iint_S (\mathbf{n}\cdot\mathbf{E})\,\mathrm{grad}\, G^h ds + i\omega\mu\sigma^* \iint_S (\mathbf{n}\times\mathbf{H})G^h ds \ .\end{aligned}$$

Transformation of the second integral is not difficult

$$\sigma^*\mathrm{rot}' \iint_S [\mathbf{n}\times\mathbf{E}]G^h ds = \sigma^* \iint_S [\mathbf{n}\times\mathbf{E}]\times\mathrm{grad}\, G^h ds \ .$$

We obtain finally

$$\mathrm{rot}'\mathbf{C}^H(\mathbf{r}') = \sigma^* \iint_S [(\mathbf{n}\cdot\mathbf{E})\,\mathrm{grad}\, G^h + (\mathbf{n}\times\mathbf{E})\times\mathrm{grad}\, G^h + i\omega\mu(\mathbf{n}\times\mathbf{H})G^h] ds \ .$$

Compare this result to (8.4.2). As is seen,

$$\mathrm{rot}'\mathbf{C}^H(\mathbf{r}') = \sigma^*\mathbf{C}^E(\mathbf{r}') \ , \tag{8.4.4a}$$

which was to be proved. In a similar way, it is proved that

$$\mathrm{rot}'\mathbf{C}^E(\mathbf{r}') = i\omega\mu\mathbf{C}^H(\mathbf{r}') \ . \tag{8.4.4b}$$

It follows from (8.4.4a) and (8.4.4b) that the functions $\mathbf{C}^E$ and $\mathbf{C}^H$ beyond the surface S satisfy the Helmholtz equation

$$\Delta\mathbf{C}^{E,H} + k^{*2}\mathbf{C}^{E,H} = \mathbf{0} \ . \tag{8.4.5}$$

2) If the densities **E**, **H** of the Stratton-Chu type integrals coincide with the boundary values of the vector fields **E** and **H**, satisfying the Maxwell equations

$$\mathrm{rot}\,\mathbf{H} = \sigma^*\mathbf{E} \ , \quad \mathrm{rot}\,\mathbf{E} = i\omega\mu\mathbf{H} \ ,$$

in the domain D right to the surface S, then

$$\mathbf{C}^E(\mathbf{r}') = \begin{cases}\mathbf{E}(\mathbf{r}') \ , & \mathbf{r}'\in D \ , \\ 0 \ , & \mathbf{r}'\in C\bar{D} \ ;\end{cases} \qquad \mathbf{C}^H(\mathbf{r}') = \begin{cases}\mathbf{H}(\mathbf{r}') \ , & \mathbf{r}'\in D \ , \\ 0 \ , & \mathbf{r}'\in C\bar{D} \ .\end{cases} \tag{8.4.6}$$

The proof directly follows from Stratton-Chu integral formulas (8.3.36).

3) If the densities **E**, **H** of the Stratton-Chu type integrals coincide with the boundary values of the vector fields **E**, **H** satisfying the Maxwell equations

$$\operatorname{rot}\mathbf{H} = \sigma^*\mathbf{E} \ , \quad \operatorname{rot}\mathbf{E} = i\omega\mu\mathbf{H}$$

in the domain $C\bar{D}$ right to the surface S and these vanish at infinity, then

$$\mathbf{C}^E(\mathbf{r}') = \begin{cases} 0 \ , & \mathbf{r}' \in D \ , \\ -\mathbf{E}(\mathbf{r}') \ , & \mathbf{r}' \in C\bar{D} \ ; \end{cases} \quad \mathbf{C}^H(\mathbf{r}') = \begin{cases} 0 \ , & \mathbf{r}' \in D \ , \\ -\mathbf{H}(\mathbf{r}') \ , & \mathbf{r}' \in C\bar{D} \ . \end{cases} \tag{8.4.7}$$

This property of the Stratton-Chu integral formulas is established analogously to formula (5.1.37) of Sect. 7.1.6 with due regard to the fact that the integrals $\mathbf{C}^E$, $\mathbf{C}^H$ around the sphere of a radius R tend to zero, provided R tends to infinity.

4) If the densities $\mathbf{E}$ and $\mathbf{H}$ of the Stratton-Chu type integrals satisfy Hölder conditions (5.2.14) at the surface S, there are limit values of the Stratton-Chu type integrals on the inner and outer sides of a smooth surface S, which are equal for $\mathbf{r}_0 \in S$ to

$$\begin{aligned} \mathbf{C}^{E_+}(\mathbf{r}_0) &= \lim_{\mathbf{r}' \to \mathbf{r}_0^+} \mathbf{C}^E(\mathbf{r}') = \mathbf{C}_0^E(\mathbf{r}_0) + \tfrac{1}{2}\mathbf{E}(\mathbf{r}_0) \ , \\ \mathbf{C}^{H_+}(\mathbf{r}_0) &= \lim_{\mathbf{r}' \to \mathbf{r}_0^+} \mathbf{C}^H(\mathbf{r}') = \mathbf{C}_0^H(\mathbf{r}_0) + \tfrac{1}{2}\mathbf{H}(\mathbf{r}_0) \ , \\ \mathbf{C}^{E_-}(\mathbf{r}_0) &= \lim_{\mathbf{r}' \to \mathbf{r}_0^-} \mathbf{C}^E(\mathbf{r}') = \mathbf{C}_0^E(\mathbf{r}_0) - \tfrac{1}{2}\mathbf{E}(\mathbf{r}_0) \ , \\ \mathbf{C}^{H_-}(\mathbf{r}_0) &= \lim_{\mathbf{r}' \to \mathbf{r}_0^-} \mathbf{C}^H(\mathbf{r}') = \mathbf{C}_0^H(\mathbf{r}_0) - \tfrac{1}{2}\mathbf{H}(\mathbf{r}_0) \ , \end{aligned} \tag{8.4.8}$$

where the singular integrals

$$\begin{aligned} \mathbf{C}_0^E(\mathbf{r}_0) &= \iint_S [(\mathbf{n}\cdot\mathbf{E})\operatorname{grad} G^h + (\mathbf{n}\times\mathbf{E})\times\operatorname{grad} G^h + i\omega\mu(\mathbf{n}\times\mathbf{H})G^h]\, ds \ , \\ \mathbf{C}_0^H(\mathbf{r}_0) &= \iint_S [(\mathbf{n}\cdot\mathbf{H})\operatorname{grad} G^h + (\mathbf{n}\times\mathbf{H})\times\operatorname{grad} G^h + \sigma^*(\mathbf{n}\times\mathbf{E})G^h]\, ds \ , \\ G^h &= G^h(\mathbf{r}_0|\mathbf{r}) = \frac{-1}{4\pi}\, \frac{\exp ik^* \ |\mathbf{r}-\mathbf{r}_0|}{|\mathbf{r}-\mathbf{r}_0|} \end{aligned} \tag{8.4.9}$$

are defined in terms of the principal Cauchy value, in accordance with the definition of Sect. 5.2.2 [formula (5.2.13)]. Thus, the Stratton-Chu type integrals undergo discontinuities on S, which are equal to

$$\mathbf{C}^{E_+}(\mathbf{r}_0) - \mathbf{C}^{E_-}(\mathbf{r}_0) = \mathbf{E}(\mathbf{r}_0) \ , \quad \mathbf{C}^{H_+}(\mathbf{r}_0) - \mathbf{C}^{H_-}(\mathbf{r}_0) = \mathbf{H}(\mathbf{r}_0) \ . \tag{8.4.10}$$

At the same time, the sums of limit values of the Stratton-Chu type integrals are equal, according to (8.4.8), to the double singular integrals

$$\mathbf{C}^{E_+}(\mathbf{r}_0) + \mathbf{C}^{E_-}(\mathbf{r}_0) = 2\mathbf{C}_0^E(\mathbf{r}_0) \ , \quad \mathbf{C}^{H_+}(\mathbf{r}_0) + \mathbf{C}^{H_-}(\mathbf{r}_0) = 2\mathbf{C}_0^H(\mathbf{r}_0) \ . \tag{8.4.11}$$

The proof of formulas (8.4.8) through (8.4.11) is just the same as for three-dimensional analogs of the Sokhotsky-Plemelj formulas (see Sect. 5.2.3). For this reason, we will confine ourselves to the explanation of these formulas at the physical level.

Take any field $\mathbf{F}$ excited by sources and rotations distributed over a closed surface S with a density q_S and $\mathbf{j}_S$, which will satisfy, beyond the surface, the Helmholtz equation $\Delta\mathbf{F} + k^{*2}\mathbf{F} = 0$. It is known that [see, for instance, (8.3.39)]

$$\mathbf{F}(\mathbf{r}') = -\iint_S \{q_S \operatorname{grad} G^h + \mathbf{j}_S \times \operatorname{grad} G^h\}\, ds \; .$$

The limit values of this field on the inner S^+ and outer S^- sides of the surface S are

$$\mathbf{F}^+(\mathbf{r}_0) = \mathbf{F}_0(\mathbf{r}_0) - \tfrac{1}{2}[q_S(\mathbf{r}_0)\mathbf{n} + \mathbf{j}_S(\mathbf{r}_0)\times\mathbf{n}] \; ,$$
$$\mathbf{F}^-(\mathbf{r}_0) = \mathbf{F}_0(\mathbf{r}_0) + \tfrac{1}{2}[q_S(\mathbf{r}_0)\mathbf{n} + \mathbf{j}_S(\mathbf{r}_0)\times\mathbf{n}] \; , \tag{8.4.12}$$

for $\mathbf{r}_0 \in S$, where $\mathbf{n}$ is the unit vector of an outward normal and the singular integral $\mathbf{F}_0(\mathbf{r}_0)$ exists in terms of the principal Cauchy value.

For a discontinuity of the field $\mathbf{F}$ at the surface S we have

$$\mathbf{F}^+(\mathbf{r}_0) - \mathbf{F}^-(\mathbf{r}_0) = -q_S(\mathbf{r}_0)\mathbf{n}(\mathbf{r}_0) - \mathbf{j}_S(\mathbf{r}_0)\times\mathbf{n}(\mathbf{r}_0) \; . \tag{8.4.13}$$

Compare the field $\mathbf{F}$ to $\mathbf{C}^E$. The vector function $\mathbf{C}^E$ exhibits properties of an electric field and its discontinuity at the surface S is caused by equivalent electric charges and magnetic currents. Hence, going from $\mathbf{F}$ over to $\mathbf{C}^E$ we should take that $q_S = q_S^e/\varepsilon^*$ and $\mathbf{j}_S = -\mathbf{j}_S^m$, whence we have, in accordance with (8.3.40), (8.4.12), and (8.4.13),

$$-q_S\mathbf{n} - \mathbf{j}_S\times\mathbf{n} = \frac{-q_S^e\mathbf{n}}{\varepsilon^*} + \mathbf{j}_S^m\times\mathbf{n} = (\mathbf{n}\cdot\mathbf{E})\mathbf{n} + (\mathbf{n}\times\mathbf{E})\times\mathbf{n} = \mathbf{E} \; ,$$

i.e.,

$$\mathbf{C}^{E+}(\mathbf{r}_0) = \mathbf{C}_0^E(\mathbf{r}_0) + \tfrac{1}{2}\mathbf{E}(\mathbf{r}_0) \; , \quad \mathbf{C}^{E-}(\mathbf{r}_0) = C_0^E(\mathbf{r}_0) - \tfrac{1}{2}\mathbf{E}(\mathbf{r}_0) \; ,$$

$$\mathbf{C}^{E+}(\mathbf{r}_0) - \mathbf{C}^{E-}(\mathbf{r}_0) = \mathbf{E} \; ,$$

which was to be proved. Similarly, limit values and discontinuities of the integral $\mathbf{C}^H$ are determined. Formulas (8.4.8), (8.4.10), (8.4.11) are electromagnetic counterparts of Sokhotsky-Plemelj formulas (5.2.30), (5.2.32), and (5.2.33).

8.4.3 Modified Stratton-Chu Type Integrals

In the Stratton-Chu type integrals, just as in the three-dimensional Cauchy integral analogs, the surface S is assumed to be closed. We can suggest, however, such a form of these integrals as will retain its properties also in the case of an open integration surface (Berdichevsky and Zhdanov 1984).

Let the surface Γ be bounded by a contour L. Consider the integrals

$$\mathbf{M}^E(\mathbf{r}') = \mathbf{C}^E(\mathbf{r}') + \frac{1}{\sigma^*}\mathbf{L}^H(\mathbf{r}') \; , \quad \mathbf{M}^H(\mathbf{r}') = \mathbf{C}^H(\mathbf{r}') + \frac{1}{i\omega\mu}\mathbf{L}^E(\mathbf{r}') \; , \tag{8.4.14}$$

where

$$\mathbf{L}^E(\mathbf{r}') = \operatorname{grad}' \oint_L G^h\mathbf{E}\cdot d\mathbf{l} \; , \quad \mathbf{L}^H(\mathbf{r}') = \operatorname{grad}' \oint_L G^h\mathbf{H}\cdot d\mathbf{l} \; . \tag{8.4.15}$$

The direction of the normal to the surface Γ and the direction of traversal of the contour L should be matched so that the unit vectors $\mathbf{n}$, $\mathbf{v}$, and $\mathbf{l}$ form a right triad (Fig. 32).

The integrals $\mathbf{M}^E$, $\mathbf{M}^H$ are *modified integrals of the Stratton-Chu type.* They differ from the integrals $\mathbf{C}^E$, $\mathbf{C}^H$ by the terms including the linear integrals $\mathbf{L}^E$, $\mathbf{L}^H$ which reflect the contributions of equivalent electric and magnetic charges accumulated on the boundary of the surface Γ. When the surface Γ is closed, $\mathbf{M}^E = \mathbf{C}^E$ and $\mathbf{M}^H = \mathbf{C}^H$, since $\mathbf{L}^E = \mathbf{L}^H = 0$. Hence, at the closed surface Γ the integrals $\mathbf{C}^{E,H}$ and $\mathbf{M}^{E,H}$ are identical.

Transform the expressions for $\mathbf{M}^E$ and $\mathbf{M}^H$. According to Eqs. (8.3.43) and (8.3.44), we have

$$\oint_L G^h \mathbf{H}\, d\mathbf{l} = \oint_L G^h \mathbf{H}\cdot(\mathbf{n}\times\mathbf{v})\,dl = -\oint_L G^h(\mathbf{n}\times\mathbf{H})\cdot\mathbf{v}\,dl = -\iint_\Gamma \mathrm{div}_S G^h(\mathbf{n}\times\mathbf{H})\,ds$$

$$= -\iint_\Gamma G^h \mathrm{div}_S(\mathbf{n}\times\mathbf{H})\,ds - \iint_\Gamma (\mathbf{n}\times\mathbf{H})\cdot\mathrm{grad}_S G^h\,ds$$

$$= \sigma^* \iint_\Gamma (\mathbf{n}\cdot\mathbf{E})G^h\,ds + \mathrm{div}' \iint_\Gamma (\mathbf{n}\times\mathbf{H})G^h\,ds\ .$$

Substituting this relation into Eq. (8.4.14) we obtain, with due reference to Eq. (8.4.2):

$$\mathbf{M}^{\mathbf{E}}(\mathbf{r}') = \mathrm{rot}' \iint_\Gamma (\mathbf{n}\times\mathbf{E})G^h\,ds + i\omega\mu \iint_\Gamma (\mathbf{n}\times\mathbf{H})G^h\,ds + \frac{\mathrm{grad}'\,\mathrm{div}'}{\sigma^*}\iint_\Gamma (\mathbf{n}\times\mathbf{H})G^h\,ds$$

$$= \mathrm{rot}' \iint_\Gamma (\mathbf{n}\times\mathbf{E})G^h\,ds + \frac{\mathrm{grad}'\,\mathrm{div}' - \Delta'}{\sigma^*}\iint_\Gamma (\mathbf{n}\times\mathbf{H})G^h\,ds$$

$$= \frac{1}{\sigma^*}\,\mathrm{rot}'\,\mathrm{rot}' \iint_\Gamma (\mathbf{n}\times\mathbf{H})G^h\,ds + \mathrm{rot}' \iint_\Gamma (\mathbf{n}\times\mathbf{E})G^h\,ds\ . \tag{8.4.16}$$

In a similar way, it is established that

$$\mathbf{M}^H(\mathbf{r}') = \frac{1}{i\omega\mu}\,\mathrm{rot}'\,\mathrm{rot}' \iint_\Gamma (\mathbf{n}\times\mathbf{E})G^h\,ds + \mathrm{rot}' \iint_\Gamma (\mathbf{n}\times\mathbf{H})G^h\,ds\ . \tag{8.4.17}$$

Compare Eqs. (8.4.16) and (8.4.17) to Eqs. (8.3.46) and (8.3.47). The modified Stratton-Chu type integrals and the integrals from the modified Stratton-Chu formulas have a similar structure. If the surface Γ is closed, the integrals $\mathbf{M}^E$, $\mathbf{M}^H$ exhibit the same properties as the integrals $\mathbf{C}^E$, $\mathbf{C}^H$ do.

It is interesting to consider the properties of the integrals $\mathbf{M}^E$, $\mathbf{M}^H$ in the case where the surface Γ is not closed.

1) The modified Stratton-Chu type integrals outside the surface Γ is bounded by a contour L satisfy the Maxwell equations. According to Eqs. (8.4.16) and (8.4.17),

$$\mathrm{rot}'\mathbf{M}^H(\mathbf{r}') = \frac{\mathrm{rot}'\,\mathrm{rot}'\,\mathrm{rot}'}{i\omega\mu}\iint_\Gamma (\mathbf{n}\times\mathbf{E})G^h\,ds + \mathrm{rot}'\,\mathrm{rot}' \iint_\Gamma (\mathbf{n}\times\mathbf{H})G^h\,ds$$

$$= -\frac{\mathrm{rot}'\,\Delta'}{i\omega\mu}\iint_\Gamma (\mathbf{n}\times\mathbf{E})G^h\,ds + \mathrm{rot}'\,\mathrm{rot}' \iint_\Gamma (\mathbf{n}\times\mathbf{H})G^h\,ds$$

$$= -\frac{\text{rot}'}{i\omega\mu} \iint_{\Gamma} (\mathbf{n}\times\mathbf{E})\Delta' G^h ds + \text{rot}'\text{rot}' \iint_{\Gamma} (\mathbf{n}\times\mathbf{H}) G^h ds$$

$$= \sigma^* \text{rot}' \iint_{\Gamma} (\mathbf{n}\times\mathbf{E}) G^h ds + \text{rot}'\text{rot}' \iint_{\Gamma} (\mathbf{n}\times\mathbf{H}) G^h ds = \sigma^* \mathbf{M}^E(\mathbf{r}') . \tag{8.4.18a}$$

In a similar way, it is proved that

$$\text{rot}'\mathbf{M}^E(\mathbf{r}') = i\omega\mu\mathbf{M}^H(\mathbf{r}') . \tag{8.4.18b}$$

This evidently suggests that the functions $\mathbf{M}^E$, $\mathbf{M}^H$ satisfy the Helmholtz equation: $\Delta'\mathbf{M}^{E,H} + k^{*2}\mathbf{M}^{E,H} = 0$.

2) Let us specify the limit values of the modified Stratton-Chu type integrals at the surface Γ bounded by a contour L. Choose a positive and a negative side of the surface Γ so that the normal vector be oriented from the negative to the positive side. According to Eqs. (8.4.8) and (8.4.14) for $\mathbf{r} \in \Gamma$, we have

$$\begin{aligned}
\mathbf{M}^{E+}(\mathbf{r}_0) &= \mathbf{C}^{E+}(\mathbf{r}_0) + \frac{1}{\sigma^*}\mathbf{L}^H(\mathbf{r}_0) = \mathbf{C}_0^E(\mathbf{r}_0) + \frac{1}{\sigma^*}\mathbf{L}^H(\mathbf{r}_0) + \frac{1}{2}\mathbf{E}(\mathbf{r}_0) , \\
\mathbf{M}^{H+}(\mathbf{r}_0) &= \mathbf{C}^{H+}(\mathbf{r}_0) + \frac{1}{i\omega\mu}\mathbf{L}^E(\mathbf{r}_0) = \mathbf{C}_0^H(\mathbf{r}_0) + \frac{1}{i\omega\mu}\mathbf{L}^E(\mathbf{r}_0) + \frac{1}{2}\mathbf{H}(\mathbf{r}_0) , \\
\mathbf{M}^{E-}(\mathbf{r}_0) &= \mathbf{C}^{E-}(\mathbf{r}_0) + \frac{1}{\sigma^*}\mathbf{L}^H(\mathbf{r}_0) = \mathbf{C}_0^E(\mathbf{r}_0) + \frac{1}{\sigma^*}\mathbf{L}^H(\mathbf{r}_0) - \frac{1}{2}\mathbf{E}(\mathbf{r}_0) , \\
\mathbf{M}^{H-}(\mathbf{r}_0) &= \mathbf{C}^{H-}(\mathbf{r}_0) + \frac{1}{i\omega\mu}\mathbf{L}^E(\mathbf{r}_0) = \mathbf{C}_0^H(\mathbf{r}_0) + \frac{1}{i\omega\mu}\mathbf{L}^E(\mathbf{r}_0) - \frac{1}{2}\mathbf{H}(\mathbf{r}_0) ,
\end{aligned} \tag{8.4.19}$$

where $\mathbf{M}^{E,H\pm}$, $\mathbf{C}^{E,H\pm}$ are the limit values of the integrals on the positive and the negative side of the surface Γ, respectively; $\mathbf{C}_0^{E,H}$ stands for singular integrals (8.4.9) in terms of the principal Cauchy value; $\mathbf{L}^{E,H}$ are the proper integrals (8.4.15).

Thus, the modified Stratton-Chu type integrals suffer discontinuities at the surface Γ:

$$\mathbf{M}^{E+} - \mathbf{M}^{E-} = \mathbf{E} \ ; \quad \mathbf{M}^{H+} - \mathbf{M}^{H-} = \mathbf{H} . \tag{8.4.20}$$

We have considered the main properties of the Stratton-Chu integral relations. All these relations are valid also for piecewise smooth surfaces, just as is the case with three-dimensional Cauchy integral analogs of Sect. 4.5.

8.4.4 Matrix Representation

The Stratton-Chu type integrals (8.4.2), (8.4.3) and (8.4.16), (8.4.17) can be written in matrix form. We will confine ourselves to integrals (8.4.2) and (8.4.3). In an arbitrary Cartesian basis $\{\mathbf{d}_x, \mathbf{d}_y, \mathbf{d}_z\}$ the vectors $\mathbf{C}^E$, $\mathbf{C}^H$ and $\mathbf{E}$, $\mathbf{H}$, are decomposed as follows:

$$\mathbf{C}^E = C_a^E \mathbf{d}_a \ , \quad \mathbf{C}^H = C_a^H \mathbf{d}_a \ , \quad \mathbf{E} = E_\beta \mathbf{d}_\beta \ ,$$

$$\mathbf{H} = H_\beta \mathbf{d}_\beta \ , \quad \mathbf{n} = n_\gamma \mathbf{d}_\gamma \ , \quad a, \beta, \gamma = x, y, z \ .$$

According to Eqs. (5.4.7) and (5.4.8), we have

$$(\mathbf{n}\cdot\mathbf{E})\,\mathrm{grad}\,G^h = E_\beta n_\gamma G^h_{,a}\delta_{\beta\gamma}\mathbf{d}_a = E_\beta n_\gamma G^h_{,\eta}\delta_{a\eta}\delta_{\beta\gamma}\mathbf{d}_a \ ;$$

$$[\mathbf{n}\times\mathbf{E}]\times\mathrm{grad}\,G^h = E_\beta n_\gamma G^h_\eta \varepsilon_{a\eta\zeta}\varepsilon_{\beta\gamma\zeta}\mathbf{d}_a \ ; \quad a, \beta, \gamma, \zeta, \eta = x, y, z \ ;$$

$$i\omega\mu(\mathbf{n}\times\mathbf{H})G^h = i\omega\mu H_\beta n_\gamma G^h \varepsilon_{\gamma\beta a}\mathbf{d}_a \ .$$

A little transformation yields

$$(\mathbf{n}\cdot\mathbf{E})\,\mathrm{grad}\,G^h + (\mathbf{n}\times\mathbf{E})\times\mathrm{grad}\,G^h + i\omega\mu(\mathbf{n}\times\mathbf{H})G^h$$

$$= [E_\beta G^h_{,\eta}(\delta_{a\eta}\delta_{\beta\gamma} + \varepsilon_{a\eta\zeta}\varepsilon_{\beta\gamma\zeta}) + i\omega\mu H_\beta G^h \varepsilon_{\gamma\beta a}]\,n_\gamma \mathbf{d}_a$$

$$= (\Delta_{a\beta\gamma\eta}E_\beta G^h_{,\eta} + i\omega\mu\varepsilon_{\gamma\beta a}H_\beta G^h)\,n_\gamma \mathbf{d}_a \ , \quad a, \beta, \gamma, \eta = x, y, z \ , \qquad (8.4.21)$$

where the four-subscript Δ symbol is expressed in terms of the Kronecker δ and ε symbols by means of formula (5.4.11). Substitution of (8.4.21) into (8.4.2) yields

$$C_a^E(\mathbf{r}') = \iint_S (\Delta_{a\beta\gamma\eta}E_\beta G^h_{,\eta} + i\omega\mu\varepsilon_{\gamma\beta a}H_\beta G^h)\,n_\gamma ds \ , \quad a, \beta, \gamma, \eta = x, y, z \ . \quad (8.4.22)$$

Similarly is determined a matrix form of the integral $\mathbf{C}^H$. According to Eq. (8.4.3),

$$G_a^H(\mathbf{r}') = \iint_S (\Delta_{a\beta\gamma\eta}H_\beta G^h_{,\eta} + \sigma^* \varepsilon_{\gamma\beta a}E_\beta G^h)\,n_\gamma ds \ , \quad a, \beta, \gamma, \eta = x, y, z \ . \quad (8.4.23)$$

The matrix formulas enable us to calculate the Stratton-Chu type integrals from the components E_β, H_β of their electric and magnetic densities.

8.4.5 Stratton-Chu Type Integrals for a Quasi-Stationary Field

By analogy to the Stratton-Chu type integrals for monochromatic fields, we can take into consideration Stratton-Chu type integrals for arbitrary time-varying fields, in particular for the model of a quasi-stationary field which is of the most practical concern in geophysical applications.

Let some vector fields $\mathbf{E}$ and $\mathbf{H}$ be specified on the inner side of a closed smooth surface S bounding the domain D, assuming that they depend on time t and satisfy the following conditions: (a) they have continuously differentiable tangential components $\mathbf{E}_\tau$, $\mathbf{H}_\tau$; (b) their normal components E_n, H_n are related to the tangential components by

$$E_n = -\frac{1}{\sigma}\,\mathrm{div}_S(\mathbf{n}\times\mathbf{H}_\tau) \ , \quad \frac{\partial H_n}{\partial t} = \frac{1}{\mu}\,\mathrm{div}_S(\mathbf{n}\times\mathbf{E}_\tau) \ , \qquad (8.4.24)$$

where σ and μ are some constants; div_S is the symbol of surface divergence.

One can readily see, analogously to Sect. 8.3.6, that relations (8.4.24) are a two-dimensional consequence of the Maxwell Eqs. (8.1.22) for a quasi-stationary field.

Consider the expressions

$$\mathbf{C}^E(\mathbf{r}', t') = \int_{-\infty}^{t'} \iint_S \left[(\mathbf{n}\cdot\mathbf{E})\,\mathrm{grad}\, G^d + (\mathbf{n}\times\mathbf{E})\times\mathrm{grad}\, G^d + \mu(\mathbf{n}\times\mathbf{H})\frac{\partial G^d}{\partial t} \right] ds\,dt \ , \tag{8.4.25a}$$

and

$$\mathbf{C}^H(\mathbf{r}', t') = \int_{-\infty}^{t'} \iint_S [(\mathbf{n}\cdot\mathbf{H})\,\mathrm{grad}\, G^d + (\mathbf{n}\times\mathbf{H})\times\mathrm{grad}\, G^d + \sigma(\mathbf{n}\times\mathbf{E})\,G^d]\, ds\,dt \ , \tag{8.4.25b}$$

where G^d is the Green fundamental function for the equation of diffusion fitted by formula (8.3.15). These expressions coincide with the integrals from Stratton-Chu formulas (8.3.18a) and (8.3.18b). The functions $\mathbf{E}$ and $\mathbf{H}$ related by (8.4.24) are "electric" and "magnetic" densities of the Stratton-Chu type integrals for a quasi-stationary field. Stratton-Chu type integrals (8.4.25a) and (8.4.25b) exhibit some properties similar to those of their monochromatic counterparts outlined in Sect. 8.4.2. Here are some of their properties.

1) The Stratton-Chu type integrals outside the surface S satisfy the equations for a quasi-stationary field

$$\mathrm{rot}'\mathbf{C}^H = \sigma\mathbf{C}^E \ , \quad \mathrm{rot}'\mathbf{C}^E = -\mu\frac{\partial \mathbf{C}^H}{\partial t} \ , \tag{8.4.26}$$

whence it follows that outside S, the fields $\mathbf{C}^E$, $\mathbf{C}^H$ satisfy the equation of diffusion

$$\Delta'\mathbf{C}^{E,H} - \mu\sigma\frac{\partial \mathbf{C}^{E,H}}{\partial t'} = \mathbf{0} \ . \tag{8.4.27}$$

2) If the densities $\mathbf{E}$, $\mathbf{H}$ of the integrals of the Stratton-Chu type coincide with the boundary values of vector fields satisfying in the domain D and at S the Maxwell Eqs.

$$\mathrm{rot}\,\mathbf{H} = \sigma\mathbf{E} \ , \quad \mathrm{rot}\,\mathbf{E} = -\mu\frac{\partial \mathbf{H}}{\partial t} \ , \quad \mathrm{div}\,\mathbf{H} = 0 \ , \quad \mathrm{div}\,\mathbf{E} = 0 \ , \tag{8.4.28}$$

then the following formulas

$$\mathbf{C}^E(\mathbf{r}', t') = \begin{cases} \mathbf{E}(\mathbf{r}', t') \ , & \mathbf{r}' \in D \ , \\ 0 \ , & \mathbf{r}' \in C\bar{D} \ , \end{cases} \qquad \mathbf{C}^H(\mathbf{r}', t') = \begin{cases} \mathbf{H}(\mathbf{r}', t') \ , & \mathbf{r}' \in D \ , \\ 0 \ , & \mathbf{r}' \in C\bar{D} \ . \end{cases} \tag{8.4.29}$$

hold good.

3) If the electric $\mathbf{E}$ and magnetic $\mathbf{H}$ densities of the Stratton-Chu type integrals are Hölder continuous at S [formula (5.2.14)], there exist limit values of these integrals on the inner and outer side of a smooth surface S, which are defined by formulas of type (8.4.8).

The proof of the above properties is similar to that presented for a monochromatic field in Sect. 8.3.2.

In conclusion, a simple physical explanation of integrals (8.4.25a) and (8.4.25b) will be given. Let extraneous electric and magnetic currents and charges be distributed over a closed surface with surface densities $\mathbf{j}_S^e$, $\mathbf{j}_S^m$, q_S^e, and q_S^m, respectively.

Everywhere outside S they excite an electromagnetic field

$$\mathbf{E}(\mathbf{r}', t') = \int_{-\infty}^{t'} \iint_S \left(\frac{q_S^e}{\varepsilon} \operatorname{grad} G^d - \mathbf{j}_S^m \times \operatorname{grad} G^d + \mu \mathbf{j}_S^e \frac{\partial G^d}{\partial t} \right) ds\, dt \ , \tag{8.4.30a}$$

$$\mathbf{H}(\mathbf{r}', t') = \int_{-\infty}^{t'} \iint_S \left(\frac{q_S^m}{\mu} \operatorname{grad} G^d + \mathbf{j}_S^e \times \operatorname{grad} G^d - \sigma \mathbf{j}_S^m G^d \right) ds\, dt \ . \tag{8.4.30b}$$

A comparison of (8.4.25a) and (8.4.25b) with (8.4.30a, b) reveals that the electromagnetic field of real sources outside S coincides with the field excited by equivalent extraneous (fictitious) currents and charges distributed over S. The density of fictitious sources is expressed in terms of normal and tangential components of the electromagnetic field

$$\mathbf{j}_S^e = -\mathbf{n} \times \mathbf{H} \ , \quad \mathbf{j}_S^m = \mathbf{n} \times \mathbf{E} \ , \quad q_S^e = -\varepsilon(\mathbf{n} \cdot \mathbf{E}) \ , \quad q_S^m = -\mu(\mathbf{n} \cdot \mathbf{H}) \ . \tag{8.4.31}$$

8.5 Extension of the Stratton-Chu Formulas to Inhomogeneous Media

The theory of Stratton-Chu type integrals elaborated in the previous subsections can be extended to the case of inhomogeneous media, which expands its applications in geophysics.

The problem is stated as follows. Consider a magnetically homogeneous medium ($\mu = \text{const}$) with specified piecewise continuous distributions of electric conductivity $\sigma(\mathbf{r})$ and permittivity $\varepsilon(\mathbf{r})$. Take a volume D bounded by a surface S. Within this model, the electromagnetic field is excited by sources located outside the domain D. It is assumed that the time dependence of the fields is fitted by the factor $\exp(-i\omega t)$, i.e., a monochromatic field is considered.

It is necessary to derive expressions for the electromagnetic field at any point $\mathbf{r}_0$ inside the domain D from its values on the inner side of the surface S.

8.5.1 Green Electromagnetic Tensors and Their Properties

Within the model concerned, the electric and the magnetic fields everywhere in the domain D satisfy the Maxwell equations

$$\nabla \times \mathbf{H} = \sigma^*(\mathbf{r})\mathbf{E}) \ , \quad \nabla \times \mathbf{E} = i\omega\mu \mathbf{H} \ , \tag{8.5.1}$$ [1]

where $\sigma^*(\mathbf{r}) = \sigma(\mathbf{r}) - i\omega\varepsilon(\mathbf{r})$ is the complex electric conductivity of a medium.

[1] Within this subsection, the operations rot, div, and grad will be denoted by the operator ∇, since it facilitates the handling of tensor objects.

By virtue of the principle of uniqueness (Harrington 1961), the electromagnetic field in a finite domain free of sources is defined unequivocally by the tangential components of the field on the boundary of S. Furthermore, initial Eqs. (8.5.1) are linear in currents. This gives grounds for representing the sought electromagnetic field at any point $\mathbf{r}'$ inside the domain D in the form (Felsen and Marcuvitz 1973)

$$\begin{aligned} \mathbf{E}(\mathbf{r}') &= -\iint_S [\hat{\mathscr{G}}_{11}(\mathbf{r}'|\mathbf{r})\mathbf{j}_S^e(\mathbf{r}) + \hat{\mathscr{G}}_{12}(\mathbf{r}'|\mathbf{r})\mathbf{j}_S^m(\mathbf{r})]\,ds \;, \\ \mathbf{H}(\mathbf{r}') &= -\iint_S [\hat{\mathscr{G}}_{21}(\mathbf{r}'|\mathbf{r})\mathbf{j}_S^e(\mathbf{r}) + \hat{\mathscr{G}}_{22}(\mathbf{r}'|\mathbf{r})\mathbf{j}_S^m(\mathbf{r})]\,ds \;, \quad \mathbf{r}' \in D \;, \end{aligned} \tag{8.5.2}$$

where $\mathbf{j}_S^e(\mathbf{r})$ and $\mathbf{j}_S^m(\mathbf{r})$ are the densities of the electric and magnetic (fictitious) fields at the surface S, respectively, and $\hat{\mathscr{G}}_{11}$, $\hat{\mathscr{G}}_{12}$, $\hat{\mathscr{G}}_{21}$, $\hat{\mathscr{G}}_{22}$ are the Green electromagnetic tensors[2] whose vector components imply the electric and magnetic fields excited at the point $\mathbf{r}'$ by electric and magnetic sources of a unit intensity located at the point $\mathbf{r}$ of the surface S (i.e., by point electric and magnetic dipoles); the integration is over the coordinate $\mathbf{r}$.

Below are given the properties of the functions $\mathscr{G}_{ij}$ $(i, j = 1, 2)$.

First, being fields of elementary sources of the electric and magnetic types, they follow the Maxwell equations:

$$\nabla' \times \hat{\mathscr{G}}_{11} = i\omega\mu\,\hat{\mathscr{G}}_{21} \;, \quad \nabla' \times \hat{\mathscr{G}}_{21} = \sigma^*(\mathbf{r}')\,\hat{\mathscr{G}}_{11} - \hat{\mathbf{I}}\delta(\mathbf{r}'-\mathbf{r}) \;; \tag{8.5.3}$$

$$\nabla' \times \hat{\mathscr{G}}_{12} = i\omega\mu\,\hat{\mathscr{G}}_{22} + \hat{\mathbf{I}}\delta(\mathbf{r}'-\mathbf{r}) \;, \quad \nabla' \times \hat{\mathscr{G}}_{22} = \sigma^*(\mathbf{r}')\,\hat{\mathscr{G}}_{12} \;, \tag{8.5.4}$$

where δ is the Dirac function and $\hat{\mathbf{I}}$ is the idem factor (see App. B).

Equations (8.5.3) and (8.5.4) suggest that $\hat{\mathscr{G}}_{11}$ and $\hat{\mathscr{G}}_{12}$ also satisfy the equations

$$\begin{aligned} \nabla' \times [\nabla' \times \hat{\mathscr{G}}_{11}] - k^{*2}(\mathbf{r}')\,\hat{\mathscr{G}}_{11} &= -i\omega\mu\hat{\mathbf{I}}\delta(\mathbf{r}-\mathbf{r}') \;, \\ \nabla' \times [\nabla' \times \hat{\mathscr{G}}_{12}] - k^{*2}(\mathbf{r}')\,\hat{\mathscr{G}}_{12} &= -i\omega\mu\nabla' \times \hat{\mathbf{I}}\delta(\mathbf{r}-\mathbf{r}') \;, \end{aligned} \tag{8.5.5}$$

while $\hat{\mathscr{G}}_{21}$ and $\hat{\mathscr{G}}_{22}$ meet the equations

$$\begin{aligned} \nabla' \times [\nabla' \times \hat{\mathscr{G}}_{21}] - k^{*2}(\mathbf{r}')\,\hat{\mathscr{G}}_{21} - \nabla'\sigma^* \times \frac{1}{\sigma^*}\,\nabla' \times \hat{\mathscr{G}}_{21} &= -\nabla' \times \hat{\mathbf{I}}\delta(\mathbf{r}-\mathbf{r}') \;, \\ \nabla' \times [\nabla' \times \hat{\mathscr{G}}_{22}] - k^{*2}(\mathbf{r}')\,\hat{\mathscr{G}}_{22} - \nabla'\sigma^* \times \frac{1}{\sigma^*}\,\nabla' \times \hat{\mathscr{G}}_{22} &= \sigma^*(\mathbf{r})\hat{\mathbf{I}}\delta(\mathbf{r}-\mathbf{r}') \;, \end{aligned} \tag{8.5.6}$$

where $k^*(\mathbf{r}') = i\omega\mu\sigma^*(\mathbf{r}')$.

Second, the tensor functions $\hat{\mathscr{G}}_{ij}$ $(i, j = 1, 2)$ exhibit symmetry and can be shown, using the Lorentz lemma, to satisfy the following reciprocal relations

---

[2] Those wishing to delve more deeply into the concept of the Green electromagnetic tensor may start with Appendix B. In addition, this Appendix presents a detailed derivation of formulas (8.5.2) relying on the Lorentz lemma for an inhomogeneous medium.

$$\hat{\mathscr{G}}_{11}(\mathbf{r}|\mathbf{r}') = \tilde{\hat{\mathscr{G}}}_{11}(\mathbf{r}'|\mathbf{r}) \ , \quad \hat{\mathscr{G}}_{22}(\mathbf{r}|\mathbf{r}') = \tilde{\hat{\mathscr{G}}}_{22}(\mathbf{r}'|\mathbf{r}) \ ,$$
$$\hat{\mathscr{G}}_{12}(\mathbf{r}|\mathbf{r}') = -\tilde{\hat{\mathscr{G}}}_{21}(\mathbf{r}'|\mathbf{r}) \ , \quad \hat{\mathscr{G}}_{21}(\mathbf{r}|\mathbf{r}') = -\tilde{\hat{\mathscr{G}}}_{12}(\mathbf{r}'|\mathbf{r}) \ , \tag{8.5.7}$$

where the swung dash denotes the operation of transposition.

The above properties of the Green electromagnetic tensors are proved in Appendix B.

8.5.2 Stratton-Chu Formulas for an Inhomogeneous Medium

It follows from formulas (8.5.3), (8.5.4), and (8.5.7) that the electric and the magnetic fields, as represented by (8.5.2), can be expressed in terms of one of the functions $\hat{\mathscr{G}}_{ij}$ $(i, j = 1, 2)$. For the sake of convenience, we will introduce tensor functions of the "electric" and "magnetic" types by the formulas

$$\hat{\mathbf{G}}^e(\mathbf{r}|\mathbf{r}') = \hat{\mathscr{G}}_{11}(\mathbf{r}|\mathbf{r}')(i\omega\mu)^{-1} \ , \quad \hat{\mathbf{G}}^m(\mathbf{r}|\mathbf{r}') = \hat{\mathscr{G}}_{21}(\mathbf{r}|\mathbf{r}')(i\omega\mu)^{-1} \ . \tag{8.5.8}$$

Then, using (8.5.3), (8.5.4), (8.5.7) and (8.5.8), we can express all the tensor functions $\hat{\mathscr{G}}_{ij}$ $(i, j = 1, 2)$ in terms of $\hat{\mathbf{G}}^e$ and $\hat{\mathbf{G}}^m$:

$$\hat{\mathscr{G}}_{11}(\mathbf{r}|\mathbf{r}') = i\omega\mu\hat{\mathbf{G}}^e(\mathbf{r}|\mathbf{r}') \ ,$$
$$\tilde{\hat{\mathscr{G}}}_{12}(\mathbf{r}|\mathbf{r}') = -\tilde{\hat{\mathscr{G}}}_{21}(\mathbf{r}'|\mathbf{r}) = -i\omega\mu\tilde{\hat{\mathbf{G}}}^m(\mathbf{r}'|\mathbf{r}) = -\tilde{\boldsymbol{\nabla}}'\times\tilde{\hat{\mathbf{G}}}^e(\mathbf{r}|\mathbf{r}') \ , \tag{8.5.9}$$
$$\hat{\mathscr{G}}_{21}(\mathbf{r}|\mathbf{r}') = i\omega\mu\hat{\mathbf{G}}^m(\mathbf{r}|\mathbf{r}') \ , \quad \tilde{\hat{\mathscr{G}}}_{22}(\mathbf{r}|\mathbf{r}') = -\boldsymbol{\nabla}\times\tilde{\hat{\mathbf{G}}}^m(\mathbf{r}'|\mathbf{r}) = -\tilde{\boldsymbol{\nabla}}'\times\tilde{\hat{\mathbf{G}}}^m(\mathbf{r}|\mathbf{r}') \ ,$$

where the operator $\boldsymbol{\nabla}'$ operates on the coordinates of the vector $\mathbf{r}'$ while the swung dash designates, just as previously, the operation of transposition.

Expressing the surface electric and magnetic currents in terms of tangential components of the magnetic and electric fields, respectively

$$\mathbf{j}^e_S = -\mathbf{n}\times\mathbf{H} \ , \quad \mathbf{j}^m_S = \mathbf{n}\times\mathbf{E} \tag{8.5.10}$$

($\mathbf{n}$ being hereinafter the unit vector of an outward pointing normal to the surface S) and substituting (8.5.9) and (8.5.10) into (8.5.2) we obtain

$$\mathbf{E}(\mathbf{r}') = \iint_S \{i\omega\mu\hat{\mathbf{G}}^e(\mathbf{r}'|\mathbf{r})[\mathbf{n}\times\mathbf{H}] + [\tilde{\boldsymbol{\nabla}}\times\tilde{\hat{\mathbf{G}}}^e(\mathbf{r}'|\mathbf{r})][\mathbf{n}\times\mathbf{E}]\}ds \ ,$$
$$\mathbf{H}(\mathbf{r}') = \iint_S \{i\omega\mu\hat{\mathbf{G}}^m(\mathbf{r}'|\mathbf{r})[\mathbf{n}\times\mathbf{H}] + [\tilde{\boldsymbol{\nabla}}\times\tilde{\hat{\mathbf{G}}}^m(\mathbf{r}'|\mathbf{r})][\mathbf{n}\times\mathbf{E}]\}ds \ , \quad \mathbf{r}\in D \ . \tag{8.5.11 a}$$

It is remarkable that formulas (8.5.11 a) for the fields $\mathbf{E}$ and $\mathbf{H}$ in the domain D can be derived directly by means of the Lorentz lemma (see App. B2), while outside the domain D we have

$$\iint_S \{i\omega\mu\hat{\mathbf{G}}^{e,m}(\mathbf{r}'|\mathbf{r})[\mathbf{n}\times\mathbf{H}] + [\tilde{\boldsymbol{\nabla}}\times\tilde{\hat{\mathbf{G}}}^{e,m}(\mathbf{r}'|\mathbf{r})](\mathbf{n}\times\mathbf{E})\}ds = 0 \ , \quad \mathbf{r}'\in C\bar{D} \ , \tag{8.5.11 b}$$

where $C\bar{D}$ is the complement of the domain $\bar{D}$ (including its boundary) with respect to the whole space.

Thus, formulas (8.5.11 a) and (8.5.11 b) are the Stratton-Chu integral formulas for an inhomogeneous medium. Note that the Stratton-Chu formulas were first extended to the case of a horizontally layered medium by E. V. Zakharov and I. V. Ilyin (1970).

8.5.3 Transition to the Model of a Homogeneous Medium

It will be shown now that in the case of a homogeneous medium, formulas (8.5.11 a) and (8.5.11 b) are reduced to Stratton-Chu formulas (8.3.36 a) and (8.3.36 b). Indeed, for $\sigma = \text{const}$ and $\varepsilon = \text{const}$, Eqs. (8.5.3) and (8.5.5) imply that

$$\Delta \hat{\mathscr{G}}_{11} + k^{*2} \hat{\mathscr{G}}_{11} - \frac{1}{\sigma^*} \nabla' \nabla' \delta(\mathbf{r}-\mathbf{r}') = i\omega\mu \hat{\mathbf{I}} \delta(\mathbf{r}-\mathbf{r}') ,$$

where $k^{*2} = i\omega\mu\sigma^* = \text{const}$, or in the operator form

$$(\Delta' + k^{*2}) \hat{\mathscr{G}}_{11} = \left(i\omega\mu \hat{\mathbf{I}} + \frac{1}{\sigma^*} \nabla' \nabla' \right) \delta(\mathbf{r}-\mathbf{r}') . \tag{8.5.12}$$

Introduce the scalar function $G(\mathbf{r}|\mathbf{r}')$ so as to have

$$\hat{\mathscr{G}}_{11} = \left(i\omega\mu \hat{\mathbf{I}} + \frac{1}{\sigma^*} \nabla' \nabla' \right) G . \tag{8.5.13}$$

This substitution fits Eq. (8.5.12), provided the function G satisfies the scalar equation

$$\Delta' G + k^{*2} G = \delta(\mathbf{r}-\mathbf{r}') \tag{8.5.14}$$

and, hence, it has the form

$$G = G^h(\mathbf{r}|\mathbf{r}') = \frac{-1}{4\pi} \frac{\exp ik^* |\mathbf{r}-\mathbf{r}'|}{|\mathbf{r}-\mathbf{r}'|} . \tag{8.5.15}$$

On the other hand, for a constant σ^*-value, Eqs. (8.5.4) and (8.5.6) suggest that

$$\Delta' \hat{\mathscr{G}}_{22} + k^{*2} \hat{\mathscr{G}}_{22} = -\frac{1}{i\omega\mu} \nabla' \nabla' \delta(\mathbf{r}-\mathbf{r}') - \sigma^* \hat{\mathbf{I}} \delta(\mathbf{r}-\mathbf{r}') ,$$

or in the operator form

$$(\Delta' + k^{*2}) \hat{\mathbf{G}}_{22} = -[\sigma^* \hat{\mathbf{I}} + (i\omega\mu)^{-1} \nabla' \nabla'] \delta(\mathbf{r}-\mathbf{r}') . \tag{8.5.16}$$

It is evident from (8.5.16) that the tensor function $\hat{\mathscr{G}}_{22}$ can be also expressed in terms of the scalar function G^h, satisfying Eq. (8.5.14):

$$\hat{\mathscr{G}}_{22} = -[\sigma^* \hat{\mathbf{I}} + (i\omega\mu)^{-1} \nabla' \nabla'] G^h . \tag{8.5.17}$$

Then, it follows from Eqs. (8.5.3) and (8.5.4), with due reference to Eqs. (8.5.13) and (8.5.17), that

$$\hat{\mathscr{G}}_{12} = -\nabla' \times \hat{\mathbf{I}} G^h \ , \qquad (8.5.18)$$

$$\hat{\mathscr{G}}_{21} = \nabla' \times \hat{\mathbf{I}} G^h \ . \qquad (8.5.19)$$

Thus, in a homogeneous medium, the electromagnetic tensor functions acquire a new property, namely symmetry:

$$\hat{\mathscr{G}}_{12}(\mathbf{r}|\mathbf{r}') = -\hat{\mathscr{G}}_{21}(\mathbf{r}|\mathbf{r}') \ . \qquad (8.5.20)$$

In the case of a homogeneous medium, the electric and the magnetic Green functions, introduced earlier according to (8.5.8), will be designated by $\hat{\mathbf{G}}_0^e$ and $\hat{\mathbf{G}}_0^m$, respectively. Then, in view of (8.5.8), (8.5.13), and (8.5.19), we write

$$\hat{\mathbf{G}}_0^e = \left(\hat{\mathbf{I}} + \frac{1}{k^{*2}} \nabla' \nabla'\right) G^h \ , \quad \hat{\mathbf{G}}_0^m = (\nabla' \times \hat{\mathbf{I}} G^h) \frac{1}{i\omega\mu} \ . \qquad (8.5.21)$$

A direct check attests to the fact that

$$\hat{\mathbf{G}}_0^e(\mathbf{r}|\mathbf{r}') = \tilde{\hat{\mathbf{G}}}_0^e(\mathbf{r}|\mathbf{r}') \ , \quad \hat{\mathbf{G}}_0^m(\mathbf{r}|\mathbf{r}') = -\tilde{\hat{\mathbf{G}}}_0^m(\mathbf{r}|\mathbf{r}') \ . \qquad (8.5.22)$$

Now express the electromagnetic tensor functions $\hat{\mathscr{G}}_{ij}$ in terms of $\hat{\mathbf{G}}_0^e$ and $\hat{\mathbf{G}}_0^m$, allowing for relations (8.5.3), (8.5.4), (8.5.7), and (8.5.20):

$$\hat{\mathscr{G}}_{11} = i\omega\mu \hat{\mathbf{G}}_0^e \ , \quad \hat{\mathscr{G}}_{21} = i\omega\mu \hat{\mathbf{G}}_0^m = \nabla' \times \hat{\mathbf{G}}_0^e \ ,$$

$$\hat{\mathscr{G}}_{12} = -\hat{\mathscr{G}}_{21} = -i\omega\mu \hat{\mathbf{G}}_0^m = -\nabla' \times \hat{\mathbf{G}}_0^e \ , \qquad (8.5.23)$$

$$\hat{\mathscr{G}}_{22} = -\nabla' \times \hat{\mathbf{G}}_0^m = -\frac{1}{i\omega\mu} [\nabla' \times (\nabla' \times \hat{\mathbf{G}}_0^e)] = -\sigma^* \hat{\mathbf{G}}_0^e \ .$$

Substituting (8.5.23) and (8.5.10) into (8.5.2) we write

$$\mathbf{E}(\mathbf{r}') = \iint_S \{i\omega\mu \hat{\mathbf{G}}_0^e(\mathbf{r}'|\mathbf{r}) \cdot (\mathbf{n} \times \mathbf{H}) + [\nabla' \times \hat{\mathbf{G}}_0^e(\mathbf{r}'|\mathbf{r})](\mathbf{n} \times \mathbf{E})\} ds \ ,$$

$$(8.5.24)$$

$$\mathbf{H}(\mathbf{r}') = \iint_S \{i\omega\mu \hat{\mathbf{G}}_0^m(\mathbf{r}'|\mathbf{r}) \cdot [\mathbf{n} \times \mathbf{H}] + [\nabla' \times \hat{\mathbf{G}}_0^m(\mathbf{r}'|\mathbf{r})] \cdot (\mathbf{n} \times \mathbf{E})\} ds \ , \quad \mathbf{r}' \in D \ ,$$

Expressions (8.5.24) are the Stratton-Chu formulas for a homogeneous medium, which are consistent with the formulas (8.5.11 a) and (8.5.11 b) obtained earlier for an inhomogeneous medium. If only the electric tensor function is employed, formulas (8.5.24) take the form

$$\mathbf{E}(\mathbf{r}') = \iint_S [i\omega\mu \hat{\mathbf{G}}_0^e \cdot (\mathbf{n} \times \mathbf{H}) + (\nabla' \times \hat{\mathbf{G}}_0^e) \cdot (\mathbf{n} \times \mathbf{E})] \, ds \ ,$$

$$(8.5.25)$$

$$\mathbf{H}(\mathbf{r}') = \iint_S \sigma^* [(\nabla' \times \hat{\mathbf{G}}_0^e) \cdot (\mathbf{n} \times \mathbf{H}) + \sigma^* \hat{\mathbf{G}}_0^e \cdot (\mathbf{n} \times \mathbf{E})] \, ds \ , \quad \mathbf{r}' \in D \ .$$

Now rewrite the right-hand sides of Eq. (8.5.24), with due account taken of (8.5.8), (8.5.13), (8.5.17) through (8.5.19):

$$\iint_S [i\omega\mu \hat{\mathbf{G}}_0^e \cdot (\mathbf{n}\times\mathbf{H}) + (\nabla' \times \hat{\mathbf{G}}_0^e)\cdot(\mathbf{n}\times\mathbf{E})]\, ds$$

$$= \iint_S \left[i\omega\mu \left(\hat{\mathbf{I}} + \frac{1}{k^{*2}} \nabla'\nabla' \right) G^h \cdot (\mathbf{n}\times\mathbf{H}) + (\nabla' \times G^h \hat{\mathbf{I}})\cdot(\mathbf{n}\times\mathbf{E}) \right] ds \; , \tag{8.5.26}$$

$$\iint_S [(\nabla' \times \hat{\mathbf{G}}_0^e)\cdot(\mathbf{n}\times\mathbf{H}) + \sigma^* \hat{\mathbf{G}}_0^e \cdot (\mathbf{n}\times\mathbf{E})]\, ds$$

$$= \iint_S \left\{ [\nabla' \times (G^h \hat{\mathbf{I}})]\cdot[\mathbf{n}\times\mathbf{H}] + \sigma^* \left(\hat{\mathbf{I}} + \frac{1}{k^{*2}} \nabla'\nabla' \right) G^h \cdot (\mathbf{n}\times\mathbf{E}) \right\} ds \; .$$

But

$$[\nabla' \times (G^h \hat{\mathbf{I}})]\cdot(\mathbf{n}\times\mathbf{E}) = \nabla' \times [G^h \hat{\mathbf{I}} \cdot (\mathbf{n}\cdot\mathbf{E})] = \nabla' \times G^h (\mathbf{n}\times\mathbf{E}) = (\mathbf{n}\times\mathbf{E}) \times \nabla G^h \tag{8.5.27}$$

and, similarly,

$$[\nabla' \times (G^h \hat{\mathbf{I}})]\cdot(\mathbf{n}\times\mathbf{H}) = (\mathbf{n}\times\mathbf{H}) \times \nabla G^h \; .$$

Moreover, it can be shown that

$$\iint_S i\omega\mu \left(\frac{1}{k^{*2}} \nabla'\nabla' G^h \right) \cdot (\mathbf{n}\times\mathbf{H})\, ds = -\iint_S (\mathbf{n}\cdot\mathbf{E}) \nabla' G^h\, ds \; ,$$

$$\iint_S \sigma^* \left(\frac{1}{k^{*2}} \nabla'\nabla' G^h \right) \cdot (\mathbf{n}\times\mathbf{E})\, ds = -\iint_S (\mathbf{n}\cdot\mathbf{H}) \nabla' G^h\, ds \; . \tag{8.5.28}$$

Indeed, allowing for Eqs. (8.5.21) and (8.5.22) as well as for the fact that the operator ∇ operates only on the coordinates of the vector $\mathbf{r}$, while ∇' operates on the coordinates of $\mathbf{r}'$, we write

$$i\omega\mu \left(\frac{1}{k^{*2}} \nabla'\nabla' G^h \right) \cdot (\mathbf{n}\times\mathbf{H}) = \frac{1}{\sigma^*} (\mathbf{n}\times\mathbf{H}) \cdot \nabla'\nabla' G^h = \frac{1}{\sigma^*} [(\mathbf{n}\times\mathbf{H})\cdot\nabla'] \nabla' G^h$$

$$= \frac{1}{\sigma^*} \nabla' [\nabla' G^h (\mathbf{n}\times\mathbf{H})] = -\frac{1}{\sigma^*} \nabla' [\nabla G^h \cdot (\mathbf{n}\times\mathbf{H})]$$

$$= -\frac{1}{\sigma^*} \nabla' \{\nabla \cdot [G^h (\mathbf{n}\times\mathbf{H})] - G^h \nabla \cdot (\mathbf{n}\times\mathbf{H})\} \; . \tag{8.5.29}$$

Let us integrate (8.5.29) over a closed surface S:

$$\iint_S i\omega\mu \left(\frac{1}{k^{*2}} \nabla'\nabla' G^h \right) \cdot (\mathbf{n}\times\mathbf{H})\, ds$$

$$= -\frac{1}{\sigma^*} \nabla' \iint_S \{\nabla \cdot [G^h (\mathbf{n}\times\mathbf{H})] - G^h \nabla \cdot (\mathbf{n}\times\mathbf{H})\} ds \; . \tag{8.5.30}$$

It follows directly from the Maxwell equations that

$$\frac{1}{\sigma^*} \nabla_S \cdot (\mathbf{n}\times\mathbf{H}) = \mathbf{n}\cdot\mathbf{E} \; . \tag{8.5.31}$$

Besides, as the surface S is closed, in view of (8.3.44) we have

$$\iint_S \{\nabla\cdot[G^h(\mathbf{n}\times\mathbf{H})]\}ds \equiv 0 \ . \tag{8.5.32}$$

A substitution of (8.5.31) and (8.5.32) into (8.5.30) yields the first of formulas (8.5.28). The second formula of (8.5.28) follows directly from the first one, provided $\mathbf{E}$ and $\mathbf{H}$ are exchanged and σ^* is replaced by $i\omega\mu$. The final form of formulas (8.5.24), taking into account (8.5.27) and (8.5.28), is as follows

$$\begin{aligned}\mathbf{E}(\mathbf{r}') &= \iint_S [i\omega\mu(\mathbf{n}\times\mathbf{H})G^h + (\mathbf{n}\cdot\mathbf{E})\nabla G^h + (\mathbf{n}\times\mathbf{E})\times\nabla G^h]\,ds \ , \\ \mathbf{H}(\mathbf{r}') &= \iint_S [(\mathbf{n}\times\mathbf{H})\times\nabla G^h + (\mathbf{n}\cdot\mathbf{H})\nabla G^h + \sigma^* G^h(\mathbf{n}\times\mathbf{E})]\,ds \ .\end{aligned} \tag{8.5.33}$$

Thus, we have arrived at Stratton-Chu formulas (8.3.36a) and (8.3.36b) for a homogeneous medium.

8.5.4 Stratton-Chu Type Integrals in an Inhomogeneous Medium and Their Properties

Let us take into consideration integral operators $\hat{F}$ and $\hat{\Psi}$ so that

$$\hat{F}^\Gamma(\hat{G})\mathbf{P} = \iint_\Gamma i\omega\mu\hat{\mathbf{G}}(\mathbf{n}\times\mathbf{P})ds \ , \quad \hat{\Psi}^\Gamma(\hat{G})\mathbf{P} = \iint_\Gamma (\tilde{\nabla}\times\tilde{\hat{\mathbf{G}}})(\mathbf{n}\times\mathbf{P})ds \ , \tag{8.5.34}$$

where $\mathbf{P}$ is an arbitrary vector field specified and integrated over a closed surface Γ; $\mathbf{n}$ is the vector of an outward pointing normal to Γ.

Let $\mathbf{e}(\mathbf{r})$ and $\mathbf{h}(\mathbf{r})$ be arbitrary piecewise differentiable vector fields specified over some closed smooth surface S bounding a finite domain D of space. Let $\hat{\mathbf{G}}^{\mathbf{e}}(\mathbf{r}|\mathbf{r}')$ and $\hat{\mathbf{G}}^{\mathbf{m}}(\mathbf{r}|\mathbf{r}')$ be tensor functions satisfying Eqs. (8.5.1) with specified parameters μ, ε, and $\sigma^*(\mathbf{r})$ (piecewise continuous in $\bar{D}$) as well as the radiation conditions at infinity (Stratton 1948).

Determine the vector fields $\mathbf{u}(\mathbf{r})$ and $\mathbf{v}(\mathbf{r})$ ($\mathbf{r}\in S$) on S so that to meet the conditions

$$\mathbf{u}_\tau = \mathbf{e}_\tau \ , \quad \mathbf{v}_\tau = \mathbf{h}_\tau \ , \quad \mathbf{v}_n = \frac{-1}{i\omega\mu}\nabla\cdot[\mathbf{n}\times\mathbf{e}] \ , \quad \mathbf{u}_n = -\frac{1}{\tilde{\sigma}^*}\nabla\cdot(\mathbf{n}\times\mathbf{h}) \ , \tag{8.5.35}$$

where the subscripts τ and $\mathbf{n}$ denote the tangential and normal components of the vectors $\mathbf{u}$, $\mathbf{v}$, $\mathbf{e}$, $\mathbf{h}$ with respect to the surface S, respectively; the parameter $\tilde{\sigma}^* = \tilde{\sigma}^*(\mathbf{r})$ coincides with $\sigma^*(\mathbf{r})$ everywhere on S, except maybe for the set of zero-measure points $\tilde{\sigma}^*(\mathbf{r}) = \sigma^*(\mathbf{r})$, $\mathbf{r}\in S$. Then, evidently the expressions

$$\mathbf{M}^e(\mathbf{r}') = \hat{F}^S(\hat{G}^e)\mathbf{v} + \hat{\Psi}^S(\hat{G}^e)\mathbf{u} \ , \quad \mathbf{M}^m(\mathbf{r}') = \hat{F}^S(\hat{G}^m)\mathbf{v} + \hat{\Psi}^S(\hat{G}^m)\mathbf{u} \tag{8.5.36}$$

make sense at any point of space not belonging to S, and can be shown to be solutions for the Maxwell Eqs. (8.5.1). Integrals (8.5.36) will be called *Stratton-Chu type integrals for an inhomogeneous medium.*

The properties of the Stratton-Chu type integrals listed in Sect. 8.4 for the homogeneous medium case are readily extended to the case of an inhomogeneous medium.

1) *The vectors* $\mathbf{M}^e$ *and* $\mathbf{M}^m$ *introduced, according to* (8.5.36), *outside the surface S satisfy Maxwell Eqs.* (8.5.1).

The validity of this claim can be checked directly. Indeed, using Eqs. (8.5.3) through (8.5.8) as well as relying on the fact that the operator ∇' operates only on the coordinates of the vector $\mathbf{r}'$, we can write

$$\nabla'\times\mathbf{M}^e(\mathbf{r}') = \iint_S \{i\omega\mu[\nabla'\times\hat{\mathbf{G}}^{\mathrm{e}}(\mathbf{r}'|\mathbf{r})]\cdot\mathbf{v}_\tau + \nabla'\times[\tilde{\nabla}\times\tilde{\hat{\mathbf{G}}}^{\mathrm{e}}(\mathbf{r}'|\mathbf{r})]\cdot\mathbf{u}_\tau\}ds$$

$$= \iint_S \{(i\omega\mu)^{-1}\hat{\mathbf{G}}^{\mathrm{m}}(\mathbf{r}'|\mathbf{r})\cdot\mathbf{v}_\tau + i\omega\mu[\nabla'\times\tilde{\hat{\mathbf{G}}}^{\mathrm{m}}(\mathbf{r}|\mathbf{r}')]\cdot\mathbf{u}_\tau\}ds$$

$$= i\omega\mu\iint_S \{i\omega\mu\hat{\mathbf{G}}^{\mathrm{m}}(\mathbf{r}'|\mathbf{r})\mathbf{v}_\tau + [\tilde{\nabla}\times\tilde{\hat{\mathbf{G}}}^{\mathrm{m}}(\mathbf{r}'|\mathbf{r})]\mathbf{u}_\tau\}ds = i\omega\mu\mathbf{M}^m(\mathbf{r}') \ , \tag{8.5.37}$$

$$\nabla'\times\mathbf{M}^m(\mathbf{r}') = \iint_S \{i\omega\mu[\nabla'\times\hat{\mathbf{G}}^{\mathrm{m}}(\mathbf{r}'|\mathbf{r})]\cdot\mathbf{v}_\tau + \nabla'\times[\tilde{\nabla}\times\tilde{\hat{\mathbf{G}}}^{\mathrm{m}}(\mathbf{r}'|\mathbf{r})]\cdot\mathbf{u}_\tau\}ds$$

$$= \sigma^*(\mathbf{r}')\iint_S \{i\omega\mu\hat{\mathbf{G}}^{\mathrm{e}}(\mathbf{r}'|\mathbf{r})\cdot\mathbf{v}_\tau + i\omega\mu\tilde{\hat{\mathbf{G}}}^{\mathrm{m}}(\mathbf{r}|\mathbf{r}')\cdot\mathbf{u}_\tau\}ds$$

$$= \sigma^*(\mathbf{r})\iint_S \{i\omega\mu\hat{\mathbf{G}}^{\mathrm{e}}(\mathbf{r}'|\mathbf{r})\cdot\mathbf{v}_\tau + [\tilde{\nabla}\times\tilde{\hat{\mathbf{G}}}^{\mathrm{e}}(\mathbf{r}'|\mathbf{r})]\cdot\mathbf{u}_\tau\}ds = \sigma^*(\mathbf{r}')\mathbf{M}^e(\mathbf{r}') \ ,$$

which was to be proved.

It is evident from (8.5.37) that outside the surface S (for $\mathbf{r}'\neq\mathbf{r}$) the vectors $\mathbf{M}^e$ and $\mathbf{M}^m$ satisfy the same equations as the electric ($\hat{\mathbf{G}}^{\mathrm{e}}$) and the magnetic ($\hat{\mathbf{G}}^{\mathrm{m}}$) tensor functions do. Indeed,

$$\nabla'\times[\nabla'\times\mathbf{M}^e(\mathbf{r}')] = i\omega\mu\nabla'\times\mathbf{M}^m(\mathbf{r}') = i\omega\mu\sigma^*(\mathbf{r}')\mathbf{M}^e(\mathbf{r}') \ ,$$

$$\nabla'\times[\nabla'\times\mathbf{M}^m(\mathbf{r}')] = \nabla'\times(\sigma^*(\mathbf{r}')\mathbf{M}^e(\mathbf{r}')) = \sigma^*(\mathbf{r})\nabla'\times\mathbf{M}^e(\mathbf{r}') + \nabla'\sigma^*(\mathbf{r}')\times\mathbf{M}^e(\mathbf{r}')$$

$$= i\omega\mu\sigma^*(\mathbf{r}')\mathbf{M}^m(\mathbf{r}') + \nabla'\sigma^*(\mathbf{r}')\times\left[\frac{1}{\sigma^*(\mathbf{r}')}\nabla'\times\mathbf{M}^m(\mathbf{r}')\right] \ ,$$

or

$$\nabla'\times[\nabla'\times\mathbf{M}^e(\mathbf{r}')] - k^{*2}(\mathbf{r}')\mathbf{M}^e(\mathbf{r}') = 0 \ ,$$

$$\nabla'\times[\nabla'\times\mathbf{M}^m(\mathbf{r}')] - k^{*2}(\mathbf{r}')\mathbf{M}^m(\mathbf{r}') - \nabla'\sigma^*\times\left[\frac{1}{\sigma^*}\nabla'\times\mathbf{M}^m(\mathbf{r}')\right] = 0 \ , \tag{8.5.38}$$

where, as previously, $k^{*2}(\mathbf{r}) = i\omega\mu\sigma^*(\mathbf{r})$.

2) *If the densities* $\mathbf{u}$ *and* $\mathbf{v}$ *of the Stratton-Chu type integrals coincide with the boundary values of the vector fields* $\mathbf{e}$ *and* $\mathbf{h}$ *satisfying in the domain D the Maxwell equations* $\nabla\times\mathbf{h} = \sigma^*(\mathbf{r})\mathbf{e}$, $\nabla\times\mathbf{e} = i\omega\mu\mathbf{h}$, *then we have*

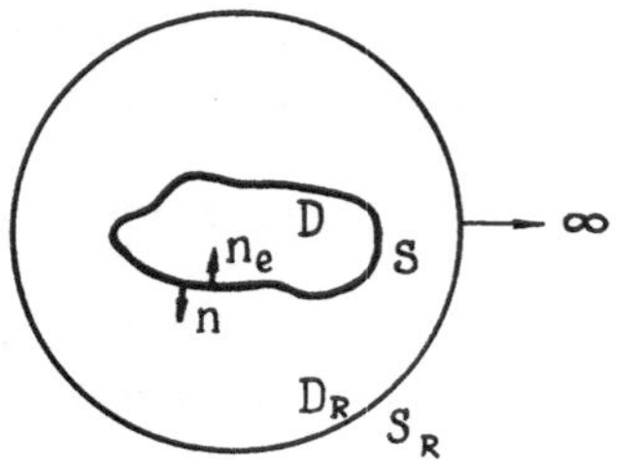

Fig. 48. Inference of property 3 of the Stratton-Chu type integrals in an inhomogeneous medium

$$\mathbf{M}^e(\mathbf{r}') = \begin{cases} \mathbf{u}(\mathbf{r}') , & \mathbf{r}' \in D , \\ 0 , & \mathbf{r}' \in C\bar{D} , \end{cases} \qquad \mathbf{M}^m(\mathbf{r}') = \begin{cases} \mathbf{v}(\mathbf{r}') , & \mathbf{r}' \in D , \\ 0 , & \mathbf{r}' \in C\bar{D} . \end{cases} \tag{8.5.39}$$

The proof follows directly from formulas (8.5.11 a) and (8.5.11 b).

3) *If the densities* $\mathbf{u}$ *and* $\mathbf{v}$ *of the Stratton-Chu type integrals coincide with the boundary values of the vector fields* $\mathbf{e}$ *and* $\mathbf{h}$ *satisfying in the domain* $C\bar{D}$ *and at the surface S the Maxwell equations* $\nabla \times \mathbf{h} = \sigma^*(\mathbf{r})\mathbf{e}$, $\nabla \times \mathbf{e} = i\omega\mu\mathbf{h}$ *and vanish at infinity, i.e.,*

$$\mathbf{e} \underset{\mathbf{r}\to\infty}{\to} \mathbf{0} , \quad \mathbf{h} \underset{\mathbf{r}\to\infty}{\to} \mathbf{0} , \tag{8.5.40}$$

then we have

$$\mathbf{M}^e(\mathbf{r}') = \begin{cases} 0 , & \mathbf{r}' \in D , \\ -\mathbf{u} , & \mathbf{r}' \in C\bar{D} , \end{cases} \qquad \mathbf{M}^m(\mathbf{r}') = \begin{cases} 0 , & \mathbf{r}' \in D , \\ -\mathbf{v} , & \mathbf{r}' \in C\bar{D} . \end{cases} \tag{8.5.41}$$

In order to prove (8.5.41), enclose the domain D within a sphere of a radius R and consider the domain D_R bounded by the surfaces S and S_R (Fig. 48). Letting the sphere radius R tend to infinity and allowing for (8.5.39) and (8.5.40), we write

$$-\mathbf{M}^e(\mathbf{r}') = \begin{cases} 0 , & \mathbf{r}' \in D , \\ \mathbf{u}(\mathbf{r}') , & \mathbf{r}' \in C\bar{D} , \end{cases} \qquad -\mathbf{M}^m(\mathbf{r}') = \begin{cases} 0 , & \mathbf{r}' \in D , \\ \mathbf{v}(\mathbf{r}') , & \mathbf{r}' \in C\bar{D} , \end{cases}$$

where the minus sign in the left-hand sides is related to the fact that at the surface S the outward normal $\mathbf{n}_e$ to the domain points in the direction opposite to the positive one adopted earlier for the normal to the same surface (Fig. 48). Thus, the validity of (8.5.41) is proved.

4) *Let the functions* $\mathbf{u}(\mathbf{r})$ *and* $\mathbf{v}(\mathbf{r})$ *be continuously differentiable along some surface S bounding a finite domain of space and satisfy along S the conditions*

$$v_n = \frac{-1}{i\omega\mu} \nabla' \cdot [\mathbf{n} \times \mathbf{u}] , \quad u_n = -\frac{1}{\tilde{\sigma}^*} \nabla' \cdot [\mathbf{n} \times \mathbf{v}] , \tag{8.5.42}$$

where $\tilde{\sigma}^* = \tilde{\sigma}^*(\mathbf{r}')$, with $\tilde{\sigma}^*(\mathbf{r}') = \sigma^*(\mathbf{r}')$, $\mathbf{r}' \in S$ (except maybe for the set of zero-measure points). *Then, on the inner and outer sides of the surface S there exist limit values of the Stratton-Chu type integrals equal (at ordinary points) to*

$$\mathbf{M}^{e+}(\mathbf{r}_0) = \lim_{\mathbf{r}' \to \mathbf{r}_0^+} \mathbf{M}^e(\mathbf{r}') = \mathbf{C}_0^E(\mathbf{r}_0) + \Delta\mathbf{C}^E(\mathbf{r}_0) + \tfrac{1}{2}\mathbf{u}(\mathbf{r}_0) , \tag{8.5.43 a}$$

$$\mathbf{M}^{m+}(\mathbf{r}_0) = \lim_{\mathbf{r}' \to \mathbf{r}_0^+} \mathbf{M}^m(\mathbf{r}') = \mathbf{C}_0^H(\mathbf{r}_0) + \Delta\mathbf{C}^H(\mathbf{r}_0) + \tfrac{1}{2}\mathbf{v}(\mathbf{r}_0) , \tag{8.5.43 b}$$

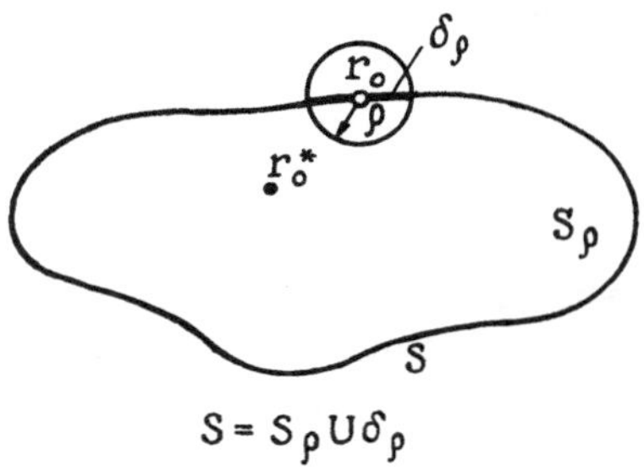

Fig. 49. Inference of property 4 of the Stratton-Chu type integrals in an inhomogeneous medium

$$\mathbf{M}^{e-}(\mathbf{r}_0) = \lim_{\mathbf{r}' \to \mathbf{r}_0^-} \mathbf{M}^e(\mathbf{r}') = \mathbf{C}_0^E(\mathbf{r}_0) + \Delta\mathbf{C}^E(\mathbf{r}_0) - \tfrac{1}{2}\mathbf{u}(\mathbf{r}_0) \ , \tag{8.5.43 c}$$

$$\mathbf{M}^{m-}(\mathbf{r}_0) = \lim_{\mathbf{r}' \to \mathbf{r}_0^-} \mathbf{M}^m(\mathbf{r}') = \mathbf{C}_0^H(\mathbf{r}_0) + \Delta\mathbf{C}^H(\mathbf{r}_0) - \tfrac{1}{2}\mathbf{v}(\mathbf{r}_0) \ , \tag{8.5.43 d}$$

where $\mathbf{C}_0^{E,H}(\mathbf{r}_0)$ are the ordinary singular integrals of the Stratton-Chu type for a homogeneous medium which are defined by the formulas

$$\mathbf{C}_0^{E,H}(\mathbf{r}_0) = \hat{F}^S[\hat{\mathbf{G}}_0^{e,m}(\mathbf{r}_0|\mathbf{r})]\mathbf{v} + \hat{\Psi}^S[\hat{\mathbf{G}}_0^{e,m}(\mathbf{r}_0|\mathbf{r})]\mathbf{u} \ , \tag{8.5.43 e}$$

$$\Delta\mathbf{C}^{E,H}(\mathbf{r}_0) = \hat{F}^S[\Delta\hat{\mathbf{G}}^{e,m}(\mathbf{r}_0|\mathbf{r})]\mathbf{v} + \hat{\Psi}^S[\Delta\hat{\mathbf{G}}^{e,m}(\mathbf{r}_0|\mathbf{r})]\mathbf{u} \ . \tag{8.5.43 f}$$

Here $\hat{\mathbf{G}}_0^{e,m}(\mathbf{r}_0|\mathbf{r})$ is the Green function of a homogeneous medium with $\sigma^* = \sigma^*(\mathbf{r}_0)$, while $\Delta\hat{\mathbf{G}}^{e,m}(\mathbf{r}_0) = \hat{\mathbf{G}}^{e,m}(\mathbf{r}_0|\mathbf{r}) - \hat{\mathbf{G}}_0^{e,m}(\mathbf{r}_0|\mathbf{r})$. Usually, the singular integrals $\mathbf{C}_0^{E,H}$ are determined in terms of the Cauchy principal value.

Let us prove the validity of (8.5.43 a) and (8.5.43 b). Take a point $\mathbf{r}_0^*$ outside the surface S, $\mathbf{r}_0^* \in D$ (Fig. 49). Evidently, the integrals $\mathbf{M}^{e,m}(\mathbf{r}_0^*)$ can be represented in the form

$$\mathbf{M}^{e,m}(\mathbf{r}_0^*) = \mathbf{C}_0^{E,H}(\mathbf{r}_0^*) + \Delta\mathbf{C}^{E,H}(\mathbf{r}_0^*) \ , \tag{8.5.44}$$

where $\mathbf{C}_0^{E,H}(\mathbf{r}_0^*)$ and $\Delta\mathbf{C}^{E,H}(\mathbf{r}_0^*)$ are the ordinary nonsingular integrals ($\mathbf{r}_0^* \notin S$, $\mathbf{r} \in S$).

Find the limits values of $\mathbf{M}^{e,m}(\mathbf{r}_0^*)$ for $\mathbf{r}_0^* \to \mathbf{r}_0$ ($\mathbf{r}_0 \in S$, $\mathbf{r}_0^* \in D$):

$$\lim_{\mathbf{r}_0^* \to \mathbf{r}_0} \mathbf{M}^{e,m}(\mathbf{r}_0^*) = \lim_{\mathbf{r}_0^* \to \mathbf{r}_0} [\mathbf{C}^{E,H}(\mathbf{r}_0^*) + \Delta\mathbf{C}^{E,H}(\mathbf{r}_0^*)] \ . \tag{8.5.45}$$

Note, as a preliminary step, that

$$\Delta\mathbf{C}^{E,H}(\mathbf{r}_0^*) = \lim_{\varrho \to 0} \{[\hat{F}^{S_\varrho}(\hat{\mathbf{G}}^{e,m}) - \hat{F}^{S_\varrho}(\hat{\mathbf{G}}_0^{e,m})]\mathbf{v} + [\hat{\Psi}^{S_\varrho}(\hat{\mathbf{G}}^{e,m}) - \Psi^{S_\varrho}(\hat{\mathbf{G}}_0^{e,m})]\mathbf{u}\}$$

$$= \lim_{\varrho \to 0} \{\hat{F}^{S_\varrho}(\Delta\hat{\mathbf{G}}^{e,m})\mathbf{v} + \hat{\Psi}^{S_\varrho}(\Delta\hat{\mathbf{G}}^{e,m})\mathbf{u}\} \ , \tag{8.5.46}$$

where $\mathbf{r} \in S$, and $S_\varrho = S \backslash \delta\varrho$ is the part of the surface S from which the part δ_ϱ is cut by a sphere of a radius ϱ circumscribed from the point $\mathbf{r}_0$ (see Fig. 49).

Then, allowing for (8.5.46), we can write

$$\lim_{\mathbf{r}_0^* \to \mathbf{r}_0} \Delta \mathbf{C}^{E,H}(\mathbf{r}_0^*) = \lim_{\mathbf{r}_0^* \to \mathbf{r}_0} \lim_{\varrho \to 0} \{\hat{F}^{S_\varrho}[\Delta \hat{\mathbf{G}}^{e,m}(\mathbf{r}_0^* | \mathbf{r})]\mathbf{v} + \hat{\Psi}^{S_\varrho}[\Delta \hat{\mathbf{G}}^{e,m}(\mathbf{r}_0^* | \mathbf{r})]\mathbf{u}\}$$

$$= \lim_{\varrho \to 0} \{\hat{F}^{S_\varrho}(\Delta \hat{\mathbf{G}}^{e,m}(\mathbf{r}_0 | \mathbf{r}))\mathbf{v} + \hat{\Psi}^{S_\varrho}(\Delta \hat{\mathbf{G}}^{e,m}(\mathbf{r}_0 | \mathbf{r}))\mathbf{u}\} , \qquad (8.5.47)$$

where it is taken into consideration that, according to Dmitriev and Farzan (1980), there is a finite limit $\lim_{\mathbf{r} \to \mathbf{r}_0} \Delta \hat{\mathbf{G}}^{e,m}(\mathbf{r}_0 | \mathbf{r})$. Hence, limit (8.5.47) exists as an ordinary improper integral

$$\lim_{\mathbf{r}_0^* \to \mathbf{r}_0} \Delta \mathbf{C}^{E,H}(\mathbf{r}_0^*) = \hat{F}^S[\Delta \hat{\mathbf{G}}^{e,m}(\mathbf{r}_0 | \mathbf{r})]\mathbf{v} + \hat{\Psi}^S[\Delta \hat{\mathbf{G}}^{e,m}(\mathbf{r}_0 | \mathbf{r})]\mathbf{u} = \Delta \mathbf{C}^{E,H}(\mathbf{r}_0) . \qquad (8.5.48)$$

The integrals $\mathbf{C}_0^{E,H}(\mathbf{r}_0^*)$ are ordinary Stratton-Chu type integrals for a homogeneous medium, therefore their limits are defined by the known expressions (8.4.8)

$$\lim_{\mathbf{r}_0^* \to \mathbf{r}_0^+} \mathbf{C}^E(\mathbf{r}_0^*) = \mathbf{C}_0^E(\mathbf{r}_0) + \tfrac{1}{2}\mathbf{u}(\mathbf{r}_0) , \quad \lim_{\mathbf{r}_0^* \to \mathbf{r}_0^+} \mathbf{C}^H(\mathbf{r}_0^*) = \mathbf{C}_0^H(\mathbf{r}_0) + \tfrac{1}{2}\mathbf{v}(\mathbf{r}_0) . \qquad (8.5.49)$$

Substituting (8.5.48) and (8.5.49) into (8.5.45) we arrive at (8.5.43 a) and (8.5.43 b), which was to be proved. In a similar way, relations (8.5.43 c) and (8.5.43 d) are proved (here $\mathbf{r}_0^* \in C\bar{D}$). It follows, in particular, from (8.5.43 a) through (8.5.43 f) that

$$\mathbf{M}_+^e - \mathbf{M}_-^e = \mathbf{u}(\mathbf{r}_0) , \quad \mathbf{M}_+^m - \mathbf{M}_-^m = \mathbf{v}(\mathbf{r}_0) , \quad \mathbf{r}_0 \in S . \qquad (8.5.50)$$

Formulas (8.5.43 a) through (8.5.43 f) and (8.5.50) can be regarded as *electromagnetic counterparts of three-dimensional Sokhotsky-Plemelj formulas* (5.2.30), (5.2.32), and (5.2.33).

Thus, we have extended the theory of the Stratton-Chu type integrals to inhomogeneous media. It is worthwile noting that to determine the Green electromagnetic tensors for an arbitrary medium is generally a very complicated problem. However, there is a variety of classes of models of inhomogeneous media for which this problem is solvable. The most important example of these models is a horizontally layered medium. Within this model, the corresponding Green tensors can be expressed explicitly by means of the Fourier or Hankel integral transforms on the elementary functions (Weidelt 1975).

8.6 Integral Transforms of Electromagnetic Fields

The body of Stratton-Chu type integrals developed in the previous subsections enables a number of theorems of complex analysis describing the properties of analytical function boundary values to be extended to electromagnetic fields. These theorems, which are extremely important for solution of various electrodynamics problems, are treated in this subsection. For the sake of simplicity, we will confine ourselves to a monochromatic electromagnetic field, although all the provisions of the theory considered below apply to a nonstationary field as well.

8.6.1 Integral Boundary Conditions for the Electromagnetic Field on the Boundary of a Homogeneous Domain

Consider a domain D in space, which is bounded by a smooth surface S. The electromagnetic parameters ε, μ and σ of a medium inside this domain are assumed to be constant and known. Let Hölder continuous vector functions $\mathbf{e}(\mathbf{r})$ and $\mathbf{h}(\mathbf{r})$ ($\mathbf{r} \in S$) having continuously differentiable tangential components be specified on S. Which conditions should be satisfied by these functions so that they might be treated as boundary values of a certain monochromatic electromagnetic field $\mathbf{E}$, $\mathbf{H}$ satisfying inside D (right to the boundary S) the equations

$$\operatorname{rot}\mathbf{H} = \sigma^* \mathbf{E} \ , \quad \operatorname{rot}\mathbf{E} = i\omega\mu\mathbf{H} \ , \tag{8.6.1}$$

where $\sigma^* = \sigma - i\omega\varepsilon$ and ω is the angular frequency of the field?

This problem will be solved relying on the theorems obtained in Sect. 5.3 for Laplace vector fields and employing the body of Stratton-Chu type integrals.

Note, first of all, that the simplest *requirement* imposed on $\mathbf{e}$ and $\mathbf{h}$ has, according to (8.3.41) and (8.3.42), the form

$$e_n = -\frac{1}{\sigma^*}\operatorname{div}_S(\mathbf{n}\times\mathbf{h}_\tau) \ , \quad h_n = -\frac{1}{i\omega\mu}\operatorname{div}_S(\mathbf{n}\times\mathbf{e}_\tau) \ , \tag{8.6.2}$$

where $\mathbf{e}_n$, $\mathbf{h}_n$ and $\mathbf{e}_\tau$, $\mathbf{h}_\tau$ are the normal and the tangential components of the vector functions $\mathbf{e}$ and $\mathbf{h}$ at the surface S, respectively. But this condition is not sufficient for the functions $\mathbf{e}$ and $\mathbf{h}$ to be the boundary values of the electromagnetic field satisfying Eqs. (8.6.1) in D.

To specify the necessary and sufficient conditions for boundary values of the electromagnetic field, Stratton-Chu type integrals (8.4.2) and (8.4.3) are written with an electric density $\mathbf{e}$ and a magnetic density $\mathbf{h}$ [in the assumption of fulfilled conditions (8.6.2)]:

$$\mathbf{C}^E(\mathbf{r}'; \mathbf{e}, \mathbf{h}) = \iint_S \{(\mathbf{n}\cdot\mathbf{e})\operatorname{grad} G^h + [\mathbf{n}\times\mathbf{e}]\times\operatorname{grad} G^{h-} + i\omega\mu[\mathbf{n}\times\mathbf{h}]\, G^h\}ds , \tag{8.6.3 a}$$

$$\mathbf{C}^H(\mathbf{r}'; \mathbf{e}, \mathbf{h}) = \iint_S [(\mathbf{n}\cdot\mathbf{h})\operatorname{grad} G^h + (\mathbf{n}\times\mathbf{h})\times\operatorname{grad} G^h + \sigma^*(\mathbf{n}\times\mathbf{e})\, G^h]\, ds . \tag{8.6.3 b}$$

In accordance with property 2 of the Stratton-Chu type integrals (see Sect. 8.4), *the necessary condition* that $\mathbf{e}$ and $\mathbf{h}$ be boundary values of a certain electromagnetic field $\mathbf{E}$, $\mathbf{H}$ excited by sources external with respect to $\bar{D}$, involves the following identities

$$\mathbf{C}^E(\mathbf{r}'; \mathbf{e}, \mathbf{h}) \equiv 0 \ , \quad \mathbf{C}^H(\mathbf{r}'; \mathbf{e}, \mathbf{h}) \equiv 0 \quad \text{for any} \quad \mathbf{r}' \in C\bar{D} \ . \tag{8.6.4}$$

Conditions (8.6.4) in conjunction with (8.6.2) are also *sufficient conditions* for $\mathbf{e}$ and $\mathbf{h}$ to be boundary values of the fields $\mathbf{E}$ and $\mathbf{H}$, defined by the formulas

$$\mathbf{E}(\mathbf{r}') = \mathbf{C}^E(\mathbf{r}'; \mathbf{e}, \mathbf{h}) \ , \quad \mathbf{H}(\mathbf{r}') = \mathbf{C}^H(\mathbf{r}'; \mathbf{e}, \mathbf{h}) \ . \tag{8.6.5}$$

To prove this claim, we will employ the electromagnetic counterparts of the Sokhotsky-Plemelj formulas (8.4.10)

$$\mathbf{C}^{E+}(\mathbf{r}_0;\ \mathbf{e},\ \mathbf{h})-\mathbf{C}^{E-}(\mathbf{r};\ \mathbf{e},\ \mathbf{h})=\mathbf{e}(\mathbf{r}_0)\ , \tag{8.6.6}$$

where $\mathbf{r}_0 \in S$. Obviously, in accordance with (8.6.4), the function $\mathbf{C}^E(\mathbf{r}';\ \mathbf{e},\ \mathbf{h})$ is extendible continuously on S from the outside of the domain D, with

$$\mathbf{C}^{E-}(\mathbf{r}_0;\ \mathbf{e},\ \mathbf{h})=\lim_{\mathbf{r}'\to\mathbf{r}_0^-}\mathbf{C}^E(\mathbf{r}';\ \mathbf{e},\ \mathbf{h})=0\ . \tag{8.6.7}$$

A substitution of (8.6.7) into (8.6.6) reveals that the function $\mathbf{E}(\mathbf{r}')$ is extendible continuously on S from the inside of the domain D, with

$$\lim_{\mathbf{r}'\to\mathbf{r}_0^+}\mathbf{E}(\mathbf{r}')=\mathbf{C}^{E+}(\mathbf{r}_0;\ \mathbf{e},\ \mathbf{h})=\mathbf{e}(\mathbf{r}_0)\ , \tag{8.6.8}$$

which was to be proved. Similarly, the validity of the relation

$$\lim_{\mathbf{r}'\to\mathbf{r}_0^+}\mathbf{H}(\mathbf{r}')=\mathbf{C}^{H+}(\mathbf{r}_0;\ \mathbf{e},\ \mathbf{h})=\mathbf{h}(\mathbf{r}_0) \tag{8.6.9}$$

is proved.

The obtained result can be formulated as the following theorem.

Theorem 8.6.1. For the Hölder continuous vector functions, with $\mathbf{e}(\mathbf{r})$ and $\mathbf{h}(\mathbf{r})$ having continuously differentiable tangential components and specified on a closed smooth surface S (bounding the domain D), to be boundary values on the inner side of this surface of a certain electromagnetic field excited by sources external to the domain D, it is necessary and sufficient that conditions (8.6.2) and (8.6.4) be met.

This theorem can be regarded as an electromagnetic counterpart of theorem 5.3.1 for boundary values of the Laplace field.

It is evident that using the body of Stratton-Chu type integrals and electromagnetic analogs of the Sokhotsky-Plemelj formulas, all the remaining theorems of Sect. 5.3.1 can be extended to the case of an electromagnetic field. Below only the formulations of these theorems are given, the proving being left to the reader.

Theorem 8.6.2. The necessary and sufficient condition that the vector functions $\mathbf{e}(\mathbf{r})$ and $\mathbf{h}(\mathbf{r})$, which are Hölder continuous on a closed smooth surface S (bounding the domain D) and which have therein differentiable tangential components, serve as boundary values on the inner side of S of an electromagnetic field $\mathbf{E}$, $\mathbf{H}$ excited by sources external to the domain D involves the following relations:

$$e_n=-\frac{1}{\sigma^*}\,\mathrm{div}_S(\mathbf{n}\times\mathbf{h}_\tau)\ ,\quad h_n=-\frac{1}{i\omega\mu}\,\mathrm{div}_S(\mathbf{n}\times\mathbf{e}_\tau)\ ; \tag{8.6.10}$$

$$\mathbf{e}(\mathbf{r}_0)=2\mathbf{C}_0^E(\mathbf{r}_0;\ \mathbf{e},\ \mathbf{h})\ ,\quad \mathbf{h}(\mathbf{r}_0)=2\mathbf{C}_0^H(\mathbf{r}_0;\ \mathbf{e},\ \mathbf{h})\ , \tag{8.6.11}$$

where $\mathbf{C}_0^E$ and $\mathbf{C}_0^H$ are the singular integrals determined in terms of the Cauchy principal value, in accordance with formulas (8.4.9); $\mathbf{n}$ being an outward pointing normal with respect to the domain D.

Theorem 8.6.3. For the vector functions $\mathbf{e}(\mathbf{r})$ and $\mathbf{h}(\mathbf{r})$, which are Hölder continuous, have continuously differentiable tangential components, and are specified on a closed smooth surface S (bounding the domain D), to be boundary values on the outer side of this surface of a certain electromagnetic field excited by sources concentrated inside D, it is necessary and sufficient that conditions (8.6.10) be satisfied and

$$\mathbf{C}^E(\mathbf{r}'; \mathbf{e}, \mathbf{h}) = 0 \ , \quad \mathbf{C}^H(\mathbf{r}'; \mathbf{e}, \mathbf{h}) \equiv 0 \quad \text{for any} \quad \mathbf{r}' \in D \ , \tag{8.6.12}$$

or

$$\mathbf{e}(\mathbf{r}_0) = -2\mathbf{C}_0^E(\mathbf{r}_0; \mathbf{e}, \mathbf{h}) \ , \quad \mathbf{h}(\mathbf{r}_0) = -2\mathbf{C}_0^H(\mathbf{r}_0; \mathbf{e}, \mathbf{h}) \ . \tag{8.6.13}$$

Relations (8.6.11) and (8.6.13) are called *integral boundary conditions for a monochromatic electromagnetic field on the boundary of a homogeneous domain.*

8.6.2 Integral Boundary Conditions for the Electromagnetic Field on the Boundary of an Inhomogeneous Domain

Now assume that inside the domain D the complex electric conductivity $\sigma^*(\mathbf{r})$ is an arbitrary continuous (right to the boundary S) function of coordinates and that $\mu = \text{const}$. Let us formulate the conditions to be met by the functions $\mathbf{e}(\mathbf{r})$ and $\mathbf{h}(\mathbf{r})$ specified on the boundary S of the domain D so that they might be treated as boundary values of some monochromatic field $\mathbf{E}, \mathbf{H}$ excited by sources external to the domain D. First of all, the functions $\mathbf{e}(\mathbf{r})$ and $\mathbf{h}(\mathbf{r})$ should be interrelated, just as is the case with a homogeneous domain D, by relations (8.6.2). These conditions, however, do not seem to be sufficient. To establish the necessary and sufficient boundary conditions, write the Stratton-Chu type integrals for an inhomogeneous medium (8.5.36) with densities equal to $\mathbf{e}(\mathbf{r})$ and $\mathbf{h}(\mathbf{r})$:

$$\begin{aligned} \mathbf{M}^e(\mathbf{r}'; \mathbf{e}, \mathbf{h}) &= \hat{F}^S(\hat{\mathbf{G}}^e)\mathbf{e} + \hat{\Psi}^S(\hat{\mathbf{G}}^e)\mathbf{h} \ , \\ \mathbf{M}^m(\mathbf{r}'; \mathbf{e}, \mathbf{h}) &= \hat{F}^S(\hat{\mathbf{G}}^m)\mathbf{e} + \hat{\Psi}^S(\hat{\mathbf{G}}^m)\mathbf{h} \ , \end{aligned} \tag{8.6.14}$$

where the integral operators $\hat{F}$ and $\hat{\Psi}$ are defined by formulas (8.5.34). Just as with the homogeneous domain D, it is easy to show, relying on formulas (8.5.50) that *the necessary and sufficient condition that the vector functions* $\mathbf{e}(\mathbf{r})$ *and* $\mathbf{h}(\mathbf{r})$, *continuously differentiable on a closed smooth surface S bounding the domain D, should serve boundary values on the inner side of S of the electromagnetic field* $\mathbf{E}, \mathbf{H}$ *which satisfies everywhere in D right to the boundary S the equations*

$$\operatorname{rot}\mathbf{H} = \sigma^*(\mathbf{r})\mathbf{E} \ , \quad \operatorname{rot}\mathbf{E} = i\omega\mu\mathbf{H} \ , \tag{8.6.15}$$

is the following identities:

$$\mathbf{M}^e(\mathbf{r}'; \mathbf{e}, \mathbf{h}) = 0 \ , \quad \mathbf{M}^m(\mathbf{r}'; \mathbf{e}, \mathbf{h}) \equiv 0 \quad \textit{for any} \quad \mathbf{r}' \in C\bar{D} \ . \tag{8.6.16}$$

By reasoning similar to that in Sect. 5.3.1 and using formulas (8.5.43a) through (8.5.43f), we can establish the conditions equivalent to (8.6.11):

$$\mathbf{e}(\mathbf{r}_0) = 2\mathbf{C}_0^E(\mathbf{r}_0;\ \mathbf{e},\ \mathbf{h}) + 2\Delta\mathbf{C}^E(\mathbf{r}_0;\ \mathbf{e},\ \mathbf{h}) \ ,$$

$$\mathbf{h}(\mathbf{r}_0) = 2\mathbf{C}_0^H(\mathbf{r}_0;\ \mathbf{e},\ \mathbf{h}) + 2\Delta\mathbf{C}^H(\mathbf{r}_0;\ \mathbf{e},\ \mathbf{h}) \ , \tag{8.6.17}$$

where the singular integrals $\mathbf{C}_0$ and $\Delta\mathbf{C}$ are fitted by formulas (8.5.43 e, f).

Relations (8.6.17) *in conjunction with* (8.6.2) *are the most convenient form of integral boundary conditions for the electromagnetic field on the boundary of an inhomogeneous domain.*

And finally, it is easy to show, relying on the findings in Sect. 8.5, that conditions (8.6.17) remain valid also for a domain D with a piecewise continuous distribution of complex electric conductivity.

8.6.3 Determination of the Electromagnetic Field from a Specified Discontinuity

Consider the following problem. Let the vector functions $\mathbf{e}(\mathbf{r})$ and $\mathbf{h}(\mathbf{r})$, which are Hölder continuous on S and have thereon continuously differentiable tangential components and which are interrelated by (8.6.2) (a two-dimensional consequence of the Maxwell equations), be specified on a closed smooth surface S. It is necessary to find the field $\mathbf{E}$, $\mathbf{H}$ satisfying Maxwell Eqs. (8.6.1) everywhere outside S, vanishing at infinity, and experiencing, on transition through the surface S, a discontinuity equal to

$$\mathbf{E}^+(\mathbf{r}_0) - \mathbf{E}^-(\mathbf{r}_0) = \mathbf{e}(\mathbf{r}_0) \ , \quad \mathbf{H}^+(\mathbf{r}_0) - \mathbf{H}^-(\mathbf{r}_0) = \mathbf{h}(\mathbf{r}_0) \ , \quad \mathbf{r}_0 \in S \ . \tag{8.6.18}$$

One can readily see that the solution to this problem, based on the electromagnetic counterparts of the Sokhotsky-Plemelj formulas (8.4.10), involves the following Stratton-Chu type integrals:

$$\mathbf{E}(\mathbf{r}') = \mathbf{C}^E(\mathbf{r}') = \iint_S \{(\mathbf{n}\cdot\mathbf{e})\,\mathrm{grad}\,G^h + [\mathbf{n}\times\mathbf{e}]\times\mathrm{grad}\,G^h + i\omega\mu[\mathbf{n}\times\mathbf{h}]\,G^h\}\,ds \ ,$$

$$\mathbf{H}(\mathbf{r}') = \mathbf{C}^H(\mathbf{r}') = \iint_S \{(\mathbf{n}\cdot\mathbf{h})\,\mathrm{grad}\,G^h + [\mathbf{n}\times\mathbf{h}]\times\mathrm{grad}\,G^h + \sigma^*[\mathbf{n}\times\mathbf{e}]\,G^h\}\,ds \ . \tag{8.6.19}$$

We should note that this solution is, just as in the case of complex analytical functions (Sect. 1.3.2) and of three-dimensional Laplace fields (Sect. 5.3.2) *unique.* The proof of this fact follows directly from the property of uniqueness of an electromagnetic field (analogous to the property of uniqueness of harmonic functions) to be considered in Sect. 9.2.

Along with representation (8.6.19), to evaluate the electromagnetic field from a specified discontinuity, we can also employ the equivalent expressions of this field in terms of modified Stratton-Chu type integrals (8.4.16) and (8.4.17).

$$\mathbf{E}(\mathbf{r}') = \mathbf{M}^E(\mathbf{r}') = \frac{1}{\sigma^*}\,\mathrm{rot}'\mathrm{rot}' \iint_S [\mathbf{n}\times\mathbf{h}]\,G^h\,ds + \mathrm{rot}' \iint_S [\mathbf{n}\times\mathbf{e}]\,G^h\,ds \ ,$$

$$\mathbf{H}(\mathbf{r}') = \mathbf{M}^H(\mathbf{r}') = \frac{1}{i\omega\mu}\,\mathrm{rot}'\mathrm{rot}' \iint_S [\mathbf{n}\times\mathbf{e}]\,G^h\,ds + \mathrm{rot}' \iint_S [\mathbf{n}\times\mathbf{h}]\,G^h\,ds \ . \tag{8.6.20}$$

Representations (8.6.20) are superior to those of (8.6.19) because the former allow explicitly only for a discontinuity of the field tangential components, since $\mathbf{n}\times\mathbf{h}=\mathbf{n}\times\mathbf{h}_\tau$, $\mathbf{n}\times\mathbf{e}=\mathbf{n}\times\mathbf{e}_\tau$.

What is more, in the case of representations (8.6.20) the surface may be open. In other words, these representations yield a solution to the following problem:

On an arbitrary closed or open smooth surface S there are specified continuously differentiable vector functions $\mathbf{e}_\tau(\mathbf{r})$ *and* $\mathbf{h}_\tau(\mathbf{r})$ *directed everywhere along the tangent line to S. It is necessary to find an electromagnetic field* **E, H** *vanishing at infinity and satisfying Eqs.* (8.6.1) *everywhere outside S, a field whose tangential components undergo a discontinuity on S equal to* $\mathbf{e}_\tau$ *and* $\mathbf{h}_\tau$, *respectively.*

We should like to stress that solution (8.6.20) for the stated problem of determining an electromagnetic field from a specified discontinuity is *unique.*

Finally, in concluding this subsection note that the above result is readily extendible to the case of a medium with an arbitrary piecewise continuous distribution of $\sigma^*(\mathbf{r})$. This can be done by means of the Stratton-Chu type integrals for inhomogeneous media [formula (8.5.36)].

8.6.4 Inversion Formulas for the Stratton-Chu Type Integrals

Consider, as usual, a domain D in space bounded by a smooth surface S. Let us specify arbitrary Hölder continuous vector functions $\mathbf{e}(\mathbf{r})$ and $\mathbf{h}(\mathbf{r})$ on S, which have continuously differentiable tangential components and which are interrelated by conditions (8.6.2). Let $\mathbf{u}(\mathbf{r}_0)$ and $\mathbf{v}(\mathbf{r}_0)$ denote singular integrals:

$$\mathbf{u}(\mathbf{r}_0)=2\mathbf{C}_0^E(\mathbf{r}_0;\,\mathbf{e},\,\mathbf{h})=2\iint_S\{(\mathbf{n}\cdot\mathbf{e})\operatorname{grad}G^h$$

$$+[\mathbf{n}\times\mathbf{e}]\times\operatorname{grad}G^h+i\omega\mu[\mathbf{n}\times\mathbf{h}]G^h\}ds\;, \tag{8.6.21 a}$$

$$\mathbf{v}(\mathbf{r}_0)=2\mathbf{C}_0^H(\mathbf{r}_0;\,\mathbf{e},\,\mathbf{h})=2\iint_S\{(\mathbf{n}\cdot\mathbf{h})\operatorname{grad}G^h$$

$$+[\mathbf{n}\times\mathbf{h}]\times\operatorname{grad}G^h+\sigma^*[\mathbf{n}\times\mathbf{e}]G^h\}ds\;,\quad \mathbf{r}_0\in S\;. \tag{8.6.21 b}$$

Now assume that the functions $\mathbf{u}(\mathbf{r}_0)$ and $\mathbf{v}(\mathbf{r}_0)$ are specified, while $\mathbf{e}(\mathbf{r})$ and $\mathbf{h}(\mathbf{r})$ are unknown. Then expressions (8.6.21 a) and (8.6.21 b) are *singular integral equations in functions* $\mathbf{e}(\mathbf{r})$ and $\mathbf{h}(\mathbf{r})$, similar to Eqs. (1.3.11) and (5.3.17) for the Cauchy-type integrals and their three-dimensional analogs. Let us solve these equations. To this end, we will use the method described earlier for inverting the Cauchy-type integrals (Sects. 1.3.3 and 5.3.3).

Note, first of all, that relying on the properties of the Stratton-Chu type integrals (Sect. 8.4.2), one is readily convinced that the functions $\mathbf{u}(\mathbf{r})$ and $\mathbf{v}(\mathbf{r})$, just as $\mathbf{e}$ and $\mathbf{h}$, are Hölder continuous on S and interrelated by conditions (8.6.2). In particular, once $\mathbf{e}$ and $\mathbf{h}$ are boundary values on S of some electromagnetic field **E, H** excited by sources external to D, we have, in accordance with theorem 8.6.2,

$$\mathbf{u}(\mathbf{r}_0)=\mathbf{r}(\mathbf{r}_0)\ ,\quad \mathbf{v}(\mathbf{r}_0)=\mathbf{h}(\mathbf{r}_0)\ ,\quad \mathbf{r}_0\in S\ . \tag{8.6.22}$$

In a general case, however, we have $\mathbf{u}\not\equiv\mathbf{e}$ and $\mathbf{v}\not\equiv\mathbf{h}$.

Let us determine $\mathbf{e}$ and $\mathbf{h}$. As a preliminary step, using the electromagnetic counterparts of the Sokhotsky-Plemelj formulas (8.4.11), we will cast formulas (8.6.21 a) and (8.6.21 b) in the form

$$\begin{aligned}&\mathbf{C}^{E+}(\mathbf{r}_0;\ \mathbf{e},\ \mathbf{h})+\mathbf{C}^{E-}(\mathbf{r}_0;\ \mathbf{e},\ \mathbf{h})=\mathbf{u}(\mathbf{r}_0)\ ,\\ &\mathbf{C}^{H+}(\mathbf{r}_0;\ \mathbf{e},\ \mathbf{h})+\mathbf{C}^{H-}(\mathbf{r}_0;\ \mathbf{e},\ \mathbf{h})=\mathbf{v}(\mathbf{r}_0)\ .\end{aligned} \tag{8.6.23}$$

Following the conclusion drawn in Sects. 1.3.3 and 5.3.3, let $\mathbf{E}(\mathbf{r})$ and $\mathbf{H}(\mathbf{r})$ denote the functions defined as follows:

$$\mathbf{E}(\mathbf{r})=\begin{cases}\mathbf{C}^E(\mathbf{r};\ \mathbf{e},\ \mathbf{h})\ , & \mathbf{r}\in D\ ,\\ -\mathbf{C}^E(\mathbf{r};\ \mathbf{e},\ \mathbf{h})\ , & \mathbf{r}\in C\bar{D}\ ;\end{cases} \tag{8.6.24a}$$

$$\mathbf{H}(\mathbf{r})=\begin{cases}\mathbf{C}^H(\mathbf{r};\ \mathbf{e},\ \mathbf{h})\ , & \mathbf{r}\in D\ ,\\ -\mathbf{C}^H(\mathbf{r};\ \mathbf{e},\ \mathbf{h})\ , & \mathbf{r}\in C\bar{D}\ .\end{cases} \tag{8.6.24b}$$

Evidently, the functions $\mathbf{E}(\mathbf{r})$ and $\mathbf{H}(\mathbf{r})$ everywhere outside S satisfy Maxwell Eqs. (8.6.1) and, according to (8.6.23), experience a discontinuity on transition through S, which is equal to

$$\mathbf{E}^+(\mathbf{r}_0)-\mathbf{E}^-(\mathbf{r}_0)=\mathbf{u}(\mathbf{r}_0)\ ,\quad \mathbf{H}^+(\mathbf{r}_0)-\mathbf{H}^-(\mathbf{r}_0)=\mathbf{v}(\mathbf{r}_0)\ . \tag{8.6.25}$$

According to the results of Sect. 8.6.3, the electromagnetic field $\mathbf{E}$, $\mathbf{H}$ is defined from a specified discontinuity using the following Stratton-Chu type integrals

$$\mathbf{E}(\mathbf{r}')=\mathbf{C}^E(\mathbf{r}';\ \mathbf{u},\ \mathbf{v})\ ,\quad \mathbf{H}(\mathbf{r}')=\mathbf{C}^H(\mathbf{r}';\ \mathbf{u},\ \mathbf{v})\ . \tag{8.6.26}$$

Applying the electromagnetic counterparts of the Sokhotsky-Plemelj formulas (8.4.11) to integrals (8.6.26), we write

$$\begin{aligned}&\mathbf{E}^+(\mathbf{r}_0)+\mathbf{E}^-(\mathbf{r}_0)=\mathbf{C}^{E+}(\mathbf{r}_0;\ \mathbf{u},\ \mathbf{v})+\mathbf{C}^{E-}(\mathbf{r}_0;\ \mathbf{u},\ \mathbf{v})=2\mathbf{C}_0^E(\mathbf{r}_0;\ \mathbf{u},\ \mathbf{v})\ ,\\ &\mathbf{H}^+(\mathbf{r}_0)+\mathbf{H}^-(\mathbf{r}_0)=\mathbf{C}^{H+}(\mathbf{r}_0;\ \mathbf{u},\ \mathbf{v})+\mathbf{C}^{H-}(\mathbf{r}_0;\ \mathbf{u},\ \mathbf{v})=2\mathbf{C}_0^H(\mathbf{r}_0;\ \mathbf{u},\ \mathbf{v})\ .\end{aligned} \tag{8.6.27}$$

On the other hand, by virtue of (8.6.24) and Sokhotsky-Plemelj formulas (8.4.10), we have

$$\begin{aligned}&\mathbf{E}^+(\mathbf{r}_0)+\mathbf{E}^-(\mathbf{r}_0)=\mathbf{C}^{E+}(\mathbf{r}_0;\ \mathbf{e},\ \mathbf{h})-\mathbf{C}^{E-}(\mathbf{r}_0;\ \mathbf{e},\ \mathbf{h})=\mathbf{e}(\mathbf{r}_0)\ ,\\ &\mathbf{H}^+(\mathbf{r}_0)+\mathbf{H}^-(\mathbf{r}_0)=\mathbf{C}^{H+}(\mathbf{r}_0;\ \mathbf{e},\ \mathbf{h})-\mathbf{C}^{H-}(\mathbf{r}_0;\ \mathbf{e},\ \mathbf{h})=\mathbf{h}(\mathbf{r}_0)\ .\end{aligned} \tag{8.6.28}$$

The left-hand sides of Eqs. (8.6.27) and (8.6.28) are equal, hence the right-hand sides are equal, too:

$$\begin{aligned}\mathbf{e}(\mathbf{r}_0)=2\mathbf{C}_0^E(\mathbf{r}_0;\ \mathbf{u},\ \mathbf{v})=2\iint_S\{&(\mathbf{n}\cdot\mathbf{u})\operatorname{grad}G^h\\ &+[\mathbf{n}\times\mathbf{u}]\times\operatorname{grad}G^h+i\omega\mu[\mathbf{n}\times\mathbf{v}]G^h\}ds\ ,\end{aligned} \tag{8.6.29a}$$

$$\mathbf{h}(\mathbf{r}_0) = 2\mathbf{C}_0^H(\mathbf{r}_0;\ \mathbf{u},\ \mathbf{v}) = 2\iint_S \{(\mathbf{n}\cdot\mathbf{v})\,\mathrm{grad}\,G^h$$

$$+[\mathbf{n}\times\mathbf{v}]\times\mathrm{grad}\,G^h + \sigma^*[\mathbf{n}\times\mathbf{u}]\,G^h\}ds\ . \tag{8.6.29b}$$

The resultant formulas (8.6.29a) *and* (8.6.29b) *yield a solution to singular integral Eqs.* (8.6.21a) *and* (8.6.21b).

It is evident that using similar reasoning we can go from formulas (8.6.29a) and (8.6.29b) over again to formulas (8.6.21a) and (8.6.21b). Thus, we arrive at a pair of mutually invertible integral transforms, which will be referred to as *Stratton-Chu integral transforms:*

$$\mathbf{u}(\mathbf{r}_0) = 2\iint_S [(\mathbf{n}\cdot\mathbf{e})\,\mathrm{grad}\,G^h + (\mathbf{n}\times\mathbf{e})\times\mathrm{grad}\,G^h + i\omega\mu(\mathbf{n}\times\mathbf{h})\,G^h]\,ds\ ,$$

$$\mathbf{v}(\mathbf{r}_0) = 2\iint_S [(\mathbf{n}\cdot\mathbf{h})\,\mathrm{grad}\,G^h + (\mathbf{n}\times\mathbf{h})\times\mathrm{grad}\,G^h + \sigma^*(\mathbf{n}\times\mathbf{e})\,G^h]\,ds\ ; \tag{8.6.30a}$$

$$\mathbf{e}(\mathbf{r}_0) = 2\iint_S [(\mathbf{n}\cdot\mathbf{u})\,\mathrm{grad}\,G^h + (\mathbf{n}\times\mathbf{u})\times\mathrm{grad}\,G^h + i\omega\mu(\mathbf{n}\times\mathbf{v})\,G^h]\,ds\ ,$$

$$\mathbf{h}(\mathbf{r}_0) = 2\iint_S [(\mathbf{n}\cdot\mathbf{v})\,\mathrm{grad}\,G^h + (\mathbf{n}\times\mathbf{v})\times\mathrm{grad}\,G^h + \sigma^*(\mathbf{n}\times\mathbf{u})\,G^h]\,ds\ . \tag{8.6.30b}$$

It is noteworthy that, by construction, the normal $\mathbf{n}$ in (8.6.30) points outside the domain D; nevertheless, just as with the three-dimensional Cauchy integral transforms, these formulas hold good even if the direction of the normal is reversed everywhere.

It is obvious that for $\omega = 0$ and $\sigma^* = 0$ (a stationary field in an insulator), formulas (8.6.30) change over into formulas (5.3.26), i.e., the Stratton-Chu integral transforms degenerate into the Cauchy integral transforms.

Along with the Stratton-Chu integral transforms (8.6.30), one can also construct a pair of mutually inversible transforms based on the Stratton-Chu type integrals for an inhomogeneous medium:

$$\mathbf{u}(\mathbf{r}_0) = 2\mathbf{C}_0^E(\mathbf{r}_0;\ \mathbf{e},\ \mathbf{h}) + 2\Delta\mathbf{C}^E(\mathbf{r}_0;\ \mathbf{e},\ \mathbf{h})\ ,$$

$$\mathbf{v}(\mathbf{r}_0) = 2\mathbf{C}_0^H(\mathbf{r}_0;\ \mathbf{e},\ \mathbf{h}) + 2\Delta\mathbf{C}^H(\mathbf{r}_0;\ \mathbf{e},\ \mathbf{h})\ ; \tag{8.6.31a}$$

$$\mathbf{e}(\mathbf{r}_0) = 2\mathbf{C}_0^E(\mathbf{r}_0;\ \mathbf{u},\ \mathbf{v}) + 2\Delta\mathbf{C}^E(\mathbf{r}_0;\ \mathbf{u},\ \mathbf{v})\ ,$$

$$\mathbf{h}(\mathbf{r}_0) = 2\mathbf{C}_0^H(\mathbf{r}_0;\ \mathbf{u},\ \mathbf{v}) + 2\Delta\mathbf{C}^H(\mathbf{r}_0;\ \mathbf{u},\ \mathbf{v})\ , \tag{8.6.31b}$$

where the singular integrals $\mathbf{C}_0^{E,H}$ and $\Delta\mathbf{C}^{E,H}$ are fitted by formulas (8.5.43e) and (8.5.43f).

8.6.5 Stratton-Chu Integral Transforms on a Plane

Rewrite formulas (8.6.30) for the case where S is a coordinate plane XY under the assumption that the functions $\mathbf{e}$, $\mathbf{h}$, and $\mathbf{u}$, $\mathbf{v}$ tend to zero at infinity. Then there exist principal values of the Stratton-Chu type integrals within infinite limits (in terms similar to those defined in Sect. 5.3.4 for the Cauchy-type integrals) and formula (8.6.30a) takes the form

$$\mathbf{u}(x_0, y_0, 0) = 2 \iint\limits_{-\infty}^{+\infty} \left[e_z \frac{\partial G^h}{\partial y} \mathbf{d}_x + e_z \frac{\partial G^h}{\partial y} \mathbf{d}_y - \left(e_x \frac{\partial G^h}{\partial x} + e_y \frac{\partial G^h}{\partial y} \right) \mathbf{d}_z \right.$$
$$\left. + i\omega\mu (h_x G^h \mathbf{d}_y - h_y G^h \mathbf{d}_x) \right] ds \; ,$$
$$\mathbf{v}(x_0, y_0, 0) = 2 \iint\limits_{-\infty}^{+\infty} \left[h_z \frac{\partial G^h}{\partial x} \mathbf{d}_x + h_z \frac{\partial G^h}{\partial y} \mathbf{d}_y - \left(h_x \frac{\partial G^h}{\partial x} + h_y \frac{\partial G^h}{\partial y} \right) \mathbf{d}_z \right.$$
$$\left. + \sigma^*(e_x G^h \mathbf{d}_y - e_y G^h \mathbf{d}_x) \right] ds \; . \tag{8.6.32}$$

Here it is taken into consideration that the normal $\mathbf{n}$ points upward.

Let the symbols $\hat{G}^h$, $\hat{G}^h_x$, and $\hat{G}^h_y$ stand for the following integral operators:

$$\hat{G}^h_x f(x, y) = \hat{G}^h_x f = 2 \iint\limits_{-\infty}^{+\infty} \frac{\partial G^h}{\partial x} f dx dy \; ,$$
$$\hat{G}^h_y f(x, y) = \hat{G}^h_y f = 2 \iint\limits_{-\infty}^{+\infty} \frac{\partial G^h}{\partial y} f dx dy \; , \tag{8.6.33}$$
$$\hat{G}^h f(x, y) = \hat{G}^h f = 2 \iint\limits_{-\infty}^{+\infty} G^h f dx dy \; .$$

Then relations (8.6.32) can be written as follows:

$$u_x = \hat{G}^h_x e_z - i\omega\mu \hat{G}^h h_y \; , \quad u_y = \hat{G}^h_y e_z + i\omega\mu \hat{G}^h h_x \; , \quad u_z = -\hat{G}^h_x e_x - \hat{G}^h_y e_y \; ; \tag{8.6.34a}$$

$$v_x = \hat{G}^h_x h_z - \sigma^* \hat{G}^h e_y \; , \quad v_y = \hat{G}^h_y h_z + \sigma^* \hat{G}^h e_x \; , \quad v_z = -\hat{G}^h_x h_x - \hat{G}^h_y h_y \; . \tag{8.6.34b}$$

Similarly, inversion formulas (8.6.30b) are written as

$$e_x = \hat{G}^h_x u_z - i\omega\mu \hat{G}^h v_y \; , \quad e_y = \hat{G}^h_y u_z + i\omega\mu \hat{G}^h v_x \; , \quad e_z = -\hat{G}^h_x u_x - \hat{G}^h_y u_y \; ; \tag{8.6.35a}$$

$$h_x = \hat{G}^h_x v_z - \sigma^* \hat{G}^h u_y \; , \quad h_y = \hat{G}^h_y v_z + \sigma^* \hat{G}^h u_x \; , \quad h_z = -\hat{G}^h_x v_x - \hat{G}^h_y v_y \; . \tag{8.6.35b}$$

Relations (8.6.34a) and (8.6.34b) are an extension of the three-dimensional Hilbert transforms (5.3.29a) and (5.3.29b) to the electromagnetic field case.

And finally, assume that $\mathbf{e}$ and $\mathbf{h}$ are boundary values in the plane $z = 0$ of the fields $\mathbf{E}$, $\mathbf{H}$ satisfying Maxwell Eqs. (8.6.1) everywhere in the lower half-space and tending to zero at infinity. Then, by virtue of theorem 8.6.2, the identities $\mathbf{e} \equiv \mathbf{u}$ and $\mathbf{h} \equiv \mathbf{v}$ hold good and relations (8.6.34a) and (8.6.34b) take the form

$$e_x = \hat{G}_x^h e_z - i\omega\mu\hat{G}^h h_y\ ,\quad e_y = \hat{G}_y^h e_z + i\omega\mu\hat{G}^h h_x\ ,\quad e_z = -\hat{G}_x^h e_x - \hat{G}_y^h e_y\ ; \tag{8.6.36a}$$

$$h_x = \hat{G}_x^h h_z - \sigma^*\hat{G}^h e_y\ ,\quad h_y = \hat{G}_y^h h_z + \sigma^*\hat{G}^h e_x\ ,\quad h_z = -\hat{G}_x^h h_x - \hat{G}_y^h h_y\ . \tag{8.6.36b}$$

In a similar way, if **e** and **h** are boundary values of the fields **E**, **H** satisfying Maxwell Eqs. (8.6.1) everywhere in the upper half-space and vanishing at infinity, then reversing the sign of the right-hand sides of (8.6.36a) and (8.6.36b) (since the direction of the normal **n** is reversed), we obtain

$$e_x = -\hat{G}_x^h e_z + i\omega\mu\hat{G}^h h_y\ ,\quad e_y = -\hat{G}_y^h e_z - i\omega\mu\hat{G}^h h_x\ ,\quad e_z = \hat{G}_x^h e_x + \hat{G}_y^h e_y\ ; \tag{8.6.37a}$$

$$h_x = -\hat{G}_x^h h_z + \sigma^*\hat{G}^h e_y\ ,\quad h_y = -\hat{G}_y^h h_z - \sigma^*\hat{G}^h e_x\ ,\quad h_z = \hat{G}_x^h e_x + \hat{G}_y^h e_y\ . \tag{8.6.37b}$$

Relations (8.6.36a) through (8.6.37b) yield simple linear relationships between various components of the electromagnetic field in a horizontal plane. For $\omega = 0$ and $\sigma^* = 0$, these relations go over into three-dimensional Hilbert transforms (5.3.31) and (5.3.32).

8.7 Techniques for Separation of the Earth's Electromagnetic Fields

8.7.1 Separation of the Electromagnetic Field into External and Internal Parts

As indicated in Sect. 3.1, the problem of separating the electromagnetic field of the Earth is one of the main problems in geomagnetism. The geophysical aspects of this problem, as applied to transient fields, are treated comprehensively in the study by Berdichevsky and Zhdanov (1984). Here we will confine ourselves to describing briefly how the problem is stated.

Consider a model within which an infinitely extended smooth surface S separates an inhomogeneous well-conducting Earth (Ω_-) from a homogeneous well-conducting atmosphere (Ω_+) (Fig. 50). The model is excited by extraneous currents distributed with a density $\mathbf{j}^e$ over a domain Q_e of the atmosphere. The time dependence of the field is expressed by the factor $\exp(-i\omega t)$ (a monochromatic field is considered). The complex electric conductivity of the at-

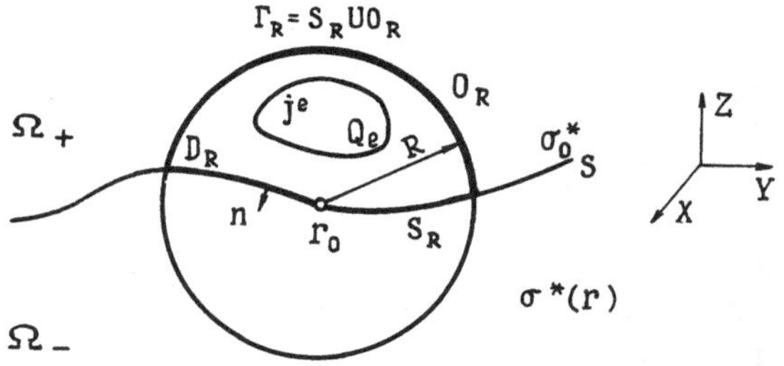

Fig. 50. Scheme explaining the method for separating the electromagnetic field into an external and an internal part: Q_e is the part of the atmosphere occupied with extrinsic currents $\mathbf{j}^e$

mosphere is denoted by σ_0^*, while that of the Earth by $\sigma^*(\mathbf{r})$. Magnetic permeability is everywhere constant and equal to μ_0. Within this model, the electromagnetic field meets the equations

$$\operatorname{rot}\mathbf{H} = \begin{cases} \sigma_0^*\mathbf{E}+\mathbf{j}^e\ , & \mathbf{r}\in\Omega_+\ ,\\ \sigma^*(\mathbf{r})\mathbf{E}\ , & \mathbf{r}\in\Omega_-\ ,\end{cases} \qquad \operatorname{rot}\mathbf{E} = i\omega\mu_0\mathbf{H}\ . \tag{8.7.1}$$

The tangential components of the electric and magnetic fields on the surface S are continuous.

Represent the electromagnetic field $\mathbf{E}$, $\mathbf{H}$ as a sum of the *external* $\mathbf{E}^e$, $\mathbf{H}^e$ and *internal* $\mathbf{E}^i$, $\mathbf{H}^i$ fields:

$$\mathbf{E} = \mathbf{E}^e+\mathbf{E}^i\ , \quad \mathbf{H} = \mathbf{H}^e+\mathbf{H}^i\ , \tag{8.7.2}$$

in which $\mathbf{E}^e$, $\mathbf{H}^e$ is the field produced by currents where the complex electric conductivity of the Earth is equal to the electric conductivity of the atmosphere:

$$\sigma^*(\mathbf{r})\equiv\sigma_0^*\ . \tag{8.7.3}$$

The external and internal fields satisfy the equations

$$\operatorname{rot}\mathbf{H}^e = \sigma_0^*\mathbf{E}^e+\mathbf{j}^e\ , \quad \operatorname{rot}\mathbf{E}^e = i\omega\mu_0\mathbf{H}^e\ ; \tag{8.7.4a}$$

$$\operatorname{rot}\mathbf{H}^i = \sigma_0^*\mathbf{E}+\mathbf{j}\ , \quad \operatorname{rot}\mathbf{E}^i = i\omega\mu_0\mathbf{H}^i\ , \tag{8.7.4b}$$

where $\mathbf{j} = [\sigma^*(\mathbf{r})-\sigma_0^*]\mathbf{E}$ are the excess currents induced in the Earth.

The problem is stated as follows: *evaluate the external and internal fields* $\mathbf{E}^e$, $\mathbf{H}^e$ *and* $\mathbf{E}^i$, $\mathbf{H}^i$ *at the Earth's surface S from the total field* $\mathbf{E}$, $\mathbf{H}$ *specified also at S.*

If the complex electric conductivity of the air is zero, $\sigma_0^* = 0$, then in view of relations (8.7.1) and (8.7.4a), (8.74b), the magnetic field in the atmosphere (outside the domain Q_e) becomes Laplacian and the fields can be separated by the Kertz-Siebert method outlined in Sect. 6.4. At the same time, employing the Stratton-Chu integrals we can elaborate a more general technique applicable to three-dimensional models with a well-conducting atmosphere. This technique will be termed a *generalized Kertz-Siebert method.*

Take a point at the Earth's surface S and circumscribe therefrom a sphere of a radius R so large as to completely enclose the source domain Q_e. Let Q_R be part of the sphere lying in Ω_+ and S_R be part of the Earth's surface found inside the sphere. Consider the Stratton-Chu type integrals over a piecewise smooth surface $\Gamma_R = S_R \cup O_R$:

$$\mathbf{C}^H(\mathbf{r}') = \iint_{\Gamma_R} [(\mathbf{n}\cdot\mathbf{H})\operatorname{grad}G_0^h+(\mathbf{n}\times\mathbf{H})\times\operatorname{grad}G_0^h+\sigma_0^*(\mathbf{n}\times\mathbf{E})G_0^h]\,ds\ ,$$

where

$$G_0^h = G^h(\mathbf{r}'|\mathbf{r}) = -\frac{1}{4\pi}\frac{\exp(ik_0^*\,|\mathbf{r}'-\mathbf{r}|)}{|\mathbf{r}'-\mathbf{r}|}\ ; \quad k_0^* = \sqrt{i\omega\mu_0\sigma_0^*}\ .$$

Represent $\mathbf{C}^H$ as a sum $\mathbf{C}^H = \mathbf{C}^{H^e}+\mathbf{C}^{H^i}$, where the integrals $\mathbf{C}^{H^e}$ and $\mathbf{C}^{H^i}$ are evaluated from the values of the external $\mathbf{H}^e$, $\mathbf{E}^e$ and internal $\mathbf{H}^i$, $\mathbf{E}^i$ electromagnetic fields:

$$\mathbf{C}^{H^e}(\mathbf{r}') = \iint_{\Gamma_R} [(\mathbf{n}\cdot\mathbf{H}^e)\operatorname{grad} G_0^h + (\mathbf{n}\times\mathbf{H}^e)\times\operatorname{grad} G_0^h + \sigma_0^*(\mathbf{n}\times\mathbf{E}^e)G_0^h]\,ds\ ,$$

$$\mathbf{C}^{H^i}(\mathbf{r}') = \iint_{\Gamma_R} [(\mathbf{n}\cdot\mathbf{H}^i)\operatorname{grad} G_0^h + (\mathbf{n}\times\mathbf{H}^i)\times\operatorname{grad} G_0^h + \sigma_0^*(\mathbf{n}\times\mathbf{E}^i)G_0^h]\,ds\ .$$

The external electromagnetic field outside Q_e is fitted by the equations

$$\operatorname{rot}\mathbf{H}^e = \sigma_0^*\mathbf{E}^e|_{\mathbf{r}\notin Q_e}\ ,\quad \operatorname{rot}\mathbf{E}^e = i\omega\mu_0\mathbf{H}^e|_{\mathbf{r}\notin Q_e}$$

and vanishes at infinity. Hence, according to (8.4.7), we have

$$\mathbf{C}^{H^e}(\mathbf{r}') = \begin{cases} \mathbf{0}\ , & \mathbf{r}'\in D_R\ ,\\ -\mathbf{H}^e(\mathbf{r}')\ , & \mathbf{r}'\in C\bar{D}_R\ ,\end{cases}$$

where D_R and $C\bar{D}_R$ are the domains which are internal and external with respect to Γ_R.

The internal electromagnetic field in $D_R \subset \Omega_+$ meets the equations

$$\operatorname{rot}\mathbf{H}^i = \sigma_0^*\mathbf{E}^i|_{\mathbf{r}\in D_R\subset\Omega_+}\ ,\quad \operatorname{rot}\mathbf{E}^i = i\omega\mu_0\mathbf{H}^i|_{\mathbf{r}\in D_R\subset\Omega_+}\ .$$

Hence, according to (8.4.6), we have

$$\mathbf{C}^{H^i}(\mathbf{r}') = \begin{cases} \mathbf{H}^i(\mathbf{r}')\ , & \mathbf{r}'\in D_R\ ,\\ \mathbf{0}\ , & \mathbf{r}'\in C\bar{D}_R\ .\end{cases}$$

Summing up $\mathbf{C}^{H^e}$ and $\mathbf{C}^{H^i}$, we find

$$\mathbf{C}^{H}(\mathbf{r}') = \begin{cases} \mathbf{H}^i(\mathbf{r}')\ , & \mathbf{r}'\in D_R\ ,\\ -\mathbf{H}^e(\mathbf{r}')\ , & \mathbf{r}'\in C\bar{D}_R\ .\end{cases}$$

Hence, one can readily obtain limit values of $\mathbf{C}^H$ at the surface S_R. Thanks to the continuity of the magnetic field, we have

$$\begin{aligned} \mathbf{C}_i^H(\mathbf{r}_0) &= \lim_{\mathbf{r}'\to\mathbf{r}_0^+} \mathbf{C}^H(\mathbf{r}') = \mathbf{H}^i(\mathbf{r}_0)\ ,\quad \mathbf{r}_0\in S_R\ ,\\ \mathbf{C}_e^H(\mathbf{r}_0) &= \lim_{\mathbf{r}'\to\mathbf{r}_0^-} \mathbf{C}^H(\mathbf{r}') = -\mathbf{H}^e(\mathbf{r}_0)\ ,\quad \mathbf{r}_0\in S_R\ . \end{aligned} \tag{8.7.5}$$

The limits values of $\mathbf{C}^H$ can be also defined by formulas (8.3.8), which are analogs of the Sokhotsky-Plemelj formulas:

$$\mathbf{C}_i^H(\mathbf{r}_0) = \mathbf{C}_0^H(\mathbf{r}_0) + \tfrac{1}{2}\mathbf{H}(\mathbf{r}_0)\ ,\quad \mathbf{C}_e^H(\mathbf{r}_0) = \mathbf{C}_0^H(\mathbf{r}_0) - \tfrac{1}{2}\mathbf{H}(\mathbf{r}_0)\ , \tag{8.7.6}$$

where the singular integral

$$\mathbf{C}_0^H(\mathbf{r}') = \iint_{\Gamma_R} [(\mathbf{n}\cdot\mathbf{H})\operatorname{grad} G_0^h + (\mathbf{n}\times\mathbf{H})\times\operatorname{grad} G_0^h + \sigma_0^*(\mathbf{n}\times\mathbf{E})G_0^h]\,ds\ ,$$

$$G_0^h = G^h(\mathbf{r}_0|\mathbf{r}) = -\frac{1}{4\pi}\,\frac{\exp(ik_0^*\,|\mathbf{r}_0-\mathbf{r}|)}{|\mathbf{r}_0-\mathbf{r}|}$$

is evaluated in terms of the Cauchy principal value. Equating (8.7.5) and (8.7.6) we obtain

$$\mathbf{H}^i(\mathbf{r}_0) = \tfrac{1}{2}\mathbf{H}(\mathbf{r}_0) + \mathbf{C}_0^H(\mathbf{r}_0)\ , \quad \mathbf{H}^e(\mathbf{r}_0) = \tfrac{1}{2}\mathbf{H}(\mathbf{r}_0) - \mathbf{C}_0^H(\mathbf{r}_0)\ . \tag{8.7.7}$$

Now proceed to the limit for $R \to \infty$. Since the field vanishes at infinity, we find

$$\lim_{R\to\infty} \iint_{O_R} [(\mathbf{n}\cdot\mathbf{H})\operatorname{grad} G_0^h + (\mathbf{n}\times\mathbf{H})\times\operatorname{grad} G_0^h + \sigma_0^*(\mathbf{n}\times\mathbf{E})G_0^h]\,ds = 0\ ,$$

whence

$$\lim_{R\to\infty} \mathbf{C}_0^H(\mathbf{r}_0) = \iint_S [(\mathbf{n}\cdot\mathbf{H})\operatorname{grad} G_0^h + (\mathbf{n}\times\mathbf{H})\times\operatorname{grad} G_0^h + \sigma_0^*(\mathbf{n}\times\mathbf{E})G_0^h]\,ds\ .$$

Hence, in accordance with (8.7.7), we have

$$\mathbf{H}^i(\mathbf{r}_0) = \tfrac{1}{2}\mathbf{H}(\mathbf{r}_0) + \iint_S [(\mathbf{n}\cdot\mathbf{H})\operatorname{grad} G_0^h + (\mathbf{n}\times\mathbf{H})\times\operatorname{grad} G_0^h + \sigma_0^*(\mathbf{n}\times\mathbf{E})G_0^h]\,ds\ ,$$

$$\mathbf{H}^e(\mathbf{r}_0) = \tfrac{1}{2}\mathbf{H}(\mathbf{r}_0) - \iint_S [(\mathbf{n}\cdot\mathbf{H})\operatorname{grad} G_0^h + (\mathbf{n}\times\mathbf{H})\times\operatorname{grad} G_0^h + \sigma_0^*(\mathbf{n}\times\mathbf{E})G_0^h]\,ds\ , \tag{8.7.8}$$

where the unit vector of the normal $\mathbf{n}$ points from the air into the Earth.

We have arrived at generalized Kertz-Siebert formulas which are an analog of formulas (6.4.12). They permit separation of the magnetic field into external and internal part in a wide class of three-dimensional models with a conducting atmosphere and an arbitrary relief of the Earth's surface. Similar formulas can be also derived for separating the electric field.

8.7.2 Separation of the Electromagnetic Field into Normal and Anomalous Parts

The problem of separating the electromagnetic field into normal and anomalous parts is a natural generalization of the above problem of determining the external and internal fields (Berdichevsky and Zhdanov 1984).

The problem is stated by using a model similar to that of Fig. 50, with the exception that the air-earth interface, i.e., the surface P is a horizontal plane $z = 0$ (Fig. 51). The Earth's complex electric conductivity everywhere outside a certain domain $Q_i \subset \Omega_-$ is a function of the vertical coordinate alone, $\sigma k^*(\mathbf{r}) = \sigma_n^*(z)$.

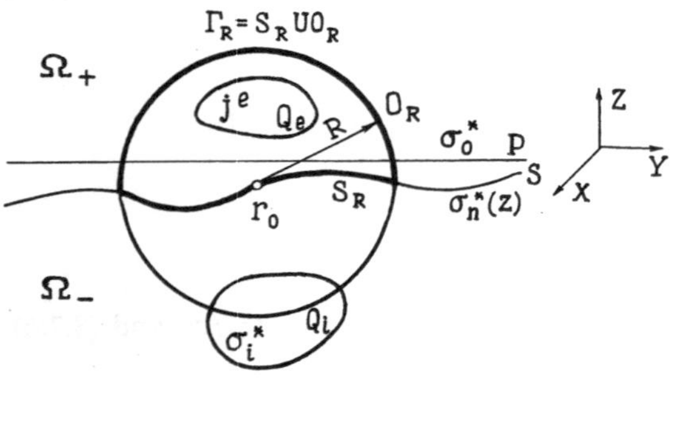

Fig. 51. Scheme explaining the method for separating the electromagnetic field into a normal and an anomalous part: Q_e is the part of the atmosphere occupied with extrinsic currents, Q_i is a deep inhomogeneous domain

Inside the inhomogeneous domain Q_i the electric conductivity is $\sigma_i^* = \sigma_n^*(z) + \sigma_a^*(\mathbf{r})$, where $\sigma_a^*(\mathbf{r})$ is an arbitrary function of coordinates. The model is excited, as previously, by extraneous currents flowing with a density $\mathbf{j}^e$ in the domain Q_e of the atmosphere.

Within the framework of this model, the electromagnetic field can be represented as a sum of the normal $\mathbf{E}^n$, $\mathbf{H}^n$ and anomalous $\mathbf{E}^a$, $\mathbf{H}^a$ fields:

$$\mathbf{E} = \mathbf{E}^n + \mathbf{E}^a \ , \quad \mathbf{H} = \mathbf{H}^n + \mathbf{H}^a \ , \tag{8.7.9}$$

where $\mathbf{E}^n$, $\mathbf{H}^n$ is the field produced by the currents $\mathbf{j}^e$ in the absence of an inhomogeneous domain Q_i, $\sigma_a^* \equiv 0$, while $\mathbf{E}^a$, $\mathbf{H}^a$ is the part of the total field that arises due to the inhomogeneity Q_i.

The normal and anomalous fields are fitted by the equations

$$\operatorname{rot} \mathbf{H}^n = \sigma_n^* \mathbf{E}^n + \mathbf{j}^e \ , \quad \operatorname{rot} \mathbf{E}^n = i\omega\mu_0 \mathbf{H}^n \ ; \tag{8.7.10a}$$

$$\operatorname{rot} \mathbf{H}^a = \sigma_n^* \mathbf{E}^a + \mathbf{j}^i \ , \quad \operatorname{rot} \mathbf{E}^a = i\omega\mu_0 \mathbf{H}^a \ , \tag{8.7.10b}$$

where $\mathbf{j}^i = \sigma_a^* \mathbf{E}$ stands for the excess currents flowing in a deep inhomogeneity Q_i.

Let the electromagnetic field $\mathbf{E}$, $\mathbf{H}$ be known at some surface S extending infinitely quasi-horizontally and either having an arbitrary relief or coinciding with the earth-air interface P (the former is the case with measurement of the electromagnetic field, say, deep down in the ocean). It is necessary to separate the field into normal and anomalous parts.

This problem can be solved using the body of Stratton-Chu type integrals in an inhomogeneous medium, with the scheme of its application identical to that of field separation into external and internal parts.

Choose a point $\mathbf{r}_0$ at the surface S and draw therefrom a sphere of a radius R so that the source domain Q_e might be inside the sphere. Let O_R be part of the sphere lying above S, while S_R be part of the surface S inside the sphere. Consider the Stratton-Chu type integrals over a closed piecewise surface $\Gamma_R = S_R \cup O_R$:

$$\mathbf{M}^{e,m}(\mathbf{r}) = \hat{F}^{\Gamma_R}(\hat{\mathbf{G}}_n^{e,m})\mathbf{H} + \hat{\Psi}^{\Gamma_R}(\hat{\mathbf{G}}_n^{e,m})\mathbf{E} \ ,$$

where $\hat{\mathbf{G}}_n^{e,m}$ stands for the electric and for the magnetic Green tensor functions of a normal horizontally layered cross section σ_n*. Represent $\mathbf{M}^{e,m}(\mathbf{r})$ as a sum: $\mathbf{M}^{e,m}(\mathbf{r}) = \mathbf{M}_n^{e,m}(\mathbf{r}) + \mathbf{M}_a^{e,m}(\mathbf{r})$ where the integrals $\mathbf{M}_n^{e,m}(\mathbf{r})$ and $\mathbf{M}_a^{e,m}(\mathbf{r})$ are determined from the values of the normal $\mathbf{E}^n$, $\mathbf{H}^n$ and anomalous $\mathbf{E}^a$, $\mathbf{H}^a$ electromagnetic fields:

$$\mathbf{M}_n^{e,m}(\mathbf{r}) = \hat{F}^{\Gamma_R}(\hat{\mathbf{G}}_n^{e,m})\mathbf{H}^n + \hat{\Psi}^{\Gamma_R}(\hat{\mathbf{G}}_n^{e,m})\mathbf{E}^n \ ,$$

$$\mathbf{M}_a^{e,m}(\mathbf{r}) = \hat{F}^{\Gamma_R}(\hat{\mathbf{G}}_n^{e,m})\mathbf{H}^a + \hat{\Psi}^{\Gamma_R}(\hat{\mathbf{G}}_n^{e,m})\mathbf{E}^a \ .$$

The normal electromagnetic field outside the domain D_R meets the conditions

$$\operatorname{rot} \mathbf{H}^n = \sigma_n^* \mathbf{E}^n \ , \quad \operatorname{rot} \mathbf{E}^n = i\omega\mu_0 \mathbf{H}^n \ , \quad \mathbf{r} \in C\bar{D}_R \ ,$$

and vanishes at infinity. Therefore, according to (8.5.41), we have

$$\mathbf{M}_n^e(\mathbf{r}) = \begin{cases} \mathbf{0}, & \mathbf{r} \in D_R, \\ -\mathbf{E}^n(\mathbf{r}), & \mathbf{r} \in C\bar{D}_R, \end{cases} \tag{8.7.11 a}$$

$$\mathbf{M}_n^m(\mathbf{r}) = \begin{cases} \mathbf{0}, & \mathbf{r} \in D_R, \\ -\mathbf{H}^n(\mathbf{r}), & \mathbf{r} \in C\bar{D}_R. \end{cases} \tag{8.7.11 b}$$

The anomalous electromagnetic field in $D_R \subset \Omega_+$ fits the equations

$$\operatorname{rot} \mathbf{H}^a = \sigma_n^* \mathbf{E}^a , \quad \operatorname{rot} \mathbf{E}^a = i\omega\mu \mathbf{H}^a , \quad \mathbf{r} \in D_R \subset \Omega_+ .$$

Hence, according to (8.5.39), we have

$$\mathbf{M}_a^e(\mathbf{r}) = \begin{cases} \mathbf{E}^a(\mathbf{r}), & \mathbf{r} \in D_R, \\ 0, & \mathbf{r} \in C\bar{D}_R; \end{cases} \tag{8.7.12 a}$$

$$\mathbf{M}_a^m(\mathbf{r}) = \begin{cases} \mathbf{H}^a(\mathbf{r}), & \mathbf{r} \in D_R, \\ 0, & \mathbf{r} \in C\bar{D}_R. \end{cases} \tag{8.7.12 b}$$

Sum up (8.7.11 a) and (8.7.12 a), as well as (8.7.11 b) and (8.7.12 b):

$$\mathbf{M}^e(\mathbf{r}) = \begin{cases} \mathbf{E}^a(\mathbf{r}), & \mathbf{r} \in D_R, \\ -\mathbf{E}^n(\mathbf{r}), & \mathbf{r} \in C\bar{D}_R, \end{cases} \tag{8.7.13 a}$$

$$\mathbf{M}^m(\mathbf{r}) = \begin{cases} \mathbf{H}^a(\mathbf{r}), & \mathbf{r} \in D_R, \\ -\mathbf{H}^n(\mathbf{r}), & \mathbf{r} \in C\bar{D}_R. \end{cases} \tag{8.7.13 b}$$

The electric and the magnetic fields can be considered, without constraints on the generality, to be continuous on the two sides of the surface (the points at the surface S where this condition is not fulfilled are eliminated from consideration). As a consequence, there are limits

$$\mathbf{M}_a^e(\mathbf{r}_0) = \lim_{\mathbf{r} \to \mathbf{r}_0^+} \mathbf{M}^e(\mathbf{r}) = \mathbf{E}^a(\mathbf{r}_0) , \tag{8.7.14 a}$$

$$\mathbf{M}_n^e(\mathbf{r}_0) = \lim_{\mathbf{r} \to \mathbf{r}_0^-} \mathbf{M}^e(\mathbf{r}) = -\mathbf{E}^n(\mathbf{r}_0) ; \tag{8.7.14 b}$$

$$\mathbf{M}_a^m(\mathbf{r}_0) = \lim_{\mathbf{r} \to \mathbf{r}_0^+} \mathbf{M}^m(\mathbf{r}) = \mathbf{H}^a(\mathbf{r}_0) , \tag{8.7.15 a}$$

$$\mathbf{M}_n^m(\mathbf{r}_0) = \lim_{\mathbf{r} \to \mathbf{r}_0^-} \mathbf{M}^m(\mathbf{r}) = -\mathbf{H}^n(\mathbf{r}_0) . \tag{8.7.15 b}$$

On the other hand, the limit values of $\mathbf{M}^{e,m}$ can be determined by formulas (8.5.43 a) through (8.5.43 f):

$$\mathbf{M}_a^e(\mathbf{r}_0) = \mathbf{C}_0^E(\mathbf{r}_0) + \Delta \mathbf{C}^E(\mathbf{r}_0) + \tfrac{1}{2}\mathbf{E}(\mathbf{r}_0) , \tag{8.7.16 a}$$

$$\mathbf{M}_n^e(\mathbf{r}_0) = \mathbf{C}_0^E(\mathbf{r}_0) + \Delta \mathbf{C}^E(\mathbf{r}_0) - \tfrac{1}{2}\mathbf{E}(\mathbf{r}_0) ; \tag{8.7.16 b}$$

$$\mathbf{M}_a^m(\mathbf{r}_0) = \mathbf{C}_0^H(\mathbf{r}_0) + \Delta \mathbf{C}^H(\mathbf{r}_0) + \tfrac{1}{2}\mathbf{H}(\mathbf{r}_0) , \tag{8.7.17 a}$$

$$\mathbf{M}_n^m(\mathbf{r}_0) = \mathbf{C}_0^H(\mathbf{r}_0) + \Delta \mathbf{C}^H(\mathbf{r}_0) - \tfrac{1}{2}\mathbf{H}(\mathbf{r}_0) , \tag{8.7.17 b}$$

where the singular integrals $\mathbf{C}_0^E$ and $\mathbf{C}_0^H$ are evaluated in terms of the Cauchy principal value.

Equating (8.7.14a) and (8.7.16a), (8.7.14b) and (8.7.16b), (8.7.15a) and (8.7.17a), (8.7.15b) and (8.7.17b), we obtain

$$\begin{aligned}
\mathbf{E}^a(\mathbf{r}_0) &= \tfrac{1}{2}\mathbf{E}(\mathbf{r}_0)+\mathbf{C}_0^E(\mathbf{r}_0)+\Delta\,\mathbf{C}^E(\mathbf{r}_0)\ ,\\
\mathbf{E}^n(\mathbf{r}_0) &= \tfrac{1}{2}\mathbf{E}(\mathbf{r}_0)-\mathbf{C}_0^E(\mathbf{r}_0)-\Delta\,\mathbf{C}^E(\mathbf{r}_0)\ ,\\
\mathbf{H}^a(\mathbf{r}_0) &= \tfrac{1}{2}\mathbf{H}(\mathbf{r}_0)+\mathbf{C}_0^H(\mathbf{r}_0)+\Delta\,\mathbf{C}^H(\mathbf{r}_0)\ ,\\
\mathbf{H}^n(\mathbf{r}_0) &= \tfrac{1}{2}\mathbf{H}(\mathbf{r}_0)-\mathbf{C}_0^H(\mathbf{r}_0)-\Delta\,\mathbf{C}^H(\mathbf{r}_0)\ .
\end{aligned}\tag{8.7.18}$$

Now proceed to the limit for $R\to\infty$. By virtue of the radiation condition at infinity, the integrals $\mathbf{C}_0^{E,H}$ and $\Delta\,\mathbf{C}^{E,H}$ taken over the half-sphere O_R tend to zero and it remains only to integrate over the surface S. Thus, formulas (8.7.18) yield a solution to the stated problem.

It is remarkable that for a normal homogeneous cross section ($\sigma_n^*(z) = \sigma^* = \text{const}$), the integrals $\Delta\,\mathbf{C}^H(\mathbf{r}_0)\equiv\Delta\,\mathbf{C}^E(\mathbf{r}_0)\equiv\mathbf{0}$ and formulas (8.7.18) go over into the formulas employed to separate the field into external and internal parts (8.7.8).

9 Analytical Continuation of the Electromagnetic Field

9.1 General Principles

The problem of *analytical continuation* lies in extending the electromagnetic field beyond the domain or surface of its initial determination. The general principles underlying analytical continuation of functions of a complex variable are outlined in Chap. 4, while those for an analytical vector field are treated in Sect. 7.1. We will apply these principles to the electromagnetic field. For the sake of simplicity, we will confine ourselves to a monochromatic field, although the results presented below also hold generally good for transient fields.

Coefficients of equations fitted by the electromagnetic field are governed by ε-, μ-, and σ-values, i.e., by the properties of a medium. This determines the specificity of the problem. The most important issue of theory is proof of the electromagnetic field analyticity and uniqueness of its analytical continuation.

9.1.1 Analytical Nature of the Electromagnetic Field

Here we will show that a time-harmonic electromagnetic field in domains of constant electromagnetic parameters of a medium is an analytical function.

Let the domain D bounded by a surface S be filled with a homogeneous medium of electric conductivity $\sigma^* = \text{const}$. The electromagnetic field is excited by sources located outside the domain D. The vector functions $\mathbf{E}$ and $\mathbf{H}$ in D and at S satisfy the Helmholtz equations

$$\Delta \mathbf{E} + k^{*2}\mathbf{E} = 0 \ , \quad \Delta \mathbf{H} + k^{*2}\mathbf{H} \equiv 0 \ ,$$

where

$$k^{*2} = i\omega\mu\sigma + \omega^2\mu\varepsilon \ .$$

Take any inner point in D and, following Sect. 7.1, circumscribe therefrom a sphere S_a of a radius a lying inside D (see Fig. 37). Employ the Stratton-Chu type integral formulas and express the electromagnetic field at some point $\mathbf{r}'$ inside S_a in terms of its value at S_a. In accordance with (8.3.36a) and (8.3.36b), we have

$$\begin{aligned} \mathbf{E}(\mathbf{r}') &= \iint_{S_a} [(\mathbf{n}\cdot\mathbf{E})\,\mathrm{grad}\,G^h + (\mathbf{n}\times\mathbf{E})\times\mathrm{grad}\,G^h + i\omega\mu(\mathbf{n}\times\mathbf{H})\,G^h]\,ds \ , \\ \mathbf{H}(\mathbf{r}') &= \iint_{S_a} [(\mathbf{n}\cdot\mathbf{H})\,\mathrm{grad}\,G^h + (\mathbf{n}\times\mathbf{H})\times\mathrm{grad}\,G^h + \sigma^*(\mathbf{n}\times\mathbf{E})\,G^h]\,ds \ , \end{aligned} \qquad (9.1.1)$$

where

$$G^h = G^h(\mathbf{r}'|\mathbf{r}) = \frac{-1}{4\pi} \frac{\exp(ik^* \, |\mathbf{r}'-\mathbf{r}|)}{|\mathbf{r}'-\mathbf{r}|} \, .$$

Now transform the Green function G^h. In accordance with (7.1.4), we have

$$|\mathbf{r}-\mathbf{r}'| = a\sqrt{1+t} \, , \tag{9.1.2}$$

where t is defined by formula (7.1.5). Within this representation,

$$G^h = \frac{-1}{4\pi a} \frac{\exp(ik^* a\sqrt{1+t})}{\sqrt{1+t}} \, , \tag{9.1.3}$$

where t can take the values from -1 to 3. It follows from (9.1.3) that for $|t|<1$ the Green function is expanded into a convergent power series

$$G^h = \sum_{n=0}^{\infty} \frac{1}{n!} G^{h(n)} t^n \, , \tag{9.1.4}$$

which is transformed, with due account taken of (7.1.5), to

$$G^h = \sum_{k=0}^{\infty} \sum_{l=0}^{\infty} \sum_{m=0}^{\infty} g_{klm}(x'-x_0)^n (y'-y_0)^l (z'-z_0)^m \, , \tag{9.1.5}$$

where the coefficients $g_{kem} = g_{kem}(\mathbf{r}'|\mathbf{r})$ are the functions of relative position of the points $\mathbf{r}'$ and $\mathbf{r} \in S$.

Substituting power series (9.1.5) into integrals (9.1.1) and allowing for the fact that this series admits termwise differentiation and integration, we obtain

$$\mathbf{E}(\mathbf{r}') = \sum_{k=0}^{\infty} \sum_{l=0}^{\infty} \sum_{m=0}^{\infty} \mathbf{e}_{klm}(x'-x_0)^k (y'-y_0)^l (z'-z_0)^m \, , \tag{9.1.6}$$

$$\mathbf{H}(\mathbf{r}') = \sum_{k=0}^{\infty} \sum_{l=0}^{\infty} \sum_{m=0}^{\infty} \mathbf{h}_{klm}(x'-x_0)^k (y'-y_0)^l (z'-z_0)^m \, . \tag{9.1.7}$$

Here

$$\mathbf{e}_{klm} = \iint_{S_a} [(\mathbf{n}\cdot\mathbf{E})\,\mathrm{grad}\, g_{klm} + (\mathbf{n}\times\mathbf{E})\times\mathrm{grad}\, g_{klm} + i\omega\mu\,[\mathbf{n}\times\mathbf{H})\, g_{klm}]\, ds \, , \tag{9.1.8a}$$

$$\mathbf{h}_{klm} = \iint_{S_a} [(\mathbf{n}\cdot\mathbf{H})\,\mathrm{grad}\, g_{klm} + (\mathbf{n}\times\mathbf{H})\times\mathrm{grad}\, g_{klm} + \sigma^*[\mathbf{n}\times\mathbf{E})\, g_{klm}]\, ds \, . \tag{9.1.8b}$$

Power series (9.1.6) and (9.1.7) converge to **E**, **H**, provided that $|t|<1$. According to (7.1.5) (7.1.9), this condition is fulfilled if $|\mathbf{r}_0-\mathbf{r}'|<\delta$ where $\delta = (\sqrt{2}-1)a$. Thus, we have indicated the neighborhood $\delta(\mathbf{r}_0)$ in which the electromagnetic field components are expanded into convergent power series. The point $\mathbf{r}_0$ has been chosen in D arbitrarily. As a result, everywhere in the domain D the electric **E** and the magnetic **H** fields are analytical functions.

It is noteworthy that the functions **E** and **H** have the same radii of convergence of the Taylor series. They are related by the Maxwell equations, hence the analyticity of one field results in the analyticity of the other.

9.1.2 Concept of Analytical Continuation of the Electromagnetic Field

As we have proved that the vectors of the electromagnetic field **E**, **H** are analytical functions, all the concepts of analytical continuation of vector fields introduced in Sect. 7.1.3 become applicable to the electromagnetic field as well.

Let us, first of all, formulate the problem of electromagnetic field continuation. Let the electromagnetic field $\mathbf{E}^0$, $\mathbf{H}^0$ be specified in a domain D_0 filled with a homogeneous medium of a complex electric conductivity σ_0^* and permeability μ_0. *Analytical continuation of the field* $\mathbf{E}^0$, $\mathbf{H}^0$ *into a larger domain D containing a medium with any variations in electric conductivity and permeability implies its extension to the field* **E**, **H**, *satisfying everywhere in D the equations* $\operatorname{rot}\mathbf{H} = \sigma_0^* \mathbf{E}$, $\operatorname{rot}\mathbf{E} = i\omega\mu_0\mathbf{H}$ *and coinciding with* $\mathbf{E}^0$, $\mathbf{H}^0$ *in* D_0.

By virtue of the uniqueness property of analytical vector fields proved in Sect. 7.1.2, it is evident that once analytical continuation of the electromagnetic field does exist, it is surely unique.

Obviously, we can apply all the concepts of vector field analytical continuation, introduced in Sect. 7.1.3, to the electromagnetic field. For instance, the term *analytical element of the electromagnetic field* is ascribed to the field **E**, **H** together with the homogeneous domain of its definition D (having constant electromagnetic parameters ε, μ, and σ), and it is denoted by (**E**, **H**, D). By analogy to the arbitrary vector field, we also deal with *chains, general and complete analytical functions of the electromagnetic field* (see Sect. 7.1.3). This question will be closely examined, using examples, in Sect. 9.1.5.

A classical method employed to construct complete analytical functions is the one suggested by Weierstrass, involving the Taylor series.

Consider a domain D_0 bounded by the surface S. It is filled with a homogeneous medium of electric conductivity σ_0^* and permeability μ_0 and is free of extraneous exciters of the field. The electric conductivity and the permeability of a medium outside D_0 may vary arbitrarily. Let the field $\mathbf{E}^0$, $\mathbf{H}^0$ be specified in D_0. To extend the field $\mathbf{E}^0$, $\mathbf{H}^0$ beyond the domain D_0, take some point $\mathbf{r}_0 \in D_0$ and, having calculated all the field derivatives, expand it into the Taylor series in power $(x-x_0)$, $(y-y_0)$, $(z-z_0)$. Thus, we represent the field $\mathbf{E}^0$, $\mathbf{H}^0$ in a sphere of convergence of the power series. Now take a point $\mathbf{r}_1$ in this sphere and assume a new Taylor series, which can bring us to a larger domain. Reiterating these expansions for more and more new points in whose neighborhood the electromagnetic field is specified, we extend gradually the domain of this field continuation and may go far beyond D_0. Upon all possible continuations along all possible chains of power series convergent spheres (along the branches of general analytical functions of the electromagnetic field) we derive functions **E**, **H** existing in the domain and called *complete analytical functions of the electric and magnetic fields.* Complete analytical functions are noncontinuable. The points at which complete analytical functions are not defined are *singular points* of the electromagnetic field. Singular points may be either isolated or form *singular lines* or *singular surfaces.* Singular points, lines, and surfaces are related to the exciters of the electromagnetic field.

Using the Taylor series, one can continue the field $\mathbf{E}^0$, $\mathbf{H}^0$ determined initially at an arbitrarily small neighborhood of the point $\mathbf{r}_0 \in D_0$.

We have considered the problem of continuing the electromagnetic field from the domain D_0 to the domain D. Once electric conductivity σ_0^* and permeability μ_0 of a medium filling the domain D_0 are known, one will be able to continue the electromagnetic field from the surface S. Let the field $\mathbf{E}^0$, $\mathbf{H}^0$ be specified on the inner side of a smooth surface S bounding the domain D_0 and let it have all tangential derivatives. Take some point $\mathbf{r}_0 \in S$. Direct differentiation yields all field derivatives in the tangential directions. Use will be made of relations following from the Maxwell equations and all the field derivatives with respect to the inward pointing normal will be calculated, with known σ_0^*- and μ_0-values. Now we can expand the field $\mathbf{E}^0$, $\mathbf{H}^0$ into the Taylor series in powers $(x-x_0)$, $(y-y_0)$, $(z-z_0)$. This series brings us to the domain D. Further expansions into the Taylor series and formation of various chains of power series convergence spheres will yield the value of the field $\mathbf{E}$, $\mathbf{H}$ in the whole domain D of existence of complete analytical functions. The same problem can be solved for the field $\mathbf{E}^0$, $\mathbf{H}^0$ specified on the inner side of any arbitrarily small part Γ of the surface S bounding the domain D_0. We should like to stress once again that, by virtue of the results obtained in Sect. 7.1.2, analytical continuation of the electromagnetic field is unique.

9.1.3 Equations of Complete Analytical Functions

In the domain D_0 of initial field determination, the complete analytical functions $\mathbf{E}$, $\mathbf{H}$ satisfy the equations

$$\operatorname{rot}\mathbf{H} = \sigma_0^*\mathbf{E} \ , \quad \operatorname{rot}\mathbf{E} = i\omega\mu_0\mathbf{H} \ , \tag{9.1.9}$$

where σ_0^* and μ_0 are the electric conductivity and permeability of a medium filling the domain D_0, respectively.

Which equations are fitted by the functions $\mathbf{E}$, $\mathbf{H}$, outside the domain D_0, i.e., where may the electric conductivity differ from σ_0^*? To answer this question, introduce functions $\mathbf{P} = \operatorname{rot}\mathbf{H} - \sigma_0^*\mathbf{E}$, $\mathbf{R} = \operatorname{rot}\mathbf{E} - i\omega\mu_0\mathbf{H}$, which are analytical in the whole domain D of their existence and are identically equal to zero in the domain D_0 of initial field determination. While proving the uniqueness of analytical continuation of the Laplace field (Sect. 7.1.2) we saw that these functions are identically zero in the whole domain of their existence. Hence, complete analytical functions satisfy everywhere Eqs. (9.1.9) with the coefficients σ_0^* and μ_0 equal to the values of electric conductivity and permeability of a medium in the domain of initial field determination. The electromagnetic field is continued together with the coefficients of the Maxwell equations that it fitted in D_0, regardless of the variations in electric conductivity or permeability of the environment. The same is true of the Helmholtz equations that follow from the Maxwell equations.

9.1.4 Principle of Continuity for the Electromagnetic Field

The problem of analytical continuation of the electromagnetic field can be also solved on the basis of *principle of continuity* extending the Penleve continuation (Sects. 4.1.7 and 7.1.5) to the electromagnetic field.

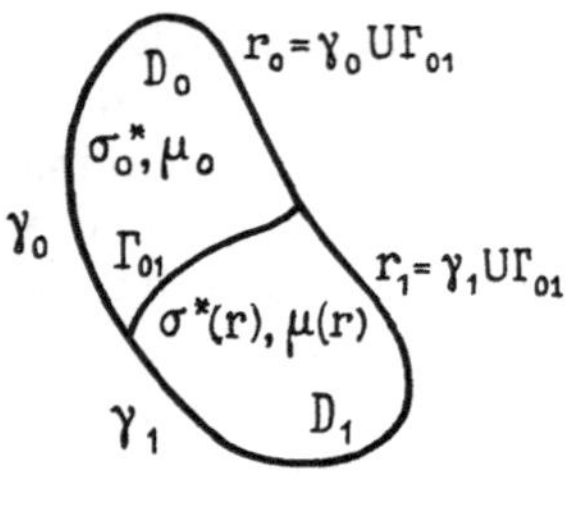

Fig. 52. Substantiation of the principle of continuity for the electromagnetic field

Consider the electromagnetic field $\mathbf{E}^0$, $\mathbf{H}^0$ specified in a domain D_0 bounded by a piecewise smooth surface Γ_0 and filled with a homogeneous medium of electric conductivity σ_0^* and permeability μ_0 (Fig. 52). Since there are no extraneous exciters of the field in D_0, we have

$$\operatorname{rot}\mathbf{H}^0 = \sigma_0^* \mathbf{E}^* \ , \quad \operatorname{rot}\mathbf{E}^0 = i\omega\mu_0 \mathbf{H}^0 \ . \tag{9.1.10}$$

Let the domain D_0 contact the domain D_1 which is bounded by a piecewise smooth surface Γ_1 having a portion Γ_{01} common with the surface Γ_0. The electric conductivity of the medium in D_1 and permeability may change arbitrarily.

To continue the field $\mathbf{E}^0$, $\mathbf{H}^0$ from D_1 to $D = D_0 \cup D_1$, it is sufficient to find a continuously differentiable field $\mathbf{E}^{(1)}$, $\mathbf{H}^{(1)}$ in $\bar{D}_1 = D_1 \cup \Gamma_1$ which satisfies the equations

$$\operatorname{rot}\mathbf{H}^{(1)} = \sigma_0^* \mathbf{E}^{(1)} \ , \quad \operatorname{rot}\mathbf{E}^{(1)} = i\omega\mu_0 \mathbf{H}^{(1)} \tag{9.1.11}$$

and has tangential components on Γ_{01} coinciding with the tangential components of the field $\mathbf{E}^0$, $\mathbf{H}^0$.

We will show that under these conditions the functions $\mathbf{E}^{(1)}$ and $\mathbf{H}^{(1)}$ permit the construction of analytical continuation of the field $\mathbf{E}^0$, $\mathbf{H}^0$, into the domain $D = D_0 \cup D_1$.

Let us start with an application of the modified Stratton-Chu integral formulas to $\mathbf{E}^0$, $\mathbf{H}^0$ and $\mathbf{E}^{(1)}$, $\mathbf{H}^{(1)}$. According to Eqs. (8.3.46), (8.3.47), (9.1.10), and (9.1.11), we have

$$\frac{1}{\sigma_0^*}\operatorname{rot}'\operatorname{rot}'\iint_{\Gamma_0}(\mathbf{n}\times\mathbf{H}^0)G_0^h ds + \operatorname{rot}'\iint_{\Gamma_0}(\mathbf{n}\times\mathbf{E}^0)G_0^h ds = \begin{cases}\mathbf{E}^0(\mathbf{r}') \ , & \mathbf{r}'\in D_0 \ , \\ \mathbf{0} \ , & \mathbf{r}'\in D_1 \ ;\end{cases} \tag{9.1.12}$$

$$\frac{1}{i\omega\mu_0}\operatorname{rot}'\operatorname{rot}'\iint_{\Gamma_0}(\mathbf{n}\times\mathbf{E}^0)G_0^h ds + \operatorname{rot}'\iint_{\Gamma_0}(\mathbf{n}\times\mathbf{H}^0)G_0^h ds = \begin{cases}\mathbf{H}^0(\mathbf{r}') \ , & \mathbf{r}'\in D_0 \ , \\ \mathbf{0} \ , & \mathbf{r}'\in D_1 \ ;\end{cases}$$

$$\frac{1}{\sigma_0^*}\operatorname{rot}'\operatorname{rot}'\iint_{\Gamma_1}(\mathbf{n}\times\mathbf{H}^{(1)})G_0^h ds + \operatorname{rot}'\iint_{\Gamma_0}(\mathbf{n}\times\mathbf{E}^{(1)})G_0^h ds = \begin{cases}\mathbf{0} \ , & \mathbf{r}'\in D_0 \ , \\ \mathbf{E}^{(1)}(\mathbf{r}') \ , & \mathbf{r}'\in D_1 \ ;\end{cases} \tag{9.1.13}$$

$$\frac{1}{i\omega\mu_0}\operatorname{rot}'\operatorname{rot}'\iint_{\Gamma_1}(\mathbf{n}\times\mathbf{E}^{(1)})G_0^h ds + \operatorname{rot}'\iint_{\Gamma_1}(\mathbf{n}\times\mathbf{H}^{(1)})G_0^h ds = \begin{cases}\mathbf{0} \ , & \mathbf{r}'\in D_0 \ , \\ \mathbf{H}^{(1)}(\mathbf{r}') \ , & \mathbf{r}'\in D_1 \ ,\end{cases}$$

where

$$G_0^h = G_0^h(\mathbf{r}'|\mathbf{r}) = \frac{-1}{4\pi}\frac{\exp(ik_0^*\,|\mathbf{r}'-\mathbf{r}|)}{|\mathbf{r}'-\mathbf{r}|}\ ; \quad k_0^* = \sqrt{i\omega\mu_0\sigma_0+\omega^2\mu\varepsilon}\ .$$

Introduce functions **E**, **H**, defined by the integral relations

$$\begin{aligned}\mathbf{E}(\mathbf{r}') &= \frac{1}{\sigma_0^*}\operatorname{rot}'\operatorname{rot}'\left[\iint_{\Gamma_0}(\mathbf{n}\times\mathbf{H}^0)G_0^h ds + \iint_{\Gamma_1}(\mathbf{n}\times\mathbf{H}^{(1)})G_0^h ds\right]\\ &\quad+\operatorname{rot}'\left[\iint_{\Gamma_0}(\mathbf{n}\times\mathbf{E}^0)G_0^h ds + \iint_{\Gamma_1}(\mathbf{n}\times\mathbf{E}^{(1)})G_0^h ds\right]\ ;\\ \mathbf{H}(\mathbf{r}') &= \frac{1}{i\omega\mu_0}\operatorname{rot}'\operatorname{rot}'\left[\iint_{\Gamma_0}(\mathbf{n}\times\mathbf{E}^0)G_0^h ds + \iint_{\Gamma_1}(\mathbf{n}\times\mathbf{E}^{(1)})G_0^h ds\right]\\ &\quad+\operatorname{rot}'\left[\iint_{\Gamma_0}(\mathbf{n}\times\mathbf{H}^0)G_0^h ds + \iint_{\Gamma_1}(\mathbf{n}\times\mathbf{H}^{(1)})G_0^h ds\right]\ .\end{aligned} \tag{9.1.14}$$

With due reference to Eqs. (9.1.12) and (9.1.13), we obtain

$$\mathbf{E}(\mathbf{r}') = \begin{cases}\mathbf{E}^0(\mathbf{r}')\ , & \mathbf{r}'\in D_0\ ,\\ \mathbf{E}^{(1)}(\mathbf{r}')\ , & \mathbf{r}'\in D_1\ ;\end{cases}\qquad \mathbf{H}(\mathbf{r}') = \begin{cases}\mathbf{H}^0(\mathbf{r}')\ , & \mathbf{r}'\in D_0\ ,\\ \mathbf{H}^{(1)}(\mathbf{r}')\ , & \mathbf{r}'\in D_1\ .\end{cases} \tag{9.1.15}$$

In Eq. (9.1.15) the integrals taken over the common part Γ_{01} cancel one another, due to the conciding tangential components of the fields $\mathbf{E}^0$, $\mathbf{H}^0$ and $\mathbf{E}^{(1)}$, $\mathbf{H}^{(1)}$. Hence,

$$\begin{aligned}\mathbf{E}(\mathbf{r}') &= \frac{1}{\sigma_0^*}\operatorname{rot}'\operatorname{rot}'\left[\iint_{\gamma_0}(\mathbf{n}\times\mathbf{H}^0)G_0^h ds + \iint_{\gamma_1}(\mathbf{n}\times\mathbf{H}^{(1)})G_0^h ds\right]\\ &\quad+\operatorname{rot}'\left[\iint_{\gamma_0}(\mathbf{n}\times\mathbf{E}^0)G_0^h ds + \iint_{\gamma_1}(\mathbf{n}\times\mathbf{E}^{(1)})G_0^h ds\right]\ ;\\ \mathbf{H}(\mathbf{r}') &= \frac{1}{i\omega\mu_0}\operatorname{rot}'\operatorname{rot}'\left[\iint_{\gamma_0}(\mathbf{n}\times\mathbf{E}^0)G_0^h ds + \iint_{\gamma_1}(\mathbf{n}\times\mathbf{E}^{(1)})G_0^h ds\right]\\ &\quad+\operatorname{rot}'\left[\iint_{\gamma_0}(\mathbf{n}\times\mathbf{H}^0)G_0^h ds + \iint_{\gamma_1}(\mathbf{n}\times\mathbf{H}^{(1)})G_0^h ds\right]\ ,\end{aligned} \tag{9.1.16}$$

where γ_0 and γ_1 are the parts of the surfaces Γ_0 and Γ_1 remaining after Γ_{01} has been eliminated.

Clearly, integration in (9.1.16) is over a closed surface $\Gamma = \gamma_0 + \gamma_1$ bounding the domain $D = D_0 \cup D_1$ and the functions **E** and **H** are modified Stratton-Chu type integrals. Consequently, in the domain D these functions satisfy the equations

$$\operatorname{rot}\mathbf{H} = \sigma_0^*\mathbf{E}\ , \quad \operatorname{rot}\mathbf{E} = i\omega\mu_0\mathbf{H}\ . \tag{9.1.17}$$

It follows from Eqs. (9.1.15) and (9.1.17) that the field **E**, **H** is the analytical continuation of the field $\mathbf{E}^0$, $\mathbf{H}^0$ into the domain $D = D_1 \cup D_0$.

9.1.5 Electromagnetic Field in the Riemann Space

Real electric and magnetic fields are described by single-valued functions. Complete analytical functions of the electromagnetic field may contain fictitious electric and magnetic fields and may be, just as in the case of an arbitrary analytical field, multiple-valued. To study such multiple-valued functions, it is necessary to resort to the concept of a Riemann space (see Sect. 7.1.3).

Take, for example, the known Sommerfeld problem on plane wave diffraction in a conducting half-plane. The model cross section is seen in Fig. 53. An infinite homogeneous medium of an electric conductivity σ contains a perfectly conducting half-plane (screen) on which a plane monochromatic wave is incident from above. Displacement currents are neglected. The magnetic permeability is everywhere μ_0. The electric field is parallel to the edge of the half-plane. It is defined by the formula (Berdichevsky and Zhdanov 1984)

$$\mathbf{E} = E_0 \left[\exp(-ik\varrho\sin\varphi) \left(1 + \frac{1}{\sqrt{i\pi}} \int_{-\infty}^{\sqrt{2k\varrho}\cos\psi/2} \exp(i\eta^2)\, d\eta \right) \right.$$

$$\left. + \exp(ik\varrho\sin\varphi) \frac{1}{\sqrt{i\pi}} \int_{-\infty}^{-\sqrt{2k\varrho}\sin\psi/2} \exp(i\eta^2)\, d\eta \right] ,$$

where ϱ and φ are the polar coordinates; $\psi = \pi/2 - \varphi$; $k = \sqrt{i\omega\mu_0\sigma}$.

One can see that the field $\mathbf{E}$ has a period of 4π over φ. As a result, it fills a two-sheeted Riemann space, with the branch line being at the edge of the half-plane. The sheets of the Riemann space corresponding to the physical and fictitious spaces are joined in the half-plane.

Let the field $\mathbf{E}$ be specified in the neighborhood of some point lying away from the well-conducting half plane. Its continuation along all possible paths not intersecting the half-plane provides a real electric field on the first sheet of the Riemann space (the first branch of a complete analytical function). Intersecting the half-plane we move to the second sheet of the Riemann space and obtain a fictitious electric field (the second branch of a complete analytical function). A complete analytical function of the electric field is two-valued.

This two-dimensional model indicates that if the electromagnetic field is continued through an inhomogeneous medium, its complete analytical functions may prove multiple-valued.

Transition to three-dimensional models would render the mathematical analysis somewhat more complicated, although the principle remains the same. A general idea of the properties of complete analytical functions of a three-dimensional electromagnetic field can be gained from Sommerfeld's theory of branched solutions (Frank and Mises 1927; Morse and Feshbach 1953).

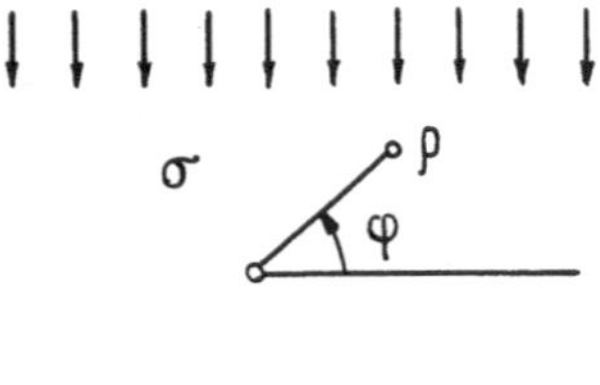

Fig. 53. Incidence of a plane wave onto a conducting half-plane

Fig. 54. Point source Q above a plane screen; Q' is the mirror image of the point source; Re is the path in real space; Im is the path in fictitious space

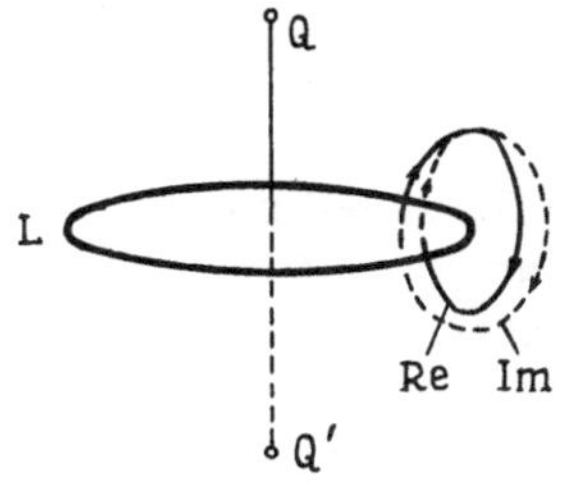

Consider, for instance, following Berdichevsky and Zhdanov (1984), a plane perfectly conducting screen bounded by an arbitrary contour L (Fig. 54). The electromagnetic field is excited by a point dipole source Q located above the screen. The tangential components of the electric field at the screen surface are zero, while the tangential components of the magnetic field suffer a discontinuity equal to the surface density of the current. The problem can be solved by the method of image sources. The real physical space is complemented by a fictitious mathematical space in which a point source Q' of an appropriate polarity is located symmetrically about the screen. Thus, we have two copies of three-dimensional space which constitute a *double Riemann space.* The real and the fictitious spaces are cut along the screen and joined so as to connect each side of the cut in one space with the opposite side of the cut in the other space. The electromagnetic field in the two copies of the Riemann space do not coincide. Analytical continuation of the electromagnetic field along all possible paths not intersecting the screen yields the first branches of complete analytical functions. These branches refer to the real physical space with a singular point in Q. Intersecting the screen we ignore the magnetic field discontinuity and get into the fictitious space containing the second branches of complete analytical functions with a singular point in Q'. This singular point is a mirror image of the singular point Q of the real field. Intersecting the screen twice we come through the two copies of the Riemann space and return to the real physical space. Thus, within the Sommerfeld model, complete analytical functions are two-valued and the contour L of the screen is a branch line of the field. Representing complete analytical functions in the double Riemann space we restore their uniqueness.

If the model includes several plane screens, intersecting each of them may give rise to a new fictitious space. The number of branch lines and of the branches of complete analytical functions increases. Thus, it is possible to construct a Riemann space consisting of two, three, and more copies.

9.2 Analytical Continuation of the Electromagnetic Field into Geoelectrical Inhomogeneities

Of great value in the theory of electromagnetic field analytical continuation is the question whether it is possible to continue the Stratton-Chu type integrals through the integration surface.

9.2.1 Analytical Continuation of the Stratton-Chu Type Integral Through the Integration Surface

The stated problem is solved, just as with the continuation of three-dimensional analogs of the Cauchy-type integrals (see Sect. 7.2), using the Cauchy-Kovalevskaya theorem.

Consider, following Sect. 8.6 [formulas (8.6.3a) and (8.6.3b)], the Stratton-Chu type integrals $\mathbf{C}^E(\mathbf{r}'; \mathbf{e}, \mathbf{h})$ and $\mathbf{C}^H(\mathbf{r}', \mathbf{e}, \mathbf{h})$ over an arbitrary closed piecewise smooth surface S. The densities of the integrals $\mathbf{e}(\mathbf{r})$ and $\mathbf{h}(\mathbf{r})$ are related, as usual, by Eqs. (8.6.2). Let us taken an analytical part Γ of the surface S and assume that the components $\mathbf{e}(\mathbf{r})$ and $\mathbf{h}(\mathbf{r})$ are fitted on Γ by analytical functions. Fix on a point $\mathbf{r}_0 \in \Gamma$. Then, according to the Cauchy-Kovalevskaya theorem (Sretensky 1946), in the neighborhood $\varepsilon(\mathbf{r}_0)$ of the point $\mathbf{r}_0$ there is an analytical vector field $\mathbf{E}^\Gamma(\mathbf{r})$ satisfying the equations

$$\operatorname{div} \mathbf{E}^\Gamma = 0 \;, \quad \Delta \mathbf{E}^\Gamma + k^{*2} \mathbf{E}^\Gamma = 0 \tag{9.2.1}$$

and the boundary conditions

$$\mathbf{E}^\Gamma|_\Gamma = \mathbf{e} \;, \quad \left.\frac{\partial \mathbf{E}^\Gamma}{\partial n}\right|_\Gamma = \operatorname{grad}_\Gamma e_n - \mathbf{n} \operatorname{div}_\Gamma \mathbf{e}_\tau - i\omega\mu(\mathbf{n}\times\mathbf{h}) \;. \tag{9.2.2}$$

Introduce into $\varepsilon(\mathbf{r}_0)$ the function

$$\mathbf{H}^\Gamma = \frac{1}{i\omega\mu} \operatorname{rot} \mathbf{E}^\Gamma \;. \tag{9.2.3}$$

Since, by definition, $\operatorname{div} \mathbf{E}^\Gamma = 0$ we have

$$\operatorname{rot} \mathbf{H}^\Gamma = \frac{1}{i\omega\mu} \operatorname{rot}\operatorname{rot} \mathbf{E}^\Gamma = \frac{1}{i\omega\mu} (\operatorname{grad}\operatorname{div} \mathbf{E}^\Gamma - \Delta \mathbf{E}^\Gamma) = \sigma^* \mathbf{E}^\Gamma \;. \tag{9.2.4}$$

Evaluate $\mathbf{H}^\Gamma$ at the surface Γ. The vector $\mathbf{H}^\Gamma$ can be decomposed into a tangential $\mathbf{H}^\Gamma_\tau$ and a normal $\mathbf{H}^\Gamma_n$ components. Fixing on the normal direction we find, with due account taken of (9.2.2) and (9.2.3),

$$\begin{aligned}
\mathbf{H}^\Gamma_\tau &= [\mathbf{n}\times\mathbf{H}^\Gamma]\times\mathbf{n} = \frac{1}{i\omega\mu} [\mathbf{n}\times\operatorname{rot}\mathbf{E}^\Gamma]\times\mathbf{n} \\
&= \frac{1}{i\omega\mu} [\operatorname{grad} E^\Gamma_n - (\mathbf{n}\cdot\operatorname{grad})\mathbf{E}^\Gamma]\times\mathbf{n} \\
&= \frac{1}{i\omega\mu} \left[\operatorname{grad}_\Gamma E^\Gamma_n + \mathbf{n}\frac{\partial E^\Gamma_n}{\partial n} - \frac{\partial \mathbf{E}^\Gamma}{\partial n}\right]\times\mathbf{n} \\
&= \frac{1}{i\omega\mu} \left[\operatorname{grad}_\Gamma E^\Gamma_n - \mathbf{n}\operatorname{div}\mathbf{E}^\Gamma_\tau - \frac{\partial \mathbf{E}^\Gamma}{\partial n}\right]\times\mathbf{n} \\
&= \frac{1}{i\omega\mu} [\operatorname{grad}_\Gamma e_n - \mathbf{n}\operatorname{div}_\Gamma \mathbf{e}_\tau - \operatorname{grad}_\Gamma e_n \\
&\quad + \mathbf{n}\operatorname{div}_\Gamma \mathbf{e}_\tau + i\omega\mu[\mathbf{n}\times\mathbf{h}]]\times\mathbf{n} = [\mathbf{n}\times\mathbf{h}]\times\mathbf{n} = \mathbf{h}_\tau \;;
\end{aligned}$$

$$H_n^\Gamma = \mathbf{n}\cdot\mathbf{H}^\Gamma = -\frac{1}{i\omega\mu}\,\mathrm{div}_\Gamma[\mathbf{n}\times\mathbf{E}^\Gamma] = -\frac{1}{i\omega\mu}\,\mathrm{div}_\Gamma[\mathbf{n}\times\mathbf{e}] = h_n \ .$$

Thus,

$$\mathbf{H}^\Gamma|_\Gamma = \mathbf{h}|_\Gamma \ . \tag{9.2.5}$$

Equations (9.2.2) through (9.2.5) imply that the functions $\mathbf{E}^\Gamma$, $\mathbf{H}^\Gamma$, satisfy the equations

$$\mathrm{rot}\,\mathbf{H}^\Gamma = \sigma^*\mathbf{E}^\Gamma \ , \quad \mathrm{rot}\,\mathbf{E}^\Gamma = i\omega\mu\mathbf{H}^\Gamma \tag{9.2.6}$$

and boundary conditions at the surface Γ:

$$\mathbf{E}^\Gamma|_\Gamma = \mathbf{e} \ , \quad \mathbf{H}^\Gamma|_\Gamma = \mathbf{h} \ . \tag{9.2.7}$$

Now employ the electromagnetic counterparts of the Sokhotsky-Plemelj formulas. According to (8.4.10),

$$\begin{aligned} \mathbf{C}^{E-}(\mathbf{r}_0;\ \mathbf{e},\ \mathbf{h}) &= \mathbf{C}^{E+}(\mathbf{r}_0;\ \mathbf{e},\ \mathbf{h}) - \mathbf{e}(\mathbf{r}_0) \ , \\ \mathbf{C}^{H-}(\mathbf{r}_0;\ \mathbf{e},\ \mathbf{h}) &= \mathbf{C}^{H+}(\mathbf{r}_0;\ \mathbf{e},\ \mathbf{h}) - \mathbf{h}(\mathbf{r}_0) \ . \end{aligned} \tag{9.2.8}$$

The latter relations can be rewritten, allowing for (9.2.7), in the form

$$\begin{aligned} \mathbf{C}^{E-}(\mathbf{r}_0;\ \mathbf{e},\ \mathbf{h}) &= \mathbf{C}^{E+}(\mathbf{r}_0;\ \mathbf{e},\ \mathbf{h}) - \mathbf{E}^\Gamma(\mathbf{r}_0) \ , \\ \mathbf{C}^{H-}(\mathbf{r}_0;\ \mathbf{e},\ \mathbf{h}) &= \mathbf{C}^{H+}(\mathbf{r}_0;\ \mathbf{e},\ \mathbf{h}) - \mathbf{H}^\Gamma(\mathbf{r}_0) \ . \end{aligned} \tag{9.2.9}$$

Let D stand for the domain bounded by S and D^+ for the common part of the neighborhood $\varepsilon(\mathbf{r}_0)$ and D, $D^+ = \varepsilon(\mathbf{r}_0) \cap D$.

Clearly, the right-hand side of (9.2.9) represents boundary conditions on Γ of electromagnetic fields satisfying Eqs. (9.2.6) everywhere in the domain D^+ adjoining Γ. The left-hand side describes the boundary conditions on Γ of electromagnetic fields satisfying Eqs. (9.2.6) everywhere outside D. As a consequence, following the principle of continuity for the electromagnetic field (Sect. 9.1.4), the right-hand side is the analytical continuation of the left-hand side through the surface Γ.

In a similar way, one can prove the possibility of analytical continuation of the Stratton-Chu type integrals through an analytical part of the surface Γ from the inside outwards. Formulas (9.2.9), just as formulas (7.2.20) for the Laplace field, suggest that the values of the Stratton-Chu type integrals continued through the surface Γ, $\mathbf{C}^{E,H\mathrm{cont}}(\mathbf{r};\ \mathbf{e},\ \mathbf{h})$ differ from the integrals themselves $\mathbf{C}^{E,H}(\mathbf{r};\ \mathbf{e},\ \mathbf{h})$ by amounts equal to the functions $[-\mathbf{E}^\Gamma(\mathbf{r})]$, $[-\mathbf{H}^\Gamma(\mathbf{r})]$:

$$\begin{aligned} \mathbf{C}^{E\mathrm{cont}}(\mathbf{r};\ \mathbf{e},\ \mathbf{h}) - \mathbf{C}^{E}(\mathbf{r};\ \mathbf{e},\ \mathbf{h}) &= -\mathbf{E}^\Gamma(\mathbf{r}) \ , \\ \mathbf{C}^{H\mathrm{cont}}(\mathbf{r};\ \mathbf{e},\ \mathbf{h}) - \mathbf{C}^{H}(\mathbf{r};\ \mathbf{e},\ \mathbf{h}) &= -\mathbf{H}^\Gamma(\mathbf{r}) \ . \end{aligned} \tag{9.2.10}$$

As a consequence, the significant conclusions reached in Sect. 7.2.3 for the Laplace field are extended to the electromagnetic field. They are: (1) the singular points $\mathbf{C}^{E,H\mathrm{cont}}(\mathbf{r};\ \mathbf{e},\ \mathbf{h})$ coincide with the singular points of the functions $\mathbf{E}^\Gamma$ and

$\mathbf{H}^{\Gamma}$, respectively; (2) the lines joining different analytical parts of the surface are generally branch lines for complete analytical functions of the Stratton-Chu type integrals.

These properties are of fundamental value in the theory of analytical continuation of the electromagnetic field and its singular points.

9.2.2 Analytical Continuation of the Electromagnetic Field into a Homogeneous Domain Bounded by an Analytical and Piecewise Analytical Surfaces

Consider a model typical of geoelectrics, within which a homogeneous Earth with a normal complex electric conductivity $\sigma_n^* = \text{const}$ contains a deep inhomogeneity D of a constant complex electric conductivity $\sigma_d^* \neq \sigma_n^*$ (Fig. 55). The Earth is bounded by a surface Σ (closed or infinitely extended), while the deep inhomogeneity is bounded by a surface S (closed, analytical, or piecewise analytical). In this model, the field is excited by external currents distributed outside the Earth (in the atmosphere).

Consider the problem of continuing analytically the electromagnetic field from the Earth's surface into the well-conducting Earth. Techniques for this continuation will be discussed in the following subsection. At this point, however, we are interested in the feasibility of electromagnetic field continuation into the inhomogeneity and in the location of singular points of this continuation. These questions are answered readily if we employ the Stratton-Chu integral formula and express the electromagnetic field $\mathbf{E}, \mathbf{H}$ in the domain $\tilde{\Omega}_-$ found between Σ and S in terms of its components at the surfaces Σ and S.

Indeed, according to (8.4.6), we have

$$\mathbf{E}(\mathbf{r}') = \mathbf{C}^{E(\Sigma)}(\mathbf{r}'; \mathbf{E}, \mathbf{H}) + \mathbf{C}^{E(S)}(\mathbf{r}'; \mathbf{E}, \mathbf{H}) ,$$

$$\mathbf{H}(\mathbf{r}') = \mathbf{C}^{H(\Sigma)}(\mathbf{r}'; \mathbf{E}, \mathbf{H}) + \mathbf{C}^{H(S)}(\mathbf{r}'; \mathbf{E}, \mathbf{H}) , \tag{9.2.11}$$

where the superscripts enclosed in parentheses with the symbols of the Stratton-Chu type integrals indicate the surface over which the corresponding integral is taken. Evidently, the integrals $\mathbf{C}^{E(\Sigma)}$ and $\mathbf{C}^{H(\Sigma)}$ are analytical everywhere in the domain inside the Earth. Therefore, seeking singular points of the fields $\mathbf{E}(\mathbf{r})$ and

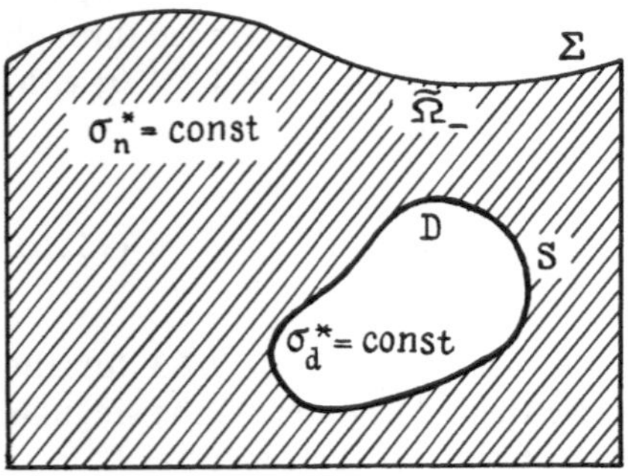

Fig. 55. Geoelectrical model within which a homogeneous Earth of a normal complex electric conductivity $\sigma_n^* = \text{const}$ contains a deep inhomogeneity D with a complex electric conductivity $\sigma_d^* \neq \sigma_n^*$

$\mathbf{H}(\mathbf{r})$ of particular interest to us are only integrals over the surface S. These integrals are present in Stratton-Chu formulas (9.2.11) if $\sigma_d^* \neq \sigma_n^*$, hence, they determine the contribution of the inhomogeneity D into the observed field. They can be related to the fictitious surface currents flowing over S. In accordance with (8.3.39) and (8.3.40), we have

$$\mathbf{C}^{E(S)}(\mathbf{r}'; \mathbf{E}, \mathbf{H}) = -\iint_S \left(\frac{q_S^e}{\sigma^*}\,\mathrm{grad}\,G^h - \mathbf{j}_S^m \times \mathrm{grad}\,G^h + i\omega\mu\,\mathbf{j}^e G^h\right) ds \;;$$

$$\mathbf{C}^{H(S)}(\mathbf{r}'; \mathbf{E}, \mathbf{H}) = -\iint_S \left(\frac{q_S^m}{\mu}\,\mathrm{grad}\,G^h + \mathbf{j}_S^e \times \mathrm{grad}\,G^h - \sigma^* \mathbf{j}^m G^h\right) ds \;, \tag{9.2.12}$$

where

$$\mathbf{j}_S^e = -(\mathbf{n} \times \mathbf{H})_S \;, \quad \mathbf{j}_S^m = (\mathbf{n} \times \mathbf{E})_S \;,$$

$$q_S^e = -\varepsilon^*(\mathbf{n} \cdot \mathbf{E})_S \;, \quad q_S^m = -\mu(\mathbf{n} \cdot \mathbf{H})_S \;. \tag{9.2.13}$$

Thus, the action of real field exciters distributed over the inhomogeneity D and at the surface S is replaced by the effect of fictitious (equivalent) exciters distributed only over S. Precisely these fictitious exciters determine the singularities of the field continued into D.

Now we will employ the results of the previous subsection, according to which the Stratton-Chu type integrals can be continued analytically through any analytical part Γ of the surface S. The singular points of the field continued in this manner coincide with those of the functions $\mathbf{E}^\Gamma$ and $\mathbf{H}^\Gamma$ related to the surface exciters by

$$E_n^\Gamma|_S = -\frac{q_S^e}{\varepsilon^*} \;, \quad H_n^\Gamma|_S = -\frac{q_S^m}{\mu} \;,$$

$$(\mathbf{n} \times \mathbf{E}_\tau^\Gamma)_S = \mathbf{j}_S^m \;, \quad (\mathbf{n} \times \mathbf{H}_\tau^\Gamma)_S = -\mathbf{j}_S^e \;. \tag{9.2.14}$$

If all the surface S is analytical, the singular points of the continued field $\mathbf{E}^\Gamma$, $\mathbf{H}^\Gamma$ are sure to lie inside D. This claim is easily proved by contradiction. If the functions $\mathbf{E}^\Gamma$, $\mathbf{H}^\Gamma$ have no singular points in D, i.e., if they satisfy Eqs. (9.2.6) in the whole domain, formulas (9.2.11) do not contain, according to (8.4.6), integrals over the surface S, which is equivalent to the absence of inhomogeneity.

In accordance with (9.2.14), the structure of an analytically continued field $\mathbf{E}^\Gamma$, $\mathbf{H}^\Gamma$ and, hence, the location of its singular points are determined by the distribution of fictitious surface currents at S, i.e., not only by the shape of an inhomogeneity, but also by the external field configuration and the Earth's surface relief.

Take, as the simplest example, the known model of electrodynamics, within which a homogeneous sphere of a radius a and electric conductivity σ_d immersed in an infinite non-conducting medium is excited by a uniform magnetic field $\mathbf{H}^n$ (Berdichevsky and Zhdanov 1984). Within a spherical coordinate system whose axis is along the primary magnetic field, for $\mathbf{r} \geqslant a$ we have

$$H_r = H^n \left(1 + \frac{a^3 D}{r^3}\right) \cos\theta \ , \quad H_\theta = -H^n \left(1 - \frac{a^3 D}{2r^3}\right) \sin\theta \ ,$$

where

$$D = \frac{3 J_{3/2}(k_d a) - k_d a J_{1/2}(k_d a)}{k_d a J_{1/2}(k_d a)} \ ;$$

$J_{1/2}$ and $J_{3/2}$ are the half-integer-order Bessel functions.

Extending these formulas to the region $r<a$, we continue the magnetic field **H** analytically into the sphere. The singular point of analytical continuation lies at the center of the sphere (third-order pole).

Similarly, one can prove that it is possible to continue the field **E**, **H** into the homogeneity D bounded by a piecewise analytical surface S. Intersecting different analytical parts of the surface S we derive, in accordance with the Cauchy-Kovalevskaya theorem, different functions $\mathbf{E}^\Gamma$, $\mathbf{H}^\Gamma$. Under such analytical continuation, the field $\mathbf{E}^\Gamma$, $\mathbf{H}^\Gamma$ inside the inhomogeneity D is multiple-valued. The lines joining different analytical parts of the surface S are generally branch lines of the field $\mathbf{E}^\Gamma$, $\mathbf{H}^\Gamma$. Each analytical part Γ of the surface may be related to its own branch of the field $\mathbf{E}^\Gamma$, $\mathbf{H}^\Gamma$ and its own copy of the Riemann space. Since the fictitious copies of the Riemann space are infinite, the singular points of the field $\mathbf{E}^\Gamma$, $\mathbf{H}^\Gamma$ may happen to be outside the inhomogeneity D. This situation is the case in the Sommerfeld problem on electromagnetic field diffraction by a plane screen (see Sect. 9.1.5).

The developed concepts provide the basis for geoelectrical interpretation of analytically continued fields. Examples of this interpretation can be found in the monograph by M. N. Berdichevsky and M. S. Zhdanov (1984).

9.3 Techniques for Analytical Continuation of the Electromagnetic Field

9.3.1 Forms of Analytical Continuation of the Electromagnetic Field in Geoelectrical Problems

A geophysicist studying electromagnetic anomalies makes use of analytical continuation either to rise over the Earth and get rid of unnecessary details associated with near-surface inhomogeneities or to penetrate deep into the Earth and, having approached a deep inhomogeneity, identify the effects indiscernible from the Earth's surface. All these problems fall into two classes:

1) continuation of the normal field into the Earth and of the anomalous field into the air;
2) continuation of the anomalous field into the Earth and of the normal field into the air.

In the problems of the first class, the field is continued into a domain devoid of exciters and, hence, of singular points. This is the simplest form of analytical continuation. An operator of this continuation is constructed directly by means of the Stratton-Chu integral.

Consider a model in which a closed Earth's surface S separates the Earth (domain Ω_-) from the atmosphere (domain Ω_+). The atmosphere is homogeneous, its complex electric conductivity σ_0^* is known. The complex electric conductivity of the Earth changes in accordance with an arbitrary law $\sigma^*(\mathbf{r}) = \sigma_n^* + \Delta\sigma^*(\mathbf{r})$ where the normal electric conductivity σ_n^* is a constant. It is taken as known. The magnetic permeability is everywhere equal to μ_0.

The model is excited, as previously, by extraneous currents flowing with a density $\mathbf{j}^e$ in the domain Q_e of the atmosphere.

Let the normal field $\mathbf{E}^n$, $\mathbf{H}^n$ be specified at the Earth's surface. Continue this field into the Earth. According to Eqs. (8.4.2), (8.4.3), (8.4.6), (8.4.16), and (8.4.17), for $\mathbf{r}' \in \Omega_-$ we have

$$\mathbf{E}^n(\mathbf{r}') = \iint\limits_S \{(\mathbf{n}\cdot\mathbf{E}^n)\operatorname{grad} G_n^h + [\mathbf{n}\times\mathbf{E}^n]\times\operatorname{grad} G_n^h + i\omega\mu_0[\mathbf{n}\times\mathbf{H}^n]G_n^h\}ds$$

$$= \frac{1}{\sigma_n^*}\operatorname{rot}'\operatorname{rot}'\iint\limits_S [\mathbf{n}\times\mathbf{H}^n]G_n^h ds + \operatorname{rot}'\iint\limits_S [\mathbf{n}\times\mathbf{E}^n]G_n^h ds \; ;$$

$$\mathbf{H}^n(\mathbf{r}') = \iint\limits_S \{(\mathbf{n}\cdot\mathbf{H}^n)\operatorname{grad} G_n^h + [\mathbf{n}\times\mathbf{H}^n]\times\operatorname{grad} G_n^h + \sigma_n^*[\mathbf{n}\times\mathbf{E}^n]G_n^h\}ds \tag{9.3.1}$$

$$= \frac{1}{i\omega\mu_0}\operatorname{rot}'\operatorname{rot}'\iint\limits_S [\mathbf{n}\times\mathbf{E}^n]G_n^h ds + \operatorname{rot}'\iint\limits_S [\mathbf{n}\times\mathbf{H}^n]G_n^h ds \; ,$$

where the unit vector of the normal $\mathbf{n}$ points from the Earth into the air and

$$G_n^h = G_n^h(\mathbf{r}'|\mathbf{r}) = \frac{-1}{4\pi}\frac{\exp(ik_n^*\,|\mathbf{r}'-\mathbf{r}|)}{|\mathbf{r}'-\mathbf{r}|} \; ; \quad k_n^* = \sqrt{i\omega\mu_0\sigma_n^*} \; .$$

A similar operator can be constructed for continuing the anomalous field $\mathbf{E}^a$, $\mathbf{H}^a$ into the air. Allowing for the fact that the Stratton-Chu integrals taken over a sphere of an infinite radius are zero, we obtain for

$$\mathbf{E}^a(\mathbf{r}') = -\iint\limits_S \{(\mathbf{n}\cdot\mathbf{E}^a)\operatorname{grad} G_0^h + [\mathbf{n}\times\mathbf{E}^a]\times\operatorname{grad} G_0^h + i\omega\mu_0[\mathbf{n}\times\mathbf{H}^a]G_0^h\}ds$$

$$= -\frac{1}{\sigma_n^*}\operatorname{rot}'\operatorname{rot}'\iint\limits_S [\mathbf{n}\times\mathbf{H}^a]G_0^h ds - \operatorname{rot}'\iint\limits_S [\mathbf{n}\times\mathbf{E}^a]G_0^h ds \; ;$$

$$\mathbf{H}^a(\mathbf{r}') = -\iint\limits_S \{(\mathbf{n}\cdot\mathbf{H}^a)\operatorname{grad} G_0^h + [\mathbf{n}\times\mathbf{H}^a]\times\operatorname{grad} G_0^h + \sigma_0^*[\mathbf{n}\times\mathbf{E}^a]G_0^h\}ds \tag{9.3.2}$$

$$= -\frac{1}{i\omega\mu_0}\operatorname{rot}'\operatorname{rot}'\iint\limits_S [\mathbf{n}\times\mathbf{E}^a]G_0^h ds - \operatorname{rot}'\iint\limits_S [\mathbf{n}\times\mathbf{H}^a]G_0^h ds \; ,$$

where the unit vector of the normal $\mathbf{n}$ points again from the Earth to the air and

$$G_0^h = G_0^h(\mathbf{r}'|\mathbf{r}) = \frac{-1}{4\pi}\frac{\exp(ik_0^*\,|\mathbf{r}'-\mathbf{r}|)}{|\mathbf{r}'-\mathbf{r}|} \; ; \quad k_0^* = \sqrt{i\omega\mu_0\sigma_0^*} \; .$$

Operators (9.3.1) and (9.3.2) are applicable to the model with an infinitely extended surface S. If the electric conductivity of the air σ_0 is assumed to be zero and the displacement currents in the air are negligible, operator (9.3.2) loses the term containing the normal component of the field $\mathbf{E}^n$ and operator (9.3.2) is meaningful only for the magnetic field.

In the second class of problems, the field is continued toward its exciters, i.e., into a domain definitely known to contain singular points (lines, surfaces) of the field. This form of analytical continuation admits no longer the utilization of operators (9.3.1) and (9.3.2). In these problems, a continuation operator can be constructed by means of the Taylor series but this approach is clearly cumbersome. Here, analytical continuation of the electromagnetic field calls for specific mathematical techniques, similar to those employed in the continuation of Laplace fields (see Sect. 7.5). These techniques will be outlined in the next subsection.

9.3.2 Problem Statement

The problem is formulated as follows. A homogeneous Earth of a known normal complex electric conductivity $\sigma_n^* = \text{const}$ and magnetic permeability μ_0 contains a deep inhomogeneity D of an unknown, constant or variable, electric conductivity $\sigma_d^* \neq \sigma_n^*$. The electromagnetic field $\mathbf{E}^0$, $\mathbf{H}^0$ (total or anomalous) has been specified at the Earth's surface Σ. It must be continued into the Earth. By definition of analytical continuation, we should find a field $\mathbf{E}$, $\mathbf{H}$ which inside a near-surface domain $\tilde{\Omega}_-$ would satisfy the equations $\operatorname{rot}\mathbf{H} = \sigma_n^*\mathbf{E}$, $\operatorname{rot}\mathbf{E} = i\omega\mu_0\mathbf{H}$, while at the Earth's surface it might coincide with $\mathbf{E}^0$, $\mathbf{H}^0$.

For known values of σ_n^* and μ_0, the Maxwell equations permit calculation of the normal field derivatives $\partial\mathbf{E}^0/\partial n$ and $\partial\mathbf{H}^0/\partial n$ from the field $\mathbf{E}^0$, $\mathbf{H}^0$. Thus, we have to solve virtually the Cauchy boundary problem for the Helmholtz equation. Proceeding to analytical continuation of the field $\mathbf{E}^0$, $\mathbf{H}^0$, we know nothing (or almost nothing) about the inhomogeneity D, i.e., about the distribution of singular points of the field $\mathbf{E}$, $\mathbf{H}$. Hence, we continue the field into some domain $\tilde{\Omega}_-$ whose lower boundary γ is not known.

To solve the problem, we introduce a provisional boundary γ separating the domain $\tilde{\Omega}_-$ from the domain that may contain singular points (lines, surfaces) of the field. For γ we choose a coordinate surface from the space of orthogonal coordinates admitting separation of variables in the Helmholtz equation. These coordinates include, for instance, rectangular, spherical, cylindrical, conical, spheroidal, ellipsoidal ones. We will confine ourselves to the construction of continuation operators relying on rectangular coordinates.

9.3.3 Continuation of the Field into a Layer

The model of a medium is depicted in Fig. 56. The Earth is bounded by a quasi-plane surface Σ oscillating near the plane $z = 0$. The normal complex electric conductivity $\sigma_n^* = \text{const}$ is assumed to be known. The inhomogeneity D is of an ar-

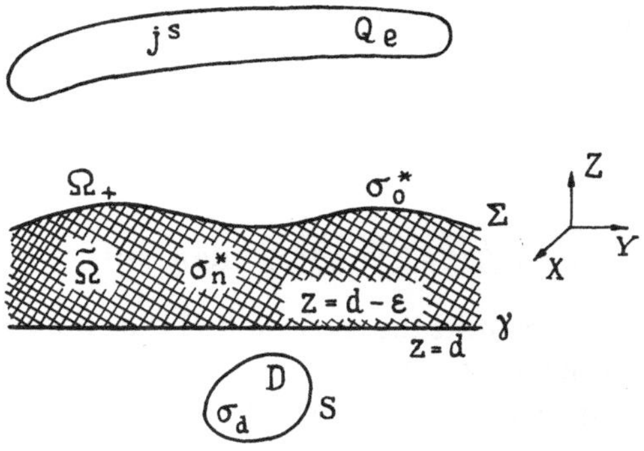

Fig. 56. Analytical continuation of the electromagnetic field into a horizontal layer

bitrary shape. Its boundary S never comes in contact with Σ. The model is excited by an extraneous current of a density $\mathbf{j}^e$ localized in the domain Q_e of the atmosphere (either conducting or nonconducting). The field $\mathbf{E}^0$, $\mathbf{H}^0$ is specified at the surface of the Earth and is to be continued into a certain domain $\tilde{\Omega}_-$ of the Earth.

For a lower boundary γ of the domain $\tilde{\Omega}_-$ take the plane $z = d - \varepsilon$, where d is a minimum depth to the inhomogeneity D, while ε is an arbitrarily small positive quantity. Evidently, this plane lies above the inhomogeneity D.

Consider a horizontal layer $\tilde{\Omega}_-$ bounded by the Earth's surface Σ and the plane γ. Express the electromagnetic field in $\tilde{\Omega}_-$ in terms of its values at Σ and γ. According to Eqs. (8.3.36a) and (8.3.36b), we have for $\mathbf{r}' \in \tilde{\Omega}_-$

$$\begin{aligned}\mathbf{E}(\mathbf{r}') = &-\iint_{\Sigma} [(\mathbf{n}\cdot\mathbf{E}^0)\operatorname{grad} G_n^h + (\mathbf{n}\times\mathbf{E}^0)\times\operatorname{grad} G_n^h + i\omega\mu_0(\mathbf{n}\times\mathbf{H})\, G_n^h]\, ds \\ &+\iint_{\gamma} [(\mathbf{n}\cdot\mathbf{E})\operatorname{grad} G_n^h + (\mathbf{n}\times\mathbf{E})\times\operatorname{grad} G_n^h + i\omega\mu_0(\mathbf{n}\times\mathbf{H}^0)\, G_n^h]\, ds \;; \\ \mathbf{H}(\mathbf{r}') = &-\iint_{\Sigma} [(\mathbf{n}\cdot\mathbf{H}^0)\operatorname{grad} G_n^h + (\mathbf{n}\times\mathbf{H}^0)\times\operatorname{grad} G_n^h + \sigma_n^*(\mathbf{n}\times\mathbf{E}^0)\, G_n^h]\, ds \\ &+\iint_{\gamma} [(\mathbf{n}\cdot\mathbf{H})\operatorname{grad} G_n^h + (\mathbf{n}\times\mathbf{H})\times\operatorname{grad} G_n^h + \sigma_n^*(\mathbf{n}\times\mathbf{E})\, G_n^h]\, ds \;,\end{aligned} \tag{9.3.3}$$

where

$$G_n^h = G_n^h(\mathbf{r}'|\mathbf{r}) = \frac{-1}{4\pi}\,\frac{\exp(ik_n^*\,|\mathbf{r}'-\mathbf{r}|)}{|\mathbf{r}'-\mathbf{r}|} \;; \quad k_n^* = \sqrt{i\omega\mu_0\sigma_n^*} \;; \quad \operatorname{Re} k_n^* > 0 \;.$$

An operator of continuation of the field $\mathbf{E}^0$, $\mathbf{H}^0$ into the domain $\tilde{\Omega}_-$ will be constructed if the integrals over the surface γ are transformed into the integrals over the Earth's surface Σ. The way to this transformation involves spectral decomposition of the Green function. The function G_n^h in integrals over the plane γ depends on the location of the points $\mathbf{r}' \in \tilde{\Omega}_-$ and $\mathbf{r} \in \gamma$. The z'- and z-axes of these points satisfy the condition $z' > z$ under which

$$G_n^h = \frac{1}{8\pi^2} \iint_{-\infty}^{\infty} \frac{f(\mathbf{r}')g(\mathbf{r})}{\eta_n}\, dk_x\, dk_y \;, \tag{9.3.4}$$

where the harmonics f, g describe quasi-plane waves

$$f(\mathbf{r}') = -\exp[i(k_x x' + k_y y')]\exp(-\eta_n z')|_{z'>z} \ ,$$

$$g(\mathbf{r}') = \exp[-i(k_x x + k_y y)]\exp(\eta_n z)|_{z'>z} \ , \qquad (9.3.5)$$

$$\eta_n = \sqrt{k_x^2 + k_y^2 - k_n^{*2}} \ , \quad \operatorname{Re}\eta_n > 0 \ .$$

Substitute the expansion of the Green function in quasi-plane waves (9.3.4) into the second integrals of formula (9.3.3). Due to the absolute and uniform convergence of these integrals:

$$\iint_\gamma [(\mathbf{n}\cdot\mathbf{E})\operatorname{grad} G_n^h + (\mathbf{n}\times\mathbf{E})\times\operatorname{grad} G_n^h + i\omega\mu_0(\mathbf{n}\times\mathbf{H})G_n^h]\,ds$$
$$= \frac{1}{8\pi^2}\iint_{-\infty}^{\infty}\frac{f}{\eta_n}\,dk_x\,dk_y\iint_\gamma \{(\mathbf{n}\cdot\mathbf{E})\operatorname{grad} g + [\mathbf{n}\times\mathbf{E}]\times\operatorname{grad} g + i\omega\mu_0[\mathbf{n}\times\mathbf{H}]g\}ds \ ;$$
$$(9.3.6)$$
$$\iint_\gamma [(\mathbf{n}\cdot\mathbf{H})\operatorname{grad} G_n^h + (\mathbf{n}\times\mathbf{H})\times\operatorname{grad} G_n^h + \sigma_n^*(\mathbf{n}\times\mathbf{H})G_n^h]\,ds$$
$$= \frac{1}{8\pi^2}\iint_{-\infty}^{\infty}\frac{f}{\eta_n}\,dk_x\,dk_y\iint_\gamma \{(\mathbf{n}\cdot\mathbf{H})\operatorname{grad} g + [\mathbf{n}\times\mathbf{H}]\times\operatorname{grad} g + \sigma_n^*[\mathbf{n}\times\mathbf{E}]g\}ds \ ,$$

where the arguments in the harmonics f, g are omitted.

Now take formulas (8.3.31) and (8.3.32) and write them for the case where extraneous currents and charges inside the integration domain D are zero:

$$\iiint_D [\mathbf{E}(\Delta P + k^{*2}P)]\,dv = \iint_S [(\mathbf{n}\cdot\mathbf{E})\operatorname{grad} P + (\mathbf{n}\times\mathbf{E})\times\operatorname{grad} P + i\omega\mu_0(\mathbf{n}\times\mathbf{H})P]\,ds \ ;$$
$$(9.3.7)$$
$$\iiint_D [\mathbf{H}(\Delta P + k^{*2}P)]\,dv = \iint_S [(\mathbf{n}\cdot\mathbf{H})\operatorname{grad} P + (\mathbf{n}\times\mathbf{H})\times\operatorname{grad} P + \sigma^*(\mathbf{n}\times\mathbf{E})P]\,ds \ .$$

Applying formulas (9.3.7) to the layer $\tilde{\Omega}_-$ for $P = g$, we obtain

$$\iiint_{\tilde{\Omega}_-} \mathbf{E}(\Delta g + k^{*2}g)\,dv = -\iint_\Sigma [(\mathbf{n}\cdot\mathbf{E}^0)\operatorname{grad} g + (\mathbf{n}\times\mathbf{E}^0)\times\operatorname{grad} g + i\omega\mu_0(\mathbf{n}\times\mathbf{H}^0)g]\,ds$$
$$-\iint_\gamma [(\mathbf{n}\cdot\mathbf{E})\operatorname{grad} g + (\mathbf{n}\times\mathbf{E})\times\operatorname{grad} g + i\omega\mu_0(\mathbf{n}\times\mathbf{H})g]\,ds \ ;$$
$$(9.3.8)$$
$$\iiint_{\tilde{\Omega}_-} \mathbf{H}(\Delta g + k^{*2}g)\,dv = -\iint_\Sigma \{(\mathbf{n}\cdot\mathbf{H}^0)\operatorname{grad} g + [\mathbf{n}\times\mathbf{H}^0]\times\operatorname{grad} g + \sigma_n^*[\mathbf{n}\times\mathbf{E}^0]g\}ds$$
$$-\iint_\gamma \{(\mathbf{n}\cdot\mathbf{H})\operatorname{grad} g + [\mathbf{n}\times\mathbf{H}]\times\operatorname{grad} g + \sigma_n^*[\mathbf{n}\times\mathbf{E}]g\}ds \ .$$

Since the harmonic g defined by Eq. (9.3.5) satisfies the Helmholtz equation, the left-hand sides of formulas (9.3.7) are equal to zero. Hence,

$$\iint_{\gamma} [(\mathbf{n}\cdot\mathbf{E})\,\mathrm{grad}\,g+(\mathbf{n}\times\mathbf{E})\times\mathrm{grad}\,g+i\omega\mu_0(\mathbf{n}\times\mathbf{H})g]\,ds$$
$$= -\iint_{\Sigma} [(\mathbf{n}\cdot\mathbf{E}^0)\,\mathrm{grad}\,g+(\mathbf{n}\times\mathbf{E}^0)\times\mathrm{grad}\,g+i\omega\mu_0(\mathbf{n}\times\mathbf{H}^0)g]\,ds\ ;$$
$$\iint_{\gamma} [(\mathbf{n}\cdot\mathbf{H})\,\mathrm{grad}\,g+(\mathbf{n}\times\mathbf{H})\times\mathrm{grad}\,g+\sigma_n^*(\mathbf{n}\times\mathbf{E})g]\,ds$$
$$= -\iint_{\Sigma} [(\mathbf{n}\cdot\mathbf{H}^0)\,\mathrm{grad}\,g+(\mathbf{n}\times\mathbf{H}^0)\times\mathrm{grad}\,g+\sigma_n^*(\mathbf{n}\times\mathbf{E}^0)g]\,ds\ . \qquad (9.3.9)$$

We have transformed the integrals over the plane γ into the integrals over the Earth's surface Σ. On substituting of Eqs. (9.3.9) into (9.3.6), and (9.3.6) into (9.3.3) and orienting the unit vector of the normal $\mathbf{n}$ from the Earth into the air we obtain *analytical continuation of the field* $\mathbf{E}^0$, $\mathbf{H}^0$ *into the horizontal layer* $\tilde{\Omega}_-$:

$$\mathbf{E}(\mathbf{r}') = \iint_{S} [(\mathbf{n}\cdot\mathbf{E}^0)\,\mathrm{grad}\,G_n^h+(\mathbf{n}\times\mathbf{E}^0)\times\mathrm{grad}\,G_n^h+i\omega\mu_0(\mathbf{n}\times\mathbf{H}^0)\,G_n^h]\,ds$$
$$-\frac{1}{8\pi^2}\iint_{-\infty}^{\infty}\frac{f}{\eta_n}\,dk_x dk_y \iint_{S} [(\mathbf{n}\cdot\mathbf{E}^0)\,\mathrm{grad}\,g$$
$$+(\mathbf{n}\times\mathbf{E}^0)\times\mathrm{grad}\,g+i\omega\mu_0(\mathbf{n}\times\mathbf{H}^0)g]\,ds\ ;$$
$$\mathbf{H}(\mathbf{r}') = \iint_{\Sigma} [(\mathbf{n}\cdot\mathbf{H}^0)\,\mathrm{grad}\,G_n^h+(\mathbf{n}\times\mathbf{H}^0)\times\mathrm{grad}\,G_n^h+\sigma_n^*(\mathbf{n}\times\mathbf{E}^0)\,G_n^h]\,ds$$
$$-\frac{1}{8\pi^2}\iint_{-\infty}^{\infty}\frac{f}{\eta_n}\,dk_x dk_y \iint_{\Sigma} [(\mathbf{n}\cdot\mathbf{H}^0)\,\mathrm{grad}\,g$$
$$+(\mathbf{n}\times\mathbf{H}^0)\times\mathrm{grad}\,g+\sigma_n^*(\mathbf{n}\times\mathbf{E}^0)g]\,ds\ . \qquad (9.3.10)$$

In practice, the most common models are those with a plane Earth's surface $z = 0$. Within these models, the Earth's surface is a coordinate one and the Green function can be expanded not only in the second but also in the first integrals of formulas (9.3.10). Since the z-axes of the points $\mathbf{r}' \in \Omega_-$ and $\mathbf{r} \in \Sigma$ in these integrals meet the condition $z' < z$, the expansion of the Green function takes the form

$$G_n^h = \frac{1}{8\pi^2}\iint_{-\infty}^{\infty}\frac{f^*(\mathbf{r}')g^*(\mathbf{r})}{\eta_n}\,dk_x dk_y\ , \qquad (9.3.11)$$

where

$$f^*(\mathbf{r}') = -\exp[-i(k_x x'+k_y y')]\exp(\eta_n z')\,|_{z'<z}\ ;$$
$$g^*(\mathbf{r}) = \exp[i(k_x x+k_y y)]\exp(-\eta_n z)\,|_{z'<z}\ . \qquad (9.3.12)$$

Substitute expansion (9.3.11) into the first integrals of formulas (9.3.10) and, discarding the atmospheric electric conductivity, take that $\mathbf{n}\cdot\mathbf{E}^0|_{\Sigma}=0$. Then, for $\mathbf{r}'\in\tilde{\Omega}_-$

$$\begin{aligned}\mathbf{E}(\mathbf{r}') = \frac{1}{8\pi^2}\Bigg\{ & \iint\limits_{-\infty}^{\infty} \frac{f^*}{\eta_n}\,dk_x dk_y \iint\limits_{\Sigma} [(\mathbf{n}\cdot\mathbf{E}^0)\,\mathrm{grad}\,g^* + (\mathbf{n}\times\mathbf{E}^0)\times\mathrm{grad}\,g^* \\ & + i\omega\mu_0(\mathbf{n}\times\mathbf{H}^0)g^*]\,ds - \iint\limits_{-\infty}^{\infty} \frac{f}{\eta_n}\,dk_x dk_y \iint\limits_{\Sigma} [(\mathbf{n}\cdot\mathbf{E}^0)\,\mathrm{grad}\,g \\ & + (\mathbf{n}\times\mathbf{E}^0)\times\mathrm{grad}\,g + i\omega\mu_0(\mathbf{n}\times\mathbf{H}^0)g]\,ds\Bigg\}\ ; \\ \mathbf{H}(\mathbf{r}') = \frac{1}{8\pi^2}\Bigg\{ & \iint\limits_{-\infty}^{\infty} \frac{f^*}{\eta_n}\,dk_x dk_y \iint\limits_{\Sigma} [(\mathbf{n}\cdot\mathbf{H}^0)\,\mathrm{grad}\,g^* + (\mathbf{n}\times\mathbf{H}^0)\times\mathrm{grad}\,g^* \\ & + \sigma_n^*(\mathbf{n}\times\mathbf{E}^0)g^*]\,ds - \iint\limits_{-\infty}^{\infty} \frac{f}{\eta_n}\,dk_x dk_y \iint\limits_{\Sigma} [(\mathbf{n}\cdot\mathbf{H}^0)\,\mathrm{grad}\,g \\ & + (\mathbf{n}\times\mathbf{H}^0)\times\mathrm{grad}\,g + \sigma_n^*(\mathbf{n}\times\mathbf{E}^0)g]\,ds\Bigg\}\ ,\end{aligned} \tag{9.3.13}$$

wherefrom a little (but rather cumbersome) transformation yields

$$\begin{aligned} E_x(\mathbf{r}') &= \frac{1}{4\pi^2}\iint\limits_{-\infty}^{\infty}\left(e_x^0\,\mathrm{ch}\,\eta_n z' + i\omega\mu_0 h_y^0\,\frac{\mathrm{sh}\,\eta_n z'}{\eta_n}\right)\exp[-i(k_x x' + k_y y')]\,dk_x dk_y\ ; \\ E_y(\mathbf{r}') &= \frac{1}{4\pi^2}\iint\limits_{-\infty}^{\infty}\left(e_y^0\,\mathrm{ch}\,\eta_n z' - i\omega\mu_0 h_x^0\,\frac{\mathrm{sh}\,\eta_n z'}{\eta_n}\right)\exp[-i(k_x x' + k_y y')]\,dk_x dk_y\ ; \\ E_z(\mathbf{r}') &= \frac{i}{4\pi^2}\iint\limits_{-\infty}^{\infty}(k_x e_x^0 + k_y e_y^0)\,\frac{\mathrm{sh}\,\eta_n z'}{\eta_n}\exp[-i(k_x x' + k_y y')]\,dk_x dk_y\ ; \end{aligned} \tag{9.3.14a}$$

$$\begin{aligned} H_x(\mathbf{r}') &= \frac{1}{4\pi^2}\iint\limits_{-\infty}^{\infty}\left[h_x^0\,\mathrm{ch}\,\eta_n z' + (\sigma_n^* e_y^0 - ik_x h_z^0)\,\frac{\mathrm{sh}\,\eta_n z'}{\eta_n}\right] \\ &\quad\times\exp[-i(k_x x' + k_y y')]\,dk_x dk_y\ ; \\ H_y(\mathbf{r}') &= \frac{1}{4\pi^2}\iint\limits_{-\infty}^{\infty}\left[h_y^0\,\mathrm{ch}\,\eta_n z' - (\sigma_n^* e_x^0 + ik_y h_z^0)\,\frac{\mathrm{sh}\,\eta_n z'}{\eta_n}\right] \\ &\quad\times\exp[-i(k_x x' + k_y y')]\,dk_x dk_y\ ; \\ H_z(\mathbf{r}') &= \frac{1}{4\pi^2}\iint\limits_{-\infty}^{\infty}\left[h_z^0\,\mathrm{ch}\,\eta_n z' + i(k_x h_x^0 + k_y h_y^0)\,\frac{\mathrm{sh}\,\eta_n z'}{\eta_n}\right] \\ &\quad\times\exp[-i(k_x x' + k_y y')]\,dk_x dk_y\ , \end{aligned} \tag{9.3.14b}$$

where $\mathbf{e}^0$, $\mathbf{h}^0$ are the spectra of the field $\mathbf{E}^0$, $\mathbf{H}^0$ defined by the relations

$$\mathbf{e}^0 = \iint\limits_{-\infty}^{\infty} \mathbf{E}^0 \exp[i(k_x x + k_y y)]\, dx\, dy \;;$$
$$\mathbf{h}^0 = \iint\limits_{-\infty}^{\infty} \mathbf{H}^0 \exp[i(k_x x + k_y y)]\, dx\, dy \;. \tag{9.3.15}$$

9.3.4 Continuation of a Two-Dimensional Electromagnetic Field

A two-dimensional continuation operator is constructed in a similar way. We start from two-dimensional Stratton-Chu formulas (8.3.48) and (8.3.49). The expansion of the two-dimensional Green function in quasi-plane waves for $\partial/\partial y = 0$ has the form

$$G_n^h = \frac{1}{4\pi} \int\limits_{-\infty}^{\infty} \frac{f(\mathbf{r}')g(\mathbf{r})}{\eta_n}\, dk_x \;, \tag{9.3.16}$$

where

$$f(\mathbf{r}') = -\exp(ik_x x'] \exp(-\eta_n z')|_{z'>z} \;;$$
$$g(\mathbf{r}) = \exp(-ik_x x)\exp(\eta_n z)|_{z'>z} \;; \quad \eta_n = \sqrt{k_x^2 - k_n^{*2}} \;; \quad \mathrm{Re}\,\eta_n > 0 \;.$$

An appropriate transformation yields analytical continuation of the field $\mathbf{E}^0$, $\mathbf{H}^0$ into the horizontal band $\tilde{\Omega}_-$. In the case of E-polarization,

$$\mathbf{E}(\mathbf{r}') = \int\limits_{\Sigma} \left[\mathbf{E}^0 \frac{\partial G_n^h}{\partial n} + i\omega\mu_0(\mathbf{n}\times\mathbf{H}^0) G_n^h \right] ds$$
$$- \frac{1}{4\pi} \int\limits_{-\infty}^{\infty} \frac{f}{\eta_n}\, dk_x \int\limits_{\Sigma} \left[\mathbf{E}^0 \frac{\partial g}{\partial n} + i\omega\mu_0(\mathbf{n}\times\mathbf{H}^0) g \right] ds \;; \tag{9.3.17a}$$

$$\mathbf{H}(\mathbf{r}') = \int\limits_{\Sigma} [(\mathbf{n}\cdot\mathbf{H}^0)\,\mathrm{grad}\, G_n^h + (\mathbf{n}\times\mathbf{H}^0)\times \mathrm{grad}\, G_n^h + \sigma_n^*(\mathbf{n}\times\mathbf{E}^0) G_n^h]\, ds$$
$$- \frac{1}{4\pi} \int\limits_{-\infty}^{\infty} \frac{f}{\eta_n}\, dk_x \int\limits_{S} [(\mathbf{n}\cdot\mathbf{H}^0)\,\mathrm{grad}\, g + (\mathbf{n}\times\mathbf{H}^0)\times \mathrm{grad}\, g + \sigma_n^*(\mathbf{n}\times\mathbf{E}^0) g]\, ds \;, \tag{9.3.17b}$$

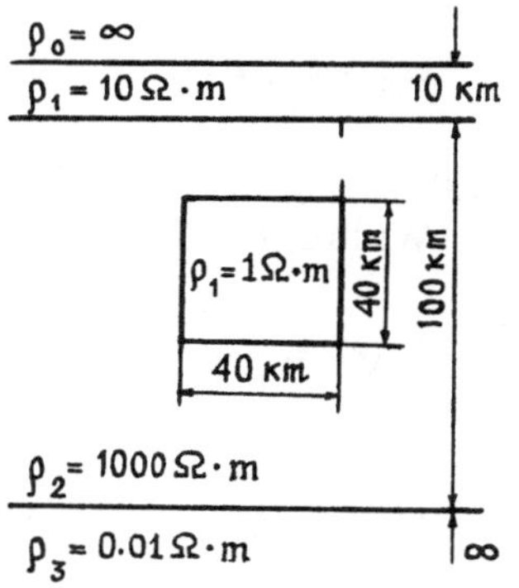

Fig. 57. Geoelectrical model

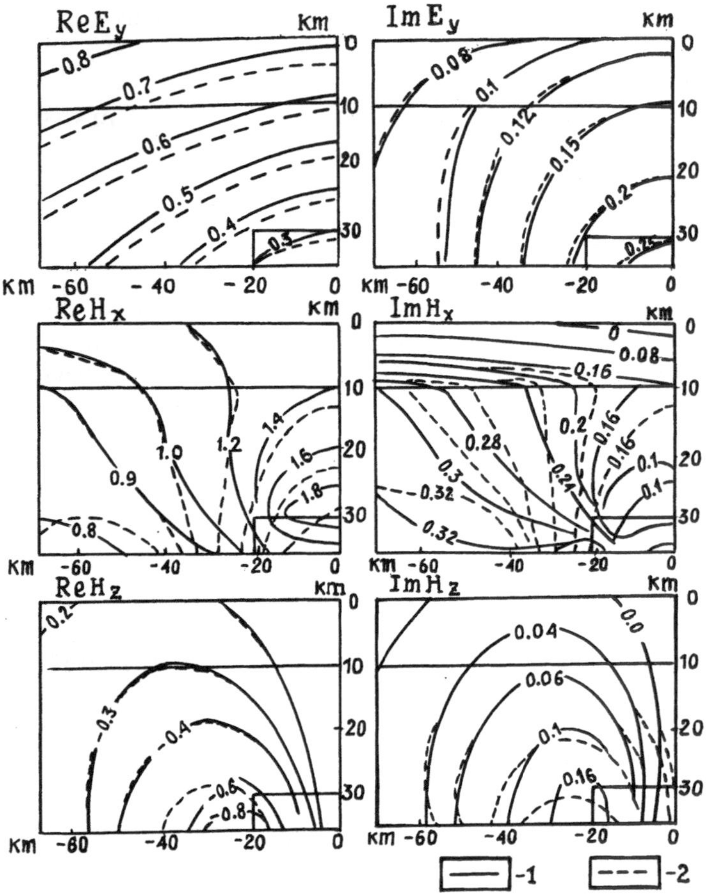

Fig. 58. Results of modeling and analytical continuation of the electromagnetic field for the model of Fig. 57 (for period $T = 2100s$); ——— finite-difference modeling; - - - analytical continuation

where the unit vector of the normal **n** points from the Earth to the air and

$$G_n^h = G_n^h(\mathbf{r}'|\mathbf{r}) = \frac{-i}{4} H_0^{(1)}(k_n^* \; |\mathbf{r}' - \mathbf{r}|)$$

($H_0^{(1)}$ being the zero-order Hankel function of the first kind).

Within the model of a plane Earth's surface

$$E_y(\mathbf{r}') = \frac{1}{2\pi} \int_{-\infty}^{\infty} \left(e_y^0 \operatorname{ch} \eta_n z' - i\omega\mu_0 h_x^0 \frac{\operatorname{sh} \eta_n z'}{\eta_n} \right) \exp(-ik_x x') dk_x \; ;$$

$$H_x(\mathbf{r}') = \frac{1}{2\pi} \int_{-\infty}^{\infty} \left(h_x^0 \operatorname{ch} \eta_n z' + \eta_n h_z^0 \frac{\operatorname{sh} \eta_n z'}{ik_x} \right) \exp(-ik_x x') dk_x \; ; \qquad (9.3.18)$$

$$H_z(\mathbf{r}') = \frac{1}{2\pi} \int_{-\infty}^{\infty} \left(h_z^0 \operatorname{ch} \eta_n z' + ik_x h_x^0 \frac{\operatorname{sh} \eta_n z'}{\eta_n} \right) \exp(-ik_x x') dk_x \ ,$$

which follows directly from (9.3.14).

In concluding this subsection we should like to note that analytical continuation of the electromagnetic field into the lower half-space is, just as in the Laplace field case, an ill-posed problem and its numerical solution calls for regularization.

This problem can be regularized, say, by regularizing filtering described comprehensively in the monograph by Berdichevsky and Zhdanov (1984). Figure 58 exemplifies the results of analytical continuation of the *E*-polarized field within the model of Fig. 57. The electromagnetic field at the Earth's surface has been calculated using finite-difference modeling. Figure 58 is an isoline map of the electromagnetic field components (dotted lines) continued on the basis of formulas (9.3.18) (with the help of a regularizing filtering algorithm). For comparison, the maps produced from the results of finite-difference modeling are also given (solid lines). Calculations have revealed that the field above an inhomogeneous square conducting inclusion is reconstructed fairly exactly.

The slight discordance between the modeling and the analytical continuation results is due to the smoothing (regularizing filtering) of the continued field. Below the level of the inhomogeneity upper edge, the continued field becomes distorted, which is attributed, by the general theory of analytical continuation, to the presence of singular points at the corners of a conducting square.

Those concerned with the applied aspects of this problem are referred to the monograph by Berdichevsky and Zhdanov (1984) where techniques for continuation into more complicated domains are treated in detail.

10 Migration of the Electromagnetic Field

10.1 Definition of the Concept of Migration

The previous chapter dealt with the theory of analytical continuation of the electromagnetic field in a well-conducting medium. The examination was confined to a monochromatic field. As indicated in our study (Zhdanov 1981), similar techniques can be also elaborated for transient electromagnetic fields. But the latter can be continued using a method somewhat different from that utilized in ordinary analytical continuation. This method is known as *migration of the electromagnetic field* (Zhdanov and Frenkel 1982a, b, 1983a, b). As will be shown below, this procedure is very much similar to the method used for seismic data interpretation and called seismoholography or seismic migration (Petrashen and Nakhamkin 1973; Timoshin 1978; Berkhout 1984; Claerbout 1985). It extends, in turn, the ideas of optical or radio holography (Caulfiled 1979: Handbook of Optical Holography) to the case of seismic (wave) fields.

It is well known that optical holography permits reconstruction of a space image of an object by using a hologram displaying not only the amplitude but also the phase structure of the electromagnetic field. It is sufficient to illuminate a hologram with a coherent light beam. In seismics, the Earth's surface, a huge space hologram, registers the amplitudes and phases of wave seismic fields rather than those of light waves. The procedure of hologram illumination by a coherent light beam is replaced by one that makes the points at the Earth's surface oscillate in the reversed time conditions $\tau = T - t$ (t being the ordinary time, and T the signal recording interval). Thereby a wave field is produced in the Earth, which propagates in reversed time and focuses at diffraction points and imaginary sources of reflected waves. Carrying out this operation numerically by a computer and drawing in space the amplitudes of reconstructed waves at the instant of their arrival at a specified point of a medium yield a seismic image of the internal structure of the Earth (this question will be treated in greater detail in Chap. 12).

A similar approach applies in principle to electromagnetic data interpretation. For instance, the electromagnetic field produced by a high-power artificial or natural source and recorded at the Earth's surface can be regarded as a huge space hologram. The problem is to reconstruct a "geoelectrical image" of the medium from this "hologram". To do this, it is sufficient, just as in seismoholography, to arrange fictitious extrinsic currents and charges at observation points, which would vary in reversed time, according to the law determined by the field measured at these points [i.e., such as would be related to the measured fields by equations like (8.4.31)]. Then the electromagnetic field induced by these sources will "light up" the deep structure of the Earth, as is the case with image reconstruction in ordinary holography. We should like to stress here that the mathematical formulation of the above procedure brings us just to the Stratton-Chu type integrals (written, however, for reversed time).

These are general considerations underlying the method of electromagnetic field migration. Now we will proceed to a systematic exposition of the migration theory.

10.1.1 Definition of a Migration Field

Let us state the following problem. At the Earth's surface Σ, over a time interval 0 to T, there is specified an electromagnetic field $\mathbf{E}^0$, $\mathbf{H}^0$, excited by a nonstationary source of a density $\mathbf{j}^Q$ which is located either on the outer side of the surface Σ or is raised above the Earth, and which is switched on at an instant $t=0$. The Earth is notable for a constant electric conductivity σ, except for some deep domain D (finite or infinitely extended) where the electric conductivity may vary arbitrarily. The magnetic permeability is everywhere equal to the value in vacuum μ_0. Displacement currents are neglected, i.e., a quasi-stationary model of the field is considered.

The problem is to find the domain of an anomalous electric conductivity from the fields $\mathbf{E}^0(\mathbf{r}, t)$ and $\mathbf{H}^0(\mathbf{r}, t)$ measured at the surface of the Earth. To solve this problem, we will do the following field transformation. We shall pass from ordinary time t to reversed time $\tau = T-t$ and specify auxiliary fields $\mathbf{F}(\mathbf{r}, t)$ and $\mathbf{R}(\mathbf{r}, t)$ at the Earth's surface by the formulas

$$\mathbf{F}_{tg}(\mathbf{r}, \tau) = \mathbf{E}^0_{tg}(\mathbf{r}, T-\tau) \ , \quad F_n(\mathbf{r}, \tau) = \frac{1}{c_1} E^0_n(\mathbf{r}, T-\tau) \ ,$$

$$\mathbf{R}_{tg}(\mathbf{r}, \tau) = \mathbf{H}^0_{tg}(\mathbf{r}, T-\tau) \ , \quad R_n(\mathbf{r}, \tau) = -H^0_n(\mathbf{r}, T-\tau) \ . \tag{10.1.1}$$

Here the subscript tg refers to the tangential component of the field.

Evidently, within this definition, the functions $\mathbf{F}$ and $\mathbf{R}$ satisfy conditions (8.2.24) (with t replaced by reversed time τ and on condition that the conductivity of the lower half-space domain adjacent to Σ is $\sigma^{m1} = c_1\sigma$):

$$\mathbf{F}_n = -\frac{1}{\sigma^{m1}} \operatorname{div}_\Gamma(\mathbf{n}\times\mathbf{R}_{tg}) \ , \quad \frac{\partial R_n}{\partial \tau} = \frac{1}{\mu_0} \operatorname{div}_\Gamma(\mathbf{n}\times\mathbf{F}_{tg}) \ . \tag{10.1.2}$$

Hence, these functions can be employed as densities of the Stratton-Chu type integrals written for reversed time:

$$\mathbf{E}^{m1}(\mathbf{r}', \tau') = \int_0^{\tau'} \iint_\Sigma \left[(\mathbf{n}\cdot\mathbf{F})\nabla G^{m1} + (\mathbf{n}\times\mathbf{F})\times\nabla G^{m1} + (\mathbf{n}\times\mathbf{R})\mu_0 \frac{\partial G^{m1}}{\partial \tau} \right] ds\,d\tau \ , \tag{10.1.3a}$$

$$\mathbf{H}^{m1}(\mathbf{r}', \tau') = \int_0^{\tau'} \iint_\Sigma [(\mathbf{n}\cdot\mathbf{R})\nabla G^{m1} + (\mathbf{n}\times\mathbf{R})\times\nabla G^{m1} + (\mathbf{n}\times\mathbf{F})\sigma^{m1} G^{m1}] \, ds\,d\tau \ , \tag{10.1.3b}$$

where

$$G^{m1}(\mathbf{r}', \tau'|\mathbf{r}, \tau) = G^{d}(\mathbf{r}', \tau'|\mathbf{r}, \tau)$$
$$= \frac{-(\mu_0\sigma^{m1})^{1/2}}{8\pi^{3/2}(\tau-\tau')^{3/2}} \exp\left(-\frac{\mu_0\sigma^{m1}\ |\mathbf{r}-\mathbf{r}'|^2}{4(\tau-\tau')}\right) \varkappa(\tau-\tau') \ . \quad (10.1.3\text{c})$$

In accordance with the main properties of the Stratton-Chu type integrals, the fields $\mathbf{E}^{m1}$ and $\mathbf{H}^{m1}$ outside Σ satisfy the Maxwell equations:

$$\operatorname{rot}\mathbf{H}^{m1} = \sigma^{m1}\mathbf{E}^{m1} \ ; \quad \operatorname{rot}\mathbf{E}^{m1} = \mu_0 \frac{\partial \mathbf{H}^{m1}}{\partial \tau'} \ . \quad (10.1.4)$$

The field $\mathbf{E}^{m1}(\mathbf{r}', \tau')$, $\mathbf{H}^{m1}(\mathbf{r}', \tau')$ will be called a *migration electromagnetic field,* while the constant quantity c_1 a migration constant.

The migration field $\mathbf{E}^{m1}$, $\mathbf{H}^{m1}$ can be made to have a simple physical meaning. Indeed, according to (8.4.31), determine the fictitious electric and magnetic charges and currents corresponding to the functions **F**, **R** and to the field $\mathbf{E}^0$, $\mathbf{H}^0$:

$$\mathbf{J}^{*E}(\mathbf{r}, \tau) = -\mathbf{n}\times\mathbf{H}^0(\mathbf{r}, T-\tau) \ , \quad \mathbf{J}^{*H}(\mathbf{r}, \tau) = \mathbf{n}\times\mathbf{E}^0(\mathbf{r}, T-\tau) \ ,$$
$$q_S^{*E}(\mathbf{r}, \tau) = -\frac{\varepsilon}{c_1}[\mathbf{n}\cdot\mathbf{E}^0(\mathbf{r}, T-\tau)] \ , \quad q_S^{*H}(\mathbf{r}, \tau) = \mu_0[\mathbf{n}\cdot\mathbf{H}^0(\mathbf{r}, T-\tau)] \ . \quad (10.1.5)$$

As a result, the expressions for $\mathbf{E}^{m1}$, $\mathbf{H}^{m1}$ can be cast in the form:

$$\mathbf{E}^{m1}(\mathbf{r}', \tau') = -\int_0^{\tau'}\iint_{\Sigma}\left(\frac{q_S^{*E}\nabla G^{m1}}{\varepsilon} - \mathbf{J}^{*H}\times\nabla G^{m1} + \mu_0\mathbf{J}^{*E}\frac{\partial G^{m1}}{\partial\tau}\right) ds\,d\tau \ ; \quad (10.1.6\text{a})$$

$$\mathbf{H}^{m1}(\mathbf{r}', \tau') = -\int_0^{\tau'}\iint_{\Sigma}\left(\frac{-q_S^{*H}\nabla G^{m1}}{\mu_0} + \mathbf{J}^{*E}\times\nabla G^{m1} - \sigma^{m1}\mathbf{J}^{*H}G^{m1}\right) ds\,d\tau \ . \quad (10.1.6\text{b})$$

Thus, the migration field is an electromagnetic field excited in a homogeneous conducting Earth by a system of fictitious currents and charges evaluated from the observed field E^0, H^0 and varying in reversed time (i.e., by those which are switched on at a finite instant of field-recording time $t = T$, operated in reversed time $\tau = T-t$, and switched off at an instant $t = 0$). Substitution of (10.1.5) into (10.1.6) yields

$$\mathbf{E}^{m1}(\mathbf{r}', \tau') = \int_0^{\tau'}\iint_{\Sigma}\left\{\frac{[\mathbf{n}\cdot\mathbf{E}^0(\mathbf{r}, T-\tau)]}{c_1}\nabla G^{m1} + [\mathbf{n}\times\mathbf{E}^0(\mathbf{r}'; T-\tau)]\times\nabla G^{m1}\right.$$
$$\left. + [\mathbf{n}\times\mathbf{H}^0(\mathbf{r}, T-\tau)]\mu_0\frac{\partial G^{m1}}{\partial\tau}\right\} ds\,d\tau \ ;$$
$$(10.1.7)$$
$$\mathbf{H}^{m1}(\mathbf{r}', \tau') = \int_0^{\tau'}\iint_{\Sigma}\{-[\mathbf{n}\cdot\mathbf{H}^0(\mathbf{r}, T-\tau)]\nabla G^{m1} + [\mathbf{n}\times\mathbf{H}^0(\mathbf{r}, T-\tau)]\times\nabla G^{m1}$$
$$+ [\mathbf{n}\times\mathbf{E}^0(\mathbf{r}, T-\tau)]\sigma^{m1}G^{m1}\}ds\,d\tau \ .$$

Replace the variable τ in the right-hand sides of expressions (10.1.7) by $T-t$, then we have

$$\mathbf{E}^{m1}(\mathbf{r}', T-t') = \int_{t'}^{T} \iint_{\Sigma} \left\{ \frac{(\mathbf{n}\cdot\mathbf{E}^0)}{c_1} \nabla \tilde{G}^{m1} + (\mathbf{n}\times\mathbf{E}^0)\times\nabla\tilde{G}^{m1} + (\mathbf{n}\times\mathbf{H}^0)\mu_0 \frac{\partial \tilde{G}^{m1}}{\partial t} \right\} ds\,dt \ , \quad (10.1.8\text{a})$$

$$\mathbf{H}^{m1}(\mathbf{r}', T-t') = \int_{t'}^{T} \iint_{\Sigma} [-(\mathbf{n}\cdot\mathbf{H}^0)\nabla\tilde{G}^{m1} + (\mathbf{n}\times\mathbf{H}^0)\times\nabla\tilde{G}^{m1} + (\mathbf{n}\times\mathbf{E}^0)\sigma^{m1}G^{m1}]\,ds\,dt \ , \quad (10.1.8\text{b})$$

where $\tilde{G}^{m1} = \tilde{G}^{m1}(\mathbf{r}', t'|\mathbf{r}, t)$ is the function conjugate to the Green fundamental function for the diffusion equation. The functions G^{m1} and $\tilde{G}^{m1}$ are related by (Morse and Feshbach 1953)

$$\tilde{G}^{m1}(\mathbf{r}', t'|\mathbf{r}, t) = G^{m1}(\mathbf{r}', -t'|r, -t) = G^{m1}(\mathbf{r}, t|r', t') \ . \quad (10.1.9)$$

The transformation of the field $\mathbf{E}^0$, $\mathbf{H}^0$ observed at Σ fitted by formulas (10.1.8) will be termed *migration of the electromagnetic field.*

It is remarkable that migration of the electromagnetic field is a stable procedure, since the operators included in the right-hand side of expressions (10.1.8) are restricted (in space L_2).

10.1.2 System of Migration Transforms of Nonstationary Electromagnetic Fields

The concept of a migration field involves some arbitrary rules for specifying auxiliary fields **F**, **R**. If the introduced transform of the electromagnetic field $\mathbf{E}^0$, $\mathbf{H}^0$ is to define some new electromagnetic field varying in reversed time τ, the auxiliary field **F**, **R** can be introduced in three ways more besides (10.1.1):

$$\begin{vmatrix} \mathbf{F}_{tg}(\mathbf{r}, \tau) = -\mathbf{E}^0_{tg}(\mathbf{r}, T-\tau) \ ; & F_n(\mathbf{r}, \tau) = (1/c_2)E^0_n(\mathbf{r}, T-\tau) \ ; \\ \mathbf{R}_{tg}(\mathbf{r}, \tau) = \mathbf{H}^0_{tg}(\mathbf{r}, T-\tau) \ ; & R_n(\mathbf{r}, \tau) = H^0_n(\mathbf{r}, T-\tau) \ ; \end{vmatrix} \quad (10.1.10)$$

$$\begin{vmatrix} \mathbf{F}_{tg}(\mathbf{r}, \tau) = -\mathbf{E}^0_{tg}(\mathbf{r}, T-\tau) \ ; & F_n(\mathbf{r}, \tau) = E^0_n(\mathbf{r}, T-\tau) \ ; \\ \mathbf{R}_{tg}(\mathbf{r}, \tau) = c_3\mathbf{H}^0_{tg}(\mathbf{r}, T-\tau) \ ; & R_n(\mathbf{r}, \tau) = H^0_n(\mathbf{r}, T-\tau) \ ; \end{vmatrix} \quad (10.1.11)$$

$$\begin{vmatrix} \mathbf{F}_{tg}(\mathbf{r}, \tau) = \mathbf{E}^0_{tg}(\mathbf{r}, T-\tau) \ ; & F_n(\mathbf{r}, \tau) = E^0_n(\mathbf{r}, T-\tau) \ ; \\ \mathbf{R}_{tg}(\mathbf{r}, \tau) = c_4\mathbf{H}^0_{tg}(\mathbf{r}, T-\tau) \ ; & R_n(\mathbf{r}, \tau) = -H^0_n(\mathbf{r}, T-\tau) \ , \end{vmatrix} \quad (10.1.12)$$

where c_2, c_3, and c_4 are some constants (migration constants).

One can readily see that each of the pairs of the auxiliary fields **F**, **R** defined by formulas (10.1.10)–(10.1.12) meets conditions (10.1.2), upon substitution therein of $\sigma^{m2} = c_2\sigma$; $\sigma^{m3} = c_3\sigma$; and $\sigma^{m4} = c_4\sigma$ for σ^{m1}, respectively. Thus, we can consider three more types of migration transform, namely migration electromagnetic fields $\mathbf{E}^{m2}$, $\mathbf{H}^{m2}$, $\mathbf{E}^{m3}$, $\mathbf{H}^{m3}$ and $\mathbf{E}^{m4}$, $\mathbf{H}^{m4}$ defined, similarly to (10.1.8), by

$$\left|\begin{array}{l}\mathbf{E}^{m2}(\mathbf{r}', T-t') = \int\limits_{t'}^{T} \iint\limits_{\Sigma} \left[\frac{(\mathbf{n}\cdot\mathbf{E})}{c_2} \nabla \tilde{G}^{m2} - (\mathbf{n}\times\mathbf{E})\times\nabla\tilde{G}^{m2} + (\mathbf{n}\times\mathbf{H})\mu_0 \frac{\partial \tilde{G}^{m2}}{\partial t} \right] ds\,dt \\ \mathbf{H}^{m2}(\mathbf{r}', T-t') = \int\limits_{t'}^{T} \iint\limits_{\Sigma} [(\mathbf{n}\cdot\mathbf{H})\nabla\tilde{G}^{m2} + (\mathbf{n}\times\mathbf{H})\times\nabla\tilde{G}^{m2} - (\mathbf{n}\times\mathbf{E})\sigma^{m2}\tilde{G}^{m2}]\, ds\,dt \end{array}\right. \tag{10.1.13}$$

$$\left|\begin{array}{l}\mathbf{E}^{m3}(\mathbf{r}', T-t') = \int\limits_{t'}^{T} \iint\limits_{\Sigma} \left[(\mathbf{n}\cdot\mathbf{E}) \nabla \tilde{G}^{m3} - (\mathbf{n}\times\mathbf{E})\times\nabla\tilde{G}^{m3} + c_3(\mathbf{n}\times\mathbf{H})\mu_0 \frac{\partial \tilde{G}^{m3}}{\partial t} \right] ds\,dt \\ \mathbf{H}^{m3}(\mathbf{r}', T-t') = \int\limits_{t'}^{T} \iint\limits_{\Sigma} [(\mathbf{n}\cdot\mathbf{H})\nabla\tilde{G}^{m3} + c_3(\mathbf{n}\times\mathbf{H})\times\nabla\tilde{G}^{m3} + (\mathbf{n}\times\mathbf{E})\sigma^{m3}\tilde{G}^{m3}]\, ds\,dt \end{array}\right. \tag{10.1.14}$$

$$\left|\begin{array}{l}\mathbf{E}^{m4}(\mathbf{r}', T-t') = \int\limits_{t'}^{T} \iint\limits_{\Sigma} \left[(\mathbf{n}\cdot\mathbf{E}) \nabla \tilde{G}^{m4} + (\mathbf{n}\times\mathbf{E})\times\nabla\tilde{G}^{m4} + c_4(\mathbf{n}\times\mathbf{H})\mu_0 \frac{\partial \tilde{G}^{m4}}{\partial t} \right] ds\,dt \\ \mathbf{H}^{m4}(\mathbf{r}', T-t') = \int\limits_{t'}^{T} \iint\limits_{\Sigma} [-(\mathbf{n}\cdot\mathbf{H})\nabla\tilde{G}^{m4} + c_4(\mathbf{n}\times\mathbf{H})\times\nabla\tilde{G}^{m4} + (\mathbf{n}\times\mathbf{E})\sigma^{m4}\tilde{G}^{m4}]\, ds\,dt \end{array}\right. \tag{10.1.15}$$

where

$$\tilde{G}^{mi} = \frac{-(\mu_0\sigma^{mi})^{1/2}}{8\pi^{3/2}(t-t')^{3/2}} \exp\left(\frac{-\mu_0\sigma^{mi}\,|\mathbf{r}-\mathbf{r}'|^2}{4(t-t')}\right) \varkappa(t-t') \;;\quad i=2,\,3,\,4 \;.$$

Transformations of the field $\mathbf{E}^0$, $\mathbf{H}^0$ observed at Σ performed by formulas (10.1.8), (10.1.13)–(10.1.15) yield a *system of migration transforms of a nonstationary electromagnetic field.*

Concluding this subsection, we should like to note that our concept of electromagnetic field migration has been introduced by analogy of the seismic migration (Berkhout 1984), so as to stress the fundamental similarity of the two techniques. As will be shown below, the migration procedure makes it possible to focus the electromagnetic field onto a geoelectrical inhomogeneity.

10.2 Properties of Migration Fields

10.2.1 Equation for a Migration Field in Direct Time

To present the properties of the migration fields $\mathbf{E}^{mi}$, $\mathbf{H}^{mi}$ $(i=1, 2, 3, 4)$ more clearly, pass again from the reversed time τ to ordinary direct time t:

$$\tilde{\mathbf{E}}^i(\mathbf{r}', t') = \mathbf{E}^{mi}(\mathbf{r}', T-t') \;;\quad \tilde{\mathbf{H}}^i(\mathbf{r}', t') = \mathbf{H}^{mi}(\mathbf{r}', T-\tau') \;.$$

The fields $\tilde{\mathbf{E}}^i$, $\tilde{\mathbf{H}}^i$ in direct time satisfy the equations

$$\operatorname{rot} \tilde{\mathbf{H}}^i = \sigma^{mi} \tilde{\mathbf{E}}^i \ , \quad \operatorname{rot} \tilde{\mathbf{E}}^i = \mu_0 \frac{\partial \tilde{\mathbf{H}}^i}{\partial t'}$$

and, hence, meet the diffusion equation for reversed time

$$\Delta(\tilde{\mathbf{E}}^i, \tilde{\mathbf{H}}^i) + \mu_0 \sigma^{mi} \frac{\partial (\tilde{\mathbf{E}}^i, \tilde{\mathbf{H}}^i)}{\partial t} = 0 \ . \tag{10.2.1}$$

While the ordinary equation of diffusion describes the development of the process of electromagnetic field propagation in an ever increasing time from the source to the observer, Eqs. (10.2.1) reflect the same process in a reverse order, i.e., from the final distribution of the field $\mathbf{E}^0$, $\mathbf{H}^0$ at the Earth's surface to its initial distribution at sources. As a result, the suggested field transforms (10.1.8a, b) describe the conversion of electromagnetic fields, diverging in real media and diffracted by geoelectrical inhomogeneities, into fields converging to the corresponding points, lines, and surfaces of diffraction. In doing so, at an instant of time $\tau' = T$ (or $t' = 0$), the fields $\mathbf{E}^{mi}(\mathbf{r}', T) = \tilde{\mathbf{E}}^i(\mathbf{r}', 0)$ and $\mathbf{H}^{mi}(\mathbf{r}', T) = \tilde{\mathbf{H}}^i(\mathbf{r}', 0)$ form, just as in seismoholography, images of field exciters related to geoelectrical inhomogeneities.

The above process is virtually equivalent to the field transformation common in ordinary optical holography. Indeed, proceeding to time spectra in formulas (10.1.8a, b) and (10.1.13)–(10.1.15) we derive, on the right, formulas identical to the ordinary Fresnel formulas (with the only difference that they are written for the diffusion process). Hence, transition to reversed time reduces to the sign reversal of the electromagnetic field phase and, accordingly, to substitution of the incoming wave superposition for the propagating wave superposition.

This point is illustrated by using several theoretical and model examples, and thereby clarifying the resolution of the method and establishing some principles of interpretation of nonstationary electromagnetic fields.

10.2.2 One-, Two-, and Three-Dimensional Migrations of Electromagnetic Source Fields

Each of the above methods used to define migration fields calls for specification of a migration constant $c_i(c_1, c_2, c_3, c_4)$. It may seem at first sight that this additional parameter of the procedure makes the very concept of migration field more complicated. But it will be shown later that the migration constant "c_i" contributes to flexibility of the introduced transform of the electromagnetic field, thereby controlling its focusing properties.

Let us examine some of the most important properties of the introduced migration fields.

Consider, sequentially, the one-, two-, and three-dimensional situations, wherein the electromagnetic field is excited, respectively, by (A) an infinite horizontal current plane, (B) an infinitely long cable, and (C) a horizontal electric dipole immersed in an infinite homogeneous conducting space of an electric conductivity σ. The current plane is parallel to the coordinate plane XOY and passes through a point z_0; the infinitely long cable and the dipole are parallel to the

Y-axis, the cable passes through a point (x_0, z_0), while the dipole is fixed in space at a point (x_0, y_0, z_0), with $z_0>0$ (to simplify the analysis in this chapter the Z-axis faces downward). The current in sources varies in accordance with the law

$$I = j\delta(t) , \tag{10.2.2}$$

where $\delta(t)$ is the Dirac delta-function; j is some constant quantity.

Let $\mathbf{E}^0$, $\mathbf{H}^0$ denote the electromagnetic field recorded over a time interval $(0, T)$ at a horizontal surface Σ coinciding with the plane XOY. We will study the properties of the fields $\mathbf{E}^{mi}$, $\mathbf{H}^{mi}$ resulting from the migration of the field $\mathbf{E}^0$, $\mathbf{H}^0$ into the lower half-space.

For the sake of brevity, the discussion is confined to migration fields of the first and third types. An analysis of the properties of all migration transforms included in a system of (10.1.8), (10.1.13)–(10.1.15) (Zhdanov and Frenkel 1985) has revealed that the fields $\mathbf{E}^{m1}$; $\mathbf{H}^{m1}$ and $\mathbf{E}^{m3}$; $\mathbf{H}^{m3}$ exhibit the most interesting and important properties necessary for practical application of electromagnetic migration to the solution of inverse geoelectrical problems.

In the one-dimensional case, formulas (10.1.8) and (10.1.14) are transformed to

$$\begin{pmatrix} \mathbf{H}_x^{m1} \\ \mathbf{H}_x^{m3} \end{pmatrix} = \int_{t'}^{T} \left[\mathbf{H}_x^0 \begin{pmatrix} -\tilde{G}^{m1}_{,z} \\ -c_3 \tilde{G}^{m3}_{,z} \end{pmatrix} + \mathbf{E}_y^0 \begin{pmatrix} \sigma^{m1} \tilde{G}^{m1} \\ \sigma^{m3} \tilde{G}^{m3} \end{pmatrix} \right] dt \tag{10.2.3}$$

$$\begin{pmatrix} \mathbf{E}_y^{m1} \\ \mathbf{E}_y^{m3} \end{pmatrix} = \int_{t'}^{T} \left[\mu_0 \mathbf{H}_x^0 \begin{pmatrix} -\tilde{G}^{m1}_{,t} \\ -c_3 \tilde{G}^{m3}_{,t} \end{pmatrix} + \mathbf{E}_y^0 \begin{pmatrix} -\tilde{G}^{m1}_{,z} \\ \tilde{G}^{m3}_{,z} \end{pmatrix} \right] dt . \tag{10.2.4}$$

In the two-dimensional case, where the field is uniform and the medium is homogeneous along the y-axis, the same components of the migration field are defined by the formulas

$$\begin{pmatrix} \mathbf{H}_x^{m1} \\ \mathbf{H}_x^{m3} \end{pmatrix} = \int_{t'}^{T} \int_{-\infty}^{\infty} \left[\mathbf{H}_x^0 \begin{pmatrix} -\tilde{G}^{m1}_{,z} \\ -c_3 \tilde{G}^{m1}_{,z} \end{pmatrix} + \mathbf{H}_z^0 \begin{pmatrix} \tilde{G}^{m1}_{,x} \\ -\tilde{G}^{m3}_{,x} \end{pmatrix} + \mathbf{E}_y^0 \begin{pmatrix} \sigma^{m1} \tilde{G}^{m1} \\ \sigma^{m3} \tilde{G}^{m3} \end{pmatrix} \right] dx dt \tag{10.2.5}$$

$$\begin{pmatrix} \mathbf{E}_y^{m1} \\ \mathbf{E}_y^{m3} \end{pmatrix} = \int_{t'}^{T} \int_{-\infty}^{\infty} \left[\mu_0 \mathbf{H}_x^0 \begin{pmatrix} -\tilde{G}^{m1}_{,t} \\ -c_3 \tilde{G}^{m3}_{,t} \end{pmatrix} + \mathbf{E}_y^0 \begin{pmatrix} -\tilde{G}^{m1}_{,z} \\ \tilde{G}^{m3}_{,z} \end{pmatrix} \right] dx dt . \tag{10.2.6}$$

And finally, in the three-dimensional case, we have

$$\begin{pmatrix} \mathbf{H}_x^{m1} \\ \mathbf{H}_x^{m3} \end{pmatrix} = \int_{t'}^{T} \iint_{-\infty}^{\infty} \left[\mathbf{H}_x^0 \begin{pmatrix} -\tilde{G}^{m1}_{,z} \\ -c_3 \tilde{G}^{m3}_{,z} \end{pmatrix} \right.$$
$$\left. + \mathbf{H}_z^0 \begin{pmatrix} -\tilde{G}^{m1}_{,x} \\ -\tilde{G}^{m3}_{,x} \end{pmatrix} + \mathbf{E}_{y(3)}^0 \begin{pmatrix} \sigma^{m1} \tilde{G}^{m1} \\ \sigma^{m3} \tilde{G}^{m3} \end{pmatrix} \right] dx dt , \tag{10.2.7}$$

$$\begin{pmatrix} \mathbf{E}_y^{m1} \\ \mathbf{E}_y^{m3} \end{pmatrix} = \int_{t'}^{T} \iint_{-\infty}^{\infty} \left[\mu_0 \mathbf{H}_x^0 \cdot \begin{pmatrix} -\tilde{G}^{m1}_{,t} \\ -c_3 \tilde{G}^{m3}_{,t} \end{pmatrix} + \mathbf{E}_{y(3)}^0 \begin{pmatrix} -\tilde{G}^{m1}_{,z} \\ \tilde{G}^{m3}_{,z} \end{pmatrix} \right] dx dt , \tag{10.2.8}$$

where

$$\tilde{G}^{mi}_{,a} = \frac{\partial \tilde{G}^{mi}}{\partial a} \ , \quad a = x, y, z, t \ ; \quad \tilde{G}^{mi} = \tilde{G}^{mi}(\mathbf{r}', t|\mathbf{r}, t) \ .$$

Considering the above model situations, wherein the field is excited by elementary sources (current plane, infinitely long cable, or electric dipole), we put in formulas (10.2.3)–(10.2.8) that

$$\mathbf{H}^0_x = -j\tilde{G}^d_{,z} \ ; \quad \mathbf{H}^0_z = j\tilde{G}^d_{,x} \ ; \quad \mathbf{E}^0_y = -j\mu_0\tilde{G}^d_{,t} \ ; \quad \mathbf{E}^0_{y(3)} = j\left(\frac{1}{\sigma}\tilde{G}^d_{yg} - \mu_0 G^d_{,t}\right) \ . \tag{10.2.9a}$$

Here

$$\tilde{G}^d = \tilde{G}^d(\mathbf{r}_0, 0|\mathbf{r}, t) \tag{10.2.9b}$$

are the functions conjugate to the Green fundamental functions for the one-, two-, and three-dimensional equations of diffusion (for models containing a current plane, an infinitely long cable, and an electric dipole, respectively) (Morse and Feshbach 1953).

10.2.3 Extreme Values of Migration Fields

Now consider the main properties of migration fields. For the sake of simplicity, we will confine ourselves to theoretical models within which the electromagnetic field in a homogeneous space is excited by elementary sources described in the previous subsection. These simplest geoelectrical situations enable us to establish several important focusing properties of migration fields. In particular, it will be shown that the points of local extreme values of migration fields coincide with the location of sources of an anomalous field (current plane, infinitely long cable, electric dipole). This yields a substantiation of the applicability of the introduced transforms of the electromagnetic field to localizing the regions of the geoelectrical section occupied with excess (anomalous) currents. The same theoretical models will be used to optimize the parameters of migration transforms. Particularly, it is demonstrated that extreme values of migration fields are focused onto the sources of an anomaly at the instant when the current is switched on in sources. It is noteworthy that fields should be migrated into a medium of an electric conductivity $\sigma^{m1} = \sigma^{m3} = \sigma$ (one-dimensional case) and $\sigma^{m1} = 0.5\,\sigma$, $\sigma^{m3} = \sigma$ (two- and three-dimensional cases), i.e., the optimal values of the migration constant are $c_1 = c_3 = 1$ for the one-dimensional case and $c_1 = 0.5$ and $c_3 = 1$ for the two- and three-dimensional cases, respectively.

The above properties of migration fields will be substantiated successively for the one-, two-, and three-dimensional case in model fields excited by a horizontal current plane, an infinitely long cable, and an electric dipole.

A. *One-dimensional case (an infinite horizontal current plane immersed in a homogeneous conducting medium of an electric conductivity* σ*).* On substitution of (10.2.9) into (10.2.3) it is evident that for migration fields $\mathbf{H}^{m1}_x$ and $\mathbf{H}^{m3}_x$ for $c_1 = c_3 = 1$ at a point $z' = z_0$ at the instant $t' = 0(\tau' = T)$ the necessary and sufficient conditions of local extreme values

$$\left.\frac{\partial \mathbf{H}_x^{mi}}{\partial z'}\right|_{\substack{z'=z_0\\ t'=0}} \equiv 0 \qquad (10.2.10\text{a})$$

$$\left.\frac{\partial^2 \mathbf{H}^{mi}}{\partial z'^2}\right|_{\substack{z'=z_0\\ t'=0}} \neq 0 \qquad (10.2.10\text{b})$$

are fulfilled. For instance, for the first type of migration field $\mathbf{H}_x^{m1}$, we have, according to (10.2.3) and (10.2.9),

$$\mathbf{H}_x^{m1}(z', T-t') = j\int_{t'}^{T} (\tilde{G}_{,z}^d \tilde{G}_{,t}^{m1} - \mu_0 \sigma^{m1} \tilde{G}_{,t}^d \tilde{G}_{,t}^{m1})\, dt \ . \qquad (10.2.11)$$

Differentiate the two parts of the last equality with respect to z' and utilize the obvious property of the Green function:

$$\frac{\partial \tilde{G}^{m1}}{\partial z'} = -\frac{\partial \tilde{G}^{m1}}{\partial z} \ .$$

Then we obtain

$$\frac{\partial \mathbf{H}_x^{m1}(z' T-t')}{\partial z'} = j\int_{t'}^{T} (\tilde{G}_{,z}^d \tilde{G}_{,tz'}^{m1} - \mu_0 \sigma^{m1} \tilde{G}_{,t}^d \tilde{G}_{,zz'}^{m1})\, dt$$

$$= -j\int_{t'}^{T} (\tilde{G}_{,z}^d \tilde{G}_{,zz}^{m1} - \mu_0 \sigma^{m1} G_{,t}^d \tilde{G}_{,zz}^{m1})\, dt \ . \qquad (10.2.12)$$

Note that for $c_1 = 1$, $t' = 0$, and $z' = z_0$, the following equations

$$\sigma^{m1} = \sigma \ ; \quad \tilde{G}^{m1}\Big|_{\substack{z'=z_0\\ t'=0}} \equiv \tilde{G}^d \qquad (10.2.13)$$

take place in view of (10.2.9b).

Substituting (10.2.13) into (10.2.12) we write

$$\left.\frac{\partial \mathbf{H}_x^{m1}}{\partial z'}\right|_{\substack{c_1=1\\ t'=0\\ z'=z_0}} = -j\int_{t'}^{T} \tilde{G}_{,z}^d (\tilde{G}_{,zz}^d - \mu_0 \sigma \tilde{G}_{,t}^d)\, dt \ . \qquad (10.2.14)$$

But the function $\tilde{G}^d$, being a conjugate to the Green function G^d, satisfies the equation [compare with Eq. (8.3.16)]

$$\tilde{G}_{,zz}^d - \mu_0 \sigma \tilde{G}_{,t}^d = \delta(z-z_0)\delta(t) \equiv 0 \quad \text{for} \quad z \neq z_0 \ , \quad t > 0 \ , \qquad (10.2.15)$$

(where z is the ordinate of the horizontal plane where the initial fields H_x^0, E_y^0 are specified, the fields which are migration-transformed).

Substitution of (10.2.15) into (10.2.14) does yield the necessary condition (10.2.10a) for a local extreme value of H_x^{m1} at the point $z' = z_0$.

It is remarkable that this property for $T \to \infty$ can be also established by direct evaluation of the integral in (10.2.11).

Indeed, for $c_1 = 1$, $t' = 0$, and $T \to \infty$ the integral in the right-hand side of (10.2.11) reduces to a form which is tabulated. Omitting simple calculations we find

$$\mathbf{H}_x^{m1}(z', t')\Big|_{\substack{c_1=1\\ t'=0}} = 2j\mu_0\sigma(z_0^2+z'^2)^{-2}(z_0^2-z'^2-2z_0z') \ , \tag{10.2.16a}$$

whence

$$\frac{\partial \mathbf{H}_x^{m1}(z', t')}{\partial z'}\Bigg|_{\substack{c_1=1\\ t'=0}} = 4j\mu_0\sigma(z_0^2+z'^2)^{-3}(z'^3-3z_0^2z'+3z_0z'^2-z_0^3)\Bigg|_{z'=z_0} \equiv 0 \ . \tag{10.2.16b}$$

Differentiating (10.2.16a) twice with respect to z' and calculating the obtained expression for $z' = z_0$ we find

$$\frac{\partial^2 \mathbf{H}_x^{m1}}{\partial z'^2}\Bigg|_{\substack{c_1=1\\ t'=0\\ z'=z_0>0}} = 3j\mu_0\sigma z_0^{-4} \neq 0 \ , \tag{10.2.16c}$$

i.e., at the point $z' = z_0$ the sufficient conditions of a local extreme value of $\mathbf{H}_x^{m1}$ are also met. Note that (10.2.16b) holds good only for $z' = z_0$.

Thus, the magnetic component of the migration field $\mathbf{H}_x^{m1}$ for $c_1 = 1$ has a single local extreme value at the point traversed by the current plane. Meanwhile, the migration electromagnetic field becomes focused at the instant $t' = 0$ (or $\tau' = T$). Precisely at this instant we are able to locate the position of field exciters.

A similar result is also valid for the same parameters of the migration procedure for $\mathbf{H}_x^{m3}$, since $\mathbf{H}_x^{m3}|_{c_3=1} = -\mathbf{H}_x^{m1}|_{c_1=1}$.

B. *Two-dimensional case (an infinitely long horizontal cable immersed in a homogeneous conducting medium of an electric conductivity σ).*

In the two-dimensional case, extreme values of the migration field (with the components of $\mathbf{H}_x^{mi}$, $\mathbf{E}_y^{mi}$, $i = 1, 3$) are to be sought in the half-space $z>0$ along the vertical axis passing through the cable (this property holds true also in the three-dimensional case for a vertical axis passing through the dipole).

To simplify the analysis, we will consider, by analogy to the one-dimensional case, expressions for migration fields at the instant of time $t' = 0$ and assume an infinite interval of field observation, i.e., $T\to\infty$. Then, substituting (10.2.9a) into (10.2.5) and using the known values of tabulated integrals, we arrive at the following chain of formulas

$$\mathbf{H}_x^{m1} = I_1 + I_2 + I_3 \ , \tag{10.2.17a}$$

$$\mathbf{H}_x^{m3} = c_3 I_1 - I_2 + I_3 \ , \tag{10.2.17b}$$

where we used the following notations:

$$I_1 = \int_0^\infty \int_{-\infty}^\infty (-\mathbf{H}_x^0 \tilde{G}_{,z}^{mi})\,dx\,dt$$

$$= \frac{-j\mu_0^2\sigma\sigma^{mi}z_0z'}{64\pi^2}\int_0^\infty t^{-4}\exp\left\{-\frac{\mu_0}{4t}\,[\sigma(z_0^2+x_0^2)+\sigma^m(z'^2+x'^2)]\right\}$$

$$\cdot \int_{-\infty}^{\infty} \exp\left\{-\frac{\mu_0}{4t}\,[x^2(\sigma+\sigma^{mi})-2x(x_0\sigma+x'\sigma^{mi})]\right\} dx\,dt$$

$$=\frac{-j\mu_0^{3/2}\sigma\sigma^{mi}(\sigma+\sigma^{mi})^{-1/2}z_0z'}{32\pi^{3/2}}\int_0^{\infty} t^{-7/2}\exp\left\{-\frac{\mu_0}{4t}B\right\} dt = A\,z_0z'B^{-5/2}\,, \tag{10.2.18}$$

here

$$A=\frac{3j\sigma\sigma^{mi}(\sigma+\sigma^{mi})^{-1/2}}{4\pi\mu_0}\,;\quad B=\sigma(x_0^2+z_0^2)+\sigma^{mi}(x'^2+z'^2)\,\frac{-(x_0\sigma+x'\sigma^{mi})^2}{\sigma+\sigma^{mi}}\,.$$

Employing similar calculation procedures we find

$$I_2=\int_0^{\infty}\int_{-\infty}^{\infty} \mathbf{H}_z^0\tilde{G}_{,x}^{mi}\,dx\,dt = A\left\{x_0x'+\frac{(x_0\sigma+x'\sigma^{mi})^2}{(\sigma+\sigma^{mi})^2}\right.$$

$$\left.+\frac{B}{3(\sigma+\sigma^{mi})}-\frac{(x'+x_0)(x_0\sigma+x'\sigma^{mi})}{\sigma+\sigma^{mi}}\right\}B^{-5/2}\,; \tag{10.2.19}$$

$$I_3=\int_0^{\infty}\int_{-\infty}^{\infty} \mathbf{E}_y^0\tilde{G}^{mi}\sigma^{mi}\,dx\,dt = A\left\{\frac{2x_0(x_0\sigma+x'\sigma^{mi})}{\sigma+\sigma^{mi}}-x_0^2\right.$$

$$\left.-z_0^2+\frac{2B}{3\sigma}-\frac{[3(x_0\sigma+x'\sigma^{mi})^2+B(\sigma+\sigma^{mi})]}{3(\sigma+\sigma^{mi})^2}\right\}B^{-5/2}\,. \tag{10.2.20}$$

Let us determine stationary points of H_x^{mi}, i.e., the points at which

$$\frac{\partial\mathbf{H}_x^{mi}}{\partial x'}=0\,,\quad \frac{\partial\mathbf{H}_x^{mi}}{\partial z'}=0\,. \tag{10.2.21}$$

Rewrite (10.2.21) with due reference to (10.2.17) – (10.2.20) for the component $\mathbf{H}_x^{m1}$:

$$\frac{\partial}{\partial x'}\left\{A\left[z_0z'+x_0x'-x_0^2-z_0^2+\frac{2B}{3\sigma}+\frac{(x'-x_0)(x_0\sigma+x'\sigma^{m1})}{\sigma+\sigma^{m1}}\right]B^{-5/2}\right\}=0\,. \tag{10.2.22}$$

We will calculate the left-hand side of (10.2.22) taking, for the sake of simplicity, that $x_0=0$ and substituting $c_1\sigma$ for σ^{m1}. The result is the following equation

$$x'\left[z_0^2+6z_0z'+7c_1z'^2+\frac{c_1(7c_1+3)}{c_1+1}x'^2\right]B^{-7/2}=0\,. \tag{10.2.23}$$

Since the braced expression in (10.2.24) is strictly positive (because $c_1>0$ and $z_0z'>0$), the unique solution of (10.2.24) as well as of (10.2.21) for the first type of migration transform is $x'=x_0=0$.

Thus, we have shown that extreme-value points for the component $\mathbf{H}_x^{m1}$ of the migration field should be sought along the vertical axis passing through the cable[1]. Hence, in our further study of the extremal properties of the fields $\mathbf{H}_x^{mi}$ we can put, without any loss of generality, $x' = x_0 = 0$ in (10.2.17)–(10.2.20). In this case, these expressions become very much simplified. Indeed,

$$
\begin{aligned}
I_1|_{x'=x_0=0} &= A z_0 z' \tilde{B}^{-5/2} \;, \\
I_2|_{x'=x_0=0} &= A[\tfrac{1}{3}\tilde{B}(\sigma+\sigma^{mi})^{-1}]\tilde{B}^{-5/2} \;, \\
I_3|_{x'=x_0=0} &= A[\tfrac{2}{3}\tilde{B}\sigma^{-1} - \tfrac{1}{3}\tilde{B}(\sigma+\sigma^{mi})^{-1} - z_0^2]\tilde{B}^{-5/2} \;,
\end{aligned}
\tag{10.2.24}
$$

here

$$\tilde{B} = \sigma z_0^2 + \sigma^{mi} z'^2 \;.$$

Substituting (10.2.24) into (10.2.20) we find

$$\mathbf{H}_x^{m1}\Big|_{\substack{x'=x_0\\ t'=0}} = \tfrac{1}{3}A\sigma^{-5/2}(2c_1 z'^2 + 3z' z_0 - z_0^2)(z_0^2 + c_1 z'^2)^{-5/2} \tag{10.2.25}$$

$$
\begin{aligned}
\mathbf{H}_x^{m3}\Big|_{\substack{x'=x_0\\ t'=0}} = & -\tfrac{1}{3}A\sigma^{-5/2}(c_3+1)^{-1}[2c_3^2 z'^2 + 3c_3(c_3+1) z_0 z' \\
& -(c_3+3)z_0^2](z_0^2 + c_3 z'^2)^{-5/2} \;.
\end{aligned}
\tag{10.2.26}
$$

The necessary conditions of a local extreme value of the function $\mathbf{H}_x^{mi}$, i.e., the conditions of the kind $\partial \mathbf{H}_x^{mi}/\partial z' = 0$, lead to the following algebraic equations

$$2c_1^2 z'^3 + 4c_1 z_0 z'^2 - 3c_1 z_0^2 z' - z_0^3 = 0 \;, \tag{10.2.27 a}$$

$$2c_3^3 z'^3 + 4c^2(c_3+1) z_0 z'^2 - c_3(3c_3+5) z_0^2 z' - c_3(c_3+1) z_0^3 = 0 \;. \tag{10.2.27 b}$$

If we demand that the local extreme values $\mathbf{H}_x^{mi}$ be located at a point through which the cable passes, one should assume $z' = z_0$ in expressions (10.2.27). Upon this substitution we arrive at an algebraic equation to solve for unknown migration constants c_1 and c_3:

$$2c_1^2 + c_1 - 1 = 0 \;, \tag{10.2.28 a}$$

$$c_3(c_3^2 - 1) = 0 \;. \tag{10.2.28 b}$$

We are interested only in the positive roots of Eqs. (10.2.28 a, b) which provide optimal values of the constants c_1 and c_3 for the first and third type of migration fields. Equations (10.2.28 a, b) yield unique positive solutions

$$c_1 = c_{1\,\mathrm{opt}} = 0.5 \;, \tag{10.2.29 a}$$

$$c_3 = c_{3\,\mathrm{opt}} = 1 \;. \tag{10.2.29 b}$$

It is noteworthy that

---

[1] In a similar way, this property is proved for the components H_x^{m3}; $\mathbf{E}_y^{m1}$; $\mathbf{E}_y^{m3}$ of a two-dimensional migration field as well as for the electric and magnetic components $\mathbf{H}_y^{mi}$, $\mathbf{E}_y^{mi}$ of the migration field in the three-dimensional case.

$$\left.\frac{\partial^2 \mathbf{H}_x^{m1}}{\partial z'^2}\right|_{\substack{c_1=0.5\\ z'=z_0\\ t'=0}} \neq 0 \; ; \quad \left.\frac{\partial^2 \mathbf{H}_x^{m3}}{\partial z'^2}\right|_{\substack{c_3=1\\ z'=z_0\\ t'=0}} \neq 0 \; .$$

Consequently, for $c_1 = 0.5$ and $c_3 = 1$, i.e., in the case of migration into a medium of an electric conductivity σ and

$$\sigma^{m1} = 0.5\,\sigma \; ; \quad \sigma^{m3} = \sigma \; ,$$

respectively, both the necessary and sufficient conditions of a local extreme value are satisfied for the migration field components H_x^{m1} and H_x^{m3} at a point $z' = z_0$ coinciding with the location of the source of an anomalous field (cable) at the instant of time when the current is switched on in the source ($t' = 0$).

Let us determine whether the obtained local extreme value of the components $\mathbf{H}_x^{m1}$ and $\mathbf{H}_x^{m3}$ of the migration field is unique. To this end, substitute the found values for $c_{1\,\mathrm{opt}}$ and $c_{3\,\mathrm{opt}}$ into (10.2.28a, b). This substitution makes (10.2.27a, b) transform to the corresponding cubic equations for the new variable $a = z'/z_0$:

$$a^3 - 4a^2 - 3a - 2 = 0 \; , \tag{10.2.30a}$$

$$a^3 + 4a^2 - 4a - 1 = 0 \; . \tag{10.2.30b}$$

Each of Eqs. (10.2.30a, b) has a single positive root: $a = 1$. In this way, we have proved that in the lower half-space the components $\mathbf{H}_x^{m1}$ and $\mathbf{H}_x^{m3}$ of the migration field for $c_1 = c_{i\,\mathrm{opt}}$ and $t' = 0$ have a unique local extreme value coinciding with the location of the anomaly source.

Now consider the behavior of the electric components of a two-dimensional migration field $\mathbf{E}_y^{m1}$, $\mathbf{E}_y^{m3}$ [formulas (10.2.6)]. Here the calculation techniques needed for deriving formulas fitting these components as well as the proof of the corresponding properties are those used in the investigation of the magnetic components of the migration field. Therefore the intermediate calculations are omitted here for the sake of brevity.

The expressions for $\mathbf{E}_y^{m1}$ and $\mathbf{E}_y^{m3}$, with $x' = x_0 = 0$, $t' = 0$, and $T \to \infty$, have the form

$$\mathbf{E}_y^{m1} = \tilde{A}_1 (z_0^2 + c_1 z'^2)^{-7/2} [c_1^2 (2c_1+1) z'^3 - c_1 (4c_1+3) z_0 z'^2$$
$$- c_1 (3c_1+4) z_0^2 z' + (c_1+2) z_0^3] \; ; \tag{10.2.31}$$

$$\mathbf{E}_y^{m3} = -\tilde{A}_3 c_3 (z_0^2 + c_3 z'^2)^{-7/2} [c_3 (2c_3+1) z'^3 + c_3 (4c_3+3) z_0 z'^2$$
$$- (3c_3+4) z_0^2 z' - (c_3+2) z_0^3] \; , \tag{10.2.32}$$

where

$$\tilde{A}_i = -3j\sigma^{-2}(c_i+1)^{-3/2}/4\pi\mu_0 \; ; \quad i = 1; \, 3.$$

Just as it was done regarding the properties of $\mathbf{H}_x^{mi}$, on the basis of the necessary conditions of local extreme values of the components $\mathbf{E}_y^{m1}$ and $\mathbf{E}_y^{m3}$ at the point $z' = z_0$, we pass from expressions (10.2.31) and (10.2.32) over to algebraic equations for unknown values of c_i

$$c_1^3 - 5c_1^2 - 3c_1 + 3 = 0 \ , \quad 7c_3^3 - 5c_3^2 - 11c_3 - 1 = 0 \ .$$

An analysis of the solutions of these equations yields the following optimal (positive) values of the migration constant, provided the extreme-value points of the electric components of the migration field E_y^{m1} and E_y^{m3} are focused onto the cable,

$$c_1 = c_{1\,\mathrm{opt}}^{(1)} \cong 0.55 \ ; \quad c_1 = c_{1\,\mathrm{opt}}^{(2)} \cong 5.45 \ ; \tag{10.2.33 a}$$

$$c_3 = c_{3\,\mathrm{opt}}^{(1)} \cong 0.09 \ ; \quad c_3 = c_{3\,\mathrm{opt}}^{(2)} \cong 1.63 \ . \tag{10.2.33 b}$$

We should also note that for all values of the migration parameters at the point $z' = z_0$, the sufficient conditions of local extreme values of the components $\mathbf{E}_y^{m1}$ and $\mathbf{E}_y^{m3}$ are fulfilled as well.

As we have implied, the value of the migration constant $c_1 = 0.5$ is optimal for the component $\mathbf{H}_y^{m1}$. It is evident from (10.2.33 a) that this value is near optimal for the electric component of the migration field $\mathbf{E}_y^{m1}$, too.

Here we should like only to stress that the component E_y^{m1} has two local extreme values (z_{y1}^e and z_{y2}^e) located in the lower half-space along the vertical axis passing through the cable. If a migration is made into a medium of $\sigma^m = 0.5\,\sigma$, the deviation of z_{y1}^e from z_0 does not exceed $0.07\,z_0$, while $z_{y2}^e = 8\,z_0$, i.e., the last point is found at a depth about an order of magnitude greater than that of the cable location. Nevertheless, availability of the second (false) extreme value of E_y^{m1} should be kept in mind while localizing the sources of anomalous fields, particularly where these sources are near the observation surface.

C. *Three-dimensional case (a horizontal electric dipole immersed in a homogeneous conducting medium).* Let us study migration of a three-dimensional field. It is worth remembering that electromagnetic fields are excited by a horizontal electric dipole parallel to the y-axis. We will calculate the migration field components H_x^{m1} and E_y^{m1} along the vertical semi-axis passing through the dipole center at $t' = 0$ and $T \to \infty$. In this case,

$$\mathbf{H}_x^{m1} = S(z_0^2 + c_1 z'^2)^{-3}(c_1 z'^2 + 2z_0 z' - z_0^2) \ ; \tag{10.2.34}$$

$$\mathbf{E}_y^{m1} = Q(z_0^2 + c_1 z'^2)^{-3}[(2c_1^3 + c_1^2)z'^3 - (5c_1^2 + 3c_1)z_0 z'^2 - (4c_1^2 + 5c_1)z_0^2 z' + (c_1 + 3)z_0^3] \ , \tag{10.2.35}$$

where S and Q are some constants determined by the parameters of a medium and a dipole.

Next, assuming that the migration field extreme-value points are coincident with the dipole location, we pass, in the known manner, from expressions (10.2.34) and (10.2.35) to the corresponding algebraic equations for the migration constants

$$2c_1^2 + c_1 - 1 = 0 \ ; \tag{10.2.36 a}$$

$$10c_1^3 - 59c_1^2 - 34c_1 + 35 = 0 \ . \tag{10.2.36 b}$$

Equations (10.2.38a, b) permit estimation of optimal values of the migration constants from the necessary conditions of focusing the migration field extreme-value points onto anomaly sources.

An analysis of the solutions of these equations leads again to the value $c_1 = c_{1\,\mathrm{opt}} = 0.5$ obtained earlier for the two-dimensional case, which is a unique exact positive solution of (10.2.38a) and is fairly close to that of (10.2.38b), equal to $c_1 = c_{1\,\mathrm{opt}}^{(1)} \cong 0.55$. Note that (10.2.36b) has another positive solution $c_1 = c_{1\,\mathrm{opt}}^{(2)} \cong 6.35$. Moreover, sufficient conditions of local extreme values of the migration fields H_x^{m1} and E_y^{m1} hold true at the point $z' = z_0$ for the obtained values of the constants $c_{i\,\mathrm{opt}}$.

Hence, if the migration field is reconstructed at the instant when the current is switched ON in the dipole and the c_i-values are $c_1 = 0.5$ for $\mathbf{H}_x^{m1}$ and $c_1 \cong 0.55$ or $c_1 \cong 6.35$ for $\mathbf{E}_{y,}^{m1}$ these components have a local extreme value at the location of the dipole. It should be borne in mind that the magnetic component has one extreme value, while the electric component has two (for each of the values $c_{1\,\mathrm{opt}}^{(1)}$ and $c_{1\,\mathrm{opt}}^{(2)}$; the extreme-value point found deeper than the dipole location is a "false" one and should be discarded).

The horizontal magnetic component H_x^{m3} retains its extremal properties in the three-dimensional case, too. Indeed, following (10.1.9), (10.2.7), and (10.2.9) we can write

$$\frac{\partial \mathbf{H}_x^{m3}}{\partial z'} = j \int_{t'}^{T} \iint_{-\infty}^{\infty} (-\mathbf{H}_x^0 \tilde{G}_{,zz'}^{m3} - \mathbf{H}_z^0 \tilde{G}_{,xz'}^{m3} + \sigma^{m3} \mathbf{E}_{y(3)}^0 \tilde{G}_{,z'}^{m3})\, dx\, dy\, dt \ . \tag{10.2.37}$$

Note that $\partial \tilde{G}^{m3}/\partial z' = -\partial \tilde{G}^{m3}/\partial z$ and besides, for $c_3 = 1$, $t' = 0$, and $z' = z_0$ the following equations

$$\sigma^{m3} = \sigma \ ; \quad \tilde{G}^{m3}\Big|_{\substack{z' = z_0 \\ t' = 0}} \equiv \tilde{G}^d \tag{10.2.38}$$

take place.

Substitution of (10.2.38) into (10.2.37) yields

$$\left.\frac{\partial \mathbf{H}_x^{m3}}{\partial z'}\right|_{\substack{c_3 = 1 \\ z' = z_0 \\ t' = 0}} = j \int_0^T \iint_{-\infty}^{\infty} (\tilde{G}_{,x}^d \tilde{G}_{,xz}^d - \tilde{G}_{,z}^d \tilde{G}_{,zz}^d - \tilde{G}_{,yy}^d G_{,z} + \mu_0 \sigma \tilde{G}_{,t}^d \tilde{G}_{,z}^d)\, dx\, dy\, dt \ . \tag{10.2.39}$$

Integrating the last expression by parts with respect to the variable x we obtain

$$\begin{aligned} \left.\frac{\partial \mathbf{H}_x^{m3}}{\partial z'}\right|_{\substack{c_3 = 1 \\ z' = z_0 \\ t' = 0}} &= j \int_0^T \iint_{-\infty}^{\infty} (-\tilde{G}_{,xx}^d \tilde{G}_{,z}^d - \tilde{G}_{,z}^d \tilde{G}_{,zz}^d - \tilde{G}_{,yy}^d \tilde{G}_{,z}^d + \mu_0 \sigma \tilde{G}_{,t}^d \tilde{G}_{,z}^d)\, dx\, dy\, dt \\ &= -j \int_0^T \iint_{-\infty}^{\infty} \tilde{G}_{,z}^d (\tilde{G}_{,xx}^d + \tilde{G}_{,yy}^d + \tilde{G}_{,zz}^d - \mu_0 \sigma \tilde{G}_{,t}^d)\, dx\, dy\, dt \\ &= -j \int_0^T \iint_{-\infty}^{\infty} \tilde{G}_{,z}^d \delta(z - z_0) \delta(t)\, dx\, dy\, dt \equiv 0 \ , \end{aligned} \tag{10.2.40}$$

because $z \neq z_0$ and $t > 0$.

Thereby for $z' = z_0$ the necessary condition of an extreme value is met. In a similar way, we can show that the sufficient condition is also fulfilled:

$$\left.\frac{\partial^2 \mathbf{H}_x^{m3}}{\partial z'^2}\right|_{\substack{c_3=1\\ z'=z_0\\ t'=0}} \neq 0 \ .$$

In addition, a more detailed analysis reveals that the point (x_0, y_0, z_0) is a unique extreme-value point of H_x^{m3} in the lower half-space.

We should like to note in conclusion that the migration transformation is a procedure resistant not only to errors in the initial electromagnetic fields involved, but also to errors in specified electromagnetic parameters of a medium. For instance, it follows from calculations that the displacement of the migration field local extreme-value point toward the observation surface (away from the cable) in the two-dimensional case is about $0.2z_0$ and $0.4z_0$ for $\sigma^{m1} = \sigma$ and $\sigma^{m1} = 2\sigma$, respectively, i.e., the value of σ^{m1} differs from an optimal conductivity of a medium equal to 0.5 by a factor of 2 and 4, respectively.

10.2.4 Migration of Theoretical and Model Electromagnetic Fields

It will be shown how the established theoretical relationships manifest themselves in numerical experiments. The anomalous fields used in the calculations were produced by (1) an infinitely long cable or a horizontal electric dipole where the current variations obey the law (10.2.2) or by a system of infinitely long cables located at the same or different depths, (2) well-conducting bodies of simple geometry, which are immersed in a homogeneous half-space of minor electric conductivity σ (the primary field source was arranged at the Earth's surface and operated in a pulsed mode; to increase the depth of investigation).

In this subsection we present the calculation results only for the first type of migration fields $\mathbf{E}^{m1}$; $\mathbf{H}^{m1}$.

Consider the first series of models with elementary sources of an anomalous field. In the two-dimensional case, calculations were based on the analytical expressions for the migration field components (10.2.20). In the three-dimensional case, however, they were computer-assisted, with the program realizing numerically the migration procedure of a three-dimensional field.

Figure 59 depicts isoline maps of the horizontal magnetic component of the migration field H_x^{m1} of two cables arranged at the same depth (Fig. 59a) and at different depths (Fig. 59b). The maps have been produced for an instant $t' = 0$ ($\tau' = T$). It is clearly seen that the position of sources is detected accurately from the extreme-value points of $\mathbf{H}_x^{m1}$.

A fairly good agreement between the predictions and practical calculations is also observed in the three-dimensional situation. Figure 60 shows the isoline maps for $\mathbf{H}_x^{m1}(c_1 = 0.5)$ constructed for an instant of time $t' = 0$ in two mutually perpendicular planes: (a) the plane normal to the dipole axis and passing through its center (Fig. 60a); (b) the horizontal plane on which the dipole is arranged (Fig. 60b). It is evident from these figures that the location of a source (dipole) in three-

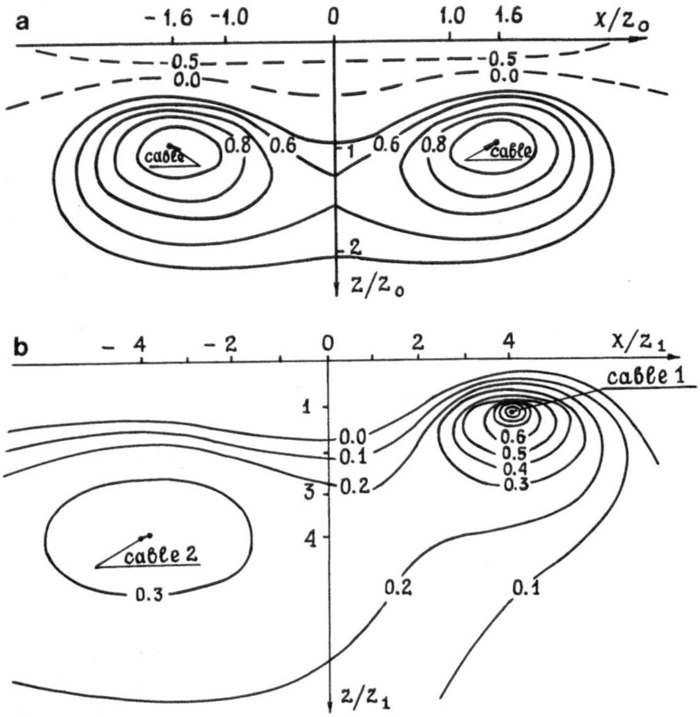

Fig. 59 a, b. Isoline maps of the migration field H_x^m of two linear currents flowing at the same depth (**a**) and at different depths (**b**). The maps are constructed for the instant $t = 0$ $(\tau' = T)$

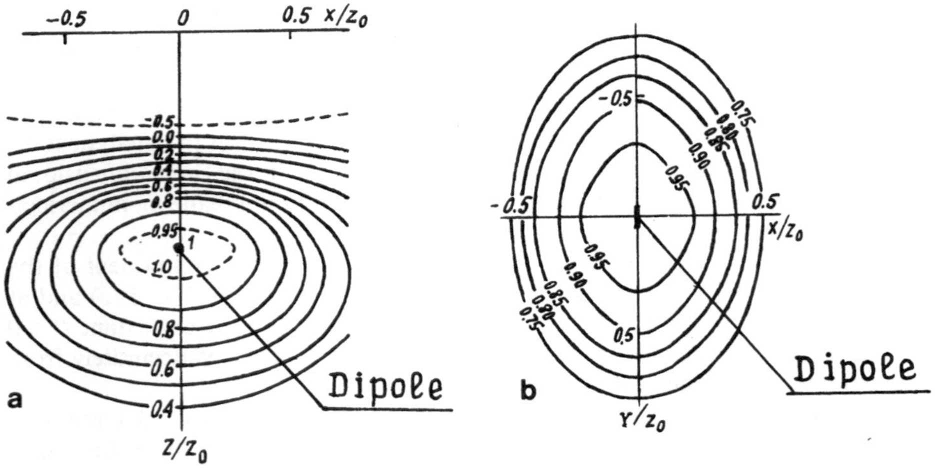

Fig. 60 a, b. Isoline maps of the migration field H_x^m of a horizontal electric dipole in two mutually perpendicular planes: **a** the plane normal to the dipole axis and passing through its center; **b** the horizontal plane on which the dipole is arranged. The maps are constructed for the instant $t' = 0$ $(\tau' = T)$

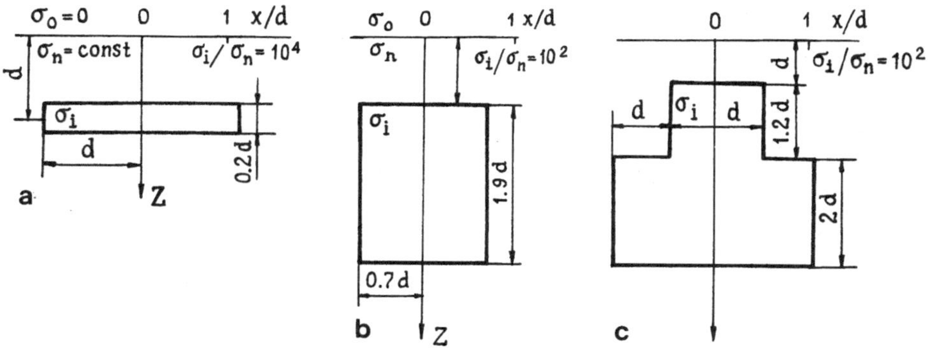

Fig. 61 a–c. Geoelectrical models used to calculate migration fields: **a** conducting horizontal band; **b** conducting rectangular insert; **c** conducting complex-shaped insert

dimensional space is definitely found from the extreme-value points of the migration field. Reconstruction of the migration field at subsequent instants of time, i.e., at $t'>0$, makes the extreme-value points shift upward; in the limit for $t'\to T$ they tend to the surface of observation. This behavior of the migration field can be explained qualitatively by the electromagnetic field diffusion in a well-conducting medium. Further along in time t', there are "switched ON" anomalous field sources located closer to the observation surface (operating in reversed time) and it is their location that is detected in the electromagnetic migration. Thus, an overall geoelectrical section (from the bottom upward) can be scanned, with the domains of excess electric conductivity seen most clearly.

Now we will dwell on the other series of models within which the initial anomalous field at the surface of a horizontally inhomogeneous Earth is calculated by the finite-difference method (Zhdanov et al. 1982). Modeling relied on simple geophysical situations (Fig. 61) and the external field was specified by a plane field whose amplitude varied in accordance with the pulsed mode law.

This permitted a quick calculation of the electromagnetic field for a number of frequencies by the finite-difference computer program. Lying within a narrow frequency range and being a smooth function of the frequency, the field spectrum was Fourier-transformed.

Models of media containing local anomaly-forming objects were employed to try out various ways of the electromagnetic field migration, allowing for the specificity of a particular model and for the properties of the migration field.

In model 1, the geoelectrical inhomogeneity is provided by a thin well-conducting horizontal layer (Fig. 61 a). At a zero instant of time ($t' = 0$), in the electromagnetic migration there show up anomalous field sources concentrated along the layer in the horizontal plane. The migration field isolines[2] enable us to define the location and shape of an anomaly-forming object (Fig. 62). Note that as t' increases, the extreme-value points of E_y^{m1} shift very slowly upward. Consequently, the excess currents localized within the layer flow into the environment

[2] Hereinafter isoline maps of the electric component of the migration field $\mathbf{E}_y^{m1}$ are presented, the migration constant being $c_1 = 0.5$.

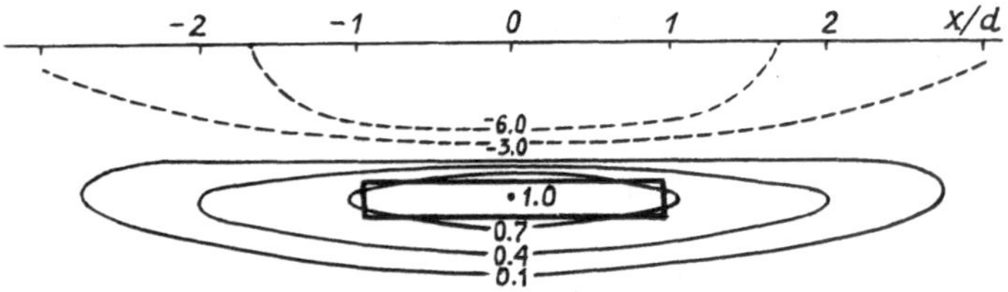

Fig. 62. Isoline map of the migration field within a conducting-band model

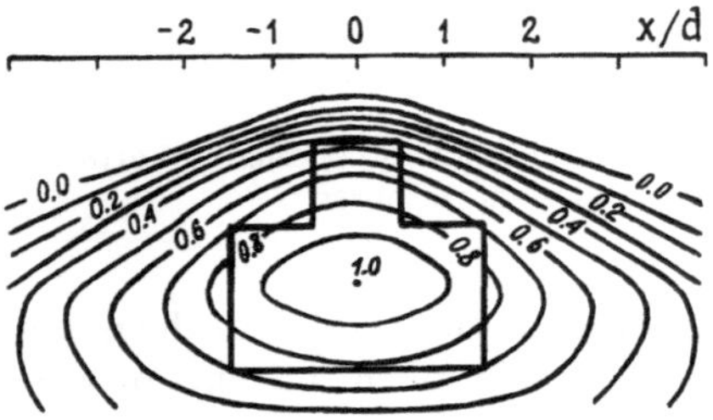

Fig. 63. Isoline map of the migration field within a complex-shaped insert model

very slowly and at a low intensity. This is evidence that the local inhomogeneity is highly conducting, its electric conductivity exceeding markedly that of the environment. In experiment, $\sigma_n = 10^{-2}\,\mathrm{ohm}^{-1}\cdot\mathrm{m}^{-1}$; $\sigma_i = 10^2\,\mathrm{ohm}^{-1}\cdot\mathrm{m}^{-1}$.

In models 2 and 3, the inhomogeneities are either a rectangular insert (Fig. 61b) or a more complex-shaped insert (Fig. 61c). The electromagnetic field migration carried out by formula (10.2.6) permits identification of the domains of the highest excess currents on the geoelectrical section at each instant of time. That is why spatial distribution of the migration field was reconstructed from layer to layer: at each level the field was reconstructed at an instant when the excess current (field source in a given layer) corresponding to this layer was switched ON. The switch-on time was defined by the expression for the depth of the skin of a quasi-stationary field (Makagonov 1977)

$$d(t) = a\sqrt{2\pi t/\mu_0\sigma_n}\;,$$

where a is some t-independent constant. The result of reconstruction for model 3 is given in Fig. 63. One can clearly see a local extreme value that is attributable to the maximum intensity of induced currents. It seems that most information is gained from reconstruction of a general spatial distribution of the migration field at the instant corresponding to the position of this extreme-value point. Indeed, as seen from Figs. 64a and 64b, the isolines of E_y^{m1} reflect rather clearly the location and shape of geoelectrical inhomogeneities. The pattern of the migration field established at other instants of time is not so indicative and may lead to erroneous results about the location and shape of an anomaly-forming object.

The above theoretical and experimental findings demonstrate that the electromagnetic field migration can provide an effective means of solving the inverse geoelectrical problem and permit reconstruction of the geoelectrical picture of a medium.

Fig. 64. a Isoline map of the migration field within a rectangular-insert model, produced by reconstructing the migration field from layer to layer; **b** Isoline map of the migration field within a complex-shaped insert model, produced by reconstructing the migration field from layer to layer

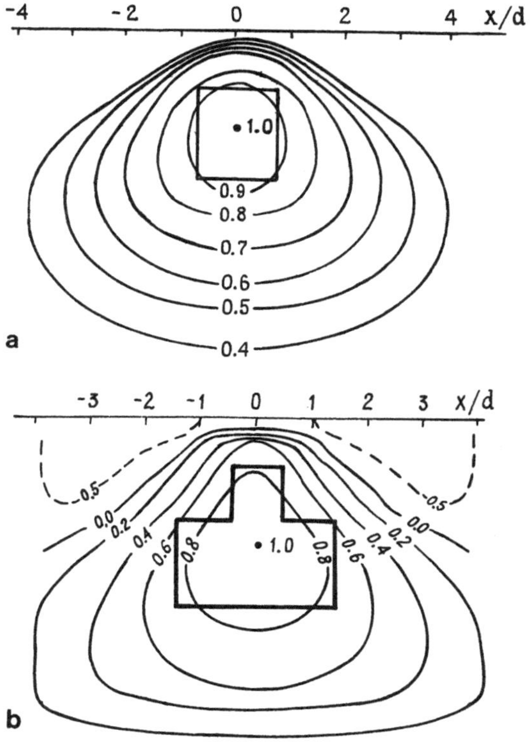

Part IV
Kirchhoff-Type Integrals in the Elastic Wave Theory

11 Kirchhoff-Type Integrals

11.1 Elastic Waves in an Isotropic Medium

Among the most important types of geophysical fields are elastic wave fields or simply wave fields. These fields are studied in seismology and in seismic prospecting and they yield extremely significant information about the deep structure of the Earth which could not be gained using other geophysical methods. In view of this, the problem of extending the body of Cauchy-type integrals to wave fields is of critical value in theoretical geophysics. In this chapter we will derive the classical Kirchhoff integral and a generalized Kirchhoff integral formula for the elastic displacement vector. The Kirchhoff-type integrals introduced on the basis of the above integral formulas are a natural extension of the Cauchy-type integrals to wave fields. They can be used to develop the theory of seismic field migration which presently underlies dynamic seismology.

For the sake of a thorough exposition of the material, the chapter starts with a consideration of the basic relations of the elastic theory (Love 1944; Landau and Lifschits 1965).

11.1.1 Stresses and Strains in Elastic Bodies

Elastic bodies exposed to applied external forces experience variations in the relative position of substance particles and have their size and shape changed, i.e., under stressed conditions elastic bodies undergo deformation. As the particles of a body change their position with respect to each other, it develops elastic forces, namely stresses opposing the deformation. The forces increase in the course of deformation and at a certain specific instant of time they can even counteract the external stress. At this point, the deformation process is discontinued and since that time a body is in a state of elastic equilibrium. As the stress is removed gradually, the elastic body regains its initial state. But if it is removed all of a sudden, particles inside the body start to oscillate. To describe these oscillations, it is necessary to quantify the relationship between the forces arising at each point of the deformed elastic body (internal stresses) and the magnitudes of deformation.

We will first describe small deformations of an elastic body. Generally speaking, the distances between individual points of a deformed elastic body are changed. Let $\mathbf{r}$ be the radius vector of some inner point M of an elastic body prior to deformation. Upon deformation, the point M is displaced by a vector $\mathbf{U}$ (the displacement vector or the deformation vector) and the radius vector $\mathbf{r}'$ of the new location of the point M is related to $\mathbf{r}$ by a simple relation

$$\mathbf{r}' = \mathbf{r} + \mathbf{U} \ . \tag{11.1.1}$$

Evidently, the displacment vector $\mathbf{U}$ depends generally upon the location of the initial point M (i.e., on the radius vector $\mathbf{r}$). In other words, a field of displacement vectors $\mathbf{U} = \mathbf{U}(\mathbf{r})$ arises in a deformed elastic body. To describe the deformation in an elastic body, we will define, following Landau and Lifschits (1965), the distance between two adjacent points prior to deformation $|d\mathbf{r}|$ and after it $|d\mathbf{r}'|$. To this end, differentiate expression (11.1.1)

$$d\mathbf{r}' = d\mathbf{r} + d\mathbf{U}$$

and calculate the square of the distance

$$|d\mathbf{r}'|^2 = |d\mathbf{r} + d\mathbf{U}|^2 = \sum_{\alpha = x, y, z} (dr_\alpha + d\mathbf{U}_\alpha)^2 \ , \tag{11.1.2}$$

where dr_α, dU_α $(\alpha = x, y, z)$ are the scalar components of the vectors $d\mathbf{r}$ and $d\mathbf{U}$.

The theory of elasticity deals, as a rule, with continuous deformations, where the displacement vectors are smooth (differentiable) functions of coordinates, hence

$$d\mathbf{U}_\alpha = \frac{\partial U_\alpha}{\partial r_\beta} dr_\beta = \mathbf{U}_{\alpha,\beta} dr_\beta \quad \alpha, \beta = x, y, z \ . \tag{11.1.3}$$[1]

Substituting (11.1.3) into (11.1.2) and raising it to the second power we write

$$|d\mathbf{r}'|^2 = |d\mathbf{r}|^2 + 2\mathbf{U}_{\alpha,\beta} dr_\alpha dr_\beta + \mathbf{U}_{\alpha,\beta} \mathbf{U}_{\alpha,\gamma} d\mathbf{r}_\beta d\mathbf{r}_\gamma \ . \tag{11.1.4}$$

Since summation in the second term of the right-hand side of (11.1.4) is over the two subscripts α and β, exchanging these subscripts we can write

$$2\mathbf{U}_{\alpha,\beta} d\mathbf{r}_\alpha d\mathbf{r}_\beta = \mathbf{U}_{\alpha,\beta} d\mathbf{r}_\alpha d\mathbf{r}_\beta + \mathbf{U}_{\beta,\alpha} d\mathbf{r}_\alpha d\mathbf{r}_\beta \ . \tag{11.1.5}$$

The subscripts in the third term of the right-hand side of (11.1.4) can be changed by replacing α by γ, β by α, and γ by β:

$$\mathbf{U}_{\alpha,\beta} \mathbf{U}_{\alpha,\gamma} dr_\beta dr_\gamma = \mathbf{U}_{\gamma,\alpha} \mathbf{U}_{\gamma,\beta} d\mathbf{r}_\alpha d\mathbf{r}_\beta \ . \tag{11.1.6}$$

Substituting (11.1.5) and (11.1.6) into (11.1.4), we write in the final form

$$|d\mathbf{r}'|^2 = |d\mathbf{r}|^2 + 2\mathbf{U}_{\alpha\beta} d\mathbf{r}_\alpha d\mathbf{r}_\beta \ , \tag{11.1.7a}$$

where

$$U_{\alpha\beta} = \tfrac{1}{2}(\mathbf{U}_{\alpha,\beta} + \mathbf{U}_{\beta,\alpha} + \mathbf{U}_{\gamma,\alpha} \mathbf{U}_{\gamma,\beta}) \ ; \tag{11.1.7b}$$

the tensor

$$\hat{\mathbf{U}} = [\mathbf{U}_{\alpha\beta}] = \begin{bmatrix} \mathbf{U}_{xx} \mathbf{U}_{xy} \mathbf{U}_{xz} \\ \mathbf{U}_{yx} \mathbf{U}_{yy} \mathbf{U}_{yz} \\ \mathbf{U}_{zx} \mathbf{U}_{zy} \mathbf{U}_{zz} \end{bmatrix}$$

[1] Following the agreement on summation introduced in Sect. 5.4, hereinafter the signs of summation over vector and tensor indexes are omitted.

is called an elastic medium deformation tensor. It is likely to be determined by the location of the point M in whose neighborhood we consider displacements of deformed points of a body, i.e., the tensor $\mathbf{U}_{\alpha\beta}$ forms a tensor field

$$\mathbf{U}_{\alpha\beta} = \mathbf{U}_{\alpha\beta}(\mathbf{r}) \ .$$

In the elastic oscillation theory one normally confines oneself to small deformations, in which case the last term in expression (11.1.7b) can be discarded as a small quantity of the second order, compared to the first two terms. In the case of small deformations, the deformation tensor takes the form

$$\mathbf{U}_{\alpha\beta} = \tfrac{1}{2}(\mathbf{U}_{\alpha,\beta} + \mathbf{U}_{\beta,\alpha}) \ . \tag{11.1.8}$$

Now consider the stresses arising inside a deformed elastic body. The total of the forces acting on an elastic body fall into two types: volume and surface. We assume that the force on an infinitesimal element of the volume dv is equal to $\mathbf{F}dv$. Here the vector $\mathbf{F}$ is a volume force related to a unit volume, i.e., the density of the volume force. Surface forces operate on elements of a surface conceived inside an elastic body or on its boundary. The value of the surface force on an element of the surface ds is equal to $\mathbf{f}ds$, where $\mathbf{f}$ is the surface force per unit area, i.e., the stresses on a given area.

Take an arbitrary domain D bounded by a surface S in a body found in a state of elastic equilibrium. The equilibrium condition of this volume can be cast in the form

$$\iiint_D \mathbf{F}(\mathbf{r})dv = \iint_S \mathbf{f}(\mathbf{r})ds \tag{11.1.9a}$$

or in the scalar form

$$\iiint_D F_\alpha(\mathbf{r})dv = \iint_S f_\alpha(\mathbf{r})ds \ . \tag{11.1.9b}$$

According to the Ostrogradsky-Gauss theorem (5.1.1), the volume integral over the scalar function is converted into a surface integral, provided this scalar function is a divergence of some vector field. As a consequence, we have

$$F_\alpha = \operatorname{div} \boldsymbol{\tau}_\alpha \ . \tag{11.1.10}$$

Substituting (11.1.10) into (11.1.9b) and employing (5.1.1), we write

$$\iint_S \boldsymbol{\tau}_\alpha \cdot \mathbf{n}\, ds = \iint_S f_\alpha ds \ . \tag{11.1.11}$$

Since equality (11.1.11) holds for any volume D and surface S selected inside an elastic body, we have

$$f_\alpha = \boldsymbol{\tau}_\alpha \cdot \mathbf{n} = \tau_{\alpha\beta} n_\beta \quad \alpha, \beta = x, y, z \ . \tag{11.1.12}$$

The tensor $\hat{\tau} = \boldsymbol{\tau}_\alpha \mathbf{d}_\alpha$ (i.e., the sum of dyadic products of the vectors $\boldsymbol{\tau}_\alpha$ and $\mathbf{d}_\alpha$) with the matrix $\tau_{\alpha\beta}$ is called a *stress tensor*.

In accordance with (11.1.10) and (11.1.12), we have

$$\mathbf{F} = \operatorname{div} \hat{\tau} \tag{11.1.13a}$$

$$\mathbf{f} = \hat{\tau} \cdot \mathbf{n} \ , \tag{11.1.13b}$$

i.e., the stress tensor enables us to define the density of volume forces and the stresses applied to a given area *ds* with a unit vector of the normal **n**.

If small deformations occur in an isotropic body, the stress tensor $\hat{\tau}$ is a linear function of the deformation tensor $\hat{\mathbf{U}}$ (and vice versa):

$$\tau_{\alpha\beta} = \frac{E}{1+\sigma}\left(\mathbf{U}_{\alpha\beta} + \frac{\sigma}{1-2\sigma}\,\mathbf{U}_{ll}\delta_{\alpha\beta}\right) \ . \tag{11.1.14}$$

Here σ and E are the elastic parameters of a medium, i.e, the Young modulus and the Poisson ratio (Landau and Lifschits 1965). This law is called Hooke's law. It is applicable virtually to any plastic deformation.

11.1.2 Equations of Motion of a Homogeneous Isotropic Elastic Medium

We should like to recall the general form of the equations of motion of a homogeneous isotropic elastic medium by the action of its own elastic forces. According to Newton's law, the density of volume forces is equal to the product of acceleration $\dfrac{\partial^2 \mathbf{U}}{\partial t^2}$ by the mass of a unit volume of a body, i.e., by its density ϱ:

$$\mathbf{F} = \varrho\,\frac{\partial^2 \mathbf{U}}{\partial t^2} \ . \tag{11.1.15}$$

Representing the volume force density in terms of the stress tensor by means of (11.1.13a), we write

$$\operatorname{div} \hat{\tau} = \varrho\,\frac{\partial^2 \mathbf{U}}{\partial t^2} \ . \tag{11.1.16}$$

Since all the deformations are assumed to be small, we can employ Hooke's law (11.1.14) and the deformation tensor expressed through the displacement vector (11.1.8). Substitution of (11.1.14) into (11.1.16) with due account of (11.1.8) yields the following form of the equation of motion of a homogeneous isotropic elastic medium:

$$C_l^2 \operatorname{grad} \operatorname{div} \mathbf{U} - C_t^2 \operatorname{rot} \operatorname{rot} \mathbf{U} = \frac{\partial^2 \mathbf{U}}{\partial t^2} \ , \tag{11.1.17}$$

where the constants C_l and C_t are expressed in terms of the density and elastic parameters of a medium (the Young modulus and the Poisson ratio) by means of the following formulas

$$C_l = \sqrt{\frac{E(1-\sigma)}{(1+\sigma)(1-2\sigma)\varrho}} \tag{11.1.18a}$$

$$C_t = \sqrt{\frac{E}{2(1+\sigma)\varrho}} \ . \qquad (11.1.18\text{b})$$

Let us take into consideration the differential operator

$$\Delta^* = C_l^2 \operatorname{grad}\operatorname{div} - C_t^2 \operatorname{rot}\operatorname{rot} \ . \qquad (11.1.19)$$

Then Eq. (11.1.17) can be rewritten as

$$\Delta^* \mathbf{U} - \frac{\partial^2 \mathbf{U}}{\partial t^2} = 0 \ . \qquad (11.1.20)$$

The operator Δ^* is called *an operator of the elastic oscillation equation.* When $C_l = C_t = C$ the operator Δ^* goes over into the Laplace operator:

$$\Delta^* = C^2 \Delta \ ,$$

while Eq. (11.1.20) turns into the wave equation

$$\Delta \mathbf{U} - \frac{1}{C^2}\frac{\partial^2 \mathbf{U}}{\partial t^2} = 0 \ . \qquad (11.1.21)$$

11.1.3 Longitudinal and Transverse Waves in a Homogeneous Isotropic Elastic Medium

Consider the field of the displacement vector. This field can be represented as a sum of rotational $\mathbf{U}^t$ and nonrotational $\mathbf{U}^l$ parts

$$\mathbf{U} = \mathbf{U}^t + \mathbf{U}^l \ , \qquad (11.1.22)$$

when the rotational part is defined by the conditions

$$\begin{cases} \operatorname{div} \mathbf{U}^t = 0 \\ \operatorname{rot} \mathbf{U}^t = \operatorname{rot} \mathbf{U} \ , \end{cases} \qquad (11.1.23)$$

while the nonrotational part is determined by the conditions

$$\begin{cases} \operatorname{div} \mathbf{U}^l = \operatorname{div} \mathbf{U} \\ \operatorname{rot} \mathbf{U}^l = 0 \ . \end{cases} \qquad (11.1.24)$$

Derive equations fitting the rotational and the nonrotational parts of the displacement vector.

For this purpose, substitute (11.1.22) into (11.1.17) [with due reference to (11.1.23) and (11.1.24)]:

$$C_l^2 \operatorname{grad}\operatorname{div} \mathbf{U}^l - C_t^2 \operatorname{rot}\operatorname{rot} \mathbf{U}^t = \frac{\partial^2 \mathbf{U}^l}{\partial t^2} + \frac{\partial^2 \mathbf{U}^t}{\partial t^2} \ . \qquad (11.1.25)$$

We will apply the operation div to the two sides of this equation. Since $\operatorname{div} \mathbf{U}^t = 0$, we obtain

$$\operatorname{div}\left(C_l^2 \Delta \mathbf{U}^l - \frac{\partial^2 \mathbf{U}^l}{\partial t^2}\right) = 0 \ . \tag{11.1.26}$$

On the other hand, the rot of the expression enclosed in parentheses is zero. Thus, the field $(C_l^2 \Delta \mathbf{U}^l - \partial^2 \mathbf{U}^l / \partial t^2)$ has neither sources nor rotations in the whole space. If there is imposed a natural physical requirement that this field should vanish at infinity, it proves to be zero in the whole space, i.e.,

$$C_l^2 \Delta \mathbf{U}^l - \frac{\partial^2 \mathbf{U}^l}{\partial t^2} = 0 \ . \tag{11.1.27}$$

Similarly, applying the operator rot to the two sides of (11.1.25) and considering that $\operatorname{rot}\mathbf{U}^l = 0$ we find

$$\operatorname{rot}\left(C_t^2 \Delta \mathbf{U}^t - \frac{\partial^2 \mathbf{U}^t}{\partial t^2}\right) = 0 \ . \tag{11.1.28}$$

Since the expression enclosed in parentheses has a zero divergence, we arrive at the following expression for the rotational field

$$C_t^2 \Delta \mathbf{U}^t - \frac{\partial^2 \mathbf{U}^t}{\partial t^2} = 0 \ . \tag{11.1.29}$$

Thus, it is evident that the rotational and the nonrotational parts of the displacement vector satisfy the wave equations with different coefficients C_t and C_l, respectively, implying, as is known, elastic velocities. Note that the waves described by the nonrotational part $\mathbf{U}^t$ of the displacement vector are referred to as *longitudinal waves,* while those corresponding to the rotational part $\mathbf{U}^t$ of the displacement vector are called *transverse waves.* Particles of an elastic body are displaced in a longitudinal wave along its direction, while in a transverse wave their direction is orthogonal to the wave propagation. It should be borne in mind that, according to expressions (11.1.18a and b), the longitudinal wave velocity always exceeds that of transverse waves: $C_l > C_t$.

11.2 Generalized Kirchhoff Integral Formula

It was shown in Chap. 8 of this book that Stratton-Chu integral formulas are a natural extension of the Cauchy integral formula to the electromagnetic theory. It turns out that the Kirchhoff integral formulas play a similar role in elastic wave theory. In this subsection we will derive these formulas shortly, relying on the Green tensor and vector formulas (Morse and Feshbach 1953).

11.2.1 Green Tensor and Vector Formulas

The Green tensor and vector formulas derive, just as in the scalar case, from the tensor statement of the Ostrogradsky-Gauss theorem [formula (B.58)]. Indeed, let us specify an auxiliary tensor field $\hat{\mathbf{\Phi}}(\mathbf{r})$:

$$\hat{\Phi}(\mathbf{r}) = C_l^2(\mathbf{U}\,\mathrm{div}\,\hat{\mathbf{P}} - \hat{\mathbf{P}}\,\mathrm{div}\,\mathbf{U}) + C_t^2(U \times \mathrm{rot}\,\hat{\mathbf{P}} + \mathrm{rot}\,\mathbf{U} \times \hat{\mathbf{P}}) \ , \tag{11.2.1}$$

where $\mathbf{U} = \mathbf{U}(\mathbf{r})$, and $\hat{\mathbf{P}} = \hat{\mathbf{P}}(\mathbf{r})$ are arbitrary vector and tensor functions, respectively, twice continuously differentiable in the domain D (up to its boundary S); C_l and C_t are arbitrary real numbers. The rules of performing algebraic and differential operations on the tensors are presented in the Sect. B.5 of Appendix B. It is noteworthy that according to Eqs. (B.52) and (B.53) div$\hat{\mathbf{P}}$ is a vector field and rot$\,\hat{\mathbf{P}}$ is a tensor field. The product $\mathbf{U}\,\mathrm{div}\,\hat{\mathbf{P}}$ should be taken as a dyad of two vectors $\mathbf{U}$ and div$\,\hat{\mathbf{P}}$, hence it forms a tensor. The product of the tensor $\hat{\mathbf{P}}$ by the scalar div$\,\mathbf{U}$ also forms a tensor. Vector products of the vector $\mathbf{U}$ and the tensor rot$\,\hat{\mathbf{P}}$ as well as of the vector rot$\,\mathbf{U}$ and the tensor $\hat{\mathbf{P}}$ are derived in accordance with formulas (B.49) and, hence, also form tensors.

Let us calculate div$\,\hat{\Phi}$. This operation is performed, following Sect. B.5 of Appendix B, using the Hamiltonian operator:

$$\mathrm{div}\,\hat{\Phi} = C_l^2\{\nabla\cdot[\mathbf{U}(\nabla\cdot\hat{\mathbf{P}})] - \nabla\cdot[\hat{\mathbf{P}}(\nabla\cdot\mathbf{U})]\}$$
$$+ C_t^2\{\nabla\cdot[\mathbf{U}\times(\nabla\times\hat{\mathbf{P}})] + \nabla\cdot[(\nabla\times\mathbf{U})\times\hat{\mathbf{P}}]\} \ . \tag{11.2.2}$$

Hereinafter the sign "·" denotes a scalar product of vectors, or of a vector by a tensor, or of a tensor into a vector, while the sign "×" refers to a vector product of vectors, or of a vector by a tensor, or of a tensor by a vector. The lack of any sign between the cofactors implies a product of a scalar by a vector or a tensor or a dyad of two vectors.

Calculate sequentially all the four summands in the right-hand side of (11.2.2):

$$\nabla\cdot[\mathbf{U}(\nabla\cdot\hat{\mathbf{P}})] = (\nabla\cdot\mathbf{U})(\nabla\cdot\hat{\mathbf{P}}) + \mathbf{U}\cdot\nabla(\nabla\cdot\hat{\mathbf{P}}) = \mathrm{div}\,\mathbf{U}\,\mathrm{div}\,\hat{\mathbf{P}} + \mathbf{U}\cdot\mathrm{grad}\,\mathrm{div}\,\hat{\mathbf{P}} \ , \tag{11.2.3}$$

$$\nabla\cdot[\hat{\mathbf{P}}(\nabla\cdot\mathbf{U})] = (\nabla\cdot\mathbf{U})(\nabla\cdot\hat{\mathbf{P}}) + \nabla(\nabla\cdot\mathbf{U}) = \mathrm{div}\,\mathbf{U}\,\mathrm{div}\,\hat{\mathbf{P}} + \mathrm{grad}\,\mathrm{div}\,\mathbf{U}\cdot\hat{\mathbf{P}} \ , \tag{11.2.4}$$

$$\nabla\cdot[\mathbf{U}\times(\nabla\times\hat{\mathbf{P}})] = (\nabla\times\mathbf{U})\cdot(\nabla\times\hat{\mathbf{P}}) - \mathbf{U}\cdot[\nabla\times(\nabla\times\hat{\mathbf{P}})]$$
$$= \mathrm{rot}\,\mathbf{U}\cdot\mathrm{rot}\,\hat{\mathbf{P}} - \mathbf{U}\cdot\mathrm{rot}\,\mathrm{rot}\,\hat{\mathbf{P}} \ , \tag{11.2.5}$$

$$\nabla\cdot[(\nabla\times\mathbf{U})\times\hat{\mathbf{P}}] = [\nabla\times(\nabla\times\mathbf{U})]\cdot\hat{\mathbf{P}} - (\nabla\times\mathbf{U})\cdot(\nabla\times\hat{\mathbf{P}})$$
$$= \mathrm{rot}\,\mathrm{rot}\,\mathbf{U}\cdot\hat{\mathbf{P}} - \mathrm{rot}\,\mathbf{U}\cdot\mathrm{rot}\,\hat{\mathbf{P}} \ . \tag{11.2.6}$$

Substituting (11.2.3) through (11.2.6) into (11.2.2) and canceling the like terms, we write

$$\mathrm{div}\,\hat{\Phi} = \mathbf{U}\cdot(C_l^2\,\mathrm{grad}\,\mathrm{div}\,\hat{\mathbf{P}} - C_t^2\,\mathrm{rot}\,\mathrm{rot}\,\hat{\mathbf{P}}) - (C_l^2\,\mathrm{grad}\,\mathrm{div}\,\mathbf{U} - C_t^2\,\mathrm{rot}\,\mathrm{rot}\,\mathbf{U})\cdot\hat{\mathbf{P}} \ . \tag{11.2.7a}$$

Or, using the operator Δ^* of the equation of elastic oscillations (11.1.19) we obtain

$$\operatorname{div}\hat{\mathbf{\Phi}} = \mathbf{U}\cdot\Delta^*\hat{\mathbf{P}} - \Delta^*\mathbf{U}\cdot\hat{\mathbf{P}} \ . \tag{11.2.7b}$$

Substituting $\hat{\mathbf{\Phi}}$ into the Ostrogradsky-Gauss tensor formula (B.58) and considering (11.2.7a, b) we write in the final form

$$\iiint_D (\mathbf{U}\cdot\Delta^*\hat{\mathbf{P}} - \Delta^*\mathbf{U}\cdot\hat{\mathbf{P}})dv = \iint_S [C_l^2\mathbf{n}\cdot(\mathbf{U}\operatorname{div}\hat{\mathbf{P}} - \hat{\mathbf{P}}\operatorname{div}\mathbf{U}) + C_t^2\mathbf{n}\cdot(\mathbf{U}\times\operatorname{rot}\hat{\mathbf{P}} + \operatorname{rot}\mathbf{U}\times\hat{\mathbf{P}})]\,ds \ . \tag{11.2.8}$$

Formula (11.2.8) is the Green tensor formula for the operator of elastic oscillations Δ^*. For $C_l = C_t = C$, formula (11.2.8) is reduced, on account of (11.1.21), to the Green tensor formula for the Laplace operator:

$$\iiint_D (\mathbf{U}\cdot\Delta\hat{\mathbf{P}} - \Delta\mathbf{U}\cdot\hat{\mathbf{P}})dv = \iint_S [\mathbf{n}\cdot(\mathbf{U}\operatorname{div}\hat{\mathbf{P}} - \hat{\mathbf{P}}\operatorname{div}\mathbf{U}) + \mathbf{n}\cdot(\mathbf{U}\times\operatorname{rot}\hat{\mathbf{P}} + \operatorname{rot}\mathbf{U}\times\hat{\mathbf{P}})]\,ds \ . \tag{11.2.9}$$

If the tensor field $\hat{\mathbf{P}}$ is replaced by the vector field $\mathbf{F}$, we arrive at the ordinary Green vector formula (Morse and Feshbach 1953):

$$\iiint_D (\mathbf{U}\cdot\Delta\mathbf{F} - \Delta\mathbf{U}\cdot\mathbf{F})dv = \iint_S [\mathbf{n}\cdot(\mathbf{U}\operatorname{div}\mathbf{F} - \mathbf{F}\operatorname{div}\mathbf{U}) + \mathbf{n}\cdot(\mathbf{U}\times\operatorname{rot}\mathbf{F} - \mathbf{F}\times\operatorname{rot}\mathbf{U})]\,ds \ . \tag{11.2.10}$$

11.2.2 Kirchhoff Integral Formulas

We will employ the Green tensor formulas (11.2.8) and (11.2.9) to elaborate the theory of integral representations of wave fields in a homogeneous and isotropic domain of space.

The equation of motion of a homogeneous isotropic medium by the action of its own restoring elastic forces is given by formulas (11.1.20) and (11.1.19), where C_l and C_t are the longitudinal and the transverse velocities, respectively.

For $C_l = C_t = C$, Eq. (11.1.20) goes over into an ordinary wave Eq. (11.1.21) fitting the longitudinal and the transverse parts of the total displacement vector (for an appropriate value of the velocity C). Below, we will study everywhere in parallel the field of the total displacement vector satisfying Eq. (11.1.20) and a simpler wave field meeting Eq. (11.1.21).

Consider the following boundary value problem: express the displacement vector field $\mathbf{U}$ in some domain D in terms of the values of U and of its normal derivative $\partial\mathbf{U}/\partial n$ on the inner side of the surface S bounding this domain. The elastic parameters of a medium, C_l and C_t, are assumed to be known.

Now let $\hat{\mathbf{P}}(r, t)$ be an arbitrary tensor field twice continuously differentiable in D, depending on time t, and absolutely integrable over the whole time axis. Then the following identity

$$\int_{-\infty}^{+\infty} [\mathbf{U}(t)\cdot\Delta * \hat{\mathbf{P}}(t) - \Delta * \mathbf{U}(t)\cdot\hat{\mathbf{P}}(t)]\,dt = \int_{-\infty}^{+\infty} \mathbf{U}(t)\cdot\left(\Delta * \hat{\mathbf{P}}(t) - \frac{\partial^2 \hat{\mathbf{P}}(t)}{\partial t^2}\right) dt \tag{11.2.11}$$

holds good.

Indeed, substituting (11.1.20) into formula (11.2.11), we obtain for the left-hand side

$$\int_{-\infty}^{+\infty} \left[(t)\cdot\Delta * \hat{\mathbf{P}}(t) - \frac{\partial^2 \mathbf{U}(t)}{\partial t^2}\cdot\hat{\mathbf{P}}(t)\right] dt \ . \tag{11.2.12}$$

Integrating the second term in integrand (11.2.12) twice by parts, we arrive at the right-hand side of (11.2.11), which was to be proved.

Now we will integrate identity (11.2.11) over the domain D and apply the Green formula (11.2.8) to the left-hand side of the identity:

$$\int_{-\infty}^{+\infty}\iint_S [C_l^2 \mathbf{n}\cdot(\mathbf{U}\,\mathrm{div}\,\hat{\mathbf{P}} - \hat{\mathbf{P}}\,\mathrm{div}\,\mathbf{U}) + C_t^2 \mathbf{n}\cdot(\mathbf{U}\times\mathrm{rot}\,\hat{\mathbf{P}} + \mathrm{rot}\,\mathbf{U}\times\hat{\mathbf{P}})]\,ds\,dt$$

$$= \int_{-\infty}^{+\infty}\iiint_D \mathbf{U}\cdot\left(\Delta * \hat{\mathbf{P}} - \frac{\partial^2 \hat{\mathbf{P}}}{\partial t^2}\right) dv\,dt \ . \tag{11.2.13}$$

Note that once $C_l = C_t = C$ and $\mathbf{U}$ is the wave field satisfying Eq. (11.1.21), formula (11.2.13) is written as follows

$$\int_{-\infty}^{+\infty}\iint_S [\mathbf{n}\cdot(\mathbf{U}\,\mathrm{div}\,\hat{\mathbf{P}} - \hat{\mathbf{P}}\,\mathrm{div}\,\mathbf{U}) + \mathbf{n}\cdot(\mathbf{U}\times\mathrm{rot}\,\hat{\mathbf{P}} + \mathrm{rot}\,\mathbf{U}\times\hat{\mathbf{P}})]\,ds\,dt$$

$$= \int_{-\infty}^{+\infty}\iiint_D \mathbf{U}\cdot\left[\Delta\hat{\mathbf{P}} - \frac{1}{C^2}\frac{\partial^2 \hat{\mathbf{P}}}{\partial t^2}\right] dv\,dt \ . \tag{11.2.14}$$

Now the function $\hat{\mathbf{P}}(\mathbf{r}, t)$ appearing in (11.2.13) will be provided by the elastic oscillation tensor of Green $\hat{\mathbf{G}}^e(\mathbf{r}', t' | \mathbf{r}, t)$ satisfying the equation

$$\Delta * \hat{\mathbf{G}}^e - \frac{\partial^2 \hat{\mathbf{G}}^e}{\partial t^2} = \hat{\mathbf{I}}\delta(\mathbf{r}-\mathbf{r}')\delta(t-t') \ , \tag{11.2.15}$$

where $\hat{\mathbf{I}}$ is the unit tensor (idem factor, see Sect. B.5).

Making the above substitution we write

$$\int_{-\infty}^{+\infty}\iint_S [C_l^2 \mathbf{n}\cdot(\mathbf{U}\,\mathrm{div}\,\hat{\mathbf{G}}^e - \hat{\mathbf{G}}^e\,\mathrm{div}\,\mathbf{U}) + C_t^2 \mathbf{n}\cdot(\mathbf{U}\times\mathrm{rot}\,\hat{\mathbf{G}}^e + \mathrm{rot}\,\mathbf{U}\times\hat{\mathbf{G}}^e)]\,ds\,dt$$

$$= \begin{cases} \mathbf{U}(\mathbf{r}', t') \ , & \mathbf{r}' \in D \\ 0 \ , & \mathbf{r}' \in C\bar{D} \ . \end{cases} \tag{11.2.16}$$

It is worth recalling what the elastic oscillation Green tensor $\hat{\mathbf{G}}^e$ is equal to. Note first of all that the tensor $\hat{\mathbf{G}}^e$ can be represented as a sum of the longitudinal $\hat{\mathbf{G}}^l$ and the transverse $\hat{\mathbf{G}}^t$ Green tensors:

$$\hat{\mathbf{G}}^e = \hat{\mathbf{G}}^l + \hat{\mathbf{G}}^t \tag{11.2.17}$$

fitted by the relations

$$\hat{\mathbf{G}}^{l}(\mathbf{r}', t'|\mathbf{r}, t) = \nabla\nabla[\hat{\mathbf{I}}(g^{(l)}(\mathbf{r}-\mathbf{r}'|t-t') - G(\mathbf{r}-\mathbf{r}')\chi(t-t'))] \tag{11.2.18a}$$

$$\hat{\mathbf{G}}^{t}(\mathbf{r}', t'|\mathbf{r}, t) = -[\nabla\times[\nabla\times\{\hat{\mathbf{I}}(g^{(t)}(\mathbf{r}-\mathbf{r}'|t-t') - G(\mathbf{r}-\mathbf{r}')\chi(t-t')\}]] \; . \tag{11.2.18b}$$

Here (1) $g^{(l,t)}$ stands for the functions derived upon double time integration of the Green functions for a scalar wave equation (with the parameters C_l and C_t, respectively):

$$g^{(l,t)}(\mathbf{r}-\mathbf{r}'|t-t') = G(\mathbf{r}-\mathbf{r}')\chi\left(\frac{|\mathbf{r}-\mathbf{r}'|}{C_{l,t}} - (t'-t)\right) \; , \tag{11.2.19}$$

where

$$\chi(t) = \int_{-\infty}^{t} \varkappa(\tau)d\tau = \begin{cases} 0 \; , & t<0 \\ t \; , & t>0 \end{cases} \; , \quad \varkappa(\tau) = \int_{-\infty}^{\tau} \delta(\tilde{\tau})d\tilde{\tau} = \begin{cases} 0 \; , & \tau<0 \\ 1 \; , & \tau>0 \; , \end{cases}$$

and $\delta(\tilde{\tau})$ is the Dirac function. (2) $G(\mathbf{r}-\mathbf{r}')$ is the Green fundamental function for the Laplace equation

$$G(\mathbf{r}-\mathbf{r}') = \frac{-1}{4\pi\,|\mathbf{r}-\mathbf{r}'|} \; .$$

Relation (11.2.16) will be called a *generalized integral formula of Kirchhoff for elastic oscillations* (a similar formula for steady-state (monochromatic) elastic oscillations was first derived by V.D. Kupradze (1950). It enables the value of the field of elastic displacements to be reconstructed everywhere inside the domain D from known values of this field and of its normal derivative at the surface S.

Now if the function $\hat{\mathbf{P}}(\mathbf{r}, t)$ in formula (11.2.14) is provided by the Green tensor for the wave equation $\hat{\mathbf{G}}^{W}(\mathbf{r}', t'|\mathbf{r}, t)$, satisfying the equation

$$\Delta\hat{\mathbf{G}}^{W} - \frac{1}{C^2}\frac{\partial^2\hat{\mathbf{G}}^{W}}{\partial t^2} = \hat{\mathbf{I}}\delta(\mathbf{r}-\mathbf{r}')\delta(t-t') \; ,$$

formula (11.2.14) permits representation

$$\int_{-\infty}^{+\infty}\iint_{S} \{\mathbf{n}\cdot(\mathbf{U}\,\mathrm{div}\,\hat{\mathbf{G}}^{W} - \hat{\mathbf{G}}^{W}\mathrm{div}\,\mathbf{U}) + \mathbf{n}\cdot[\mathbf{U}\times\mathrm{rot}\,\hat{\mathbf{G}}^{W} + \mathrm{rot}\,\mathbf{U}\times\hat{\mathbf{G}}^{W}]\}ds\,dt$$

$$= \begin{cases} \mathbf{U}(\mathbf{r}', t') \; , & \mathbf{r}'\in D \\ 0 \; , & \mathbf{r}'\in C\bar{D} \; . \end{cases} \tag{11.2.20}$$

Expression (11.2.17) is the Kirchhoff integral formula for the wave equation cast in the tensor form. Indeed the Green tensor for the wave equation $\hat{\mathbf{G}}^{W}$ is:

$$\hat{\mathbf{G}}^{W}(\mathbf{r}', t'|\mathbf{r}, t) = \hat{\mathbf{I}}G^{W}(\mathbf{r}', t'|\mathbf{r}, t) \; , \tag{11.2.21a}$$

where $\hat{\mathbf{G}}^{W}$ is the Green fundamental function for wave Eq. (11.1.21) defined by the formula (Morse and Feshbach 1953):

$$G^W(\mathbf{r}', t'|\mathbf{r}, t) = -\frac{1}{4\pi\,|\mathbf{r}-\mathbf{r}'|}\,\delta\left[\frac{|\mathbf{r}-\mathbf{r}'|}{C}-(t'-t)\right] . \qquad (11.2.21\,\mathrm{b})^2$$

Substitute (11.2.21) into (11.2.20) allowing for the following formulas for the divergence and rot of the Green tensor:

$$\mathrm{div}\,\hat{\mathbf{G}}^W = \nabla\cdot\hat{\mathbf{I}}G^W = \nabla G^W\cdot\hat{\mathbf{I}} = \nabla G^W = \mathrm{grad}\,G^W ,$$

$$\mathrm{rot}\,\hat{\mathbf{G}}^W = \nabla\times\hat{\mathbf{I}}G^W = \nabla G^W\times\hat{\mathbf{I}} = \mathrm{grad}\,G^W\times\hat{\mathbf{I}} .$$

As a result, a little transformation yields

$$\int_{-\infty}^{+\infty}\iint_S \{(\mathbf{n}\cdot\mathrm{grad}\,G^W)\mathbf{U}+(\mathbf{n}\times\mathrm{grad}\,G^W)\times\mathbf{U}+(\mathbf{n}\times\mathrm{rot}\,\mathbf{U})G^W-\mathbf{n}\,\mathrm{div}\,\mathbf{U}\,G^W\}ds\,dt$$

$$= \begin{cases}\mathbf{U}(\mathbf{r}', t') , & \mathbf{r}'\in D\\ 0 , & \mathbf{r}'\in C\bar{D} .\end{cases} \qquad (11.2.22)$$

Let us transform the left-hand side of formula (11.2.22), using the vector identity

$$(\mathbf{a}\cdot\mathbf{b})\mathbf{c}+(\mathbf{a}\times\mathbf{b})\times\mathbf{c} = (\mathbf{a}\cdot\mathbf{c})\mathbf{b}+(\mathbf{a}\times\mathbf{c})\times\mathbf{b} .$$

We obtain

$$\int_{-\infty}^{+\infty}\iint_S [(\mathbf{n}\cdot\mathbf{U})\,\mathrm{grad}\,G^W+(\mathbf{n}\times\mathbf{U})\times\mathrm{grad}\,G^W+(\mathbf{n}\times\mathrm{rot}\,\mathbf{U}-\mathbf{n}\,\mathrm{div}\,\mathbf{U})G^W]\,ds\,dt$$

$$= \begin{cases}\mathbf{U}(\mathbf{r}', t') , & \mathbf{r}'\in D\\ 0 , & \mathbf{r}'\in C\bar{D} .\end{cases} \qquad (11.2.23)$$

We will take the differential operator in formula (11.2.23), operating on the Green function G^W outside the integral sign, using formulas (8.3.22 a)

$$\int_{-\infty}^{+\infty}\left\{-\mathrm{grad}'\iint_S (\mathbf{n}\cdot\mathbf{U})G^W ds+\mathrm{rot}'\iint_S (\mathbf{n}\times\mathbf{U})G^W ds\right.$$

$$\left.+\iint_S (\mathbf{n}\times\mathrm{rot}\,\mathbf{U}-\mathbf{n}\,\mathrm{div}\,\mathbf{U})G^W ds\right\} dt = \begin{cases}\mathbf{U}(\mathbf{r}', t') , & \mathbf{r}'\in D\\ 0 , & \mathbf{r}'\in C\bar{D} .\end{cases} \qquad (11.2.24)$$

Now substituting (11.2.21 b) into (11.2.24) and integrating with respect to t, we find

$$\frac{1}{4\pi}\,\mathrm{grad}'\iint_S \frac{(\mathbf{n}\cdot\tilde{\mathbf{U}})}{|\mathbf{r}-\mathbf{r}'|}\,ds-\frac{1}{4\pi}\,\mathrm{rot}'\iint_S \frac{(\mathbf{n}\times\tilde{\mathbf{U}})}{|\mathbf{r}-\mathbf{r}'|}\,ds$$

$$-\frac{1}{4\pi}\iint_S \frac{\mathbf{n}\times\mathrm{rot}\,\tilde{\mathbf{U}}-\mathbf{n}\,\mathrm{div}\,\tilde{\mathbf{U}}}{|\mathbf{r}-\mathbf{r}'|}\,ds = \begin{cases}\mathbf{U}(\mathbf{r}', t') , & \mathbf{r}'\in D\\ 0 , & \mathbf{r}'\in C\bar{D} ,\end{cases} \qquad (11.2.25)$$

where $\tilde{\mathbf{U}}$ is the wave field at retarded instants of time (retarded field):

---

[2] Expression (11.2.21 b) is derived from (8.3.19) on exchanging $\sqrt{\mu\varepsilon}$ by $1/C$ in the latter and with due reference to the known property of δ-functions: $\delta(x-x') = \delta(x'-x)$.

$$\tilde{\mathbf{U}} = \int_{-\infty}^{+\infty} \mathbf{U}(t)\delta\left(\frac{|\mathbf{r}-\mathbf{r}'|}{C} - (t'-t)\right) dt = \mathbf{U}\left(t' - \frac{|\mathbf{r}-\mathbf{r}'|}{C}\right) . \tag{11.2.26}$$

Expression (11.2.25) is the Kirchhoff integral formula for the wave field expressed in terms of retarded fields.

11.2.3 Kirchhoff Integral Formulas for a Scalar Wave Field

The classical Kirchhoff integral formula for the scalar field is usually derived by direct integration of the scalar wave equation (Stratton 1941; Morse and Feshbach 1953). At the same time it is useful to see how this formula stems from general vector integrals of Kirchhoff. To solve this problem, resort again to expression (11.2.24) and assume that the wave field $\mathbf{U}$ is a field of longitudinal waves:

$$\mathbf{U} = \mathbf{U}^l = \operatorname{grad}\varphi , \tag{11.2.27}$$

where φ is the scalar potential satisfying the scalar wave equation

$$\Delta\varphi - \frac{1}{C^2}\frac{\partial^2\varphi}{\partial t^2} = 0 . \tag{11.2.28}$$

Substituting (11.2.27) into (11.2.24) we write

$$\int_{-\infty}^{+\infty} \left\{ -\operatorname{grad}' \iint_S (\mathbf{n}\cdot\operatorname{grad}\varphi) G^W ds + \operatorname{rot}' \iint_S [\mathbf{n}\times\operatorname{grad}\varphi] G^W ds - \iint_S \mathbf{n}\Delta\varphi G^W ds \right\} dt = \begin{cases} \operatorname{grad}\varphi(\mathbf{r}', t') , & \mathbf{r}'\in D \\ 0 , & \mathbf{r}'\in CD . \end{cases} \tag{11.2.29}$$

Let us evaluate the third integral in the left-hand side of (11.2.29), using integration by parts with respect to t and a scalar wave equation for φ and G^W:

$$\int_{-\infty}^{+\infty}\iint_S \mathbf{n}\Delta\varphi G^W ds\,dt = \int_{-\infty}^{+\infty}\iint_S \mathbf{n}\frac{1}{C^2}\frac{\partial^2\varphi}{\partial t^2} G^W ds\,dt = \int_{-\infty}^{+\infty}\iint_S \mathbf{n}\varphi\frac{1}{C^2}\frac{\partial^2 G^W}{dt^2} ds\,dt$$

$$= \int_{-\infty}^{+\infty} \mathbf{n}\varphi\Delta G^W ds\,dt . \tag{11.2.30}$$

Now we will transform the integrand in the right-hand side of (11.2.30) by changing the differentiation of the Green function $G^W(\mathbf{r}-\mathbf{r}'|t-t')$ from the variable $\mathbf{r}$ to the variable $\mathbf{r}'$:

$$\mathbf{n}\varphi\Delta G^W = \mathbf{n}\varphi\Delta' G^W = \Delta'(G^W\varphi\mathbf{n}) = \operatorname{grad}'\operatorname{div}'(G^W\varphi\mathbf{n}) - \operatorname{rot}'\operatorname{rot}'(G^W\varphi\mathbf{n})$$

$$= \operatorname{grad}'(\operatorname{grad}' G^W\cdot\varphi\mathbf{n}) - \operatorname{rot}'(\operatorname{grad}' G^W\times\varphi\mathbf{n})$$

$$= -\operatorname{grad}'(\operatorname{grad} G^W\cdot\varphi\mathbf{n}) + \operatorname{rot}'(\operatorname{grad} G^W\times\varphi\mathbf{n})$$

$$= -\operatorname{grad}'(\operatorname{grad} G^W\cdot\varphi\mathbf{n}) + \operatorname{rot}'[\operatorname{rot}(G^W\varphi\mathbf{n}) - G^W\operatorname{rot}(\varphi\mathbf{n})] ,$$

where the prime denotes differentiation with respect to $\mathbf{r}'$. Substituting the last formula into (11.2.30) we obtain

$$\int_{-\infty}^{+\infty}\iint_S \mathbf{n}\Delta\varphi G^W ds\,dt = -\mathrm{grad}'\int_{-\infty}^{+\infty}\iint_S (\mathrm{grad}\,G^W\cdot\mathbf{n})\varphi\, ds\,dt$$

$$+\mathrm{rot}'\int_{-\infty}^{+\infty}\iint_S \mathrm{rot}(G^W\varphi\mathbf{n})\,ds\,dt - \mathrm{rot}'\int_{-\infty}^{+\infty}\iint_S G^W \mathrm{rot}(\varphi\mathbf{n})\,ds\,dt\ . \qquad (11.2.31)$$

The second integral in the right-hand side of (11.2.31) is zero, according to the Stokes theorem (5.1.11), since S is a closed surface:

$$\iint_S \mathrm{rot}(G^W\varphi\mathbf{n})\,ds = \iint_S [\mathrm{grad}(G^W\varphi)\times\mathbf{n}]\,ds = 0\ . \qquad (11.2.32\mathrm{a})$$

Transformation similar to that of (6.1.20) can render the third integral to the form

$$\iint_S G^W\mathrm{rot}(\varphi\mathbf{n})\,ds = -\iint_S (\mathbf{n}\times\mathrm{grad}\,\varphi)G^W ds\ . \qquad (11.2.32\mathrm{b})$$

With due account taken of (11.2.32a) and (11.2.32b), formula (11.2.31) is rewritten as follows:

$$\int_{-\infty}^{+\infty}\iint_S \mathbf{n}\Delta\varphi G^W ds\,dt = -\mathrm{grad}'\int_{-\infty}^{+\infty}\iint_S (\mathrm{grad}\,G^W\cdot\mathbf{n})\varphi\, ds\,dt$$

$$+\mathrm{rot}'\int_{-\infty}^{+\infty}\iint_S (\mathbf{n}\times\mathrm{grad}\,\varphi)G^W ds\,dt\ . \qquad (11.2.33)$$

Substituting (11.2.33) into (11.2.29) we find

$$\mathrm{grad}'\int_{-\infty}^{+\infty}\iint_S \left(\varphi\frac{\partial G^W}{\partial\mathbf{n}} - G^W\frac{\partial\varphi}{\partial\mathbf{n}}\right)ds\,dt = \begin{cases}\mathrm{grad}'\varphi(\mathbf{r}',t')\ , & \mathbf{r}'\in D\\ 0\ , & \mathbf{r}'\in C\bar{D}\ .\end{cases} \qquad (11.2.34)$$

Hence, integrating with respect to t (allowing for expression (11.2.21b) for the Green function G^W) and omitting the symbol grad$'$, we write in the final form

$$\frac{1}{4\pi}\iint_S\left(\frac{1}{C|\mathbf{r}-\mathbf{r}'|}\frac{\partial|\mathbf{r}-\mathbf{r}'|}{\partial\mathbf{n}}\frac{\partial\tilde{\varphi}}{\partial t} - \frac{\partial}{\partial\mathbf{n}}\frac{1}{|\mathbf{r}-\mathbf{r}'|}\tilde{\varphi} + \frac{1}{|\mathbf{r}-\mathbf{r}'|}\frac{\partial\tilde{\varphi}}{\partial\mathbf{n}}\right)ds$$

$$= \begin{cases}\varphi(\mathbf{r}',t')\ , & \mathbf{r}'\in D\\ 0\ , & \mathbf{r}'\in C\bar{D}\ ,\end{cases} \qquad (11.2.35)$$

where $\tilde{\varphi} = \varphi\left(t' - \dfrac{|\mathbf{r}-\mathbf{r}'|}{C}\right)$ is the retarded potential. The obtained relation is precisely the classical integral formula of Kirchhoff for the scalar field φ.

11.2.4 Kirchhoff Integral Formulas in Matrix Notation

In what follows, it will be convenient, while solving certain problems, to use Kirchhoff integral formulas in matrix notation. The technique of this notation is identical to those in Sects. 5.4 and 8.4.4. Therefore, at this point we will only briefly outline the transition to matrix representations.

Note first of all that elastic oscillation Eq. (11.1.20) and wave Eq. (11.1.21) in matrix notation have the form

$$\Delta^*_{\alpha\beta\gamma\eta}\mathbf{U}_{\beta,\gamma\eta}-\frac{\partial^2\mathbf{U}_\alpha}{\partial t^2}=0 \ , \tag{11.2.36}$$

$$\Delta_{\alpha\beta\gamma\eta}\mathbf{U}_{\beta,\gamma\eta}-\frac{1}{C^2}\frac{\partial^2\mathbf{U}_\alpha}{\partial t^2}=0 \ , \tag{11.2.37}$$

where the subscript separated by a comma denotes, as usual, differentiation with respect to the corresponding coordinate, while the four-valence tensors $\Delta^*_{\alpha\beta\gamma\eta}$ and $\Delta_{\alpha\beta\gamma\eta}$ are formed by combining the symmetric $\delta_{\alpha\beta}$ and skew-symmetric $\varepsilon_{\alpha\beta\gamma}$. Kronecker symbols [cf. (5.4.11)]

$$\Delta^*_{\alpha\beta\gamma\eta}=C_l^2\delta_{\alpha\eta}\delta_{\beta\gamma}+C_t^2\varepsilon_{\alpha\eta\zeta}\varepsilon_{\beta\gamma\zeta} \ , \tag{11.2.38 a}$$

$$\Delta_{\alpha\beta\gamma\eta}=\delta_{\alpha\eta}\delta_{\beta\gamma}+\varepsilon_{\alpha\eta\zeta}\varepsilon_{\beta\gamma\zeta} \ . \tag{11.2.38 b}$$

Accordingly, the generalized Kirchhoff integral formula (11.2.16) is written as

$$\int_{-\infty}^{+\infty}\iint_S \Delta^*_{\alpha\beta\gamma\eta}(\mathbf{U}_\alpha G^e_{\mu\beta,\gamma}-\mathbf{U}_{\beta,\gamma}G^e_{\mu\alpha})n_\eta\,ds\,dt=\begin{cases}\mathbf{U}_\mu(\mathbf{r}',t') \ , & \mathbf{r}'\in D\\ 0 \ , & \mathbf{r}'\in C\bar{D} \ ,\end{cases} \tag{11.2.39}$$

where $G^e_{\mu\alpha}$ is the matrix of the Green tensor $\hat{\mathbf{G}}^e$ for elastic oscillations:

$$G^e_{\mu\alpha}=G^l_{\mu\alpha}+G^t_{\mu\alpha} \ . \tag{11.2.40}$$

Here $G^l_{\mu\alpha}$ and $G^t_{\mu\alpha}$ are the matrices of the longitudinal and transverse Green tenors defined, according to (11.2.18a, b), by the relations

$$G^l_{\mu\alpha}(\mathbf{r}',t'|\mathbf{r},t)=\delta_{\mu\beta}\delta_{\alpha\gamma}[g^l_{,\beta\gamma}(\mathbf{r}-\mathbf{r}'|t-t')-G_{,\beta\gamma}(\mathbf{r}-\mathbf{r}')\chi(t-t')] \tag{11.2.41 a}$$

$$G^t_{\mu\alpha}(\mathbf{r}',t'|\mathbf{r},t)=\varepsilon_{\alpha\gamma\zeta}\varepsilon_{\mu\beta\zeta}[g^{(t)}_{,\beta\gamma}(\mathbf{r}-\mathbf{r}'|t-t')-G_{,\beta\gamma}(\mathbf{r}-\mathbf{r}')\chi(t-t')] \ , \tag{11.2.41 b}$$

where the adopted designations are the same as in formulas (11.2.18a, b).

Substituting (11.2.31a, b) into (11.2.30) and contracting the corresponding tensors, we obtain

$$G^e_{\mu\alpha}=g^{(l)}_{,\mu\alpha}-g^{(t)}_{,\mu\alpha}+\delta_{\mu\alpha}(g^{(t)}_{,\zeta\zeta}-G_{,\zeta\zeta}\cdot\chi)=g^{(l)}_{,\mu\alpha}-g^{(t)}_{,\mu\alpha}+\frac{\delta_{\mu\alpha}}{C_S^2}G^{W(t)} \ , \tag{11.2.42}$$

where $G^{W(t)}$ is the Green function for a scalar wave equation with the parameter C_t:

$$G^{W(t)}=\frac{-1}{4\pi\,|\mathbf{r}-\mathbf{r}'|}\,\delta\left(\frac{|\mathbf{r}-\mathbf{r}'|}{C_t}-(t'-t)\right) . \tag{11.2.43}$$

Obviously, expression (11.2.42) is the simplest and the most convenient representation of the Green tensor for elastic oscillations.

For $C_l = C_t = C$ the elastic oscillation tensor of Green goes over into the Green tensor for the wave equation $\hat{\mathbf{G}}^W$:

$$G^e_{ma} = \frac{1}{C^2} G^W_{ma} = \frac{1}{C^2} \delta_{ma} \Gamma^n , \tag{11.2.44}$$

whereas formula (11.2.39) is reduced to the integral formula of Kirchhoff (11.2.20) in matrix notation

$$\int_{-\infty}^{+\infty} \iint_S \Delta_{\alpha\beta\gamma\eta} (\mathbf{U}_a G^W_{\mu\beta,\gamma} - \mathbf{U}_{\beta,\gamma} G^W_{\mu a}) \mathbf{n}_\eta \, ds \, dt = \begin{cases} \mathbf{U}_\mu(\mathbf{r}', t) , & \mathbf{r}' \in D \\ 0, & \mathbf{r}' \in C\bar{D} . \end{cases}$$

The above matrix notations of the Kirchhoff integral formulas prove to be very convenient in the solution of problems of analytical continuation and migration of elastic fields.

11.3 Kirchhoff-Type Integrals

Analogously to the Cauchy- and Stratton-Chu type integrals, Kirchhoff-type integrals can be taken into consideration. These integrals are convenient to use in connection with the problems of analytical continuation and migration of elastic fields to be treated in the following chapter.

11.3.1 Kirchhoff-Type Integrals for Wave Fields

To define the concept of a Kirchhoff-type integral, we will turn again to integral formula (11.2.24). It can be seen from this formula that to evaluate the wave field inside the domain D, one has to know not only the field itself on the boundary of this domain but also the divergence $\operatorname{div} \mathbf{U}$ and the tangential components of the $\operatorname{rot} \mathbf{U}$. It will be shown that the values $\operatorname{div} \mathbf{U}$ and $(\mathbf{n} \times \operatorname{rot} \mathbf{U})$ can be readily calculated from the values of the field $\mathbf{U}$ and its normal derivative $\partial \mathbf{U}/\partial \mathbf{n}$ specified at the surface S.

Indeed, transform the vector derivative

$$\mathbf{n} \times \operatorname{rot} \mathbf{U} = \mathbf{n} \times (\nabla \times \mathbf{U}) = \nabla (\mathbf{n} \cdot \mathbf{U}) - (\mathbf{n} \cdot \nabla) \mathbf{U} = \nabla (\mathbf{n} \cdot \mathbf{U}) - \frac{\partial \mathbf{U}}{\partial \mathbf{n}}$$

$$= (\mathbf{n} \times \nabla) \times \mathbf{U} + \mathbf{n} (\nabla \cdot \mathbf{U}) - \frac{\partial \mathbf{U}}{\mathbf{n}} = [\mathbf{n} \times \nabla] \times \mathbf{U} + \mathbf{n} \operatorname{div} \mathbf{U} - \frac{\partial \mathbf{U}}{\partial \mathbf{n}} . \tag{11.3.1}$$

Hence,

$$\mathbf{n} \times \operatorname{rot} \mathbf{U} - \mathbf{n} \operatorname{div} \mathbf{U} = (\mathbf{n} \times \nabla) \times \mathbf{U} - \frac{\partial \mathbf{U}}{\partial \mathbf{n}} . \tag{11.3.2}$$

The expression $(\mathbf{n} \times \nabla) \times \mathbf{U}$ can be written in terms of the surface gradient and divergence. Indeed,

$$[\mathbf{n}\times\nabla]\times\mathbf{U} = (\mathbf{n}\times\nabla_S)\times\mathbf{U} = -\mathbf{n}(\nabla_S\cdot\mathbf{U})+\nabla_S(\mathbf{n}\cdot\mathbf{U}) = \mathrm{grad}_S\mathbf{U}_n-\mathbf{n}\,\mathrm{div}_S\mathbf{U}_\tau\ , \tag{11.3.3}$$

where $\mathbf{U} = \mathbf{U}_n+\mathbf{U}_\tau$, $\mathbf{U}_n$, and $\mathbf{U}_\tau$ are the normal and the tangential components of the field $\mathbf{U}$; ∇_S, grad_S, and div_S are the operators of surface differentiation defined by formulas (5.2.1).

Substituting (11.3.3) into (11.3.2) and subsequently into (11.2.24) and using expression (11.2.21 b) of the Green fundamental function for the wave equation, we find

$$\frac{1}{4\pi}\,\mathrm{grad}'\iint_S \frac{\mathbf{n}\cdot\tilde{\mathbf{U}}}{|\mathbf{r}-\mathbf{r}'|}\,ds-\frac{1}{4\pi}\,\mathrm{rot}'\iint_S \frac{\mathbf{n}\times\tilde{\mathbf{U}}}{|\mathbf{r}-\mathbf{r}'|}\,ds$$
$$-\frac{1}{4\pi}\iint_S \frac{\mathrm{grad}_S\tilde{\mathbf{U}}_n-\mathbf{n}\,\mathrm{div}_S\tilde{\mathbf{U}}_\tau-\partial\tilde{\mathbf{U}}/\partial\mathbf{n}}{|\mathbf{r}-\mathbf{r}'|}\,ds = \begin{cases}\mathbf{U}(\mathbf{r}',t')\ , & \mathbf{r}'\in D\\ 0\ , & \mathbf{r}'\in C\bar{D}\end{cases} \tag{11.3.4}$$

where $\tilde{\mathbf{U}}$ is the retarded wave field fitted by formula (11.2.26). The first two integrals in (11.3.4) are exactly the three-dimensional Cauchy-type integral (5.2.4) written for a retarded wave field $\tilde{\mathbf{U}}$:

$$\mathbf{C}^S(\mathbf{r}',\tilde{\mathbf{U}}) = \frac{1}{4\pi}\,\mathrm{grad}'\iint_S \frac{\mathbf{n}\cdot\tilde{\mathbf{U}}}{|\mathbf{r}-\mathbf{r}'|}\,ds-\frac{1}{4\pi}\,\mathrm{rot}'\iint_S \frac{\mathbf{n}\times\tilde{\mathbf{U}}}{|\mathbf{r}-\mathbf{r}'|}\,ds\ . \tag{11.3.5}$$

Thus, we arrive at the following very convenient representation of the Kirchhoff integral formula which is equivalent to (11.2.25):

$$\mathbf{C}^S(\mathbf{r}',\tilde{\mathbf{U}})-\frac{1}{4\pi}\iint_S \frac{\mathrm{grad}_S\tilde{\mathbf{U}}_n-\mathbf{n}\,\mathrm{div}_S\tilde{\mathbf{U}}_\tau-\partial\tilde{\mathbf{U}}/\partial\mathbf{n}}{|\mathbf{r}-\mathbf{r}'|}\,ds = \begin{cases}\mathbf{U}(\mathbf{r}',t')\ , & \mathbf{r}'\in D\\ 0\ , & \mathbf{r}'\in CD\ .\end{cases} \tag{11.3.6}$$

Formula (11.3.6) shows clearly that to evaluate the wave field inside the domain D it is necessary to know this field and its normal derivative on the inner side of the surface S bounding the domain D.

The obtained expression can be used to define the concept of the Kirchhoff-type integral for a wave field. Indeed, using the approach employed repeatedly in our book, consider the expression in the left-hand side of formula (11.3.6), where instead of the functions $\mathbf{U}$ and $\partial\mathbf{U}/\partial\mathbf{n}$ there appear arbitrary nonstationary vector fields $\mathbf{W}(\mathbf{r},t)$ and $\mathbf{V}(\mathbf{r},t)$ specified at the surface S (the field $\mathbf{W}$ is also assumed to be differentiable at S):

$$\mathbf{K}^W(\mathbf{r}',\mathbf{W},\mathbf{V}) = C^S(\mathbf{r}',\tilde{\mathbf{W}})-\frac{1}{4\pi}\iint_S \frac{\mathrm{grad}_S\tilde{\mathbf{W}}_n-\mathbf{n}\,\mathrm{div}_S\tilde{\mathbf{W}}_\tau-\tilde{\mathbf{V}}}{|\mathbf{r}-\mathbf{r}'|}\,ds\ , \tag{11.3.7}$$

where $\tilde{\mathbf{W}}$ and $\tilde{\mathbf{V}}$ are the values of the fields at retarded instants of time:

$$\tilde{\mathbf{W}} = \mathbf{W}\left(t'-\frac{|\mathbf{r}-\mathbf{r}'|}{C}\right)\ ,$$
$$\tilde{\mathbf{V}} = \mathbf{V}\left(t'-\frac{|\mathbf{r}-\mathbf{r}'|}{C}\right)\ . \tag{11.3.8}$$

One is readily convinced that everywhere outside the surface S the vector function $\mathbf{K}^W$ satisfies the wave equation

$$\Delta \mathbf{K}^W - \frac{1}{C^2}\frac{\partial^2 \mathbf{K}^W}{\partial t^2} = 0 \ .$$

At the same time, once the functions $\mathbf{W}$ and $\mathbf{V}$ coincide with the boundary value of the wave field $\mathbf{U}$ and its normal derivative $\partial \mathbf{U}/\partial \mathbf{n}$ at the surface S, we have, according to (11.3.6),

$$\mathbf{K}^W\left(\mathbf{r}', \mathbf{U}, \frac{\partial \mathbf{U}}{\partial \mathbf{n}}\right) = \begin{cases} \mathbf{U}(\mathbf{r}', t') \ , & \mathbf{r}' \in D \\ 0 \ , & \mathbf{r}' \in C\tilde{D} \ . \end{cases} \tag{11.3.9}$$

Taking these simple properties of the function $\mathbf{K}^W$ into account, we naturally call it *a Kirchhoff-type integral for a wave field,* while the functions $\mathbf{W}$ and $\mathbf{V}$ are termed *its vector densities.*

11.3.2 Kirchhoff-Type Integrals for Elastic Displacement Fields

Now turn to the generalized Kirchhoff integral formula (11.2.16). We will transform the integrand in this formula, allowing for (11.3.1) and (11.3.3), in the following way:

$$\begin{aligned} & C_l^2(\mathbf{n}\cdot\mathbf{U}\,\mathrm{div}\,\hat{\mathbf{G}}^e - \mathbf{n}\cdot\hat{\mathbf{G}}^e\,\mathrm{div}\,\mathbf{U}) + C_t^2\mathbf{n}\cdot[\mathbf{U}\times\mathrm{rot}\,\hat{\mathbf{G}}^e + \mathrm{rot}\,\mathbf{U}\times\hat{\mathbf{G}}^e] \\ &= C_l^2(\mathbf{n}\cdot\mathbf{U})\,\mathrm{div}\,\hat{\mathbf{G}}^e + C_t^2[\mathbf{n}\times\mathbf{U}]\cdot\mathrm{rot}\,\hat{\mathbf{G}}^e + \{C_t^2[\mathbf{n}\times\mathrm{rot}\,\mathbf{U}] - C_l^2\mathbf{n}\,\mathrm{div}\,\mathbf{U}\}\hat{\mathbf{G}}^e \\ &= C_l^2(\mathbf{n}\cdot\mathbf{U})\,\mathrm{div}\,\hat{\mathbf{G}}^e + C_t^2[\mathbf{n}\times\mathbf{U}]\cdot\mathrm{rot}\,\hat{\mathbf{G}}^e \\ &\quad + \left\{C_l^2[\mathbf{n}\times\nabla]\times\mathbf{U} - C_l^2\frac{\partial \mathbf{U}}{\partial \mathbf{n}} + \mathbf{n}(C_t^2 - C_l^2)\,\mathrm{div}\,\mathbf{U}\right\}\hat{\mathbf{G}}^e \\ &= C_l^2(\mathbf{n}\cdot\mathbf{U})\,\mathrm{div}\,\hat{\mathbf{G}}^e + C_t^2[\mathbf{n}\times\mathbf{U}]\cdot\mathrm{rot}\,\hat{\mathbf{G}}^e + \left\{C_l^2\left(\mathrm{grad}_S\mathbf{U}_n - \mathbf{n}\,\mathrm{div}_S\mathbf{U}_\tau - \frac{\partial \mathbf{U}}{\partial \mathbf{n}}\right)\right. \\ &\quad \left. + \mathbf{n}(C_t^2 - C_l^2)\,\mathrm{div}_S\mathbf{U}_\tau + \mathbf{n}(C_t^2 - C_l^2)\frac{\partial \mathbf{U}}{\partial \mathbf{n}}\right\}\hat{\mathbf{G}}^e \ . \end{aligned} \tag{11.3.10}$$

Substituting (11.3.10) into (1.2.16) we arrive at the following representation of the Kirchhoff integral formula

$$\begin{aligned} & \int_{-\infty}^{+\infty}\iint_S \left\{C_l^2(\mathbf{n}\cdot\mathbf{U})\,\mathrm{div}\,\hat{\mathbf{G}}^e + C_t^2[\mathbf{n}\times\mathbf{U}]\cdot\mathrm{rot}\,\hat{\mathbf{G}}^e + \left[C_l^2\left(\mathrm{grad}_S\mathbf{U}_n - \mathbf{n}\,\mathrm{div}_S\mathbf{U}_\tau - \frac{\partial \mathbf{U}}{\partial \mathbf{n}}\right)\right.\right. \\ &\quad \left.\left. + \mathbf{n}(C_t^2 - C_l^2)\,\mathrm{div}_S\mathbf{U}_\tau + \mathbf{n}(C_t^2 - C_l^2)\frac{\partial \mathbf{U}_n}{\partial \mathbf{n}}\right]\hat{\mathbf{G}}^e\right\} ds\,dt \\ &= \begin{cases} \mathbf{U}(\mathbf{r}', t') \ , & \mathbf{r}' \in D \\ 0 \ , & \mathbf{r}' \in CD \ . \end{cases} \end{aligned} \tag{11.3.11}$$

As can be seen, to evaluate the field of elastic displacements inside the domain D, it is necessary to know both the field itself and its normal derivative on the inner side of the surface S bounding the domain D.

Following the previous subsection, let us construct an integral expression equal to the left-hand side of (11.3.11), in which instead of the functions $\mathbf{U}$ and $\partial\mathbf{U}/\partial\mathbf{n}$ there appear arbitrary non-stationary vector fields $\mathbf{W}(\mathbf{r}; t)$ and $\mathbf{V}(\mathbf{r}; t)$:

$$\mathbf{K}^e(\mathbf{r}', \mathbf{W}, \mathbf{V}) = \int_{-\infty}^{+\infty}\iint_S \{C_l^2(\mathbf{n}\cdot\mathbf{W})\operatorname{div}\hat{\mathbf{G}}^e + C_t^2[\mathbf{n}\times\mathbf{W}]\cdot\operatorname{rot}\hat{\mathbf{G}}^e$$
$$+[C_l^2(\operatorname{grad}_S W_n - \mathbf{n}\operatorname{div}_S\mathbf{W}_\tau - \mathbf{V}) + \mathbf{n}(C_t^2 - C_l^2)\operatorname{div}_S\mathbf{W}_\tau$$
$$+\mathbf{n}(C_t^2 - C_l^2)\mathbf{U}_n]\hat{\mathbf{G}}^e\}ds\,dt\ . \tag{11.3.12}$$

Direct checking will reveal that the vector field $\mathbf{K}^e$ satisfies everywhere outside the surface S the elastic oscillation equation

$$\Delta^*\mathbf{K}^e - \frac{\partial^2\mathbf{K}^e}{\partial t^2} = 0\ .$$

Following item 11.3.1, the vector function $\mathbf{K}^e$ will be called a *Kirchhoff-type integral for fields of elastic displacements*, while the functions $\mathbf{W}$ and $\mathbf{V}$ will be termed its *vector densities*. Evidently, if the functions $\mathbf{W}$ and $\mathbf{V}$ coincide with the boundary values of the displacement vector field $\mathbf{U}$ and its normal derivative $\partial\mathbf{U}/\partial\mathbf{n}$ at the surface S, then we have, according to (11.3.11),

$$\mathbf{K}^e(\mathbf{r}', \mathbf{U}, \partial\mathbf{U}/\partial\mathbf{n}) = \begin{cases}\mathbf{U}(\mathbf{r}', t')\ , & \mathbf{r}'\in D\\ 0\ , & \mathbf{r}'\in C\bar{D}\ .\end{cases} \tag{11.3.13}$$

Analogously to the three-dimensional Cauchy-type and Stratton-Chu type integrals, a sufficiently comprehensive theory of Kirchhoff-type integrals can be developed, including calculation of singular integrals (for $\mathbf{r}'\in S$), construction of analogs of the Sokhotsky-Plemelj formulas, etc. But in view of the limited scope of this book, we will not dwell on these issues here, giving so the reader an opportunity to repeat the reasoning and proofs outlined in detail in the previous chapters.

12 Continuation and Migration of Elastic Wave Fields

It was already mentioned in Chap. 10 that one of the most effective methods used to interpret seismic data is that of seismic migration or so-called seismoholography (Hemon 1971; Timoshin 1972, 1978; Petrashen and Nakhamkin 1973; Vasiliev 1975; Clearbout 1976, 1985; Berkhout 1980, 1984). Behind these methods is the idea of time reversal of an elastic oscillation field and continuation of the reversed wave field toward sources. For instance, in diffraction transformation the Fresnel-Huygens principle for a time-reversed field is employed (Timoshin 1978). In migration by the Clearbout method, a boundary-value problem for wave equation is numerically solved. The physical prerequisite for developing all these methods is the fact that the distribution of transformed amplitudes of elastic waves over a medium yields, just as in the case of optical holography, an image of the studied medium. However, as a rule, it is not the true field but only a certain mapping of it that is reconstructed in a medium. At the same time, in many seismic prospecting and seismology problems it is of interest to analyze the real distribution of the elastic wave field in the interior of the Earth. This analysis facilitates, on the one hand, a more complete understanding of the space-time structure of the wave field, and, as will be shown later, it is highly instrumental in the solution of inverse problems of dynamic seismics, on the other hand. In this connection it is necessary to develop methods for faithful reconstruction of the wave distribution in an elastic medium, methods allied to those of analytical continuation of the gravitational, magnetic, and electromagnetic fields considered in the previous chapters.

Below, there is a solution to the problem of continuing the field of elastic oscillations specified at a certain surface (the Earth's surface) into the lower half-space. The solution of this problem is based on the generalized Kirchhoff integral formula for elastic oscillations given in matrix notation [formula (11.2.39)]. Since under certain assumptions a seismic field in a medium can be described approximately by the wave equation (Vasiliev and Golosov 1971), along with the theory of continuation of an overall displacement vector, a simpler theory of wave field continuation satisfying the Fresnel-Huygens principle and the wave equation is elaborated. It is noteworthy that practically all the above-mentioned previous works on the theory of wave field migration operate precisely with this simplified model of the seismic field.

12.1 Analytical Continuation of Elastic Wave Fields

The problem of continuing the field of elastic displacements is very similar to those of analytical continuation of the gravitational, magnetic, and electromagnetic fields considered in Chaps. 7 and 9. The property of analyticity established earlier for electromagnetic fields is also preserved for the field of elastic displacements and for the wave field in regions of constant elastic parameters of a medium. Therefore, the results related to the concept of analytical continuation, equations for complete analytical functions, and the principle of continuity, obtained for the electromagnetic field can be extended sequentially to these fields. In view of the limited volume of this book, we are not going to dwell on these questions once again, but confine ourselves instead to the integral methods of wave field continuation, which are of great value when applied to the dynamic seismic problems (Zhdanov and Matusevich 1984).

12.1.1 Continuation of the Elastic Displacement Field into the Upper and Lower Half-Spaces in a Homogeneous Isotropic Medium

Let us specify an arbitrary piecewise-smooth surface S passing through a point at infinity in a homogeneous isotropic space with parameters C_l and C_t (Fig. 45):

$$S:\{Z=z(x,\ y)\ ,\quad -\infty<x<+\infty\ ,\quad -\infty<y<+\infty\}$$

$$0\geqslant z(x,\ y)\geqslant -h\ ,\quad h>0\ .$$

The surface S separates the whole space into two half-spaces: Ω^+ (upper) and Ω^- (lower). Let components of the displacement vector U_α and its spatial derivatives $U_{\alpha,\beta}$ be specified on the surface S at any instant of time. It is also known that the sources of elastic oscillations are found below the horizontal plane

$$P:\{z=-h\}\ ,\quad h>h_0$$

and at infinity the radiation condition is fulfilled (Kupradze 1950). The problem is to determine the magnitude of the vector U everywhere in the region of the lower half-space Ω^- lying above the plane P. This problem will be called a problem of continuation of the elastic displacement field into the lower half-space.

First consider the simpler problem of continuing elastic waves into the upper half-space Ω^+. This problem is readily solved, using formula (11.2.39). Indeed, to evaluate the field at some point $\mathbf{r}'\in\Omega^+$, it is sufficient to apply formula (11.2.39) to the surface formed by part of the sphere $O_R(\mathbf{r}')$ of a rather large radius R with its center at the point $\mathbf{r}'$ belonging to Ω^+, and by part S_R of the surface S cut from the latter by the sphere O_R. Then, letting $R\to\infty$ in the obtained expression and considering that the integral taken around the surface of the sphere tends to zero, on account of the radiation condition, we find in the final form

$$U_\mu(\mathbf{r}',\ t')=\int_{-\infty}^{\infty}\iint_S \Delta^*_{\alpha\beta\eta\gamma}[U_\alpha(t)G^e_{\mu\beta,\gamma}(t'-t)-U_{\beta,\gamma}(\tau)G^e_{\mu\alpha}(t'-t)]\,n_\eta\,ds\,dt\ . \tag{12.1.1}$$

Equation (12.1.1) admits a solution for the problem of continuing an overall displacement vector into the upper half-space.

Now return to the problem of continuing elastic waves into the lower half-space Ω^- right to the plane P. We will employ the known expansion of the Green scalar function for the wave equation (Morse and Feshbach 1953):

$$-4\pi G^{w(l,t)}(\mathbf{r}-\mathbf{r}'|t') = \frac{1}{|\mathbf{r}-\mathbf{r}'|}\,\delta\left(\frac{|\mathbf{r}-\mathbf{r}'|}{C_{l,t}}-t'\right)$$

$$= \frac{1}{4\pi^2}\iiint_{-\infty}^{\infty} \frac{\exp\,[i\underset{(l,t)}{\mathbf{\Omega}^{\pm}}\cdot(\mathbf{r}'-\mathbf{r})]}{\sqrt{v^2+w^2-\dfrac{\omega^2}{C_{l,t}^2}}}\, e^{-i\omega t}\, dv\,dw\,d\omega \ , \quad (12.1.2)$$

where

$$\underset{(l,t)}{\mathbf{\Omega}^{\pm}} = \left(v,\ w,\ \pm i\sqrt{v^2+w^2-\frac{\omega^2}{C_{l,t}^2}}\right) = (\Omega_x,\ \Omega_y,\ \underset{(l,t)}{\Omega_z^{\pm}}) \ ;$$

$$\underset{(l,t)}{\mathbf{\Omega}^{\pm}}\cdot(\mathbf{r}'-\mathbf{r}) = v(x'-x)+w(y'-y)\pm i\sqrt{v^2+w^2-\frac{\omega^2}{C_{l,t}^2}}\,(z'-z) \ ;$$

the positive sign is chosen under the condition $z'>z$, while the negative sign is taken when $z'<z$. Substitution of (12.1.2) into (11.2.42) yields decomposition of the Green tensor:

$$G^e_{\mu\alpha}(\mathbf{r}'-\mathbf{r})|t) = \frac{-i}{16\pi^3}\int_{-\infty}^{\infty}\frac{e^{-i\omega t}}{\omega^2}\iint_{-\infty}^{\infty}[\Gamma^{\pm(l)}_{\mu\alpha}e^{i\underset{(l)}{\mathbf{\Omega}^{\pm}}\cdot(\mathbf{r}'-\mathbf{r})}+\Gamma^{\pm(t)}_{\mu\alpha}e^{i\underset{(t)}{\mathbf{\Omega}^{\pm}}\cdot(\mathbf{r}'-\mathbf{r})}]$$

$$\cdot dv\,dw\,d\omega \ ; \quad (12.1.3)$$

where the matrices $\Gamma^{\pm(l)}_{\mu\alpha}$ and $\Gamma^{\pm(t)}_{\mu\alpha}$ are defined by the equations:

$$\Gamma^{\pm(l)}_{\mu\alpha} = \frac{\underset{(l)}{\Omega^{\pm}_{\mu}}\,\underset{(l)}{\Omega^{\pm}_{a}}}{\underset{(l)}{\Omega^{+}_{z}}} \ ; \quad \Gamma^{\pm(t)}_{\mu\alpha} = \frac{\dfrac{\omega^2}{C_t^2}\delta_{\mu\alpha}-\underset{(t)}{\Omega^{\pm}_{\mu}}\,\underset{(t)}{\Omega^{\pm}_{a}}}{\underset{(t)}{\Omega^{+}_{z}}} \ . \quad (12.1.4)$$

Further transforms are similar to the calculations presented in Sects. 4.5 and 9.3. Let $\mathbf{r}'$ be an arbitrary point of the lower half-space lying above P (i.e., $z'>-h$). Take such a small value $\varepsilon>0$ that $-h+\varepsilon<z'$ and draw a horizontal plane $\Sigma(z=-h+\varepsilon)$. According to (12.1.1), we have

$$U_\mu(\mathbf{r}',\ t') = \int_{-\infty}^{\infty}\iint_{\Sigma}\Delta^*_{\alpha\beta\eta\gamma}[U_\alpha(t)G^e_{\mu\beta,\gamma}(t'-t)-U_{\beta,\gamma}(t)G^e_{\mu\alpha}(t'-t)]\,n_\eta\,ds,\,dt \ .$$

$$(12.1.5)$$

Substitute formula (12.1.3) (provided that $z'>z$) into (12.1.5) and, since all integrals converge uniformly by the corresponding parameters, exchange the order of integration:

$$U_\mu(\mathbf{r}', t') = \frac{-i}{16\pi^3} \iiint_{-\infty}^{\infty} \frac{1}{\omega^2} \left\{ \exp[i(\underset{(l)}{\boldsymbol{\Omega}} \cdot \mathbf{r}')] \int_{-\infty}^{\infty} \iint_{\Sigma} \Delta^*_{\alpha\beta\eta\gamma} \right.$$

$$\cdot [U_\alpha(t) \frac{\partial}{\partial r_\gamma} (\Gamma^{(l)}_{\mu\beta} \exp[-i(\underset{(l)}{\boldsymbol{\Omega}} \cdot \mathbf{r})] \exp[-i\omega(t'-t)])$$

$$- U_{\beta,\gamma}(t) \Gamma^{(l)}_{\mu\alpha} \exp[-i(\underset{(l)}{\boldsymbol{\Omega}} \cdot \mathbf{r})] \exp[-i\omega(t'-t)] n_\eta ds dt$$

$$+ \exp[i(\underset{(t)}{\boldsymbol{\Omega}} \cdot \mathbf{r}')] \int_{-\infty}^{\infty} \iint_{\Sigma} \Delta^*_{\alpha\beta\eta\gamma} [U_\alpha(t) \frac{\partial}{\partial r_\gamma} (\Gamma^{(t)}_{\mu\beta} \exp[-i(\underset{(t)}{\boldsymbol{\Omega}} \cdot \mathbf{r})]$$

$$\cdot \exp[-i\omega(t'-t)] - U_{\beta,\gamma}(t) \cdot \Gamma^{(t)}_{\mu\alpha}$$

$$\left. \cdot \exp[-i(\underset{(t)}{\boldsymbol{\Omega}} \cdot \mathbf{r})] \exp[-i\omega(t'-t)] n_\eta ds dt \right\} dv dw d\omega \; . \qquad (12.1.6)^1$$

Analogously to the above case, construct a sphere $O_R(\mathbf{r}')$. Part of the surface of this sphere enclosed between S and Σ will be designated by C_R. Consider the closed surface γ_R formed by C_R and by the corresponding parts of the surfaces$S$ and Σ cut therefrom by the sphere $O_R(\mathbf{r}')$. One can readily see that inside γ_R for any R the following relations

$$\Delta^*_{\alpha\beta\eta\gamma} \frac{\partial^2}{\partial r_\gamma \partial r_\eta} (\Gamma^{(l,t)}_{\mu\beta} \cdot e^{-i\underset{(l,t)}{\boldsymbol{\Omega}} \cdot \mathbf{r}} e^{-i\omega(t'-t)}) - \frac{\partial^2}{\partial t^2} (\Gamma^{(l,t)}_{\mu\alpha} \cdot e^{-i\underset{(l,t)}{\boldsymbol{\Omega}} \cdot \mathbf{r}} \cdot e^{-i\omega(t'-t)}) = 0$$

$$(12.1.7)$$

are met. Hence, we have, according to formula (11.2.29),

$$\int_{-\infty}^{\infty} \iint_{\gamma_R} \Delta^*_{\alpha\beta\eta\gamma} \left\{ U_\alpha(t) \frac{\partial}{\partial r_\gamma} [\Gamma^{(l,t)}_{\mu\beta} \exp(-i\underset{(l,t)}{\boldsymbol{\Omega}} \cdot \mathbf{r}) \exp[-i\omega(t'-t)]] \right.$$

$$\left. - U_{\beta,\gamma}(t) \Gamma^{(l,t)}_{\mu\alpha} \exp(-i\underset{(l,t)}{\boldsymbol{\Omega}} \cdot \mathbf{r}) \exp[-i\omega(t'-t)] \right\} \tilde{n}_\eta ds' dt = 0 \; ,$$

$$(12.1.8)$$

where $\tilde{\mathbf{n}}$ is the unit vector of an outward normal to the surface γ_R. Letting $R \to \infty$ in expression (12.1.8), we find that the integral over C_R disappears due to the radiation condition and we have in the limit

$$\int_{-\infty}^{\infty} \iint_{\Sigma} \Delta^*_{\alpha\beta\eta\gamma} [\ldots] \tilde{n}_\eta ds dt = \int_{-\infty}^{\infty} \iint_{S} \Delta^*_{\alpha\beta\eta\gamma} [\ldots] n_\eta ds dt \; , \qquad (12.1.9)$$

where $\mathbf{n}$ is the unit vector of the normal to the surface S pointing to the lower half-space. This passage to the limit is uniform over $\boldsymbol{\Omega}$, hence, it is also possible under the sign of the outer integral in expression (12.1.6). Transformation of (12.1.6) in this way yields

[1] In expressions (12.1.6) through (12.1.13) the positive sign with the quantities Γ and Ω corresponding to the condition $Z' > Z$ is omitted, for the sake of a simplified representation.

$$U_\mu(\mathbf{r}',t') = \frac{-1}{16\pi^3}\Delta^*_{\alpha\beta\eta\gamma}\iiiint_{-\infty}^{\infty}\frac{\exp[-i\omega(t'-t)]}{\omega^2}\cdot\Big\{\exp(i\underset{(l)}{\boldsymbol{\Omega}}\cdot\mathbf{r}')\iint_S[U_\alpha(t)\underset{(l)}{\Omega_\gamma}\Gamma^{(l)}_{\mu\beta}$$

$$-iU_{\beta,\gamma}(t)\Gamma^{(l)}_{\mu\alpha}]\cdot\exp(-i\underset{(l)}{\boldsymbol{\Omega}}\cdot\mathbf{r})n_\eta ds+\exp(i\underset{(t)}{\boldsymbol{\Omega}}\cdot\mathbf{r}')\iint_S[U_\alpha(t)\underset{(t)}{\Omega_\gamma}\Gamma^{(t)}_{\mu\beta}$$

$$-iU_{\beta,\gamma}(\tau)\Gamma^{(t)}_{\mu\alpha}]\exp(-i\underset{(t)}{\boldsymbol{\Omega}}\cdot\mathbf{r})n_\eta ds\Big\}\,dv\,dw\,d\omega\,dt\,. \tag{12.1.10}$$

Proceed to spectral representations in expression (12.1.10)

$$\mathbf{u}(\mathbf{r}',\omega)=\int_{-\infty}^{\infty}\mathbf{U}(\mathbf{r}',\tau)e^{i\omega t}dt\,,\quad \mathbf{U}(\mathbf{r}',t)=\frac{1}{2\pi}\int_{-\infty}^{\infty}\mathbf{u}(\mathbf{r}',\omega)e^{-i\omega t}d\omega\,. \tag{12.1.11}$$

Then expression (12.1.10) takes the form

$$\mathbf{u}_\mu(\mathbf{r}',\omega)=\frac{-1}{8\pi^2\omega^2}\Delta^*_{\alpha\beta\eta\gamma}\iint_{-\infty}^{\infty}\Big\{\exp(i\underset{(l)}{\boldsymbol{\Omega}}\cdot\mathbf{r}')\iint_S[\mathbf{u}_\alpha(\mathbf{r},\omega)\underset{(l)}{\Omega_\gamma}\Gamma^{(l)}_{\mu\beta}$$

$$-i\mathbf{u}_{\beta,\gamma}(\mathbf{r},\omega)\Gamma^{(l)}_{\mu\alpha}]\exp(-i\underset{(l)}{\boldsymbol{\Omega}}\cdot\mathbf{r})n_\eta ds+\exp(i\underset{(t)}{\boldsymbol{\Omega}}\cdot\mathbf{r}')$$

$$\cdot\iint_S[\mathbf{u}_\alpha(\mathbf{r},\omega)\underset{(t)}{\Omega_\gamma}\Gamma^{(t)}_{\mu\beta}-i\mathbf{u}_{\beta,\gamma}(\mathbf{r},\omega)\Gamma^{(t)}_{\mu\alpha}]\exp(-i\underset{(t)}{\boldsymbol{\Omega}}\cdot\mathbf{r})n_\eta ds\Big\}\,dv\,dw\,. \tag{12.1.12}$$

Expression (12.1.12) provides a solution to the problem of continuing the field of elastic displacements into the lower half-space. Setting $C_l = C_t = C$ in (12.1.1) and (12.1.12), we derive formulas for continuing the wave field into the upper and lower half-spaces. In this case, the tensors $\Delta^*_{\alpha\beta\eta\gamma}$ and $G^e_{\mu\alpha}$ in (12.1.1) can be merely replaced by $\Delta_{\alpha\beta\eta\gamma}$ and $G^w_{\mu\alpha}$, respectively, while expression (12.1.12) can be simplified

$$\mathbf{u}_\mu(\mathbf{r}',\omega)=\frac{-1}{8\pi^2C^2}\iint_{-\infty}^{\infty}\frac{e^{i\boldsymbol{\Omega}\cdot\mathbf{r}'}}{\Omega_z}\iint_S[\Delta_{\alpha\mu\eta\gamma}\mathbf{u}_\alpha(\mathbf{r},\omega)\Omega_\gamma$$

$$-i\Delta_{\mu\beta\eta\gamma}u_{\beta,\gamma}(\mathbf{r},\omega)]e^{-i\boldsymbol{\Omega}\cdot\mathbf{r}}ds\,dv\,dw\,. \tag{12.1.13}$$

12.1.2 Integral Formulas for Continuation of the Elastic Displacement Field into the Lower Half-Space for a Two-Layer Medium

Consider the case where the parameters of the medium Ω^+ and Ω^- are different, i.e., where the surface S is an interface of media having differing acoustical properties: C_{l_0} and C_{t_0} in the media Ω^+, while C_l, C_t in Ω^-. The problem is to continue the field specified on the lower side of the surface S into the lower half-space to the horizontal plane Σ. The region enclosed between the surfaces Σ and S will be designated by T. Assume that $\mathbf{r}' \in T$. We will apply relation (11.2.39) to the region T:

$$\mathbf{u}_\mu(\mathbf{r}', t') = \int_{-\infty}^{\infty} \iint_{\Sigma} \Delta^*_{\alpha\beta\eta\gamma}[\mathbf{u}_\alpha(t) G^e_{\mu\beta,\gamma}(t'-t) - \mathbf{u}_{\beta,\gamma}(t) G^e_{\mu\alpha}(t'-t)]\,\tilde{n}_\eta\, ds\, dt$$

$$- \int_{-\infty}^{\infty} \iint_{S} \Delta^*_{\alpha\beta\eta\gamma}[\mathbf{u}_\alpha(t) G^e_{\mu\beta,\gamma}(t'-t) - \mathbf{u}_{\beta,\gamma}(t) G^e_{\mu\alpha}(t'-t)]\, n_\eta\, ds\, dt\ , \quad (12.1.14)$$

where $\tilde{\mathbf{n}}$ and $\mathbf{n}$ are the unit vectors of normals to the surfaces Σ and S, respectively, pointing to the lower half-space.

Transform expression (12.1.14) analogously to expression (12.1.5), using decomposition (12.1.3) and spectral representations (12.1.4):

$$\mathbf{u}_\mu(\mathbf{r}', \omega) = -\Delta^*_{\alpha\beta\eta\gamma} \frac{1}{8\pi^2\omega^2} \iint_{-\infty}^{\infty} \Big\{ \exp(i\underset{(l)}{\mathbf{\Omega}^+}\cdot\mathbf{r}') \iint_S [\mathbf{u}_\alpha(\mathbf{r}, \omega) \underset{(l)}{\Omega^+_\gamma} \Gamma^{+(l)}_{\mu\beta}$$

$$- i\mathbf{u}_{\beta,\gamma}(\mathbf{r}, \omega) \Gamma^{+(l)}_{\mu\alpha}] \exp(-i\underset{(l)}{\mathbf{\Omega}^+}\cdot\mathbf{r}) n_\eta ds + \exp(i\underset{(S)}{\mathbf{\Omega}^+}\cdot\mathbf{r}') \iint_S$$

$$\cdot [\mathbf{u}_\alpha(\mathbf{r}, \omega) \underset{(t)}{\Omega^+_\gamma} \Gamma^{+(t)}_{\mu\beta} - i\mathbf{u}_{\beta\gamma}(\mathbf{r}, \omega) \Gamma^{+(t)}_{\mu\alpha}] \exp(-i\underset{(S)}{\mathbf{\Omega}^+}\cdot\mathbf{r}) n_\eta ds \Big\}\, dv\, dw$$

$$+ \Delta^*_{\alpha\beta\eta\gamma} \frac{1}{8\pi^2\omega^2} \iint_{-\infty}^{\infty} \Big\{ \exp(i\underset{(l)}{\mathbf{\Omega}^-}\cdot\mathbf{r}') \iint_S [\mathbf{u}_\alpha(\mathbf{r}, \omega) \underset{(l)}{\Omega^-_\gamma} \Gamma^{-(l)}_{\mu\beta}$$

$$- i\mathbf{u}_{\beta,\gamma}(\mathbf{r}, \omega) \Gamma^{-(l)}_{\mu\alpha}] \exp(-i\underset{(l)}{\mathbf{\Omega}^-}\cdot\mathbf{r}) n_\eta ds + \exp(i\underset{(t)}{\mathbf{\Omega}^-}\cdot\mathbf{r}')$$

$$\cdot \iint_S [\mathbf{u}_\alpha(\mathbf{r}, \omega) \underset{(t)}{\Omega^-_\gamma} \Gamma^{-(t)}_{\mu\beta} - i\mathbf{u}_{\beta,\gamma}(\mathbf{r}, \omega) \Gamma^{-(t)}_{\mu\alpha}]$$

$$\cdot \exp(-i\underset{(S)}{\mathbf{\Omega}^-}\cdot\mathbf{r}) n_\eta ds \Big\}\, dv\, dw\ . \quad (12.1.15)$$

If $C_l = C_t = C$, then (12.1.15) is simplified

$$\mathbf{u}_\mu(\mathbf{r}', \omega) = \frac{-1}{8\pi^2 C^2} \iint_{-\infty}^{\infty} \frac{e^{i\Omega^+\cdot\mathbf{r}'}}{\mathbf{\Omega}^+_z} \iint_S [\Delta_{\alpha\mu\eta\gamma} \mathbf{u}_\alpha(\mathbf{r}, \omega) \Omega^+_\gamma - i\Delta_{\alpha\beta\eta\gamma} \mathbf{u}_{\beta,\gamma} \mathbf{u}_{\beta,\gamma}(r, \omega)]$$

$$\cdot e^{-i\mathbf{\Omega}^+\cdot\mathbf{r}} n_\eta ds\, dv\, dw + \frac{1}{8\pi^2 c^2} \iint_{-\infty}^{\infty} \frac{e^{i\mathbf{\Omega}^-\cdot\mathbf{r}'}}{\Omega^+_z} \iint_S [\Delta_{\alpha\mu\eta\gamma} \mathbf{u}_\alpha(\mathbf{r}, \omega) \Omega^-_\gamma$$

$$- i\Delta_{\mu\beta\eta\gamma} \mathbf{u}_{\beta,\gamma}(\mathbf{r}, \omega)] e^{-i\mathbf{\Omega}^-\cdot\mathbf{r}} n_\eta ds\, dv\, dw\ . \quad (12.1.16)$$

Formulas (12.1.15) and (12.1.16) make it possible to evaluate the field of elastic displacements everywhere in the lower half-space right to the horizontal plane $\{z = -h\}$ below which there are elastic wave sources, knowing its values and its partial derivatives at the surface S. It is noteworthy that, unlike the seismoholography techniques, it is the true field in a medium that is reconstructed. The spatial pattern of wave distribution in an elastic medium obtained by means of field continuation may be used as a basis for constructing an image of the studied medium.

12.1.3 Continuation of Elastic Displacement Fields Specified in a Horizontal Plane

Relations (12.1.15) and (12.1.16) are used to continue three-dimensional fields from an arbitrary surface S into the lower half-space. In the processing of practical observation results, however, of the most critical importance is the case where the observation surface S is a horizontal plane. Here, general integral formulas of continuation can be markedly simplified.

Let us introduce the following designations:

$$l_a(V, W)_z = \frac{1}{2\pi}\iint_{-\infty}^{\infty} \mathbf{u}_a(x, y, z)e^{-i(vx+wy)}dxdy \ ,$$

$$l_{a\beta}(V, W)_z = \frac{1}{2\pi}\iint_{-\infty}^{\infty} \mathbf{u}_{a,\beta}(x, y, z)e^{-i(vx+wy)}dxdy \ , \tag{12.1.17}$$

$$l_a(V, W)_0 = l_a \ .$$

Evidently,

$$l_{a\hat{\mu}}(V, W)_z = i\Omega_{\hat{\mu}} l_a(V, W)_z \ ; \quad \hat{\mu} = x, y \ . \tag{12.1.18}$$

If the surface S is a horizonal plane: S: $\{z = 0\}$, formula (12.1.15) written in designations (12.1.17) takes the form

$$\begin{aligned}\mathbf{u}_{\hat{\mu}}(\mathbf{r}', \omega) = \frac{1}{2\pi}\iint_{-\infty}^{\infty} e^{i(vx'+wy')} \Bigg\{ & l_{\hat{\mu}} ch(\theta_t\zeta) + \frac{\Omega_{\hat{\mu}}}{\underset{(l)}{\Omega_z^+}} l_z sh(\theta_l\zeta) \\ & + i\Omega_{\hat{\mu}} \frac{C_l^2}{\omega^2} \sum_{\beta = x, y, z} l_{\beta\beta}[ch(\theta_t\zeta) - ch(\theta_l\zeta)] \\ & + i\Omega_{\hat{\mu}} \frac{C_t^2}{\omega^2} [(l_{xz} - l_{zx})\Omega_x + (l_{yz} - l_{zy})\Omega_y] \cdot \left[\frac{1}{\underset{(t)}{\Omega_z^+}} sh(\theta_t\zeta) - \frac{1}{\underset{(l)}{\Omega_z^+}} sh(\theta_l\zeta) \right] \\ & - i sh(\theta_t\zeta) \frac{1}{\underset{(t)}{\Omega_z^+}} (l_{\hat{\mu}z} - l_{z\hat{\mu}}) \Bigg\} dvdw \ ; \quad \hat{\mu} = x, y \ .\end{aligned} \tag{12.1.19a}$$

$$\begin{aligned}\mathbf{u}_z(\mathbf{r}', \omega) = \frac{1}{2\pi}\iint_{-\infty}^{\infty} e^{i(vx'+wy')} \Bigg\{ & l_z ch(\theta_l\zeta) - \Bigg[\frac{\Omega_x}{\underset{(t)}{\Omega_z^+}} l_x + \frac{\Omega_y}{\underset{(t)}{\Omega_z^+}} l_y \\ & + i \frac{C_l^2}{\underset{(t)}{\Omega_z^+}} \frac{1}{C_t^2} \sum_{\beta = x, y, z} l_{\beta\beta} \Bigg] sh(\theta_t\zeta) + i \frac{C_t^2}{\omega^2} [(l_{xz} - l_{zx})\Omega_x + (l_{yz} - l_{zy})\Omega_y] \\ & \cdot [ch(\theta_t\zeta) - ch(\theta_l\zeta)] \Bigg\} dvdw \ ,\end{aligned} \tag{12.1.19b}$$

where

$$\theta_{l,t}=\sqrt{v^2+w^2-\frac{\omega^2}{C_{l,t}^2}}\ ,\quad \zeta=z-z'>0\ .$$

For $C_l=C_t=C$, i.e., on transition from the general equation of elastic oscillations (11.1.20) to the wave Eq. (11.1.21), we have

$$\mathbf{u}_{\hat\mu}(\mathbf{r}',\omega)=\frac{1}{2\pi}\iint_{-\infty}^{\infty} e^{i(vx'+wy')}\left\{l_{\hat\mu}ch(\theta\zeta)+\frac{\Omega_{\hat\mu}}{\Omega_z^+}l_z sh(\theta\zeta)\right.$$

$$\left.-i(l_{\hat\mu z}-l_{z\hat\mu})\frac{sh(\theta\zeta)}{\Omega_z^+}\right\}dv\,dw\ ,\quad \hat\mu=x,y\ ; \qquad (12.1.20\text{a})$$

$$\mathbf{u}_z(\mathbf{r}',\omega)=\frac{1}{2\pi}\iint_{-\infty}^{\infty} e^{i(vx'+wy')}\left\{l_z ch(\theta\zeta)\right.$$

$$\left.-\left[\Omega_x l_x+\Omega_y l_y+i\sum_{\beta=x,y,z} l_{\beta\beta}\right]\frac{sh(\theta\zeta)}{\Omega_z^+}\right\}dv\,dw\ , \qquad (12.1.20\text{b})$$

where l_a is related to $\mathbf{u}_a$ by expressions similar to (12.1.17) and

$$\theta=\sqrt{v^2+w^2-\frac{\omega^2}{C^2}}\ .$$

For applied seismic problems of great importance is the case where the surface S is an earth-air interface, i.e., an acoustically soft boundary. Then the following conditions

$$\sum_{\mu=x,y,z} l_{\mu\mu}=0\ ;\quad l_{\hat a}=0\ ;\quad l_{\hat a z}=-l_{z\hat a}\ ;\quad \hat a=xy \qquad (12.1.21)$$

are fulfilled at the surface S. Allowing for (12.1.21) (which is also valid for wave field spectra) we bring (12.1.19) and (12.1.20) to the forms:

$$\mathbf{u}_{\hat\mu}(\mathbf{r}',\omega)=\frac{1}{2\pi}\iint_{-\infty}^{\infty} e^{i(vx'+wy')}\left\{\frac{\Omega_{\hat\mu}}{\underset{(l)}{\Omega_z^+}}\left[2\frac{C_t^2}{\omega^2}(\underset{(t)}{\Omega_z^+})^2-1\right]\cdot sh(\theta_l\zeta)\right.$$

$$\left.-2\frac{C_t^2}{\omega^2}\Omega_{\hat\mu}\underset{(t)}{\Omega_z^+}sh(\theta_t\zeta)\right\}l_z dv\,dw\ ,\quad \hat\mu=x,y\ ; \qquad (12.1.22\text{a})$$

$$\mathbf{u}_z(\mathbf{r}',\omega)=\frac{1}{2\pi}\iint_{-\infty}^{\infty} e^{i(vx'+wy')}\left\{\left[2(\underset{(t)}{\Omega_z^+})^2\frac{C_t^2}{\omega^2}-1\right]ch(\theta_l\zeta)\right.$$

$$\left.+2\left[1-(\underset{(t)}{\Omega_z^+})^2\frac{C_t^2}{\omega^2}\right]ch(\theta_t\zeta)\right\}l_z dv\,dw\ ; \qquad (12.1.22\text{b})$$

$$\mathbf{u}_{\hat\mu}(\mathbf{r}',\omega)=\frac{1}{2\pi}\iint_{-\infty}^{\infty} e^{i(vx'+wy')}\frac{\Omega_{\hat\mu}}{\Omega_z^+}sh(\theta\zeta)l_z dv\,dw\ ,\quad \hat\mu=x,y\ ; \qquad (12.1.23\text{a})$$

$$\mathbf{u}_z(\mathbf{r}', \omega) = \frac{1}{2\pi} \iint_{-\infty}^{\infty} e^{i(vx'+wy')}\ ch(\theta\zeta) l_z dv dw\ . \tag{12.1.23 b}$$

Formulas (12.1.22 a, b) permit continuation of the overall displacement vector $\mathbf{U} = \{U_\alpha\}$ into the lower half-space, knowing values of its vertical components $\mathbf{u}_2$ measured at the Earth's surface. Similarly, formulas (12.1.23 a, b) allow continuation of the wave field knowing the values of its vertical component u_z. Here it is necessary to know the velocities of longitudinal C_l and transverse C_t waves in a medium or the C-parameter of the wave equation.

12.1.4 Elaboration of Regularizing Algorithms for Wave Field Continuation

The above integral formulas (12.1.15) and (12.1.16) suggested for analytical continuation of the field into the lower half-space admit only a formal solution to the problem, i.e., their implementation calls for exact and continuous values of the field at the Earth's surface. In practice, however, only approximate values of the field at a discrete network of observation points are available. As a result, direct application of numerical analogs of formulas (12.1.15) and (12.1.16) involves an error growing exceptionally rapidly with increasing depth of continuation, as well as a failure of the whole wave picture, which is due to the well-known instability of the downward analytical continuation procedure. This effect has been thoroughly studied in the theory of continuation of potential and electromagnetic fields (Zhdanov 1977; Berdichevsky and Zhdanov 1984), where effective regularizing algorithms have been synthesized. Their main point is that the unstable prodecure proving an exact solution to a problem is approximated by a set of stable procedures depending on the scalar parameter λ, called a regularization parameter (Tikhonov and Arsenin 1974). It is necessary here that with λ tending to zero the procedures of this set should more or less indefinitely approach an exact procedure. Hence, relying on a sufficiently low level of error of starting data, we can, in principle, obtain a solution which would be arbitrarily close to an exact one.

The above considerations will be illustrated by using, as an example, the continuation of a two-dimensional scalar wave field into the lower half-space in a two-layer medium. Let the line $z = 0$ be an earth-air interface. In this case formula (12.1.16) is transformed as follows

$$U(x', z', t) = \frac{1}{2\pi} \iint_{-\infty}^{\infty} l(v, 0, \omega) ch\left(\sqrt{v^2 - \frac{\omega^2}{C^2}} \cdot z'\right) e^{i(vx+\omega t)} dv d\omega\ , \tag{12.1.24}$$

where

$$l(v, 0, \omega) = \frac{1}{2\pi} \iint_{-\infty}^{\infty} U(x, 0, t) e^{-i(vx+\omega t)} dx dt\ ,$$

is the two-dimensional spectrum of the studied field $U(x', z', t)$ recorded at the surface of the Earth. Formula (12.1.24) implies that the procedure by which this formula can be realized is not stable. This is accounted for by the fact that for

$v > \frac{\omega}{C}$ the function $ch\left(\sqrt{v^2 - \frac{\omega^2}{C^2}}\, z'\right)$ increases infinitely with $z' \to \infty$. Hence in this case, at a sufficiently great depth of calculation z', the error involved in the calculation of the spectrum $l(v, z', \omega) = l(v, 0, \omega)ch\left(\sqrt{v^2 - \frac{\omega^2}{C^2}}\, z'\right)$ may be arbitrarily large, while the calculation error of $l(v, 0, \omega)$ may be arbitrarily small. This fact makes formula (12.1.24) absolutely impracticable for calculating continued values of a studied field, i.e., in practice the spectrum $l(v, 0, \omega)$ is determined numerically and, as a consequence, is always related to some uncertainty.

In this case, the regularizing algorithm can be elaborated as follows.

Consider a set of procedures

$$\tilde{U}_\lambda(x', z', t) = \frac{1}{2\pi} \iint_{-\infty}^{\infty} l(v, 0, \omega) P_\lambda(v, \omega)\, ch\left(\sqrt{v^2 - \frac{\omega^2}{C^2}} \cdot z'\right) e^{i(vx' + \omega t)}\, dv\, d\omega \;, \tag{12.1.25}$$

where

$$P_\lambda(v, \omega) = \begin{cases} 1 \;; & \text{if} \quad |l(v,0,\omega)| \geqslant \lambda \\ 0 \;; & \text{if} \quad |l(v,0,\omega)| < \lambda \;. \end{cases}$$

Evidently, when the regularization parameter λ tends to zero the value $U_\lambda(x', z', t)$ approaches indefinitely an exact value $U(x', z', t)$. To realize the above scheme, the procedures specified by formula (12.1.25) should prove stable at any λ. The latter condition follows from the fact that the spectrum $l(v, 0, \omega) \to 0$ when $(v^2 + \omega^2) \to \infty$ and the function $l(v, 0, \omega) P_x(v, \omega)$ is, therefore, nonzero only in some limited region.

Thus, formula (12.1.25) admits a numerical solution for the stated problem with satisfactory accuracy by adjusting the value of the regularization parameter λ to the error of determination of the spectrum $l(v, 0, \omega)$.

Let us illustrate realization of the above algorithm by considering the following problem. A source of elastic oscillations whose power varies according to the law

$$f(t) = \frac{\sin(\omega t - 2\pi)}{(\omega t - 2\pi)} \;; \quad t \in [0, T] \;; \quad \omega = 2\pi f \;, \quad f = 100 H_z$$

operates at a point $\mathbf{r}_0 = (x_0 = 0\,\mathrm{m},\ z_0 = -500\,\mathrm{m})$ of a homogeneous half-space $z \leqslant 0$ over a time interval $[0, T = 0.03\,\mathrm{s}]$. The induced longitudinal waves propagate in a medium at a velocity $C = 400\,\mathrm{m/s}$. A field of longitudinal waves $U(x, t)$ is recorded at the earth-air interface coinciding with the line $z = 0$, with a base $[-1600\,\mathrm{m}, 1600\,\mathrm{m}]$. The recording time is 0.5 s. It is known (Morse and Feshbach 1953) that in this case the field $U(x, t)$ on the observation line can be calculated by the formula

$$U(\mathbf{r}', t') = 4C \int_0^T \frac{f(t) \cdot \varkappa \left[(t' - t) - \frac{|\mathbf{r}' - \mathbf{r}_0|}{C} \right]}{\sqrt{C^2 (t' - t)^2 - |\mathbf{r}' - \mathbf{r}^0|^2}}\, dt \;, \tag{12.1.26}$$

where

$$\mathbf{r}' = (x', y'),\ x' \in [-1600\,\mathrm{m},\ 1600\,\mathrm{m}],\ y' = 0$$

$$\varkappa(t) = \begin{cases} 1\ ; & \text{if} \quad t>0 \\ 0\ ; & \text{if} \quad t<0\ . \end{cases}$$

It is necessary to reconstruct the field $U(\mathbf{r}', t)$ at points of the lower half-space, i.e., for $z' < 0$.

Figures 65 and 66 show the results of field continuation from the Earth's surface down to levels $z' = -20$ m, -100 m, and -200 m. Calculations were performed using discrete analogs of formulas (12.1.25) and (12.1.26). As a glance at the figure shows, the procedure specified by formula (12.1.25) is not stable. Even at the level $z' = -100$ m the wave structure of the field disappears completely (Fig. 65b). At the same time, the use of regularizing procedure (12.1.26) permits a satisfactory reconstruction of the field at all the three levels (Fig. 66).

Concluding this subsection we should like to note that the suggested regularizing algorithm can be applied successfully to implement the integral methods of analytical continuation of the gravitational, magnetic, and electromagnetic fields (Sects. 4.5, 7.5, and 8.1).

12.2 Migration of Wave Fields on the Basis of Analytical Continuation

The conventional procedure for migrating elastic wave fields is based, as noted earlier, on application of the Kirchhoff integral formulas to time-inverted wave fields $U(T-\tau)$ (T being the field recording interval). Meanwhile, there is reason to expect that a more comprehensive image of a medium can be obtained by reconstructing the true picture of elastic wave distribution in a medium, using the analytical continuation methods (Zhdanov and Matusevich 1984). Indeed, the distribution of the singularities of a reconstructed field is closely related to the nature of studied inhomogeneities and to the location of the diffraction points and reflecting boundaries. Take, for instance, a situation where the section contains a reflecting boundary on which a reflected wave forms at the instant of arrival of a direct wave. The interference of the two waves results in a larger (smaller) amplitude of oscillations taking place at points of the reflecting boundary, as compared to adjacent points. A similar effect is observed when the section includes diffracting objects. This circumstance permits application of the wave field continuation procedure to the solution of migration problems. Indeed, assume that the field of elastic oscillations is reconstructed at each point of the section at the instant of arrival of the direct wave. Then the lines and regions exhibiting anomalous amplitudes of oscillations in the resultant model will indicate the location of reflecting boundaries and diffracting objects.

The above operation will be referred to as production of a migration section. Mathematically, it can be expressed as follows

$$S(x, z) = \int_{-\infty}^{\infty} \delta[t-\tau(x, z)]\, U(x, z, t)\, dt\ ,$$

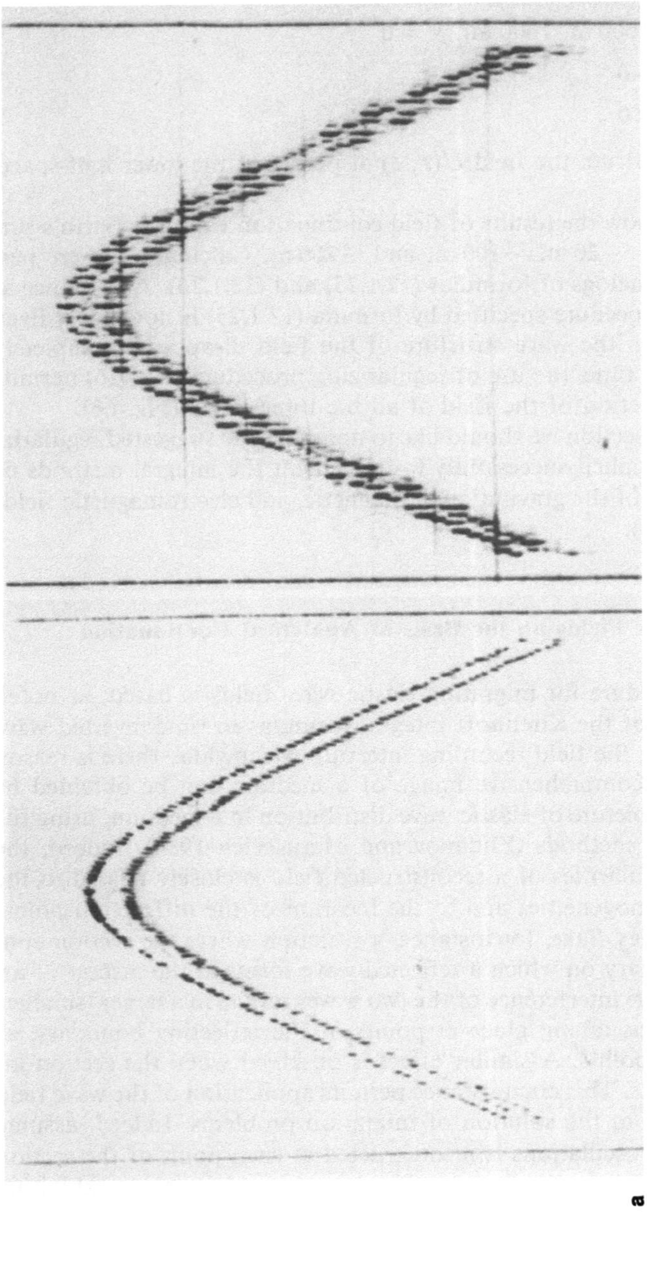

b

a

Fig. 65a–d. Continuation of the spherical wave field without regularization procedure: **a** $z' = 0$ m; **b** $z' = -20$ m; **c** $z' = -100$ m; **d** $z' = -200$ m

d

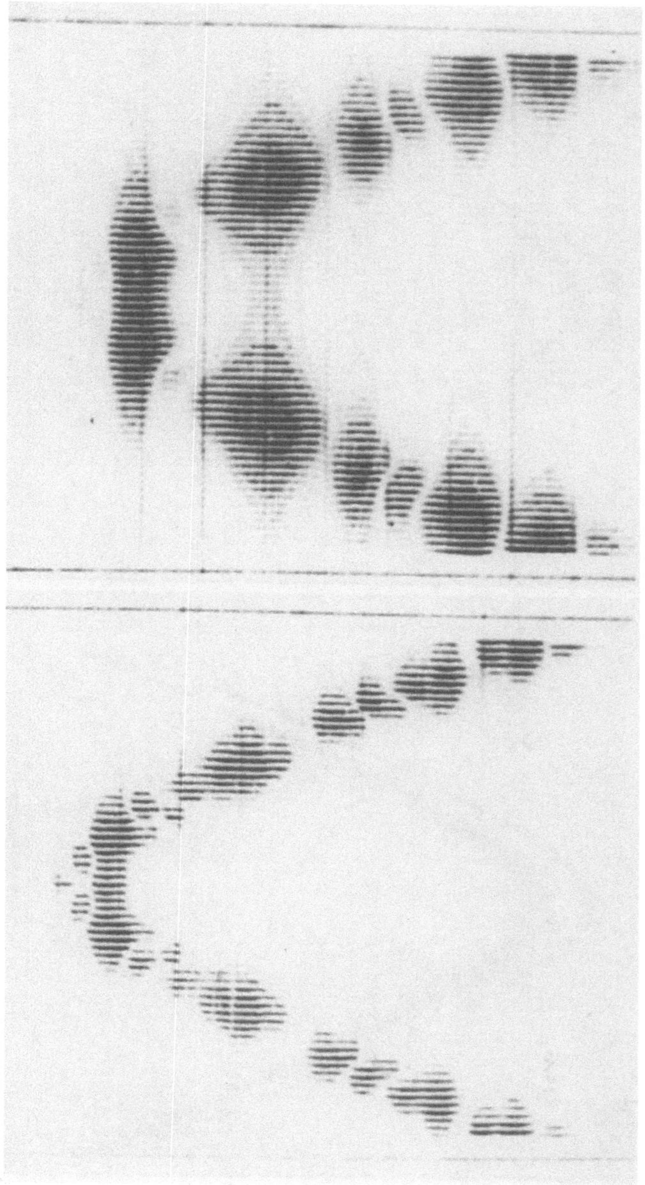

c

Fig. 65c, d

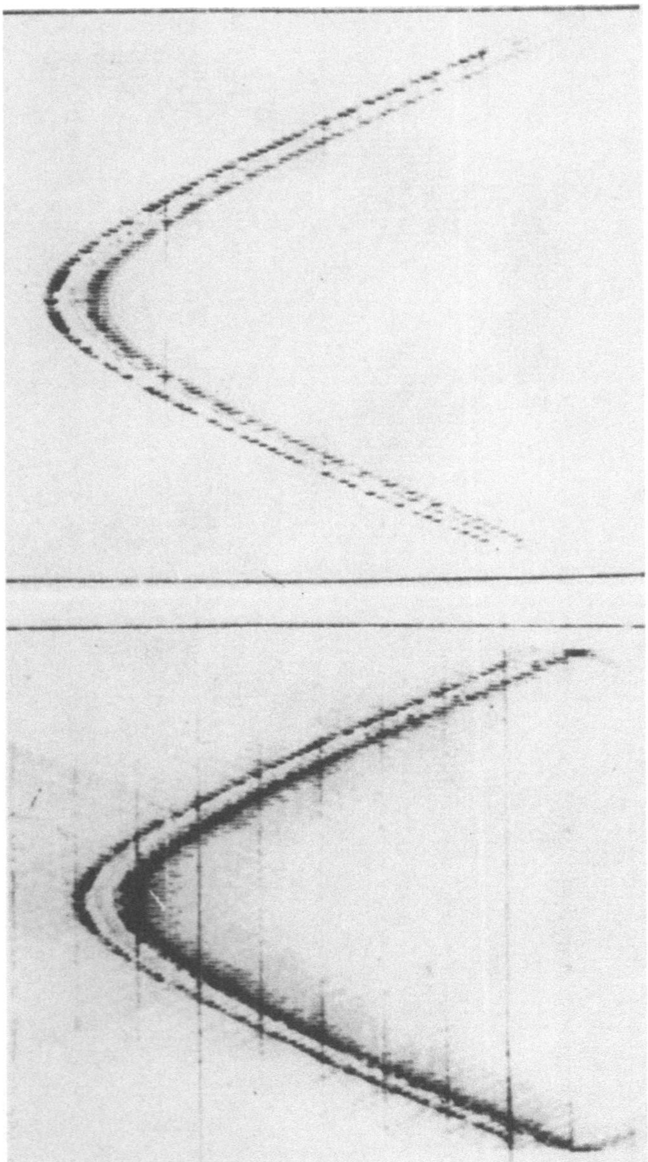

a b

Fig. 66a–d. Continuation of the spherical wave field with regularization procedure: **a** $z' = 0$ m; **b** $z' = -20$ m; **c** $z' = -100$ m; **d** $z' = -200$ m

d

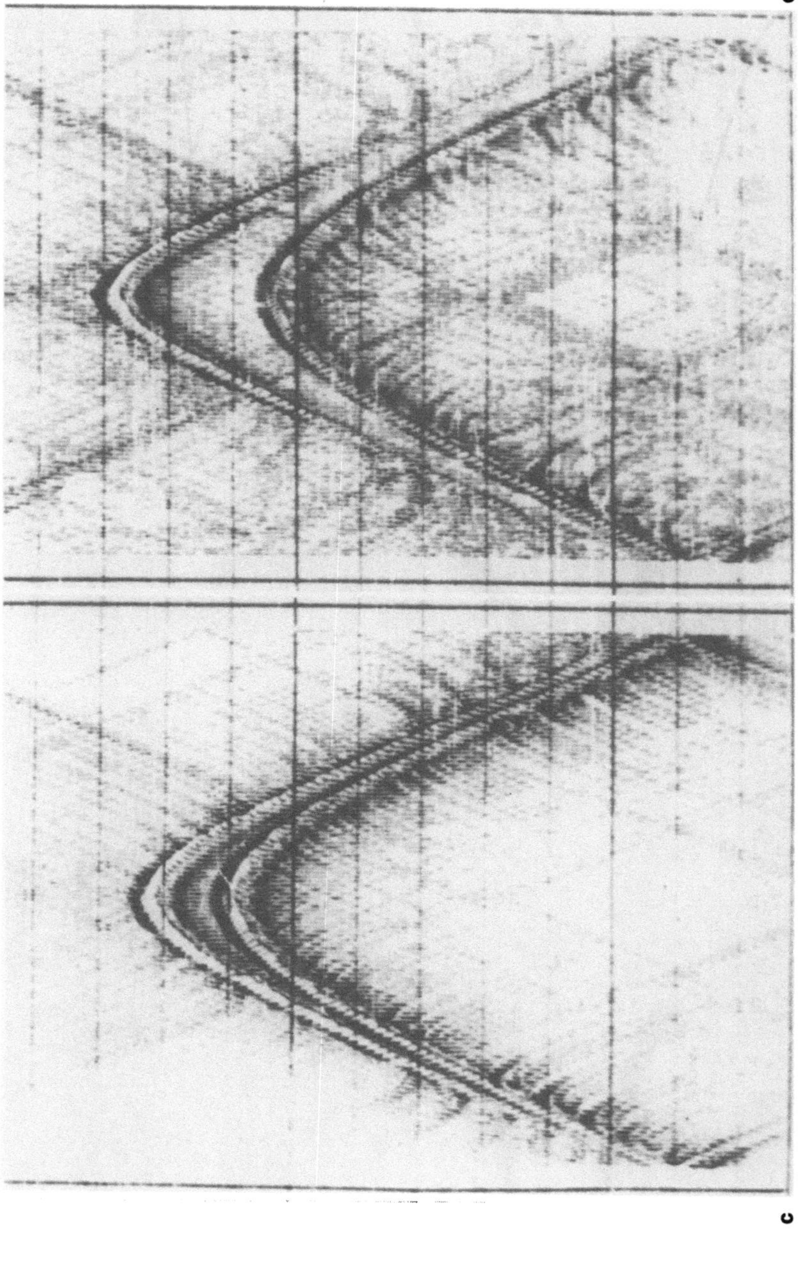

c

Fig. 66c, d

a

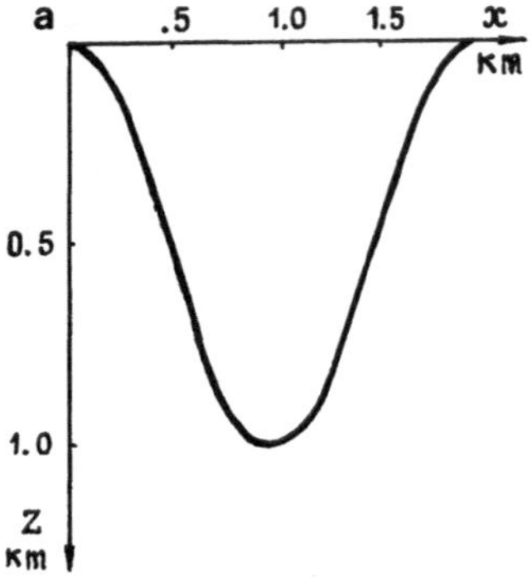

b

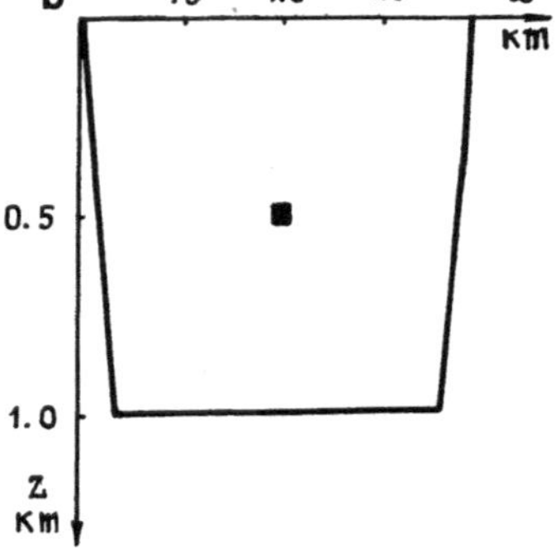

Fig. 67a, b. Media models containing: **a** curvilinear reflector; **b** diffracting point and steeply dipping reflectors

a

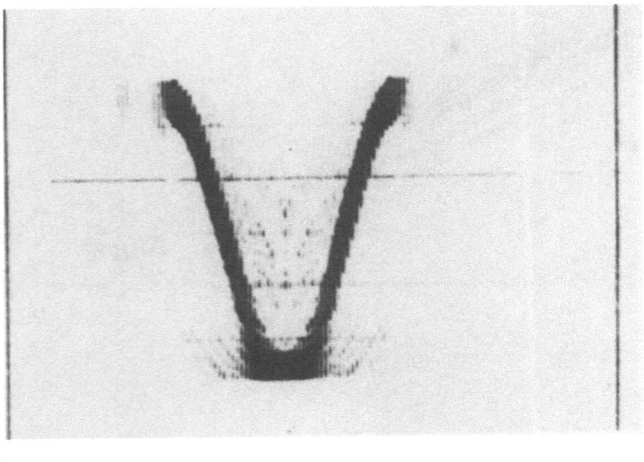

b

Fig. 68a, b. Time section (**a**) and migration (**b**) for the model of Fig. 67a

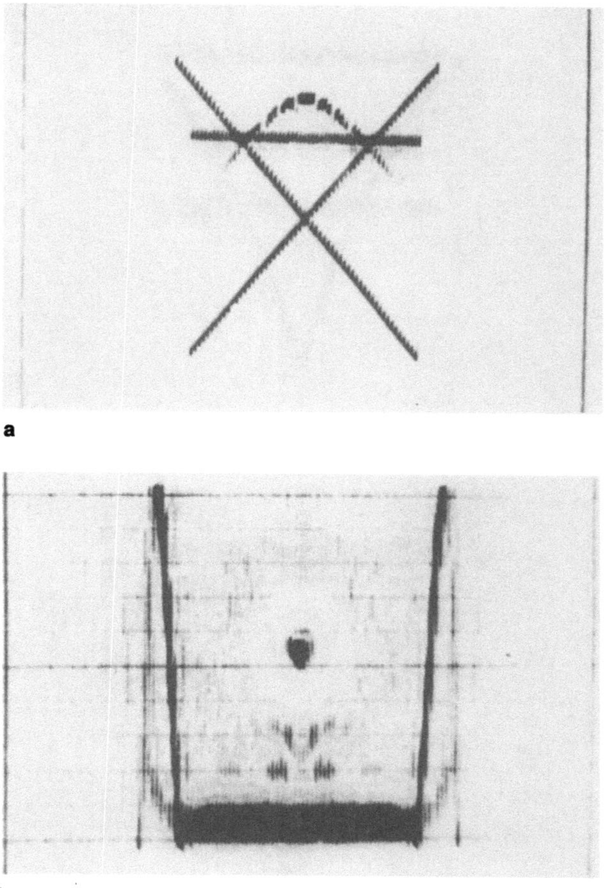

Fig. 69a, b. Time section (a) and migration (b) for the model of Fig. 67b

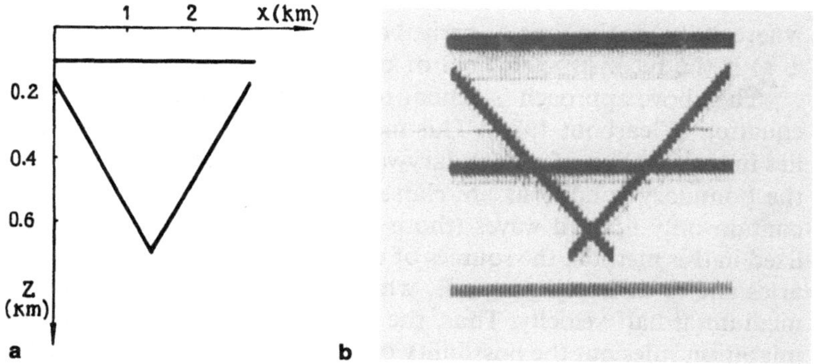

Fig. 70a, b. Three-layer media model (a) and its time section (b)

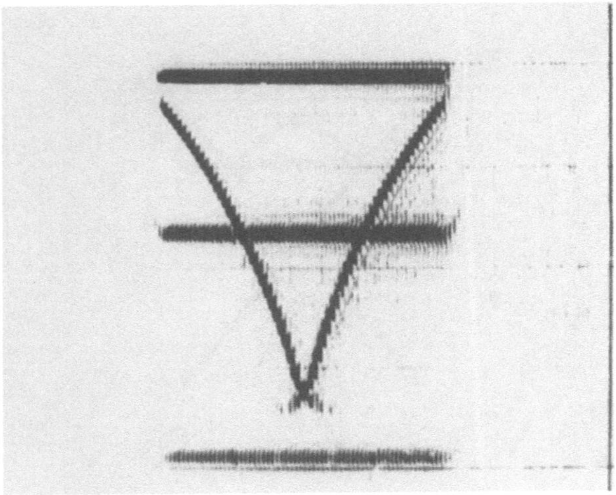

a

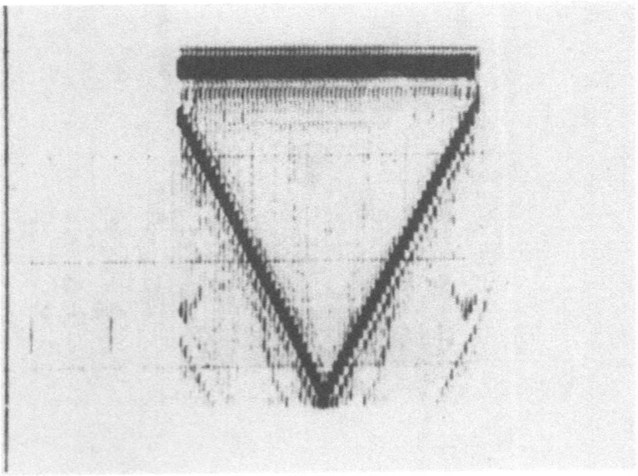

b

Fig. 71 a, b. Migration for the model of Fig. 70 at a mean velocity (**a**) and for layered models (**b**)

where $\tau(x, z)$ is the time of arrival of the direct wave at the point $\mathbf{r} = (x, z)$; $U(x, z, t)$ is the reconstructed field of elastic oscillations.

The above approach is similar to that employed in the migration of the wave equation (Clearbout 1985). This method has been so far used extensively and it lies in the solution of a boundary-value problem for the wave equation. However, the boundary conditions are chosen artificially so that the wave process might contain only upward waves (those propagating upward). Within the model utilized in this method, the sources of upward waves are located on reflecting boundaries and in diffracting regions, while the waves themselves propagate through a medium at half velocity. Thus, the model basic to the method of wave equation migration rules out the possibility of multiple reflections that may give rise to fictitious reflecting boundaries in the migration section. An exact continuation of the field will contribute to the reconstruction of the true process of elastic wave propagation in a medium, in which case multiple waves do not pass the layers where they have been formed and, hence, they cannot generate any fictitious reflecting boundaries.

Figures 67 through 71 depict numerical model examples illustrating the potential of the migration algorithm based on analytical continuation of the wave field. Figure 67 shows two models: (a) one with a curvilinear reflecting boundary and (b) the other with a diffracting point and highly dipping (around 78°) reflecting boundaries. Figures 68 and 69 concern the time sections (Figs. 68a and 69a) calculated for these models and the results of their migration (Figs. 68b and 69b). One can readily see that the configuration of the reflecting boundaries is reconstructed fairly satisfactorily. It is remarkable that the Clearbout migration imposing limitations on the depth to a reflecting boundary does not make it possible to reconstruct the inclination of such a steeply dipping reflecting boundary.

Of particular concern is the migration procedure where the field is continued in a layered model. Figures 70 and 71 illustrate the advantages of this migration making it superior to any known migration algorithm involving the model of mean velocities. Figure 70 is a model of a three-dimensional medium (Fig. 70a) containing a horizontal boundary inviting both wave refraction and multiple reflections, as is seen on the time section. Figure 71 illustrates the procedure of migration of the time section, Fig. 71a being the result of migration obtained using conventional procedures (within the model of mean velocity). Figure 71b shows the result of migration on the basis of analytical continuation within the layered model. As can be seen, approximation of the layered section by the mean velocity model tends to distort the configuration of true boundaries and produces fictitious boundaries as a result of multiple reflections. The migration algorithm for a layered model based on analytical continuation is free from the shortcomings mentioned above.

Appendix A
Space Analogs of the Cauchy-Type Integrals and the Quaternion Theory

The space analogs of the Cauchy-type integral established in Chap. 5 can be also introduced on the basis of the quaternion theory suggested by W. Hamilton. We will cite only the main provisions of this theory which are necessary for describing the Cauchy integral formulas and their related expressions in terms of quaternions.[1]

A.1 Quaternions and Operations Thereon

Consider a four-dimensional Euclidean space E_4. The elements of this space are four-vectors designated, unlike the ordinary vectors of three-dimensional space, by bars below letters: $\underline{\mathbf{a}}$, $\underline{\mathbf{b}} \in E_4$.

Let us designate the unit vectors of an arbitrary orthonormal basis E_4 by $\underline{\mathbf{i}}_0$, $\underline{\mathbf{i}}_x$, $\underline{\mathbf{i}}_y$, and $\underline{\mathbf{i}}_z$. Then any four-vector $\underline{\mathbf{a}} \in E_4$ can be written as a linear combination of unit vectors

$$\underline{\mathbf{a}} = a_0\underline{\mathbf{i}}_0 + \sum_{a=x,y,z} a_a\underline{\mathbf{i}}_a \ , \tag{A.1}$$

where a_0, a_x, a_y, and a_z are the scalar components of $\underline{\mathbf{a}}$ on a given basis.

Along with the space E_4, we can consider its three-dimensional subspace $E_3 \subset E_4$ spread over the unit vectors $\underline{\mathbf{i}}_x$, $\underline{\mathbf{i}}_y$, and $\underline{\mathbf{i}}_z$. The linear combination $\sum_{a=x,y,z} a_a\underline{\mathbf{i}}_a$ forms a vector from E_3, which will be denoted by $\mathbf{a}$:

$$\mathbf{a} = \sum_{a=x,y,z} a_a\underline{\mathbf{i}}_a \ . \tag{A.2}$$

This vector is a projection of the four-vector $\underline{\mathbf{a}}$ into E_3. Thus, any four-vector can be represented as a sum:

$$\underline{\mathbf{a}} = a_0\underline{\mathbf{i}}_0 + \mathbf{a} \ , \tag{A.3a}$$

and in particular

$$\underline{\mathbf{i}}_a = \mathbf{i}_a \ , \quad a = x, y, z \ . \tag{A.3b}$$

Turning to the four-dimensional Euclidean space E_4 is justified because it is possible in principle to introduce here the operation of vector multiplication,

[1] Appendix A has been prepared in collaboration with W. L. Vasilevsky and M. V. Shapiro.

which has very good properties similar to those of multiplication of complex numbers.

Indeed, let us define the operation of multiplication of four-vectors as follows. First, formulate the rules of multiplication of the basis unit vectors

$$\underline{\mathbf{i}}_0 * \underline{\mathbf{i}}_\alpha = \underline{\mathbf{i}}_\alpha * \underline{\mathbf{i}}_0 = \underline{\mathbf{i}}_\alpha \ , \quad \alpha = x,\ y,\ z \ , \tag{A.4a}$$

$$\underline{\mathbf{i}}_0 * \underline{\mathbf{i}}_0 = \underline{\mathbf{i}}_0^2 = \underline{\mathbf{i}}_0 \ ; \tag{A.4b}$$

$$\begin{cases} \underline{\mathbf{i}}_x * \underline{\mathbf{i}}_y = -\underline{\mathbf{i}}_y * \underline{\mathbf{i}}_x = \underline{\mathbf{i}}_z \\ \underline{\mathbf{i}}_y * \underline{\mathbf{i}}_z = -\underline{\mathbf{i}}_z * \underline{\mathbf{i}}_y = \underline{\mathbf{i}}_x \\ \underline{\mathbf{i}}_z * \underline{\mathbf{i}}_x = -\underline{\mathbf{i}}_x * \underline{\mathbf{i}}_z = \underline{\mathbf{i}}_y \end{cases} \tag{A.5a}$$

$$\underline{\mathbf{i}}_\alpha * \underline{\mathbf{i}}_\alpha = \underline{\mathbf{i}}_\alpha^2 = -\underline{\mathbf{i}}_0 \ , \quad \alpha = x,\ y,\ z \ . \tag{A.5b}$$

Thus, the unit vector $\underline{\mathbf{i}}_0$ in a given operation of multiplication serves as a unity, while the unit vectors $\underline{\mathbf{i}}_x$, $\underline{\mathbf{i}}_y$, and $\underline{\mathbf{i}}_z$ act as imaginary unities. Employing ordinary vector operations on the vectors $\mathbf{i}_x$, $\mathbf{i}_y$, and $\mathbf{i}_z$, considered in the three-dimensional subspace E_3, relations (A.5a, b) can be rewritten as

$$\underline{\mathbf{i}}_\alpha * \underline{\mathbf{i}}_\beta = -(\mathbf{i}_\alpha \cdot \mathbf{i}_\beta)\underline{\mathbf{i}}_0 + [\mathbf{i}_\alpha \times \mathbf{i}_\beta] \ , \tag{A.6}$$

where $\alpha,\ \beta = x,\ y,\ z$.

Allowing for rules (A.4a, b), (A.6), and formula (A.3a), the product of any two four-vectors can be cast as

$$\underline{\mathbf{a}} * \underline{\mathbf{b}} = (a_0\underline{\mathbf{i}}_0 + \mathbf{a}) * (b_0\underline{\mathbf{i}}_0 + b) = (a_0 b_0 - \mathbf{a}\cdot\mathbf{b})\underline{\mathbf{i}}_0 + a_0\mathbf{b} + b_0\mathbf{a} + [\mathbf{a}\times\mathbf{b}] \ . \tag{A.7}$$

One can readily see that the above-defined operation of four-vector multiplication possesses the properties of associativity and distributivity with respect to the operations of summation and multiplication of vectors by scalar factors. But this operation is noncommutative, since generally we have

$$\underline{\mathbf{a}} * \underline{\mathbf{b}} \neq \underline{\mathbf{b}} * \underline{\mathbf{a}} \ .$$

Let us make some definitions.

Definition A.1. The four-dimensional Euclidean space E_4 with an additional operation of multiplication defined according to (A.7) is designated by the letter $\mathscr{H}$ and its elements are called quaternions.

Definition A.2. If $a_0 = 0$ the quaternion $\underline{\mathbf{a}} = \mathbf{a}$ is called entirely imaginary. If $a_x = a_y = a_z = 0$, the quaternion $\underline{\mathbf{a}} = a_0\underline{\mathbf{i}}_0$ is called entirely real.

Thus, entirely imaginary quaternions can be identified with the vectors $\mathbf{a} \in E_3$, while the entirely real with the scalars a_0. In other words, quaternions combine scalars and vectors, which is particularly convenient for applications. In what follows the scalar a_0 will be called a real (scalar) part of the quaternion $\underline{\mathbf{a}}$, while the vector $\mathbf{a}$ will be termed an imaginary (vector) part of the quaternion $\underline{\mathbf{a}}$.

Since $\underline{\mathbf{i}}_0$ plays the part of an ordinary unit (as we saw earlier), it can be omitted from expression (A.3a) for the quaternion, i.e., the latter can be represented as a sum of its scalar and vector parts:

$$\underline{\mathbf{a}} = a_0 + \mathbf{a} \; . \tag{A.8}$$

Definition A.3. The quaternions $\underline{\mathbf{a}} = a_0 + \mathbf{a}$ and $\bar{\underline{\mathbf{a}}} = a_0 - \mathbf{a}$ are called conjugates.

One is readily convinced that the product of conjugate quaternions is an entirely real quaternion:

$$\underline{\mathbf{a}} * \bar{\underline{\mathbf{a}}} = \bar{\underline{\mathbf{a}}} * \underline{\mathbf{a}} = (a_0 + \mathbf{a}) * (a_0 - \mathbf{a}) = a_0^2 + \mathbf{a} \cdot \mathbf{a} = a_0^2 + a_x^2 + a_y^2 + a_z^2 = |\underline{\mathbf{a}}|^2 \; .$$

Hence, if $a \neq 0$ we can determine the inverse quaternion

$$\underline{\mathbf{a}}^{-1} = \frac{1}{|\underline{\mathbf{a}}|^2} \bar{\underline{\mathbf{a}}} \; . \tag{A.9}$$

According to (A.7), the product of two entirely imaginary quaternions $\underline{\mathbf{c}}$ and $\underline{\mathbf{d}}$, i.e., the quaternion product of two vectors $\mathbf{c}$ and $\mathbf{d}$, generates a four-vector (quaternion containing both the imaginary and the real parts):

$$\mathbf{c} * \mathbf{d} = -\mathbf{c} \cdot \mathbf{d} + \mathbf{c} \times \mathbf{d} \; .$$

Thus, multiplying three-dimensional vectors according to the quaternion laws we arrive inevitably (except for the obvious case $\mathbf{c} \perp \mathbf{d}$) at four-dimensional vectors (quaternions). Hence it is evident that in an ordinary three-dimensional space E_3 it is impossible to define a multiplication exhibiting the above very convenient properties (say, invertibility of any nonzero element).

A.2 Monogenic Functions

Let D be a domain of the three-dimensional Euclidean space E_3. Consider the quaternion functions of the vector argument defined on D: $\underline{\mathbf{f}} = \underline{\mathbf{f}}(\mathbf{r})$, $\mathbf{r} \in D$, $\underline{\mathbf{f}} \in \mathscr{H}$. It turns out that there exist a variety of classes of such functions, which are rather exact analogs of complex analytical functions. We will consider a class of these functions known as monogenic.

First define the operation of quaternion differentiation. Evidently, the vector Hamiltonian ∇ can be identified with the entirely imaginary quaternion Hamiltonian whose multiplication at the quaternion function $\underline{\mathbf{f}}(\mathbf{r})$ is equivalent to simultaneously taking the gradient of the real part of $\underline{\mathbf{f}}$ and of the divergence and rotor of the imaginary part of $\underline{\mathbf{f}}$:

$$\nabla * \underline{\mathbf{f}} = \nabla * (f_0 + \mathbf{f}) = \nabla f_0 - \nabla \cdot \mathbf{f} + \nabla \times \mathbf{f} = \mathrm{grad} f_0 - \mathrm{div} f + \mathrm{rot}\, \mathbf{f} \; . \tag{A.10}$$

In particular, Eq. (4.1.1) of an arbitrary vector field is written in terms of quaternions as follows:

$$\nabla * \mathbf{F} = -q + \mathbf{j} \; . \tag{A.11}$$

Now let us give the following definition.

Definition A.4. Let the real and the imaginary parts of the quaternion function $\underline{\mathbf{m}} = m_0 + \mathbf{m}$ be differentiable scalar $m_0(\mathbf{r})$ and vector $\mathbf{m}(\mathbf{r})$ functions of the radius vector $\mathbf{r} \in D$. If everywhere in D

$$\nabla * \underline{\mathbf{m}} = 0 \tag{A.12}$$

the quaternion function $\underline{\mathbf{m}}(\mathbf{r})$ is called monogenic in D.

Expand expression (A.12), using (A.10):

$$\nabla * \underline{\mathbf{m}} = \operatorname{grad} m_0 - \operatorname{div} \mathbf{m} + \operatorname{rot} \mathbf{m} = 0 \ . \tag{A.13}$$

Thus, the monogeneticity condition for the scalar and vector parts of the quaternion function $\underline{\mathbf{m}}$ is written in the form:

$$\operatorname{div} \mathbf{m} = 0 \ , \quad \operatorname{grad} m_0 + \operatorname{rot} \mathbf{m} = 0 \ . \tag{A.14}$$

If m_0 and $\mathbf{m}$ are functions twice differentiable in D, condition (A.14) implies that m_0 and $\mathbf{m}$ satisfy in D the Laplace equations:

$$\Delta m_0 = 0 \ , \quad \Delta \mathbf{m} = 0 \ , \quad \mathbf{r} \in D \ . \tag{A.15}$$

Thus, all the scalar components of the monogenic function $\underline{\mathbf{m}}(\mathbf{r})$ twice continuously differentiable in D are harmonic functions in D.

Particularly, the set of entirely imaginary monogenic functions $\underline{\mathbf{F}}(\mathbf{r})$ coincides with the set of Laplace fields in D.

For quaternion functions the relation that can be treated as a quaternion analog of the Ostrogradsky-Gauss theorem is valid.

Let a quaternion function $\underline{\mathbf{f}}(\mathbf{r})$, whose scalar and vector components are continuously differentiable everywhere in D right to its boundary, be specified inside the domain $D \subset E_3$ with a piecewise smooth boundary S. Then, according to the vector statements of the Ostrogradsky-Gauss theorem (5.1.1), (5.1.5), (5.1.7) and the rules of quaternion multiplication (A.7) and differentiation (A.10) the following formula

$$\iiint_D \nabla * \underline{\mathbf{f}}\, dv = \iiint_D \{\operatorname{grad} f_0 - \operatorname{div} \mathbf{f} + \operatorname{rot} \mathbf{f}\} dv = \iint_S \{f_0 \mathbf{n} - \mathbf{n} \cdot \mathbf{f} + \mathbf{n} \times \mathbf{f}\} ds = \iint_S \mathbf{n} * \underline{\mathbf{f}}\, ds \ ,$$

($\mathbf{n}$ being the unit vector of an outward normal to S) holds good.

The relation

$$\iiint_D \nabla * \underline{\mathbf{f}}\, dv = \iint_S \mathbf{n} * \underline{\mathbf{f}}\, ds \tag{A.16}$$

is called the *quaternion statement of the Ostrogradsky-Gauss theorem.*

In particular, with monogenic functions we find that for any closed piecewise smooth surface S

$$\iint_S \mathbf{n} * \underline{\mathbf{m}}\, ds = 0 \ . \tag{A.17}$$

This point can be regarded as a quaternion analog of the Cauchy integral theorem for complex analytical functions fitted by formula (2.4.7).

A.3 Quaternion Notation of Space Analogs of the Cauchy-Type Integral

A little calculation omitted here, because of its cumbersome nature, reveals that the quaternion analog of formulas (8.2.9) is the following integral identity

$$\iiint\limits_D [(\nabla * \underline{g}) * \underline{f} + \underline{g} * (\nabla * \underline{f})]\, ds = \iint\limits_S \underline{g} * \mathbf{n} * \underline{f}\, ds \ , \tag{A.18}$$

where $\underline{g}$ and $\underline{f}$ are the quaternion functions, having twice continuously differentiable scalar and vector components everywhere in the domain D right to its boundary. Note that integral relation (A.18) can be derived directly from the Stokes multidimensional formula (Spivak 1965).

We will point in formula (A.18) that $\underline{f} = \underline{m}$ is a function monogenic in D, while $\underline{g}$ is an entirely imaginary quaternion equal to the gradient of the Green fundamental function for the Laplace equation

$$\underline{g} = \operatorname{grad} G(\mathbf{r}-\mathbf{r}') = -\frac{1}{4\pi} \operatorname{grad} \frac{1}{|\mathbf{r}-\mathbf{r}'|} \ . \tag{A.19}$$

Then, according to (A.10), we have

$$\nabla * \underline{g} = \nabla * \operatorname{grad} G = -\operatorname{div}\operatorname{grad} G = -\delta(\mathbf{r}-\mathbf{r}') \ . \tag{A.20}$$

As a result, we obtain

$$\frac{1}{4\pi} \iint\limits_S \operatorname{grad} \frac{1}{|\mathbf{r}-\mathbf{r}'|} * \mathbf{n} * \underline{m}\, ds = \begin{cases} \underline{m}(\mathbf{r}') \ , & \mathbf{r}' \in D \\ 0 \ , & \mathbf{r}' \in CD \ . \end{cases} \tag{A.21}$$

Relation (A.21) is a quaternion analog for monogenic functions of the Cauchy integral formula for complex analytical functions (1.1.1). In particular, for entirely imaginary monogenic function $\underline{m} = \mathbf{F}$ ($\mathbf{F}$ being the Laplace field in D) the expression (A.21) turns exactly into formula (5.1.2). Indeed, in accordance with the rules of quaternion multiplication (A.7), we have

$$\begin{aligned} \operatorname{grad} \frac{1}{|\mathbf{r}-\mathbf{r}'|} * \mathbf{n} * \mathbf{f} &= \operatorname{grad} \frac{1}{|\mathbf{r}-\mathbf{r}'|} * (-\mathbf{n}\cdot\mathbf{F} + \mathbf{n}\times\mathbf{F}) = -\operatorname{grad} \frac{1}{|\mathbf{r}-\mathbf{r}'|} (\mathbf{n}\cdot\mathbf{F}) \\ &\quad - \left[\operatorname{grad} \frac{1}{|\mathbf{r}-\mathbf{r}'|} \cdot (\mathbf{n}\times\mathbf{F}) \right] + \operatorname{grad} \frac{1}{|\mathbf{r}-\mathbf{r}'|} \times (\mathbf{n}\times\mathbf{F}) \ . \end{aligned} \tag{A.22}$$

Substituting (A.22) into (A.21) we write

$$\begin{aligned} &-\frac{1}{4\pi} \iint\limits_S \left[(\mathbf{n}\cdot\mathbf{F}) \operatorname{grad} \frac{1}{|\mathbf{r}-\mathbf{r}'|} + (\mathbf{n}\times\mathbf{F}) \times \operatorname{grad} \frac{1}{|\mathbf{r}-\mathbf{r}'|} \right] ds \\ &\qquad -\frac{1}{4\pi} \iint\limits_S \operatorname{grad} \frac{1}{|\mathbf{r}-\mathbf{r}'|} \cdot (\mathbf{n}\times\mathbf{F})\, ds = \begin{cases} \mathbf{F}(\mathbf{r}') \ , & \mathbf{r}' \in D \\ 0 \ , & \mathbf{r}' \in CD \ . \end{cases} \end{aligned} \tag{A.23}$$

In the right-hand side of Eq. (A.23) there is an entirely imaginary quaternion, hence,

$$-\frac{1}{4\pi}\iint_S \operatorname{grad}\frac{1}{|\mathbf{r}-\mathbf{r}'|}\cdot(\mathbf{n}\times\mathbf{F})\,ds=0\ . \tag{A.24}$$

Condition (A.24) can be derived also directly. Indeed, transform the integrand in (A.24):

$$-\operatorname{grad}\frac{1}{|\mathbf{r}-\mathbf{r}'|}\cdot(\mathbf{n}\times\mathbf{F})=\left(\operatorname{grad}\frac{1}{|\mathbf{r}-\mathbf{r}'|}\times\mathbf{F}\right)\cdot\mathbf{n}$$

$$=\operatorname{rot}\frac{\mathbf{F}}{|\mathbf{r}-\mathbf{r}'|}\cdot\mathbf{n}-\frac{1}{|\mathbf{r}-\mathbf{r}'|}\operatorname{rot}\mathbf{F}\cdot\mathbf{n}\ . \tag{A.25}$$

The second term in (A.25) is zero, since $\mathbf{F}$ is the Laplace field. The integral of the first term is also zero, according to the Stokes formula (5.1.8), since S is a closed surface.

With due account taken of (A.24), formula (A.23) turns into (5.1.26), which was to be proved.

Thus, the quaternion notation of the Cauchy integral formula (5.1.26) for vector monogenic functions $\mathbf{F}$ (i.e., for Laplace fields) is

$$-\frac{1}{4\pi}\iint_S \operatorname{grad}\frac{1}{|\mathbf{r}-\mathbf{r}'|}*\mathbf{n}*\mathbf{F}\,ds=\begin{cases}\mathbf{F}(\mathbf{r}')\ , & \mathbf{r}'\in D\\ 0\ , & \mathbf{r}'\in CD\ .\end{cases} \tag{A.26}$$

Formula (A.26) can be taken as a basis for determining the Cauchy-type integrals. Indeed, let an arbitrary vector field $\boldsymbol{\varphi}$, satisfying the condition

$$\iint_S \operatorname{grad}\frac{1}{|\mathbf{r}-\mathbf{r}'|}\cdot(\mathbf{n}\times\boldsymbol{\varphi})\,ds=0\ , \tag{A.27}$$

be specified on S. Note that for vector fields continuously differentiable on S condition (A.27) follows, in view of (A.25), from condition (5.2.2)

$$\operatorname{rot}_S\boldsymbol{\varphi}_\tau\equiv 0\ , \tag{A.28}$$

where $\boldsymbol{\varphi}_\tau$ is the tangential component of the field $\boldsymbol{\varphi}$. Then the space analog of the Cauchy-type integral (5.2.3) can be represented in the quaternion form as follows

$$C^S(\mathbf{r}';\,\boldsymbol{\varphi})=-\frac{1}{4\pi}\iint_S \operatorname{grad}\frac{1}{|\mathbf{r}-\mathbf{r}'|}*\mathbf{n}*\boldsymbol{\varphi}\,ds\ . \tag{A.29}$$

Relations (A.26) and (A.29) enable all the basic provisions of the theory of Cauchy-type integral space analogs (developed in Chap. 2) to be rewritten in terms of quaternions.

For instance, the three-dimensional integral transforms of Cauchy (5.3.26) for Hölder functions on S take the form

$$\boldsymbol{\Psi}(\mathbf{r}_0) = -\frac{1}{2\pi}\iint\limits_S \operatorname{grad}\frac{1}{|\mathbf{r}-\mathbf{r}'|}*\mathbf{n}*\boldsymbol{\varphi}\,ds\ ,$$
$$\boldsymbol{\varphi}(\mathbf{r}_0) = -\frac{1}{2\pi}\iint\limits_S \operatorname{grad}\frac{1}{|\mathbf{r}-\mathbf{r}'|}*\mathbf{n}*\boldsymbol{\Psi}\,ds\ . \tag{A.30}$$

Representation (6.2.20) of the gravitational field of a homogeneous body is written, with due reference to (5.1.28b), as follows

$$\mathbf{g}(\mathbf{r}') = \tfrac{1}{3}\gamma\varrho_0 \iint\limits_S \frac{(\mathbf{r}-\mathbf{r}_0)*\mathbf{n}*(\mathbf{r}-\mathbf{r}_0)}{|\mathbf{r}-\mathbf{r}_0|^3}\ , \tag{A.31}$$

and so on.

Appendix B
Green Electromagnetic Functions for Inhomogeneous Media and Their Properties

B.1 Field Equations

The body of Green functions provides an important means for analyzing electromagnetic fields. The Green electromagnetic functions, which can be regarded as a response of the field to excitation by a point source, make it possible to describe the electric or the magnetic field observed at some point of space in terms of integrals taken over the volume occupied by electric or magnetic sources or in terms of integrals over some closed surface bounding a domain devoid of sources. The possibility of such representations is related to the validity of the principle of superposition for linear fields. Thus, the electric and the magnetic fields at a point $\mathbf{r}'$ (outside the source domain) can be generally represented in the form (Felsen and Marcuvitz 1973)

$$\mathbf{E}(\mathbf{r}') = -\iiint_V \{\hat{\mathscr{G}}_{11}(\mathbf{r}'|\mathbf{r})\cdot\mathbf{j}^e(\mathbf{r}) + \hat{\mathscr{G}}_{12}(\mathbf{r}'|\mathbf{r})\cdot\mathbf{j}^m(\mathbf{r})\}dv \ ,$$

$$\mathbf{H}(\mathbf{r}') = -\iiint_V \{\hat{\mathscr{G}}_{21}(\mathbf{r}'|\mathbf{r})\cdot\mathbf{j}^e(\mathbf{r}) + \hat{\mathscr{G}}_{22}(\mathbf{r}'|\mathbf{r})\cdot\mathbf{j}^m(\mathbf{r})\}dv \ , \qquad \text{(B.1)}[1]$$

where $\mathbf{j}^e$ and $\mathbf{j}^m$ are the volume densities of electric and magnetic currents, respectively, and $\hat{\mathscr{G}}_{ij}(i,\ j = 1,\ 2)$ are the Green electromagnetic tensor functions; $\mathbf{r}$ and $\mathbf{r}'$ are the coordinates of source and observation points; V is the domain occupied by sources.

It is evident from Eq. (B.1) that $\hat{\mathscr{G}}_{11}$, $\hat{\mathscr{G}}_{12}$ and $\hat{\mathscr{G}}_{21}$, $\hat{\mathscr{G}}_{22}$ imply the sense of the electric and the magnetic fields, respectively, excited at the point $\mathbf{r}'$ by the electric and magnetic dipoles located at the point $\mathbf{r}$.

Write the equations fitting the electromagnetic functions $\hat{\mathscr{G}}_{ij}$, introduced according to (B.1).

In this case, the equations have the form

$$\nabla'\times\mathbf{H} = \sigma^*\mathbf{E} + \mathbf{j}^e \ , \quad \nabla'\times\mathbf{E} = i\omega\mu\mathbf{H} - \mathbf{j}^m \ . \qquad \text{(B.2)}$$

Substituting Eq. (B.1) into Eq. (B.2) and taking into account that the operator ∇' operates only on the coordinates of the vector $\mathbf{r}'$ we write

[1] The negative sign with the integral is taken for the sake of convenience of choosing the Green electromagnetic tensor functions.

$$\iiint_V [(\nabla'\times \hat{\mathscr{G}}_{21})\cdot\mathbf{j}^e+(\nabla'\times \hat{\mathscr{G}}_{22})\cdot\mathbf{j}^m]\,dv=\sigma^* \iiint_V (\hat{\mathscr{G}}_{11}\cdot\mathbf{j}^e+\hat{\mathscr{G}}_{12}\cdot\mathbf{j}^m)\,dv-\mathbf{j}^e \;, \tag{B.3}$$

$$\iiint_V [(\nabla'\times \hat{\mathscr{G}}_{11})\cdot\mathbf{j}^e+(\nabla'\times \hat{\mathscr{G}}_{12})\cdot\mathbf{j}^m]\,dv=i\omega\mu \iiint_V (\hat{\mathscr{G}}_{21}\cdot\mathbf{j}^e+\hat{\mathscr{G}}_{22}\cdot\mathbf{j}^m)\,dv+\mathbf{j}^m \;,$$

or

$$\iiint_V [(\nabla'\times \hat{\mathscr{G}}_{21})\cdot\mathbf{j}^e+(\nabla'\times \hat{\mathscr{G}}_{22})\cdot\mathbf{j}^m-\sigma^*\hat{\mathscr{G}}_{11}\cdot\mathbf{j}^e-\sigma^*\hat{\mathscr{G}}_{12}\cdot\mathbf{j}^m+\hat{\mathbf{I}}\delta(\mathbf{r}-\mathbf{r}')]\,dv=0 \;, \tag{B.4}$$

$$\iiint_V [(\nabla'\times \hat{\mathscr{G}}_{11})\cdot\mathbf{j}^e+(\nabla'\times \hat{\mathscr{G}}_{12})\cdot\mathbf{j}^m$$
$$-i\omega\mu\, \hat{\mathscr{G}}_{21}\cdot\mathbf{j}^e-i\omega\mu\,\hat{\mathscr{G}}_{22}\cdot\mathbf{j}^m-\mathbf{j}^m\delta(\mathbf{r}-\mathbf{r}')]\,dv=0 \;,$$

where $\delta(\mathbf{r}-\mathbf{r}')$ is the Dirac function, which is known to possess the following property

$$\iiint_V \mathbf{A}(\mathbf{r})\delta(\mathbf{r}-\mathbf{r}')\,dv=\mathbf{A}(\mathbf{r}') \;, \quad \mathbf{r}'\in V \;. \tag{B.5}$$

Since (B.4) is expected to hold good for any value of $\mathbf{j}^e$ and $\mathbf{j}^m$, we take alternately $\mathbf{j}^e=0$ and $\mathbf{j}^m=0$ in (B.4). The result is the equation that should be satisfied by the Green electromagnetic tensor functions

$$\nabla'\times \hat{\mathscr{G}}_{11}=i\omega\mu\,\hat{\mathscr{G}}_{21} \;, \quad \nabla'\times \hat{\mathscr{G}}_{21}=\sigma^*\hat{\mathscr{G}}_{11}-\hat{\mathbf{I}}\delta(\mathbf{r}-\mathbf{r}') \;; \tag{B.6}$$

$$\nabla'\times \hat{\mathscr{G}}_{12}=i\omega\mu\,\hat{\mathscr{G}}_{22}+\hat{\mathbf{I}}\delta(\mathbf{r}-\mathbf{r}') \;, \quad \nabla'\times \hat{\mathscr{G}}_{22}=\sigma^*\hat{\mathscr{G}}_{12} \;, \tag{B.7}$$

where $\hat{\mathbf{I}}$ is the idem factor. It follows from (B.6) and (B.7) that $\hat{\mathscr{G}}_{11}$ and $\hat{\mathscr{G}}_{12}$ satisfy also the equations

$$\begin{aligned}\nabla'\times(\nabla'\times \hat{\mathscr{G}}_{11})-k^2\hat{\mathscr{G}}_{11}&=-i\omega\mu\hat{\mathbf{I}}\delta(\mathbf{r}-\mathbf{r}') \;,\\ \nabla'\times(\nabla'\times \hat{\mathscr{G}}_{12})-k^2\hat{\mathscr{G}}_{12}&=i\omega\mu\nabla'\times\hat{\mathbf{I}}\delta(\mathbf{r}-\mathbf{r}') \;,\end{aligned} \tag{B.8}$$

while the functions $\hat{\mathscr{G}}_{21}$ and $\hat{\mathscr{G}}_{22}$ meet the equations

$$\begin{aligned}\nabla'\times(\nabla'\times \hat{\mathscr{G}}_{21})-k^2\hat{\mathscr{G}}_{21}-\nabla'\sigma^*\times\left(\frac{1}{\sigma^*}\nabla'\times \hat{\mathscr{G}}_{21}\right)&=-\nabla'\times\hat{\mathbf{I}}\delta(\mathbf{r}-\mathbf{r}') \;,\\ \nabla'\times(\nabla'\times \hat{\mathscr{G}}_{22})-k^2\hat{\mathscr{G}}_{22}-\nabla'\sigma^*\times\left(\frac{1}{\sigma^*}\nabla'\times \hat{\mathscr{G}}_{22}\right)&=\sigma^*\hat{\mathbf{I}}\delta(\mathbf{r}-\mathbf{r}') \;.\end{aligned} \tag{B.9}$$

B.2 Lorentz Lemma for an Inhomogeneous Medium

The Green electromagnetic tensor functions $\hat{\mathscr{G}}_{ij}$ $(ij=1,2)$ for isotropic space, introduced according to (B.1), are known (Felsen and Marcuvitz 1973) to satisfy the reciprocal relations, which, in turn, reflect the properties of symmetry of the electromagnetic field in a given inhomogeneous medium. To derive these relations, it is natural to use the Lorentz lemma. The latter will be formulated for the case of electromagnetic fields excited by electric ($\mathbf{j}^e$) and magnetic ($\mathbf{j}^m$) sources in an inhomogeneous isotropic medium of complex electric conductivity $\sigma^*(\mathbf{r})$ and magnetic permeability $\mu = \text{const}$. It is assumed, as previously, that the time dependence of the fields is fitted by the factor $\exp(-i\omega t)$[2].

Let some domain V of a medium with parameters $\sigma^*(\mathbf{r})$ and μ contain two sets of sources ($\mathbf{j}_a^e$, $\mathbf{j}_a^m$) and ($\mathbf{j}_b^e$, $\mathbf{j}_b^m$) exciting electromagnetic fields of the same frequency ω. The field excited by the a-type sources is designated by $\mathbf{E}^a$, $\mathbf{H}^a$, while the field produced by the b-type sources is denoted by $\mathbf{E}^b$, $\mathbf{H}^b$. The corresponding equations have the form

$$\nabla\times\mathbf{H}^a = \sigma^*(\mathbf{r})\mathbf{E}^a + \mathbf{j}_a^e , \tag{B.10}$$

$$\nabla\times\mathbf{E}^a = i\omega\mu\mathbf{H}^a - \mathbf{j}_a^m , \tag{B.11}$$

$$\nabla\times\mathbf{H}^b = \sigma^*(\mathbf{r})\mathbf{E}^b + \mathbf{j}_b^e , \tag{B.12}$$

$$\nabla\times\mathbf{E}^b = i\omega\mu\mathbf{H}^b - \mathbf{j}_b^m . \tag{B.13}$$

Multiplying (B.10) scalarly by $\mathbf{E}^b$ and (B.13) by $\mathbf{H}^a$ and adding up the resultant products we have

$$\nabla\cdot[\mathbf{H}^a\times\mathbf{E}^b] = \sigma^*(\mathbf{r})(\mathbf{E}^a\cdot\mathbf{E}^b) + (\mathbf{j}_a^e\cdot\mathbf{E}^b) - i\omega\mu(\mathbf{H}^a\cdot\mathbf{H}^b) + (\mathbf{j}_b^m\cdot\mathbf{H}^a) . \tag{B.14}$$

Interchanging the indices a and b we obtain

$$\nabla\cdot[\mathbf{H}^b\times\mathbf{E}^a] = \sigma^*(\mathbf{r})(\mathbf{E}^b\cdot\mathbf{E}^a) + (\mathbf{j}_b^e\cdot\mathbf{E}^a) - i\omega\mu(\mathbf{H}^b\cdot\mathbf{H}^a) + (\mathbf{j}_a^m\cdot\mathbf{H}^b) . \tag{B.15}$$

A subtraction of (B.14) from (B.15) yields

$$\nabla\cdot[(\mathbf{E}^b\times\mathbf{H}^a) - (\mathbf{E}^a\times\mathbf{H}^b)] = (\mathbf{E}^a\cdot\mathbf{j}_b^e) + (\mathbf{H}^b\cdot\mathbf{j}_a^m) - (\mathbf{E}^b\cdot\mathbf{j}_a^e) - (\mathbf{H}^a\cdot\mathbf{j}_a^m) . \tag{B.16}$$

Integrating (B.16) over the whole domain V and employing the Ostrogradsky-Gauss theorem we derive

$$\iint_S [(\mathbf{E}^b\times\mathbf{H}^a) - (\mathbf{E}^a\times\mathbf{H}^b)]\cdot ds = \iiint_V [(\mathbf{E}^a\cdot\mathbf{j}_b^e) + (\mathbf{H}^b\cdot\mathbf{j}_a^m) - (\mathbf{E}^b\cdot\mathbf{j}_a^e) - (\mathbf{H}^a\cdot\mathbf{j}_b^m)]\, dv , \tag{B.17}$$

where S is the surface bounding the domain V, and the vector of the normal to the surface points outside the domain V.

Now let the diameter of this domain tend to infinity. Then, allowing for the radiation conditions for the fields $\mathbf{E}^a$, $\mathbf{H}^a$ and $\mathbf{E}^b$, $\mathbf{H}^b$ we can write

$$\iiint_{V\infty} [(\mathbf{E}^a\cdot\mathbf{j}_b^e) + (\mathbf{H}^b\cdot\mathbf{j}_a^m) - (\mathbf{E}^b\cdot\mathbf{j}_a^e) - (\mathbf{H}^a\cdot\mathbf{j}_b^m)]\, dv = 0 , \tag{B.18}$$

where integration is carried out over the whole space.

Relation (B.18) is a mathematical expression for the reciprocity principle in our case.

---

[2] Note that the Lorentz lemma also holds good in a more general case of nonstationary electromagnetic fields in anisotropic inhomogeneous media (Felsen and Marcuvitz 1973, vol. 1).

B.3 Reciprocal Relations

We will use the Lorentz lemma in form (B.18) to derive reciprocal relations for the Green electromagnetic tensor functions in an inhomogeneous medium.

$$\text{a)}\quad \hat{\mathscr{G}}_{11}(\mathbf{r}|\mathbf{r}') = \tilde{\hat{\mathscr{G}}}_{11}(\mathbf{r}'|\mathbf{r}) \ , \tag{B.19}$$

where the swung dash denotes the operation of transposition.

Let electric dipoles with moments **a** and **b** be arranged successively at points with radii vectors **r**′ and **r**″, respectively:

$$\mathbf{j}_a^e = \mathbf{a}\delta(\mathbf{r}-\mathbf{r}') \ , \quad \mathbf{j}_b^e = \mathbf{b}\delta(\mathbf{r}-\mathbf{r}'') \ , \tag{B.20}$$

where **a** and **b** are arbitrary constant vectors, δ is the Dirac function; **r** is the radius vector of the observation point, and it is assumed that there are no magnetic-type sources:

$$\mathbf{j}_a^m = 0 \ , \quad \mathbf{j}_b^m = 0 \ . \tag{B.21}$$

Then, according to (B.1), we can write

$$\mathbf{E}^a = -\hat{\mathscr{G}}_{11}(\mathbf{r}|\mathbf{r}')\cdot\mathbf{a} \ , \quad \mathbf{E}^b = -\hat{\mathscr{G}}_{11}(\mathbf{r}|\mathbf{r}'')\cdot\mathbf{b} \ . \tag{B.22}$$

Substitution of (B.20) through (B.22) into (B.18) yields

$$\iiint\limits_{V\infty} \{-[\hat{\mathscr{G}}_{11}(\mathbf{r}|\mathbf{r}')\cdot\mathbf{a}]\cdot\mathbf{b}\delta(\mathbf{r}-\mathbf{r}'')+[\hat{\mathscr{G}}_{11}(\mathbf{r}|\mathbf{r}'')\cdot\mathbf{b}]\cdot\mathbf{a}\delta(\mathbf{r}-\mathbf{r}')\}dv = 0 \ , \tag{B.23}$$

where integration is over the observation point coordinate.

With due account taken of the property of the δ-function, we write $\{\mathbf{b}\cdot[\hat{\mathscr{G}}_{11}(\mathbf{r}''|\mathbf{r}')]\cdot\mathbf{a}\} = \{\mathbf{a}\cdot[\hat{\mathscr{G}}_{11}(\mathbf{r}'|\mathbf{r}'')]\cdot\mathbf{b}\}$. This directly implies that $\hat{\mathscr{G}}_{11}(\mathbf{r}''|\mathbf{r}') = \tilde{\hat{\mathscr{G}}}_{11}(\mathbf{r}'|\mathbf{r}'')$ or in the previous designations $\hat{\mathscr{G}}_{11}(\mathbf{r}|\mathbf{r}') = \tilde{\hat{\mathscr{G}}}_{11}(\mathbf{r}'|\mathbf{r})$, which was to be proved (the swung dash denotes the operation of transposition).

$$\text{b)}\quad \hat{\mathscr{G}}_{22}(\mathbf{r}|\mathbf{r}') = \tilde{\hat{\mathscr{G}}}_{22}(\mathbf{r}'|\mathbf{r}) \ . \tag{B.24}$$

Let magnetic dipoles with moments **a** and **b** be arranged successively at points with radii vectors **r**′ and **r**″, respectively:

$$\mathbf{j}_a^m = \mathbf{a}\delta(\mathbf{r}-\mathbf{r}') \ , \quad \mathbf{j}_b^m = \mathbf{b}\delta(\mathbf{r}-\mathbf{r}'') \ . \tag{B.25}$$

Assume that no electric sources are available in the two cases:

$$\mathbf{j}_a^e = 0 \ , \quad \mathbf{j}_b^e = 0 \ . \tag{B.26}$$

Then, in accordance with (B.1), we have

$$\mathbf{H}^a = -\hat{\mathscr{G}}_{22}(\mathbf{r}|\mathbf{r}')\cdot\mathbf{a} \ , \quad \mathbf{H}^b = -\hat{\mathscr{G}}_{22}(\mathbf{r}|\mathbf{r}'')\cdot\mathbf{b} \ . \tag{B.27}$$

Substituting (B.25) through (B.27) into (B.18), write

$$\iiint\limits_{V\infty} \{-[\hat{\mathscr{G}}_{22}(\mathbf{r}|\mathbf{r}'')\cdot\mathbf{b}]\cdot\mathbf{a}\delta(\mathbf{r}-\mathbf{r}')+[\hat{\mathscr{G}}_{22}(\mathbf{r}|\mathbf{r}')\cdot\mathbf{a}]\cdot\mathbf{b}\delta(\mathbf{r}-\mathbf{r}'')\}dv = 0 \ . \tag{B.28}$$

It follows from (B.28) that

$$[\mathbf{b}\cdot\hat{\mathscr{G}}_{22}(\mathbf{r}''|\mathbf{r}')]\cdot\mathbf{a} = [\mathbf{a}\cdot\hat{\mathscr{G}}_{22}(\mathbf{r}'|\mathbf{r}'')]\cdot\mathbf{b}$$

or in other designations

$$\hat{\mathscr{G}}_{22}(\mathbf{r}|\mathbf{r}') = \tilde{\hat{\mathscr{G}}}_{22}(\mathbf{r}'|\mathbf{r}) \ ,$$

which was to be proved.

c) $$\hat{\mathscr{G}}_{12}(\mathbf{r}|\mathbf{r}') = -\tilde{\hat{\mathscr{G}}}_{21}(\mathbf{r}'|\mathbf{r}) \ . \tag{B.29}$$

Let an electric dipole with a moment **a** be placed at a point with a radius vector **r**′, while a magnetic dipole with a moment **b** is arranged at a point with a radius vector **r**″. Assuming no other sources in this domain, we have

$$\mathbf{j}_a^e = \mathbf{a}\delta(\mathbf{r}-\mathbf{r}') \ , \quad \mathbf{j}_a^m = 0 \ ; \tag{B.30}$$

$$\mathbf{j}_b^e = 0 \ , \quad \mathbf{j}_b^m = \mathbf{b}\delta(\mathbf{r}-\mathbf{r}'') \ . \tag{B.31}$$

Then, according to (B.1), we have

$$\mathbf{H}^a = -\hat{\mathscr{G}}_{21}(\mathbf{r}|\mathbf{r}')\cdot\mathbf{a} \ , \quad \mathbf{E}^b = -\hat{\mathscr{G}}_{12}(\mathbf{r}|\mathbf{r}'')\cdot\mathbf{b} \ . \tag{B.32}$$

Substitution of (B.30) through (B.32) into (B.18) yields

$$\iiint\limits_{V\infty} \{-[\hat{\mathscr{G}}_{12}(\mathbf{r}|\mathbf{r}'')\cdot\mathbf{b}]\cdot\mathbf{a}\delta(\mathbf{r}-\mathbf{r}')-[\hat{\mathscr{G}}_{21}(\mathbf{r}|\mathbf{r}')\cdot\mathbf{a}]\cdot\mathbf{b}\delta(\mathbf{r}-\mathbf{r}'')\}dv = 0 \ .$$

Hence we obtain $\mathbf{a}\cdot[\hat{\mathscr{G}}_{12}(\mathbf{r}'|\mathbf{r}'')\cdot\mathbf{b}] = -\mathbf{b}\cdot[\hat{\mathscr{G}}_{21}(\mathbf{r}''|\mathbf{r}')\cdot\mathbf{a}]$ or in other designations $\hat{\mathscr{G}}_{12}(\mathbf{r}|\mathbf{r}') = \tilde{\hat{\mathscr{G}}}_{21}(\mathbf{r}'|\mathbf{r})$, which was to be proved.

Thus, we have shown that the Green electromagnetic tensor functions for an inhomogeneous medium satisfy the following reciprocal relations

$$\hat{\mathscr{G}}_{ij}(\mathbf{r}|\mathbf{r}') = (-1)^{i+j}\tilde{\hat{\mathscr{G}}}_{ji}(\mathbf{r}'|\mathbf{r}) \quad (i, j = 1, 2) \ . \tag{B.33}$$

For a homogeneous medium, relations (B.33) are supplemented by two more

$$\hat{\mathscr{G}}_{11}(\mathbf{r}|\mathbf{r}') = \tilde{\hat{\mathscr{G}}}_{11}(\mathbf{r}|\mathbf{r}') \ , \tag{B.34}$$

$$\hat{\mathscr{G}}_{21}(\mathbf{r}|\mathbf{r}') = -\hat{\mathscr{G}}_{21}(\mathbf{r}|\mathbf{r}') \ . \tag{B.35}$$

B.4 Integral Representations of the Electric and Magnetic Fields

We will derive formulas for the electric and magnetic fields in a domain devoid of sources, which will be similar to the Stratton-Chu formulas for a homogeneous medium. To this end, we will employ the Lorentz lemma in form (B.17) as well as the rules of operation on tensor functions presented in Sect. 2.2.

Consider two electromagnetic fields: (1) $\mathbf{E}^a$ and $\mathbf{H}^a$, produced by arbitrary sources located beyond a finite domain V of space and (2) $\mathbf{E}^b$, $\mathbf{H}^b$ produced by the following source concentrated inside V:

$$\mathbf{j}_b^e = \mathbf{a}\delta(\mathbf{r}-\mathbf{r}') \ , \quad \mathbf{j}_b^m = 0 \ . \tag{B.36}$$

The fields $\mathbf{E}^a$, $\mathbf{H}^a$ and $\mathbf{E}^b$, $\mathbf{H}^b$ satisfy inside V the following equations, respectively:

$$\nabla' \times \mathbf{H}^a = \sigma^* \mathbf{E}^a \ , \quad \nabla' \times \mathbf{E}^a = i\omega\mu \mathbf{H}^a \ , \tag{B.37}$$

$$\nabla' \times \mathbf{H}^b = \sigma^* \mathbf{E}^b + \mathbf{a}\delta(\mathbf{r}-\mathbf{r}') \ , \quad \nabla' \times \mathbf{E}^b = i\omega\mu \mathbf{H}^b \ . \tag{B.38}$$

Note that, according to (B.1), we have

$$\mathbf{E}^b = -\hat{\mathscr{G}}_{11} \cdot \mathbf{a} \ , \quad \mathbf{H}^b = -\hat{\mathscr{G}}_{21} \cdot \mathbf{a} \ . \tag{B.39}$$

According to (B.17), we have

$$\iint_S \{[\mathbf{E}^a \times (\hat{\mathscr{G}}_{21}(\mathbf{r}'|\mathbf{r}) \cdot \mathbf{a}) - (\hat{\mathscr{G}}_{11}(\mathbf{r}'|\mathbf{r}) \cdot \mathbf{a}) \times \mathbf{H}^a]\} \cdot ds' = \iiint_V \mathbf{E}^a \cdot \mathbf{a}\delta(\mathbf{r}'-\mathbf{r})dv' \tag{B.40}$$

or without the superscript a

$$\iint_S \{(\mathbf{n}\times\mathbf{E}) \cdot [\hat{\mathscr{G}}_{21}(\mathbf{r}'|\mathbf{r}) \cdot \mathbf{a}] + [\hat{\mathscr{G}}_{11}(\mathbf{r}'|\mathbf{r}) \cdot \mathbf{a}] \times (\mathbf{n}\times\mathbf{H}) ds = \mathbf{a} \cdot \mathbf{E}(\mathbf{r}) \ .$$

Interchanging $\mathbf{r}$ and $\mathbf{r}'$ and allowing for the reciprocal relations we write

$$\iint_S \{[\mathbf{a} \cdot \hat{\mathscr{G}}_{11}(\mathbf{r}'|\mathbf{r})] \cdot (\mathbf{n}\times\mathbf{h}) - [\mathbf{a} \cdot \hat{\mathscr{G}}_{12}(\mathbf{r}'|\mathbf{r})] \cdot (\mathbf{n}\times\mathbf{E})\} ds = \mathbf{a} \cdot \mathbf{E}(\mathbf{r}') \ .$$

Hence it follows, with due account taken of the arbitrariness of the vector $\mathbf{a}$, that

$$\iint_S [\hat{\mathscr{G}}_{11}(\mathbf{r}'|\mathbf{r}) \cdot (\mathbf{n}\times\mathbf{H}) - \hat{\mathscr{G}}_{12}(\mathbf{r}'|\mathbf{r}) \cdot (\mathbf{n}\times\mathbf{E})] ds = \mathbf{E}(\mathbf{r}') \ .$$

Designating $-(\mathbf{n}\times\mathbf{H}) = \mathbf{j}_S^e$, $(\mathbf{n}\times\mathbf{E}) = \mathbf{j}_S^m$ and allowing for the fact that the right-hand side of (B.40) is zero for $\mathbf{r} \notin V$, we write in the final form

$$-\iint_S [\hat{\mathscr{G}}_{11}(\mathbf{r}'|\mathbf{r}) \cdot \mathbf{j}_S^e(\mathbf{r}) + \hat{\mathscr{G}}_{12}(\mathbf{r}'|\mathbf{r}) \cdot \mathbf{j}_S^m(\mathbf{r})] ds = \begin{cases} \mathbf{E}(\mathbf{r}') \ , & \mathbf{r}' \in V \ , \\ 0 \ , & \mathbf{r}' \notin V \ . \end{cases} \tag{B.41}$$

Similarly taking sources $\mathbf{j}_b^e = 0$, and $\mathbf{j}_b^m = \mathbf{a}\delta(\mathbf{r}-\mathbf{r}')$ inside V and considering that in this case $\mathbf{E}^b = -\hat{\mathscr{G}}_{12} \cdot \mathbf{a}$, $\mathbf{H}^b = -\hat{\mathscr{G}}_{22} \cdot \mathbf{a}$, we write successively:

$$\iint_S -\{[\mathbf{E}^a \times (\hat{\mathscr{G}}_{22}(\mathbf{r}'|\mathbf{r}) \cdot \mathbf{a})] + [(\hat{\mathscr{G}}_{12}(\mathbf{r}'|\mathbf{r}) \cdot \mathbf{a}] \times \mathbf{H}^b]\} \cdot ds' = \iiint_V \mathbf{H}^a \cdot \mathbf{a}\delta(\mathbf{r}'-\mathbf{r})dv' \ ;$$

$$-\iint_S \{(\mathbf{n}\times\mathbf{E})[\hat{\mathscr{G}}_{22}(\mathbf{r}'|\mathbf{r}) \cdot \mathbf{a}] + [\hat{\mathscr{G}}_{12}(\mathbf{r}'|\mathbf{r}) \cdot \mathbf{a}] \cdot (\mathbf{n}\times\mathbf{H})\} \cdot ds' = \mathbf{a} \cdot \mathbf{H}(\mathbf{r}) \ ;$$

$$\iint_S \{-[\mathbf{a} \cdot \hat{\mathscr{G}}_{12}(\mathbf{r}'|\mathbf{r})] \cdot (\mathbf{n}\times\mathbf{H}) + [\mathbf{a} \cdot \hat{\mathscr{G}}_{22}(\mathbf{r}'|\mathbf{r})] \cdot (\mathbf{n}\times\mathbf{E})\} \cdot ds = \mathbf{a} \cdot \mathbf{H}(\mathbf{r}') \ ; \tag{B.42}$$

$$\iint_S \{\hat{\mathscr{G}}_{12}(\mathbf{r}'|\mathbf{r}) \cdot (\mathbf{n}\times\mathbf{H}) - \hat{\mathscr{G}}_{22}(\mathbf{r}'|\mathbf{r}) \cdot (\mathbf{n}\times\mathbf{E})\} \cdot ds = \mathbf{H}(\mathbf{r}') \ ;$$

$$-\iint_S \{\hat{\mathscr{G}}_{12}(\mathbf{r}'|\mathbf{r}) \cdot \mathbf{j}_S^e + \hat{\mathscr{G}}_{22}(\mathbf{r}'|\mathbf{r}) \cdot \mathbf{j}_S^m\} ds = \begin{cases} \mathbf{H}(\mathbf{r}') \ , & \mathbf{r}' \in V \ , \\ 0 \ , & \mathbf{r}' \notin V \ , \end{cases}$$

which was to be proved.

B.5 Some Formulas and Rules of Operations on Dyadic Tensor Functions

First, define the "external" (dyadic) product of the two vectors **a** and **b** (or a dyad) by the formula

$$\hat{\mathbf{A}} = \mathbf{ab} = \begin{bmatrix} a_x b_x & a_x b_y & a_x b_z \\ a_y b_x & a_y b_y & a_y b_z \\ a_z b_x & a_z b_y & a_z b_z \end{bmatrix} , \tag{B.43}$$

where $\mathbf{a} = (a_x, a_y, a_z)$; $\mathbf{b} = (b_x, b_y, b_z)$.

Note that for the tensors represented in form (B.43) the following rules of operations (Tai 1971, p. 46–48)

$$\mathbf{c} \cdot \hat{\mathbf{A}} = (\mathbf{c} \cdot \mathbf{a})\mathbf{b} = \mathbf{b}(\mathbf{c} \cdot \mathbf{a}) = \mathbf{b}(\mathbf{a} \cdot \mathbf{c}) , \tag{B.44}$$

$$\hat{\mathbf{A}} \cdot \mathbf{c} = \mathbf{a}(\mathbf{b} \cdot \mathbf{c}) = (\mathbf{b} \cdot \mathbf{c})\mathbf{a} = (\mathbf{c} \cdot \mathbf{b})\mathbf{a} , \tag{B.45}$$

$$\tilde{\hat{\mathbf{A}}} = \mathbf{ba} , \tag{B.46}$$

hold good, where the swung dash denotes, as previously, the operation of transposition,

$$\hat{\mathbf{A}} \cdot \mathbf{c} = \mathbf{c} \cdot \tilde{\hat{\mathbf{A}}} , \tag{B.47}$$

$$\hat{\mathbf{A}} \times \mathbf{c} = \mathbf{a}(\mathbf{b} \times \mathbf{c}) , \tag{B.48}$$

$$\mathbf{c} \times \hat{\mathbf{A}} = (\mathbf{c} \times \mathbf{a})\mathbf{b} . \tag{B.49}$$

Let $\mathbf{d}_a$ stand for the unit vectors within the Cartesian system of coordinates. Then any tensor

$$\hat{\mathbf{G}} = [G_{\beta a}] = \begin{bmatrix} G_{xx} & G_{xy} & G_{xz} \\ G_{yx} & G_{yy} & G_{yz} \\ G_{zx} & G_{zy} & G_{zz} \end{bmatrix} \tag{B.50}$$

can be obviously written in the form

$$\hat{\mathbf{G}} = \sum_{a = x, y, z} \mathbf{G}_a \mathbf{d}_a , \tag{B.51}$$

where $\mathbf{G}_a$ denotes the columns in representation (B.50). Then, by definition, we have

$$\operatorname{div} \hat{\mathbf{G}} = \nabla \cdot \hat{\mathbf{G}} = \sum_{a = x, y, z} (\nabla \cdot \mathbf{G}_a)\mathbf{d}_a = \sum_{a = x, y, z} \operatorname{div} G_a d_a , \tag{B.52}$$

$$\begin{aligned} \operatorname{rot} \hat{\mathbf{G}} = \nabla \times \hat{\mathbf{G}} &= \sum_{a = x, y, z} [\nabla \times \mathbf{G}_a]\mathbf{d}_a = \sum_{a = x, y, z} \operatorname{rot} \mathbf{G}_a \mathbf{d}_a \\ &= \mathbf{d}_x \left(\frac{\partial \tilde{\mathbf{G}}_z}{\partial y} - \frac{\partial \tilde{\mathbf{G}}_y}{\partial z} \right) + \mathbf{d}_y \left(\frac{\partial \tilde{\mathbf{G}}_x}{\partial z} - \frac{\partial \tilde{\mathbf{G}}_z}{\partial x} \right) + \mathbf{d}_z \left(\frac{\partial \tilde{\mathbf{G}}_y}{\partial x} - \frac{\partial \tilde{\mathbf{G}}_x}{\partial y} \right) , \end{aligned} \tag{B.53}$$

where $\tilde{\mathbf{G}}_\beta$ denotes the vector lines in representation (B.50).

It also follows from (B.46), (B.51), and (B.53) that

$$\tilde{\nabla}\times\hat{\mathbf{G}} = \sum_{\beta=x,y,z} \mathbf{d}_\beta(\nabla\times\tilde{\mathbf{G}}_\beta) \ , \tag{B.54}$$

$$\nabla\times\tilde{\hat{\mathbf{G}}} = \sum_{\beta=x,y,z} (\nabla\times\tilde{\mathbf{G}}_\beta)\mathbf{d}_\beta \ , \tag{B.55}$$

$$\tilde{\nabla}\times\tilde{\hat{\mathbf{G}}} = \sum_{\alpha=x,y,z} \mathbf{d}_\alpha(\nabla\times\mathbf{G}_\alpha) \ . \tag{B.56}$$

And finally, prove the following formula for dyadic functions depending on two variables $\mathbf{r}$ and $\mathbf{r}'$:

$$\nabla\times[\hat{\mathbf{G}}(\mathbf{r}|\mathbf{r}')\cdot\mathbf{a}(\mathbf{r}')] = [\nabla\times\hat{\mathbf{G}}(\mathbf{r}|\mathbf{r}')]\cdot\mathbf{a}(\mathbf{r}') \ , \tag{B.57}$$

taking into consideration that the operator ∇ operates only on the coordinates of the vector $\mathbf{r}$:

$$\begin{aligned}
\nabla\times(\hat{\mathbf{G}}\cdot\mathbf{a}) &= \nabla\times\left(\sum_{\alpha=x,y,z} \mathbf{G}_\alpha\mathbf{d}_\alpha\cdot\mathbf{a}\right) = \sum_{\alpha=x,y,z} [\nabla\times(\mathbf{G}_\alpha\mathbf{d}_\alpha)\mathbf{a})] \\
&= \sum_{\alpha=x,y,z} \{\nabla\times[\mathbf{G}_\alpha(\mathbf{d}_\alpha\cdot\mathbf{a})]\} = \sum_{\alpha=x,y,z} [\nabla\times(\mathbf{G}_\alpha a_\alpha)] \\
&= \sum_{\alpha=x,y,z} (\nabla a_\alpha\times\mathbf{G}_\alpha + a_\alpha\nabla\times\mathbf{G}_\alpha) = \sum_{\alpha=x,y,z} a_\alpha(\nabla\times\mathbf{G}_\alpha) \\
&= \sum_{\alpha=x,y,z} (\nabla\times\mathbf{G}_\alpha)(\mathbf{d}_\alpha\cdot\mathbf{a}) = \sum_{\alpha=x,y,z} [(\nabla\times\mathbf{G}_\alpha)\mathbf{d}_\alpha]\cdot\mathbf{a} = (\nabla\times\hat{\mathbf{G}})\cdot\mathbf{a} \ ,
\end{aligned}$$

which was to be proved.

We recall also that, according to (B.50) and (B.51), the idem factor $\hat{\mathbf{I}}$ (unit tensor) is written as

$$\hat{\mathbf{I}} = \sum_{\alpha=x,y,z} \mathbf{d}_\alpha\mathbf{d}_\alpha = \delta_{\alpha\beta}\mathbf{d}_\alpha\mathbf{d}_\beta \ ,$$

$$\hat{\mathbf{I}}_{\alpha\beta} = \begin{bmatrix} 1 & 0 & 0 \\ 0 & 1 & 0 \\ 0 & 0 & 1 \end{bmatrix} \ .$$

B.6 Tensor Statements of the Ostrogradsky-Gauss Theorem

Let $\hat{\mathbf{G}} = \hat{\mathbf{G}}(\mathbf{r})$ be a tensor field differentiable continuously everywhere in the domain D right to its boundary S. We will evaluate a volume integral of $\operatorname{div}\hat{\mathbf{G}}$ over the domain D, using formula (B.52) and the first vector statement of the Ostrogradsky-Gauss theorem (5.1.1):

$$\iiint_D \operatorname{div}\hat{\mathbf{G}}\,dv = \sum_{a=x,y,z} \iiint_D \operatorname{div}\mathbf{G}_a dv\,\mathbf{d}_a = \sum_{a=x,y,z} \iint_S \mathbf{n}\cdot\mathbf{G}_a ds\,\mathbf{d}_a$$

$$= \iint_S \mathbf{n}\cdot\left(\sum_{a=x,y,z} \mathbf{G}_a\mathbf{d}_a\right) ds = \iint_S \mathbf{n}\cdot\hat{\mathbf{G}}ds\ .$$

Thus we have derived the following important formula

$$\iiint_D \operatorname{div}\hat{\mathbf{G}}\,dv = \iint_S \mathbf{n}\cdot\hat{\mathbf{G}}ds\ , \tag{B.58}$$

which is called the *first tensor statement of the Ostrogradsky-Gauss theorem.*

Now we will evaluate the integral of $\operatorname{rot}\hat{\mathbf{G}}$, using formula (B.53) and the third vector statement of the Ostrogradsky-Gauss theorem (5.1.7):

$$\iiint_D \operatorname{rot}\hat{\mathbf{G}}\,dv = \sum_{a=x,y,z} \iiint_D \operatorname{rot}\mathbf{G}_a dv\,\mathbf{d}_a = \sum_{a=x,y,z} \iint_S [\mathbf{n}\times\mathbf{G}_a] ds\,\mathbf{d}_a$$

$$= \iint_S \left[\mathbf{n}\times\left(\sum_{a=x,y,z} \mathbf{G}_a\mathbf{d}_a\right)\right] ds = \iint_S \mathbf{n}\times\hat{\mathbf{G}}ds\ . \tag{B.59}$$

The resultant formula is termed the *second tensor statement of the Ostrogradsky-Gauss theorem.*

References

Bateman H (1946) Some integral equation of potential theory. J Appl Phys 17, 2:91–108

Berdichevsky MN, Zhdanov MS (1981) Interpretation of anomalies of the transient electromagnetic field of the earth. Nedra, Moscow, 327 pp

Berdichevsky MN, Zhdanov MS (1984) Advanced theory of deep geomagnetic sounding. Elsevier, Amsterdam, 408 pp

Berkhout AJ (1980) Seismic migration. Elsevier, Amsterdam, 339 pp

Berkhout AJ (1984) Seismic migration B. Practical aspects. Elsevier, Amsterdam, 274 pp

Bitsadze AV (1953) Space analog of the Cauchy-type integral and some of its applications. Izv Akad Nauk SSSR Ser Mat 17:525–538

Bitsadze AV (1972) Fundamentals of theory of analytical functions of a complex variable. Nauka, Moscow, 263 pp

Brackx F, Delanghe K, Sommen F (1982) Clifford analysis – Pitman advanced publishing program. Res Notes Math, Boston 76:308 pp

Chapman S, Bartels J (1940) Geomagnetism, vol. 2, Sect 22.10. Oxford Univ Press, pp 544–1049

Caulfiled HJ (ed) (1979) Handbook of optical holography. Academic Press, London New York, 638 pp

Cherednichenko VG (1978) Determination of the density of a body from a given potential. Dokl Akad Nauk SSSR 240, 5:721–725

Claerbout JF (1976) Fundamentals of geophysical data processing. McGraw-Hill, New York, 274 pp

Claerbout JF (1985) Imaging the earth's interior. Blackwell, New York

Dmitriev VI, Farzan RH (1980) Method for calculating an anomalous electromagnetic field from a local inhomogeneity. In: Mathematical models of electromagnetic sounding and mapping in geophysics. Loránd Eötvös Univ, Budapest, pp 95–111

Dmitriev VI, Zhdanov MS (1982) Methods of solution of inverse geophysical problems. In: Computing mathematics and facilities in prospecting geophysics. Nedra, Moscow, pp 89–105

Felsen LB, Marcuvitz N (1973) Radiation and scattering of waves. Prentice-Hall, Englewood Cliffs, 888 pp

Frank Ph, Mises R (1927) Die Differential- und Integralgleichungen der Mechanik und Physik. Riemann-Webers, Braunschweig, 916 pp

Fueter R (1935) Über die analytische Darstellung der regulären Funktionen einer Quaternionenvariablen. Comm Math Helv 8:371–378

Fueter R (1937) Die Theorie der regulären Funktionen einer Quaternionenvariablen. Congr Int Math, Oslo, B1, pp 75–91

Fuks BA, Shabat BV (1964) Functions of a complex variable and some of their applications, 3rd edn. Nauka, Moscow, 387 pp

Gakhov FD (1963) Boundary-value problems, 2nd edn. Fizmatghiz, Moscow, 649 pp

Gazdag J (1978) Wave equation migration with the phase shift method. Geophysics 43, 7:1342–1351

Golizdra G Ya (1966) Singular points of analytical continuation of the gravitational field and their relationship to the shape of disturbing masses. In: Additional chapters of the course in gravimetric and magnetic prospecting. Nauka, Novosibirsk, pp 273–388

Harrington RF (1961) Time harmonic electromagnetic fields. McGraw-Hill, New York, 480 pp

Hartman O (1963) Behandlung lokaler erdmagnetischer Felder als Randwertaufgabe der Potentialtheorie. Abh Akad Wiss Göttingen Math Phys Kl 9:3–50

Hemon Ch (1971) Automatic conversion of seismic sections. In: Interpretation of seismic prospecting data by mathematical methods. Discussion Symp, Moscow, pp 59–78

Herglotz G (1914) Über die analytische Fortsetzung des Potentials ins Innere der anziehenden Massen. Teubner, Leipzig, pp 1–52

Hönl H, Maue AW, Westpfahl K (1961) Theorie der Beugung. Springer, Berlin Göttingen Heidelberg, 428 pp

Kantor IL, Solodovnikov AS (1973) Hypercomplex numbers. Nauka, Moscow, 144 pp

Kertz W (1954) Modelle für erdmagnetische induzierte electrische Ströme im Untergrund. Nachr Akad Wiss Göttingen Math Phys Kl Abt IIa:101

Kolbenheyer T (1976) Solution of space direct gravimetrical problem by means of the Bitsadze-type integrals. Contrib Geophys Inst Slovak Acad Sci 6, 1:9–17

Kolbenheyer T (1978) Analytical continuation of an external gravitational field of a three-dimensional body into its interior. In: Geophysikalische Interpretations-Methoden. Veda, Slowak Akad Wiss, pp 89–100

Koshlyakov NS, Gliner EB, Smirnov MM (1962) Governing differential equations of mathematical physics. Fizmatghiz, Moscow, 767 pp

Kupradze VD (1950) Boundary-value problems in theory of oscillations and integral equations. Gostekhizdat, Moscow Leningrad, 280 pp

Landau LD, Lifschits EM (1965) Theory of elasticity. Nauka, Moscow, 203 pp

Love AEH (1944) A treatise on the mathematical theory of elasticity, 4th edn. Dover, New York, 643 pp

Makagonov PP (1977) Skin effect in steplike variations of the electromagnetic field. Izv Vuzov Geol Razvedka 7:134–137

Moisil Gr C (1931) Sur les quaternions monogènes. Bull Sci Math (Cluj) 55

Moisil Gr C (1976) Opera matematica, vol 1. Acad RSR Bucukesti, 376 pp

Moisil Gr C, Theodoresco N (1931) Functions holomorphes dans l'espace. Mathematica 5:142–153

Morse Ph M, Feshbach H (1953) Methods of theoretical physics. McGraw-Hill, New York, 997 pp

Muskhelishvily NI (1954) Some main problems of mathematical theory of elasticity, 4th edn. Akad Nauk SSSR, Moscow, 647 pp

Muskhelishvily NI (1962) Singular integral functions, 2nd edn. Fizmatghiz, Moscow, 509 pp

Naumov VV, Grigor'ev Yu M (1982) Laurent series for the Moisil-Theodoresco system. In: Dynamics of solid media, Iss 54. Hydrodyn Inst, SO Akad Nauk SSSR, Novosibirsk, pp 115–126

Nedyalkov IP (1965) Separation of potential fields. Izv Akad Nauk SSSR Fiz Zemly 12:31–44

Noho K (1982) Holomorphic functions of a quaternion variable. Bull Fukuoka Univ Educ, vol 32, Pt III, pp 21–37

Petrashen GI, Nakhamkin SA (1973) Continuation of wave fields in seismic prospecting problems. Nauka, Leningrad, 170 pp

Puzyrev NN (1963) Theory of interpretation of seismic point observations. Gheol Gheofiz 9:66–81

Rashevsky PK (1956) Course in differential geometry. Gostekhizdat, Moscow, 420 pp

Ryabinkin LA, Napalkov Yu V, Znamensky VV, Voskresensky Yu N, Rapoport MB (1962) Theory and practice of seismic CDR method. Trans Moscow Inst Petrochem Gas Indust, vol 39. Gostoptekhizdat, Moscow, 295 pp

Siebert M (1958) Die Zerlegung eines lokalen erdmagnetischen Feldes in äußeren und inneren Anteil mit Hilfe des zweidimensionalen Fouriertheorems. Abh Akad Wiss Göttingen Math Phys Kl 4:33

Siebert M (1962) Die Zerlegung eines zweidimensionalen Magnetfeldes in äußeren und inneren Anteil mit Hilfe der Cauchyschen Integralformel. Z Geophys 28:231–236

Siebert M (1964) Ein Verfahren zur unmittelbaren Bestimmung der vertikalen Leitfähigkeitsverteilung im Rahmen der erdmagnetischen Tiefensonderung. Nachr Akad Wiss Göttingen Math Phys Kl 2:25–35

Siebert M, Kertz W (1957) Zur Zerlegung eines lokalen erdmagnetischen Feldes in äußeren und inneren Anteil. Nachr Akad Wiss Göttingen Math Phys Kl Abt IIa:87–112

Sitarova A (1977) Analytical continuation of the gravitational field into attracting masses through an arbitrary second-order surface, vol 8. In: Proc Geophys Inst, Slovak Acad Sci, pp 93–103

Spivac M (1965) Calculus on manifolds. Benjamin, New York Amsterdam, 162 pp

Sretensky LN (1946) Theory of the Newton potential. OGhIZ, Moscow, 318 pp

Stolt RH (1978) Migration by Fourier transforms. Geophysics 43, 1:23–48

Strakhov VN (1969a) Theory of approximate solution of linear ill-posed problems in the Hilbert space and its application to prospecting geophysics, P 1. Izv Akad Nauk SSSR Fiz Zemly 8:30–53

Strakhov VN (1969b) Theory of approximate solution of linear ill-posed problems in the Hilbert space, P 2. Izv Akad Nauk SSSR Fiz Zemly 9:64–96

Strakhov VN (1970a) Theory of a plane inverse problem of magnetic potential under variable magnetization. Izv Akad Nauk SSSR Fiz Zemly 3:44–58

Strakhov VN (1970b) State-of-the art and problems in mathematical theory of interpretation of gravitational and magnetic anomalies. Izv Akad Nauk SSSR Fiz Zemly 5:112–119

Strakhov VN (1970c) Some aspects of the plane inverse problem of magnetic potential. Izv Akad Nauk SSSR Fiz Zemly 9:31–41

Strakhov VN (1970d) Some aspects of the plane gravitational problem. Izv Akad Nauk SSSR Fiz Zemly 12:32–44

Strakhov VN (1970e) Analytical continuation of two-dimensional potential fields into arbitrary domains of the lower half-space adjacent to the OX-axis. Izv Akad Nauk SSSR Fiz Zemly 6:35–52

Strakhov VN (1971a) Network methods of analytical continuation of potential fields using conformal grids. Izv Akad Nauk SSSR Fiz Zemly 10:44–59

Strakhov VN (1971b) Theory of analytical continuation of two-dimensional potential fields by the conformal grid method. Izv Akad Nauk SSSR Fiz Zemly 11:34–39

Strakhov VN (1974) State-of-the art and problems in geological interpretation of gravitational and magnetic data. In: Soviet prospecting geophysics on the boundary of the 1970s. Nedra, Moscow, pp 113–121

Stratton JA (1941) Electromagnetic theory. McGraw-Hill, New York, 615 pp

Tai CT (1971) Dyadic Green's functions in electromagnetic theory. Scrahton XVII The Intext monograph series in electrical engineering. Consulting, ed DK Cheng, 246 pp

Tikhonov AN, Arsenin V Ya (1974) Methods for solving ill-posed problems. Nauka, Moscow, 223 pp

Timan AF, Trofimov VN (1968) Introduction to theory of harmonic functions. Nauka, Moscow, 207 pp

Timoshin Yu V (1972) Fundamentals of diffraction conversion of seismic recordings. Nedra, Moscow, 263 pp

Timoshin Yu V (1978) Pulsed seismic holography. Nedra, Moscow, 286 pp

Tsirulsky AV (1963) Some properties of the complex logarithmic potential of a homogeneous domain. Izv Akad Nauk SSSR Ser Gheofiz 7:1072–1075

Tsirulsky AV (1964) Relationship of analytical continuation of a logarithmic potential to establishment of the boundaries of a disturbing domain. Izv Akad Nauk SSSR Ser Gheofiz 11:1646–1693

Tsirulsky AV (1969) Uniqueness of solution to the inverse problem of potential theory. Izv Akad Nauk SSSR Fiz Zemly 6:60–65

Tsirulsky AV, Sirotin MI (1964) Analytical continuation of logarithmic potential. Izv Akad Nauk SSSR Ser Gheofiz 1:105–109

Vargova E (1977) Solution of the direct inhomogeneous magnetic problem by means of analytical four-vectors. Contrib Geophys Inst Slovak Acad Sci 8:63–67

Vasiliev SA (1975) Possibility of seismic field continuation towards the source inside a layered homogeneous medium. Izv Akad Nauk SSSR Fiz Zemly 9:28–39

Vasiliev SA, Golosov VP (1971) Interpretation of seismograms by the Huygens-Fresnel method and wave equation. In: Applied geophysics, Iss 64. Nedra, Moscow, pp 59–65

Vekua IN (1959) Generalized analytical functions. Fizmatghiz, Moscow, 628 pp

Vestine EH, Davids N (1945) Analysis and interpretation of geomagnetic anomalies. Terr Mag 50:21

Vinogradov VS (1964) One analog of the Cauchy-Riemann system in a four-dimensional space. Dokl Akad Nauk SSSR 154, 1:16–19

Waters KH (1978) Reflection seismology. Wiley-Interscience, New York, 377 pp
Wavre R (1934) Sur le problème inverse de la théorie de potential et les functions harmoniques multiformes. Comment Math Helv 6:317–327
Weaver JT (1964) On the separation of local geomagnetic fields into external and internal parts. Z Geophys 30: 29–36
Weidelt P (1975) Electromagnetic induction in three-dimensional structures. J Geophys 41, 1:85–109
Zakharov EV, Ilyin IV (1970) Integral representations of electromagnetic fields in an inhomogeneous layered medium. Izv Akad Nauk SSSR Fiz Zemly 8:62–71
Zakharov EV, Pimenov Yu V (1982) Numerical analysis of radiowave diffraction. Radio Svyaz, Moscow, 184 pp
Zamorev AA (1941) Investigation of the two-dimensional inverse problem of potential theory. Izv Akad Nauk SSSR Ser Gheogr Gheofiz 4/5:487–500
Zamorev AA (1942) Definition of the shape of a body from the derivatives of an external gravitational potential. Izv Akad Nauk SSSR Ser Gheogr Gheofiz 1/2:48–54
Zhdanov MS (1970) Relationship of singular points of the gravitational and magnetic potentials to the shape of the surface of contact. Gheol Gheofiz 6:119–122
Zhdanov MS (1971) Development of theory of analytical continuation in curvilinear domains (two-dimensional potential fields). Izv Akad Nauk SSSR Fiz Zemly 5:114–121
Zhdanov MS (1973a) Analytical continuation of two-dimensional potential fields. Gheolog Gheofiz 8:93–97
Zhdanov MS (1973b) Properties of the gravitational potential of a three-dimensional homogeneous body. Gheolog Gheofiz 12:96–101
Zhdanov MS (1973c) Separation of transient electromagnetic fields of the Earth. Izv Akad Nauk SSSR Fiz Zemly 6:43–54
Zhdanov MS (1974) Theory of interpretation of gravitational anomalies determined by three space coordinates. Izv Akad Nauk SSSR Fiz Zemly 9:32–46
Zhdanov MS (1975a) Gravitational field of a three-dimensional layered medium. Gheolog Gheofiz 6:112–120
Zhdanov MS (1975b) Theory of interpretation of deep electromagnetic anomalies using the analytical continuation methods. Izv Akad Nauk SSSR Fiz Zemly 9:59–73
Zhdanov MS (1976a) Summary of selected lectures on the specialized course in gravimetric and magnetic prospecting. Publ Moscow Inst Petrochem Gas Indust, 74 pp
Zhdanov MS (1976b) Methods of transformation and interpretation of anomalies of the gravitational, magnetic, and transient electromagnetic fields of the Earth. Doctorate in physics and mathematics, Moscow State Univ, 337 pp
Zhdanov MS (1977) Numerical methods for transforming transient electromagnetic fields. Trans Moscow Inst Petrochem Gas Indust, vol 35, Moscow, pp 67–77
Zhdanov MS (1979) Application of the Cauchy-type integral analogs to theory of interpretation of geopotential fields. Preprint IZMIRAN 26 (255), Moscow, 30 pp
Zhdanov MS (1980a) Cauchy integral analogs for the separation and continuation of electromagnetic fields within conducting matter. Geophys Surv 4:115–136
Zhdanov MS (1980b) Use of Cauchy integral analogs in the geopotential field theory. Ann Geophys 36, 4:447–458
Zhdanov MS (1981) Continuation of nonstationary electromagnetic fields in geoelectrical problems. Izv Akad Nauk SSSR Fiz Zemly 12:60–69
Zhdanov MS, Frenkel MA (1982a) Application of reversed continuation of nonstationary electromagnetic fields in the solution of inverse geoelectrical problems. In: Mathematical methods in geoelectrics. IZMIRAN, Moscow, pp 111–120
Zhdanov MS, Frenkel MA (1982b) The solution of the inverse problems on the basis of the analytical continuation of transient EM-field in the reversed time. 6th Workshop EM-induction in the Earth and Moon. Victoria (Can), Univ Victoria, p 10
Zhdanov MS, Frenkel MA (1983a) Solution of inverse problems by analytical continuation of a transient electromagnetic field in reversed time. J Geomag Geoelectr 35:747–765
Zhdanov MS, Frenkel MA (1983b) The electromagnetic migration method in solution of inverse geoelectrical problems. Dokl Akad Nauk SSSR 271, 3:589–594
Zhdanov MS, Matusevich V Yu (1984) Restoration of the spatial pattern of wave propagation in an elastic medium. Ann Geophys 2, 1:1–16
Zhdanov MS, Golubev NG, Spichak VV, Varentsov Iv M (1982) The construction of effective methods for electromagnetic modeling. Geophys J R Astron Soc 68:589–607

Subject Index